CHILTON'S
MECHANICS' HANDBOOK

VOLUME **4** AUTOMATIC TRANSMISSION REPAIR

Managing Editor	Kerry A. Freeman, S.A.E.
Senior Editor	Richard J. Rivele, S.A.E.

OFFICERS

President	William A. Barbour
Executive Vice President	James Miades
Vice President & General Manager	John P. Kushnerick

CHILTON BOOK COMPANY
Chilton Way, Radnor, Pa. 19089

Manufactured in USA
© 1981 Chilton Book Company
ISBN 0-8019-7060-1 (paperback)
ISBN 0-8019-7062-4 (hardcover)

Library of Congress Catalog
Card No. 80-70353

567890   0987654

Contents

GENERAL INFORMATION

COMPONENT REMOVAL AND OVERHAUL

METRIC NOTICE

Certain parts are dimensioned in the metric system. Many fasteners are metric and should not be replaced with a customary inch fastener. It is important to note that during any maintenance procedure or repair, the metric fastener should be salvaged for reassembly. If the fastener is not reusable, then the equivalent fastener should be used.

A mismatched or incorrect fastener can result in component damage or possibly, personal injury.

SAFETY NOTICE

Proper service and repair procedures are vital to the safe, reliable operation of all motor vehicles, as well as the personal safety of those performing repairs. This manual outlines procedures for servicing and repairing vehicles using safe, effective methods. The procedures contain many NOTES, CAUTIONS and WARNINGS which should be followed along with standard safety procedures to eliminate the possibility of personal injury or improper service which could damage the vehicle or compromise its safety.

It is important to note that repair procedures and techniques, tools and parts for servicing motor vehicles, as well as the skill and experience of the individual performing the work, vary widely. It is not possible to anticipate all of the conceivable ways or conditions under which vehicles may be serviced, or to provide cautions as to all of the possible hazards that may result. Standard and accepted safety precautions and equipment should be used when handling toxic or flammable fluids, and safety goggles or other protection should be used during cutting, grinding, chiseling, prying, or any other process that can cause material removal or projectiles.

Some procedures require the use of tools specially designed for a specific purpose. Before substituting another tool or procedure, you must be completely satisfied that neither your personal safety, nor the performance of the vehicle will be endangered.

PRINCIPLES OF OPERATION

The automatic transmission is a complex unit, designed to deliver the engine power to the drive wheels in such a manner as to provide smooth power application through automatic gear ratio changes, dictated by engine speed, power requirements and road speed. The transmission gear ratios must be provided to multiply available engine torque to power the vehicle when starting, hill climbing and passing while still maintaining a gear ratio to avoid high engine RPM at high road speeds.

To provide a fully automatic transmission, a medium must be used to transmit the engine power to the transmission through an automatic clutch unit. The medium used is oil, because it is non-compressible and because of its power transmitting qualities when put in motion. The oil is commonly called ''Automatic Transmission Fluid (ATF)'' or ''Fluid''.

Along with its power transmitting abilities, the fluid is used to control the internal hydraulic system of the transmission and to lubricate the bearings, bushing and gears while still controlling the friction qualities of the clutches and the bands. In addition, the fluid serves as a heat transfer medium to carry the heat generated by the transmission to either an air-cooled or a separate water-cooled heat exchanger.

Although the design and construction differs among automatic transmission

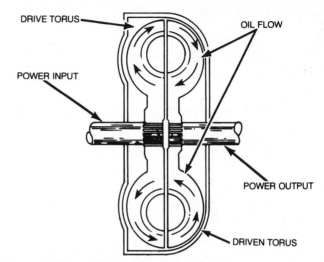

Cross-section of a fluid coupling and the oil flow direction

manufacturers, the mode of operation and the basic operating elements remain the same. The following basic operating elements are used in the construction of a typical automatic transmission.

1. Fluid coupling or torque converter.
2. Single or compound planetary gear sets.
3. Clutch packs and/or brake bands designed to operate in automatic transmission fluid.
4. Hydraulic valve controls and fluid passages.
5. Overrunning clutch units.

6. Manual, electric or vacuum-operated shift control.

To gain a better understanding of the operation of the automatic transmission elements, the following explanations are given.

FLUID COUPLING AND TORQUE CONVERTER
Fluid Coupling

The fluid coupling is used to transfer the engine torque through an impeller, bolted

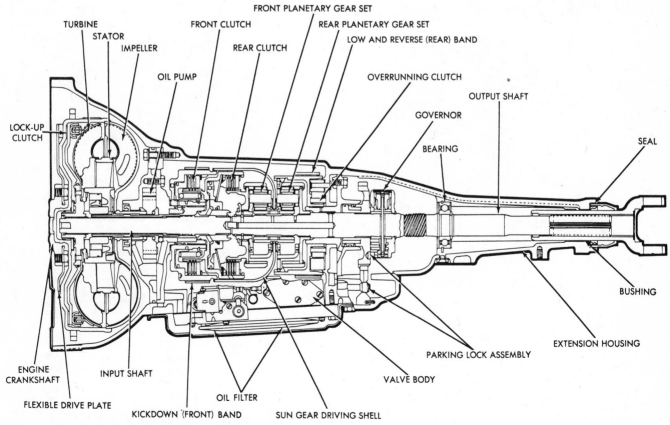

Cutaway view of a typical automatic transmission

to the engine crankshaft and turning whenever the engine operates, to a turbine, mounted on the transmission input shaft, through the use of automatic transmission fluid. Both the impeller and the turbine are constructed with vanes, so that when the impeller is speeded up by the increase of engine RPM, the automatic transmission fluid is directed from the impeller vanes to strike the turbine blades and cause the turbine to turn in proportion to engine speed, which in turn transfers the engine torque to the transmission input shaft. With the vehicle at idle, the engine torque is automatically interrupted by fluid slippage, due to low engine RPM and the loss of available engine torque to act upon the automatic transmission fluid.

The torque loss is greatest with this type of coupling and is not used without a torque multiplier in today's vehicles.

Torque Converter

The torque converter is a fluid coupler, combined with a torque multiplier to increase the engine torque to the transmission at times of vehicle start-up through the lower gear ratios, passing and hill climbing and to function as a fluid coupler when high engine and road speeds are attained and the need for a torque multiplier is not required.

The operation of the fluid coupler and

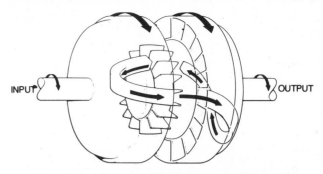

Fluid circulation through the elements of the torque converter

the torque converter are typical in many respects. The automatic transmission fluid is moved by the impeller which is bolted to and driven by the engine crankshaft. The moving fluid is directed to the turbine, attached to the transmission input shaft, and forces it to turn. A third unit, called a stator or torque multiplier, is placed between the impeller and the turbine and operates by the force of the automatic transmission fluid being directed to it.

As the impeller begins to move by starting the engine, the automatic transmission fluid is directed from the impeller vanes to the turbine vanes, which are curved opposite to the vanes of the impeller, and due to the force of the fluid, causes the turbine to rotate.

To achieve torque multiplication, the automatic transmission fluid is directed back to the impeller to accelerate the speed of the fluid and thereby increase the force on the turbine. The direction of the redirected fluid force is opposite the direction of the impeller and if allowed to strike the impeller vanes, would slow the impeller down. The third unit, a stator, is used to again redirect the fluid force from the turbine, through the stator vanes and then back to the impeller vanes in the direction of impeller rotation. A one-way clutch is used in the stator assembly to prevent the fluid force from rotating the stator opposite the impeller and turbine rotation.

As the fluid is forced from the turbine, reversed through the stator and into the

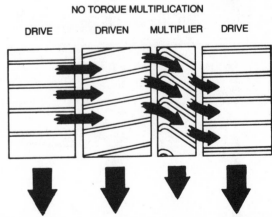

Fluid flow through the torque multiplier (stator) to increase the engine torque

Cutaway view of a torque converter assembly

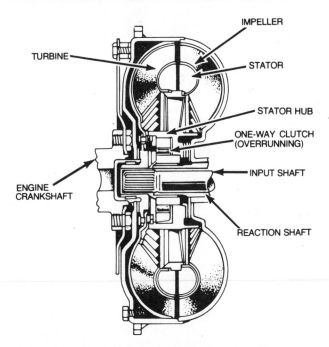

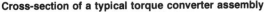

Cross-section of a typical torque converter assembly

impeller vanes, the fluid force is increased to approximately twice the energy force it had to begin with, thereby exerting a greater force on the turbine and giving torque multiplication as the fluid is forced back to the turbine blades and the fluid routing is begun again.

Torque Multiplication

Torque multiplication exists when more torque is applied to the turbine through redirected fluid force from the impeller, which is operating at a lesser torque. This condition is commonly called VORTEX FLOW. Vortex flow exists when the fluid is directed from the impeller, to the turbine, back through the stator vanes to change the fluid force direction, and to the impeller and is again directed to the turbine at a greater force.

It must be remembered that the torque multiplication decreases as the turbine speed increases in relation to the impeller. As the turbine speed catches up to the speed of the impeller, the angle of the fluid force leaving the turbine constantly changes and will strike the back of the stator vanes, causing the one-way clutch to unlock so that the impeller turbine and stator turn together. This is called COUPLING PHASE or when the turbine speed is 90% of the impeller speed.

Since the vortex flow is governed by the difference between the impeller and the turbine speeds, a constant adjustment of the torque multiplication is done automatically and will occur when the turbine rotates less than 90% of impeller speed.

Lock-up Torque Converter (Chyrsler Corp.)

To improve the fuel economy, reduce the fluid operating temperature and to

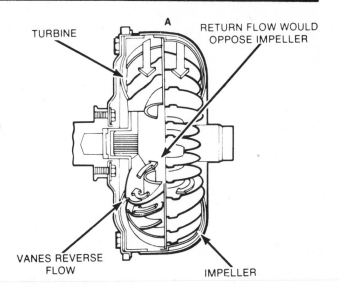

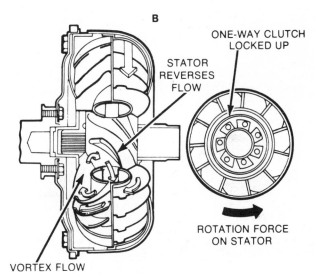

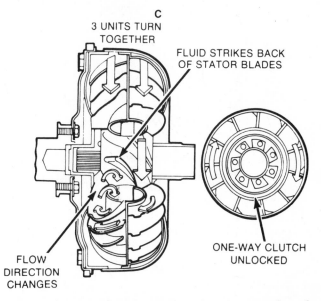

Types of fluid circulation created during the operation of the torque converter: A—Rotating Flow; B—Vortex Flow; C—Phase Coupling Flow

overcome fluid slippage in direct drive, a turbine to impeller lock-up device is incorporated in the torque converter to obtain 100% engine to transmission torque application. The lock-up torque converter is basically the same as the conventional torque converter in that a turbine, impeller and stator are used. In addition, an internal lock-up piston is used to lock the turbine to the impeller in direct drive.

The lock-up system is comprised of three major components.
1. Converter lock-up mechanism
   a. Sliding clutch piston
   b. Torsion springs
   c. Clutch friction material
2. Lock-up module attached to the valve body
3. Switch valve in the valve body

The clutch friction material is attached to the front cover, the clutch piston is mounted in the turbine, while the torsion springs are located on the forward side of the turbine.

NOTE: The torsion springs dampen engine firing impulses and absorb the shock loads that occur during the lock-up.

As the vehicle speed attains approximately 35-40 MPH and the transmission shifts into direct drive, pressured fluid is channeled through the transmission input shaft and into an area between the clutch piston and the turbine. The fluid pressure forces the piston against the front cover friction material and locks the turbine to the impeller, giving 100% engine torque transfer to the transmission.

As the vehicle speed drops below approximately 35 MPH, the fluid pressure is released and the piston disengages from the front cover friction material. The torque converter now operates in the conventional manner.

CAUTION: *The lock-up torque converter is not interchangeable with the conventional torque converter. The internal transmission parts prohibit the interchange. Do not attempt to service, repair or flush the unit; replace as a unit.*

Automatic Overdrive Torque Converter (Ford Motor Co.)

The torque converter used with the new type automatic overdrive transmission is similar in operation to other torque converters, but has an added spring damper assembly and an input shaft used for direct and overdrive gear ratios. When the engine is directed through the added input shaft, the torque converter is by-passed and a mechanical lock-up occurs.

Converter Clutch Unit (General Motors Corp.)

The lock-up clutch mechanism is incorporated in the torque converter assembly and provides a mechanical direct drive coupling of the engine to the planetary gears, to eliminate converter slippage.

The converter clutch operation is controlled by the transmission manual valve being positioned in the DRIVE detent and by a series of controls, both electric and hydraulic, to engage the converter clutch above a predetermined vehicle speed, with the transmission in the third speed or direct gear ratio.

Other Lock-up Converters

Various other automatic transmission manufacturers are engaged in the development of lock-up converter clutches to improve fuel economy and to reduce the fluid operating temperatures. Centrifugal force, oil pressure electrical and vacuum switch

controls are some of the methods being tested for reliability and dependability.

PLANETARY GEAR ASSEMBLY

To obtain the necessary gear combinations needed for gear reduction, direct drive, reverse, neutral and overdrive, planetary gear sets are used in the automatic transmission. The planetary gear set is always in mesh and consists of pinion gears mounted on a planetary carrier by shafts or pins, a sun gear or center gear and a ring gear or outer gear with internal teeth. Each of the three parts of the planetary gear set can be a drive or driven member, depending upon the gear ratio needed. All automatic transmissions use two or more planetary gear sets, with either a common or separate sun gear, to obtain the three or four speeds needed to move the vehicle and bring it to cruising speed.

Principles of Operation

To avoid confusion, the operating principle of a single planetary gear set will be examined with the comparison of the compound gear set to follow.

Gear Reduction

To obtain gear reduction, the sun gear or center is held stationary while the ring gear or outer gear is driven which causes the planetary gears and carrier to rotate or walk around the sun gear in the same direction as the ring gear turns, but not as fast. This causes the engine torque to be multiplied since the output shaft is not turning at the same RPM as the input shaft.

Direct Drive

To obtain direct drive, the entire planetary gear set must rotate as a unit by having both the sun gear and ring gear locked together and being driven from the same torque input through the planetary carrier.

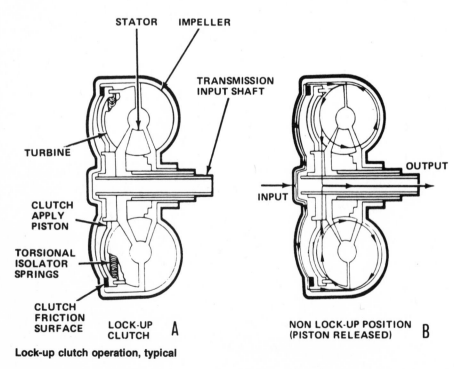
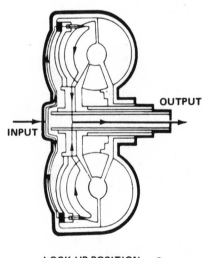

Lock-up clutch operation, typical

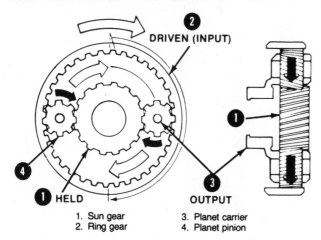

Planetary gear set in reduction

1. Sun gear
2. Ring gear
3. Planet carrier
4. Planet pinion

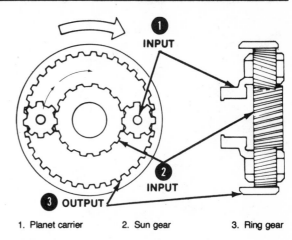

Planetary gear set in direct (no reduction)

1. Planet carrier   2. Sun gear   3. Ring gear

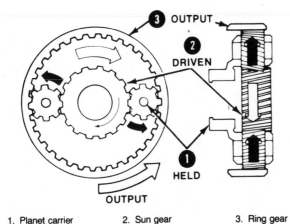

Planetary gear set in reverse

1. Planet carrier   2. Sun gear   3. Ring gear

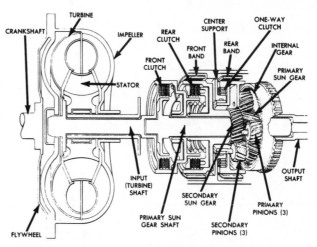

Planetary gear set in overdrive

1. Sun gear   2. Planet carrier   3. Ring gear

Reverse

To change the input power flow to a reverse rotation at the output shaft, the pinion carrier is held stationary, while the sun gear is driven, causing the planetary pinions to turn in the opposite direction, which causes the ring gear to be driven in the opposite direction as well, thereby providing a reverse gear to the transmission.

Neutral

Obtaining the neutral position where a torque input is directed into the transmission, but no torque output is developed at the output shaft, results from having none of the driving members coupled to the planetary gear set.

Overdrive

An overdrive gear ratio is possible by holding the sun gear and driving the pinion carrier, which causes the pinion gears to move around the stationary sun gear and drive the ring gear at a faster speed than the input shaft.

Compound Planetary Gear Assembly

Since the planetary gear set only provides one reduction and one direct drive in the same direction of rotation, it is necessary to use two planetary units connected together in series to obtain the three and four forward speeds. The front and rear planetary gear assemblies, which comprise the compound planetary unit, are similar in gear arrangement, but normally differ in size to obtain a different percentage of the reduction.

The compound planetary gear assembly is normally interconnected by a common sun gear or by long and short pinion gears, intermeshed to each other and both capable of being a drive or driven member. Each pinion grouping is meshed to a sun gear (the primary and secondary) which are of different sizes to allow power flow through different stages of gear reduction.

Typical planetary gear arrangement coupled to the other components of the gear train

General Information

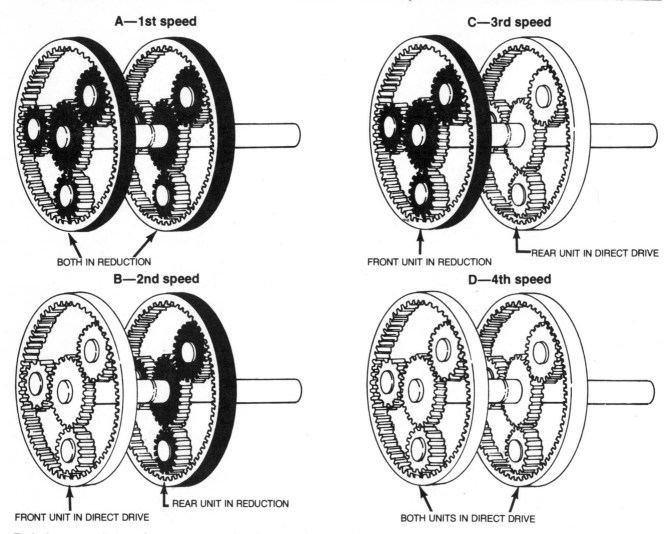

A—1st speed

BOTH IN REDUCTION

B—2nd speed

FRONT UNIT IN DIRECT DRIVE

REAR UNIT IN REDUCTION

C—3rd speed

FRONT UNIT IN REDUCTION

REAR UNIT IN DIRECT DRIVE

D—4th speed

BOTH UNITS IN DIRECT DRIVE

Typical compound planetary gear set usage in a four speed automatic transmission to obtain different gear ratios

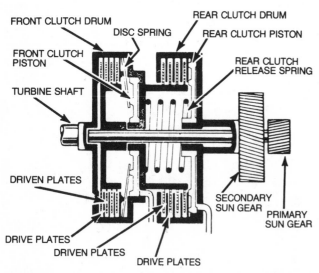

FRONT CLUTCH DRUM
REAR CLUTCH DRUM
DISC SPRING
REAR CLUTCH PISTON
FRONT CLUTCH PISTON
REAR CLUTCH RELEASE SPRING
TURBINE SHAFT
DRIVEN PLATES
SECONDARY SUN GEAR
PRIMARY SUN GEAR
DRIVE PLATES
DRIVEN PLATES
DRIVE PLATES

Typical clutch pack arrangement in the automatic transmission gear train

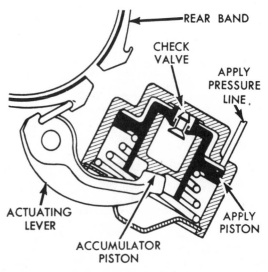

REAR BAND
CHECK VALVE
APPLY PRESSURE LINE
ACTUATING LEVER
ACCUMULATOR PISTON
APPLY PISTON

Typical brake band servo using a large servo lever (© Ford Motor Co.)

CLUTCHES AND BANDS

In order for the planetary gear assemblies to function, a combination of clutches and bands are used to hold, drive or turn with a member of the planetary gear assembly, to provide the phases of the necessary gear ratios.

The clutches and bands are applied and released by both hydraulic and spring pressure, acting upon the surface of pistons, positioned in such a manner as to operate the bands or clutches quickly and securely.

The servo pistons are designed to have a small or a large surface area for the hydraulic pressures to act upon, depending upon the force needed to apply the brake band or clutch pack. The servo levers are designed to assist the hydraulic pressures where needed, by having a force ratio, for example, of from 2.0:1 to 3.0:1, such as used with the reverse servo.

It is most important to maintain the clutch packs and brake bands with the units for which they were designed, due to the type of materials used during their manufacture. Different designs in the friction coefficient (holding power) of the material and the applying surface are used for a specific purpose: to control the quality of the shift and the holding abilities of the brake bands and clutch packs. When brake bands and clutch packs are interchanged into units not designed for them, the shift quality is changed and possible premature transmission failure can occur.

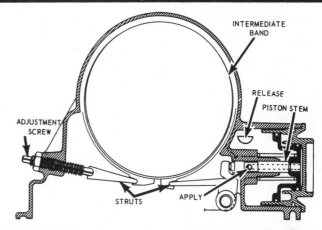

Typical brake band servo using small servo lever with adjustment screen on the opposite side of the band locating lug (© Ford Motor Co.)

Clutches

The clutch pack normally consists of a specific number of composition lined and steel plates, assembled in an alternate sequence, so as to avoid metal to metal contact. The location of the clutches are either within a drive or driven drum, attached to either the ring gear or sun gear of the planetary gear set, or arranged between the transmission case and a rotating member. Upon demand by hydraulic pressure, the clutch pack will slow and lock the companion member to obtain a specific gear ratio.

Bands

The brake band is normally a split circle with band anchors located opposite each other at the split, while other bands are of the "wraparound" type with the band anchors off-set from each other. The bands are usually made from spring steel and have a composition lining bonded to the inside surface. As the operating servo is applied by fluid pressure, the band tightens on the rotating encased drum, slowing and stopping it to provide a change in the gear ratios.

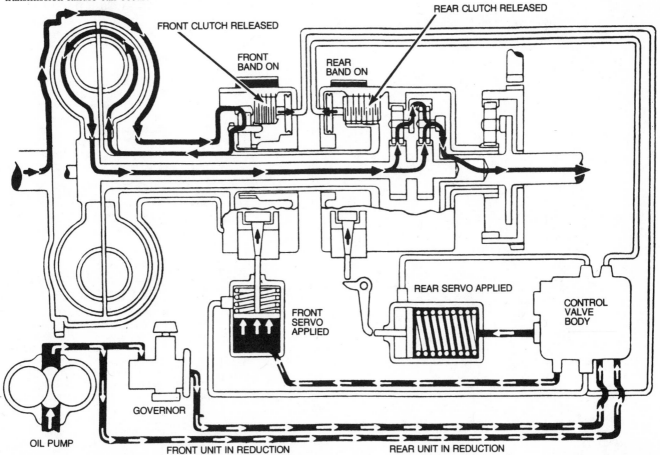

Power flow through a typical automatic transmission, using clutches, bands and planetary gear sets, with servo application shown

First Gear

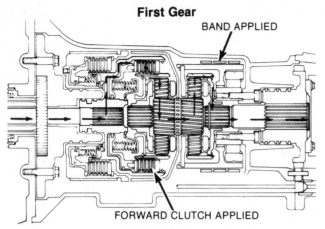

High Gear

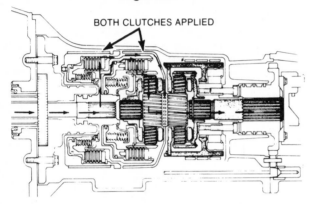

Neutral and Park (no power transfer)

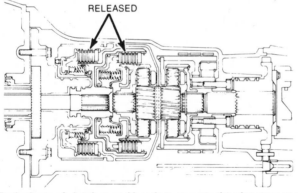

Typical use of clutches and bands to operate the planetary gear sets to obtain different gear ratios

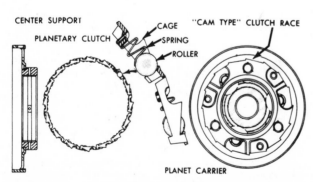

Components and roller/spring position of roller type overrunning clutch (© Ford Motor Co.)

OVERRUNNING CLUTCH

The overrunning clutch is used to allow a rotating part to turn in one direction, but not in another.

Two types of overrunning clutches are used: the roller type and the sprag type. Each type performs the same function, with only the physical appearance and lock-up procedure being different.

Roller Type Overrunning Clutch

The roller type overrunning clutch has a series of spring-loaded roller bearings set in a cam-type notch of the roller carrier, located between an inner and outer race.

By holding one race and rotating the second race in a specified direction, the rollers act against the spring tension and move into the large cam cavity of the cam notch by the action of the rotating race on the roller bearing, allowing the clutch unit to overrun or freewheel. Turning the moveable race in the opposite direction causes the spring tension to act upon the rollers, along with the race rotation, to move the roller bearings into the smaller part of the cam notch, thereby locking the held race, the roller carrier and the moveable race together by a jamming action.

Sprag Type Overrunning Clutch

The sprag type overrunning clutch uses locking lugs, shaped in appearance like flattened roller bearings. The individual sprags are placed between the inner and outer races and are held in place by two springs positioned in the sprag notches, located at each end of the sprag. When the rotating race is turned in the proper direction, the reaction on the retaining springs by the rotating races causes the sprags to jam between the inner and outer races. When turned in the opposite direction, the sprags are released and the unit is in a freewheeling or overrunning mode.

NOTE: Either type overrunning clutch unit can be found in the converters or the internal gear train, depending upon the manufacturer.

MANUAL, ELECTRIC OR VACUUM CONTROLS

The gear selection and shifting cannot be entirely automatic, but must be controlled by the operator to guide the vehicle through traffic conditions at varied speeds and in different gear reductions.

The operator selects a gearing range by manually moving a selector lever, connected by linkage or cable, to the manual valve within the transmission. As the manual valve is moved, specified pressure passages are opened to apply or release certain elements within the transmission to obtain the proper gear ratios, so that movement of the vehicle can begin.

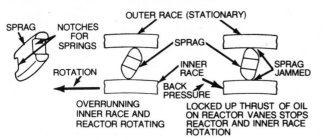

Sprag operation to lock and unlock rotating member

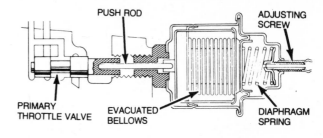

Cross-section of vacuum diaphragm (modulator) unit

To control the shifting from one gear ratio to another as the vehicle speed increases, either a mechanical connection between the carburetor throttle opening linkage and the transmission throttle valve is used to control the throttle pressure, or a vacuum diaphragm, operated by engine vacuum, is used to control the throttle pressure.

To provide a downshift control when periods of quick acceleration are needed, such as moving uphill or passing another vehicle, an electric solenoid or manual linkage connected to the accelerator or to the valve body is used to move a downshift valve, located within the control valve assembly.

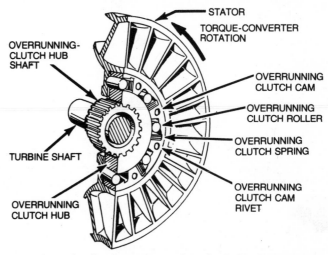

Cutaway view of roller type overrunning clutch (© Chrysler Corp.)

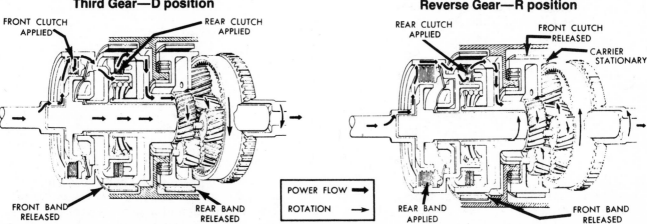

First gear—D position
Second Gear—D position
Third Gear—D position
Reverse Gear—R position

Typical power flow through rough gear ratios using clutch, band and planetary gear applications (© Ford Motor Co.)

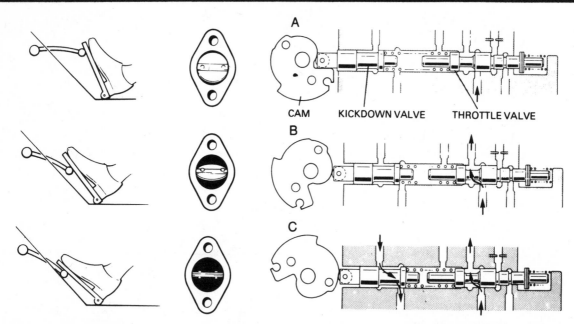

Comparison of a cable operated downshift valve in conjunction with the throttle opening in the idling mode, the throttle partical open and the kickdown valve positioned to open passages and allow the fluid to pass: A—Idling; B—Throttle partly open; C—Kickdown valve

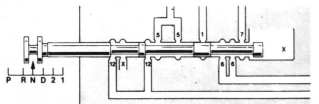

A—Neutral—Line pressure (1) is blocked by spool valve land

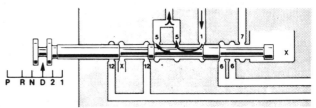

D—Drive (D)—Line pressure (1) has access to line 5. All other lines are open to exhaust ports.

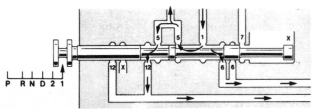

B—First Gear—Line pressure (1) has access to lines 5, 6, and 12. Adjacent exhaust port is closed. Note: Line 7 is open to exhaust port (X).

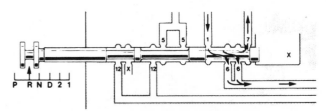

E—Reverse (R)—Line pressure has access to lines 6 and 7. Lines 5 and 12 are open to exhaust ports.

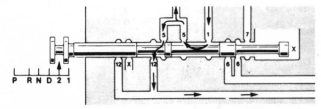

C—Second Gear—Line pressure (1) has access to lines 5 and 12. The adjacent exhaust port is closed. Lines 6 and 7 are open to other exhaust ports.

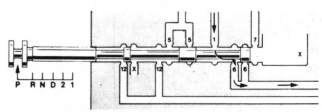

F—Park (P)—Line pressure (1) has access to only line 6. Lines 5, 7 and 12 are open to exhaust ports. Mechanical connection operates the parking pawl.

Control functions of the manual valve

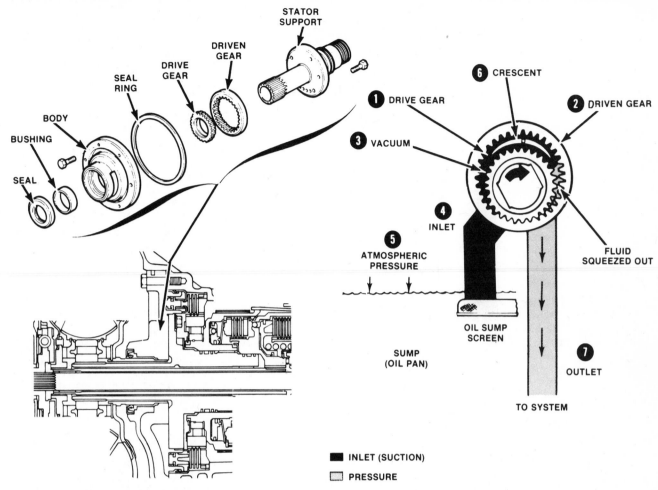

Oil pump operation, location and exploded view (© Ford Motor Co.)

COMBINING THE AUTOMATIC TRANSMISSION ELEMENTS

When all of the previously discussed elements are combined into the transmission case, we have a power transfer unit that can direct the needed torque from the engine to the drive wheels by automatically changing the gear ratios to compensate for engine load, vehicle speed and power requirements by torque multiplication at lower speeds and the change to direct (or overdrive) at higher engine and vehicle speeds.

THE HYDRAULIC SYSTEM
Oil Pump

The oil pump is the heart of the automatic transmission hydraulic system. Without it, fluid pressure is not attained nor can the automatic transmission function to change gear ratios or provide lubrication to the rotating parts. A steady flow of hydraulic fluid is required and to perform this function, a ''positive displacement'' pump is used.

POSITIVE DISPLACEMENT PUMP

Positive displacement means that the pump is designed to deliver a definite quantity of fluid as long as the pump is turning and fluid is supplied to the inlet, the volume being in proportion to the pump drive speed. The positive displacement pumps are designed to deliver more fluid pressure than the transmission needs and a pressure regulator or relief valves are used to vent the excess fluid pressure into the reservoir sump and to maintain a specific operating control pressure.

PUMP EFFICIENCY

Efficiency of the pump is the actual output in respect to the fluid displacement. Should the pump be able to displace, for example, 100 cubic inches of fluid per revolution of the pump, the efficiency of the pump is considered to be 100%. Should the internal parts of the pump become worn or damaged and the pump becomes able to deliver only 80 cubic inches of fluid, the efficiency of the pump is considered to be only 80%. It should be noted that the loss of pump efficiency does not mean the pressure is lowered, as the pressure will continue to build up as long as fluid is being pumped and not leaking off in another part of the hydraulic system. The loss of pump efficiency will affect the application of the pressure and the application of the working components by slowing their movements down.

PUMP OPERATION

Both rotary blades and internal gears are used in the positive displacement pumps, with the internal gear type most widely used in the automatic transmissions. The drive gear is operated by the converter impeller hub, which is bolted to the engine crankshaft and turns whenever the engine is operated. The drive gear is in mesh with the driven gear and whenever the drive gear is rotated, the driven gear is rotated also. As the drive and driven gears separate from mesh, a vacuum is created. The pump inlet for fluid is positioned at this point and the atmospheric pressure forces the fluid into the pump to fill the vacuum void. At the point of greatest separation, a crescent is used to trap the fluid as the gears turn past the inlet. The spaces between the crescent and the gear teeth form chambers for the fluid to be carried towards the outlet. As the gears come together again, the crescent

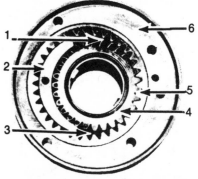

1. Low pressure area
2. Crescent
3. High pressure area
4. Drive gear
5. Driven gear
6. Pump body

Oil pump body with a crescent

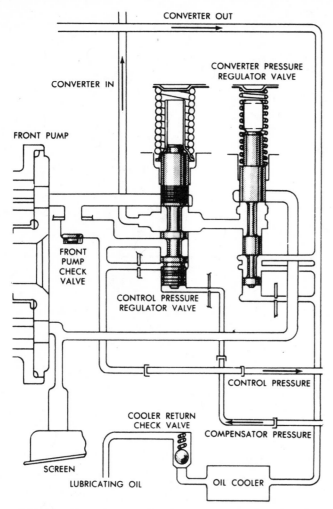

Typical control pressure regulator valve and converter pressure regulator valve used with a positive displacement oil pump (© Ford Motor Co.)

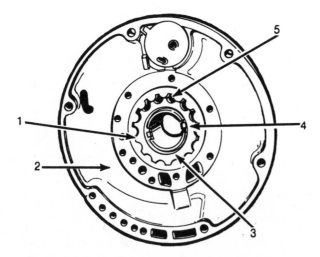

1. High pressure area
2. Pump body
3. Drive gear
4. Driven gear
5. Low pressure area

Oil pump body without a crescent

blocks the fluid from getting back to the inlet and the gear teeth squeeze the fluid and force it into the hydraulic system under pressure.

NOTE: Early pump bodies were smaller and did not require a crescent. As the demand for greater pump efficiency and fluid displacement were needed as the automatic transmission evolved, the low pressure chamber in the pump body became too large for proper fluid movement, so that a crescent was installed to block the fluid and to maintain a direction of flow.

Hydraulic Valve Controls and Fluid Passages

The hydraulic control system is supplied a constant flow of pressurized fluid from the oil pump. The pressurized fluid is used to charge the torque converter, the cooling and lubricating system, the control valve assembly, the governor system and the fluid passages in the transmission case.

The fluid pressure is regulated to maintain a specific fluid pressure, either before reaching the control valve assembly or within the assembly. Fluid pressure is directed manually by the operator, through mechanical linkages to specific passages controlling the application and release of the various bands and clutch packs to engage the initial gear ratio so that the vehicle can be moved. As the vehicle speed is increased, the shift valves within the control valve assembly are moved by governor and modulated throttle pressures by overcoming the holding control pressure and therefore automatically changing the gear ratio in relation to the speed of the vehicle and the load on the engine, by directing pressurized fluid to apply or release selected bands and clutch packs.

The automatic transmission operates on a balanced fluid/spring pressure equalization method. The spring tension is varied by its length, size and strength. It can sometimes be varied by an adjustable screw. The fluid pressure is varied by the governor assembly, the throttle valve (which is controlled by either mechanical linkage or a vacuum diaphragm), and by the vehicle road and engine speed.

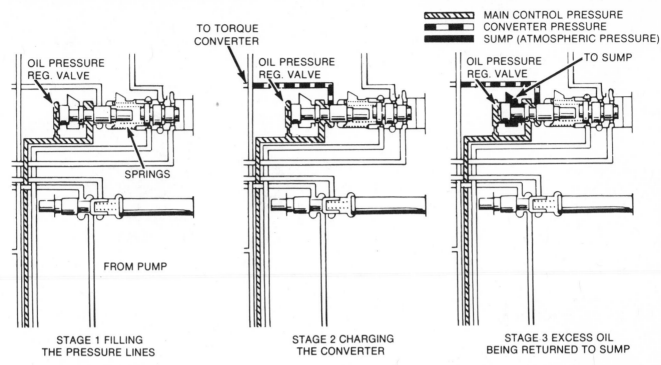

MAIN CONTROL PRESSURE
CONVERTER PRESSURE
SUMP (ATMOSPHERIC PRESSURE)

STAGE 1 FILLING
THE PRESSURE LINES

STAGE 2 CHARGING
THE CONVERTER

STAGE 3 EXCESS OIL
BEING RETURNED TO SUMP

The use of oil pressure and spring pressure opposing each other to form the balanced equalization method (© Chrysler Corp.)

The purpose of the spring tension is to assist regulated main control pressure in opposing the governor and modulated throttle pressures to move the shift valves until predetermined vehicle speeds have been reached. As the vehicle speed is decreased and drops below the predetermined shift point, the spring tension and regulated main control pressure act together as a force and move the valve back into its original position. This causes a change in the gear ratio by automatically shifting from one ratio to another.

The spring tension also controls fluttering of the control valves during periods of low fluid pressures and when opposing pressures are the same, gives the valves a more positive movement or holds the valves in a specific position. In many cases, springs are used on both sides of the valves, each having different tensions.

Any combination of valve control variations can be obtained to control the timing and quality of the shift, by the manufacturer, in the use of different spring tensions and varied oil pressures at different RPM, since spring tensions increase as the spring is compressed and the hydraulic pressure varies with the speed of the oil pump.

Valves

Different types of valves are used in the hydraulic circuits to direct or cut off the fluid flow from one area to another and to regulate the amount of fluid flow through a passage.

DIRECTIONAL VALVES

Directional valves permit fluid flow in one direction and not in the reverse direction, unless a metering effect is desired. The directional valves are normally called check valves and can be either single-acting (allows fluid to flow in one direction only) or double-acting (allows the fluid to flow from one of two passages into the third passage, but restricts the reverse flow of fluid back into one of the two original passages).

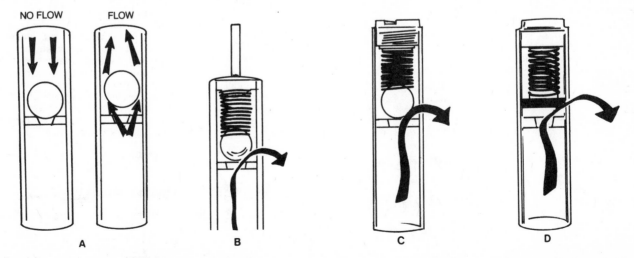

NO FLOW FLOW

A   B   C   D

Cross-section of a typical ball check valve, allowing fluid flow in one direction only (A). Cross-section of spring loaded ball check valve (B). Cross-section of adjustable spring loaded check valve (C). Cross-section of poppet valve using disc instead of a ball as a checking component (D). (© Chrysler Corp.)

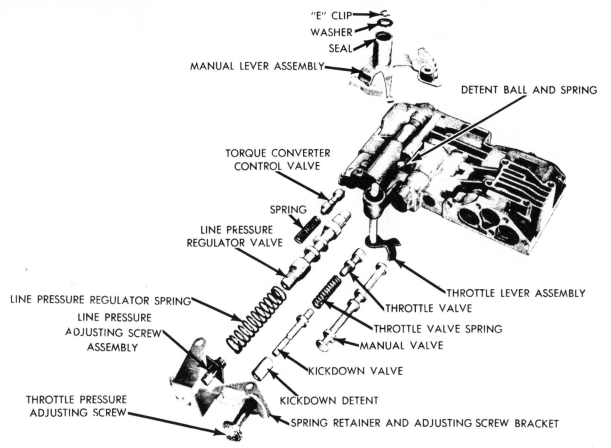

Typical control valve section—Exploded view showing valves and springs

The operation of the check valves is normally automatic and is controlled by fluid pressure or spring pressure. When the check valve is operated by fluid pressure, the ball is moved from its seat when the fluid pressure flows through the seat orifice. When the pressure direction is reversed, the fluid pressure moves the ball back to the seat and closes the orifice to prevent the fluid from passing. This type of check valve is not quick acting, since the ball is allowed to float. The spring-operated check valves have the check ball spring loaded to allow the fluid pressure to build up to a specific pressure before overcoming the spring pressure and lifting the ball from its seat. When the fluid pressure is released, the spring pressure quickly forces the ball to its seat and closes the orifice to stop the reverse flow of fluid.

A means of controlling the positive displacement oil pump pressure is needed and this type of check valve is used. It is nor-

mally called a pressure relief valve or a pressure regulator valve.

As the fluid pressure is increased and opposes the ball or valve spring pressure, the ball or valve is positioned in a hovering manner over the seat orifice and is balanced between the spring pressure and the fluid pressure, with the ball or valve moving in the direction of less pressure, either against fluid pressure or spring pressure, to maintain a specific amount of fluid pressure for the automatic transmission operation.

The spring pressure is adjustable on certain check valves and is used to recalibrate the fluid pressure without replacement of the springs. This type of check valve is called an adjustable pressure relief valve.

Another form of the check valve is called the poppet valve type, which uses a disc, cone or other form of stopper in place of the ball, to stop the fluid flow in the reverse direction.

SPOOL VALVES

The majority of valves used in the valve body and governor body are called "spool valves" because of their appearance to thread spools of different sizes, mated together. These valves are precision machined to fit in the close tolerance valve and governor body bores. The spool valves have lands to slide on and grooves to allow the fluid to pass through and move from passage to passage.

This type of valve can be operated either manually or automatically. The manual valve operation is an example of the manually operated spool valve, while the spool valves moved by fluid pressure opposing spring pressure is considered automatic. A third type of spool valve can be moved by fluid pressure alone, acting upon either end of the valve, causing the valve to move back and forth without any spring pressure resistance, and is called a shuttle valve.

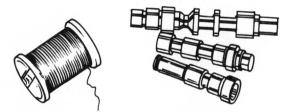

Comparison between thread spool and spool valves, hence the name "spool" (© Chrysler Corp.)

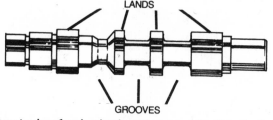

Spool valve showing lands and grooves (© Chrysler Corp.)

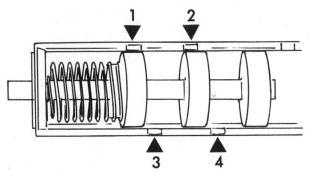

Function of vave lands holding pressure at ports 1 and 2 to prevent the pressure from moving to ports 3 and 4 (© Chrysler Corp.)

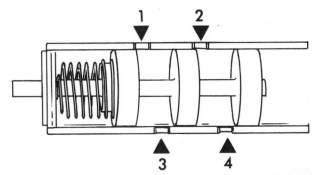

Movement of valve against spring pressure to open ports 1 and 2, allowing pressure to cross and exit through ports 3 and 4 (© Chrysler Corp.)

DIFFERENTIAL SPOOL VALVES

On many spool valves, the lands will be of different sizes. The reason is that the fluid pressure in a closed system is the same on all interior parts of the system (Pascal's Law) and that when the force of the fluid pressure is acting upon two moveable surfaces, the larger surface will have the greatest reaction. This pressure reaction is designed into the valve bodies to take advantage of the fluid pressure and to have the valve movements occur at specific pressure levels and specific torque requirement needs.

GENERAL TRANSMISSION SERVICE AND INSPECTION

Diagnosing automatic transmission problems is simplified following a definite procedure and understand the basic operation of the individual transmission that is being inspected or serviced. Do not attempt to "short-cut" the procedure or take for granted that another technician has performed the adjustments or the critical checks. It may be an easy task to locate a defective or burned-out unit, but the technician must be skilled in locating the primary reason for the unit failure and must repair the malfunction to avoid having the same failure occur again.

Each automatic transmission manufacturer has developed a diagnostic procedure for their individual transmissions. Although the operation of the units are basically the same, many differences will appear in the construction, method of unit application and the hydraulic control systems.

The same model transmissions can be installed in different makes of vehicles and are designed to operate under different load stresses, engine applications and road conditions. Each make of vehicle will have specific adjustments or use certain outside manual controls to operate the individual unit, but may not interchange with another transmission/vehicle application from the same manufacturer.

The identification of the transmission is most important so that the proper preliminary inspections and adjustments may be done and if in need of a major overhaul, the

correct parts may be obtained and installed to avoid costly delays.

Engine Performance

When engine performance has declined due to the need of an engine tune-up or a system malfunction, the operation of the transmission is greatly affected. Rough or slipping shifts and overheating of the transmission and fluid can occur, which can develop into serious internal transmission problems. Complete the adjustments or repairs to the engine before the road test is attempted or transmission adjustments made.

General Diagnosis

A sequence for diagnosis of the transmission is needed to proceed in an orderly manner. A suggested sequence is as follows:
1. Inspect and correct the fluid level.
2. Inspect and adjust the throttle or kick-down linkage.
3. Inspect and adjust the manual linkage.
4. Install one or more oil pressure gauges to the transmission as instructed in the individual transmission sections.
5. Road test the vehicle.

NOTE: During the road test, use all the selector ranges while noting any differences in operation or changes in oil pressures, so that the unit or hydraulic circuit can be isolated that is involved in the malfunction.

Inspection of the Fluid Level

Most automatic transmissions are designed to operate with the fluid level between the ADD OR ONE PINT and FULL marks on the dipstick indicator, with the fluid at normal operating temperature. The normal operating temperature is attained by operating the engine-transmission assembly for at least 8 to 15 miles of driving or its equivalent. The fluid temperature should be in the range of 150° to 200° F when normal operating temperature is attained.

NOTE: If the vehicle has been operated for long periods at high speed or in extended city traffic during hot weather, an accurate fluid level check cannot be made until the fluid cools, normally 30 minutes after the vehicle has been parked, due to fluid heat in excess of 200° F.

The transmission fluid can be checked during two ranges of temperature.

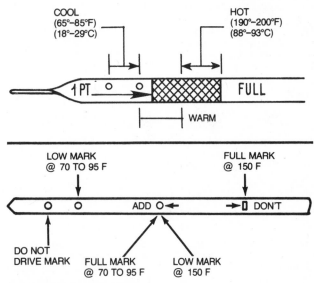

Fluid level check indicators—Typical

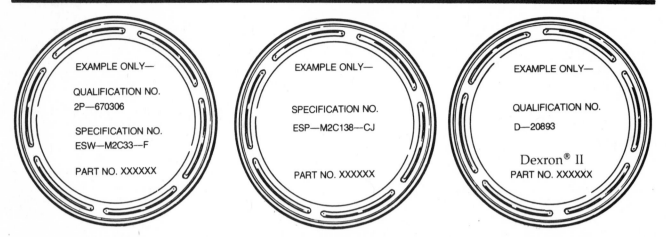

Comparison of type F, Dexron® II and type CJ fluid containers and their identifying code numbers (© Ford Motor Co.)

1. Transmission at normal operating temperature.
2. Transmission at room temperature.

During the checking procedure and adding of fluid to the transmission, it is most important not to overfill the reservoir to avoid foaming and loss of fluid through the breather, which can cause slippage and transmission failure.

TRANSMISSION AT NORMAL OPERATING TEMPERATURE

(150° to 200° F.—Dipstick hot to the touch)

1. With the vehicle on a level surface, engine idling, wheels blocked or parking brake applied, move the gear selector lever through all the ranges to fill the passages with fluid.
2. Place the selector lever in the Park position and remove the dipstick from the transmission. Wipe clean and reinsert the dipstick to its full length into the dipstick tube.
3. Remove the dipstick and observe the fluid level mark on the dipstick stem. The fluid level should be between the ADD and the FULL marks. If necessary, add fluid through the filler tube to bring the fluid level to its proper height.
4. Reinstall the dipstick and be sure it is sealed to the dipstick filler tube to avoid the entrance of dirt or water.

TRANSMISSION AT ROOM TEMPERATURE

(65° to 95° F.—Dipstick cool to touch)

CAUTION: *The automatic transmissions are sometimes overfilled because the fluid level is checked when the transmission has not been operated and the fluid is cold and contracted. As the transmission is warmed to normal operating temperature, the fluid level can change as much as ¾ inch.*

1. With the vehicle on a level surface, engine idling, wheels blocked or parking brake applied, move the selector lever through all the ranges to fill the passages with fluid.

2. Place the selector lever in the Park position and remove the dipstick from the transmission. Wipe clean and re-insert it back into the dipstick tube.
3. Remove the dipstick and observe the fluid level mark on the dipstick stem. The fluid should be directly below the FULL indicator.

NOTE: Most dipsticks will have either one mark or two marks, such as dimples or holes in the stem of the dipstick, to indicate the cold level, while others may be marked HOT or COLD levels.

4. Add enough fluid, as necessary, to the transmission, but do not overfill.

CAUTION: *This operation is most critical, due to the expansion of the fluid under heat.*

Fluid Type Specifications

The automatic transmission fluid is used for numerous functions such as a power-transmitting fluid in the torque converter, a hydraulic fluid in the hydraulic control system, a lubricating agent for the gears, bearings and bushings, a friction-controlling fluid for the bands and clutches and a heat transfer medium to carry the heat to an air or cooling fan arrangement.

Because of the varied automatic transmission designs, different frictional characteristics of the fluids are required so that one fluid cannot assure freedom from chatter or squawking from the bands and clutches. Operating temperatures have increased sharply in many new transmissions and the transmission drain intervals have been extended or eliminated completely. It is therefore most important to install the proper automatic transmission fluid into the automatic transmission designed for its use.

TYPES OF AUTOMATIC TRANSMISSION FLUID

TYPE A

This fluid was developed and recommended for use in all automatic transmissions manufactured before 1956.

Container Identification number—AQ-ATF-XXX (X = digits in the Qualification number)

TYPE A—SUFFIX A

This fluid supersedes the Type A fluid and was recommended for use in all General Motors, Chrysler and American Motors automatic transmissions manfactured between 1956 and 1967.

Container Identification number—AQ-ATF-XXXX

DEXRON®

This fluid supersedes the Type A-Suffix A fluid and was recommended for use in all General Motors, Chrysler, American Motors and certain imported vehicles automatic transmissions, manufactured since 1967. This fluid was recommended for use where the Type A and Type A-Suffix A fluids were originally specified. This fluid was designed to maintain its frictional and other characteristics longer under severe service and operation.

Container Identification number—B-XXXX

DEXRON® II

This fluid supersedes the Dexron® type fluid and meets a more severe set of performance requirements, such as improved high temperature oxidation resistance and low temperature fluidity. The Dexron® II is recommended for use in all General Motors, Chrysler, American Motors and certain imported vehicles automatic transmissions. This fluid can replace all Dexron® fluid with a B- number designation.

Container Identification number—D-XXXXX

DEXRON® II SERIES D

This fluid was developed and used, beginning with the 1977 model year and is used in place of the regular Dexron® II fluids.

The container identification is with a ''D'' prefix to the qualification number on the top of the container.

TYPE F

Ford Motor Company began developing its own specifications for automatic transmission fluid in 1959 and again updated its specifications in 1967, requiring fluid with different frictional characteristics and identified as Type F fluid.

Beginning with the 1977 model year, a new Type F (CJ) fluid was specified for use with the C-6 and newly introduced Jatco model PLA-A transmissions. This new fluid is not interchangable with the original Type F fluid.

Prior to 1967, all Ford automatic transmissions use fluids in containers marked with qualification number IP-XXXXXX, meeting Ford specification number ESW-M2C33-D.

Fluids in containers marked with qualification number 2P-XXXXXX meets Ford specification ESW-M2C33-F and is used in the Ford automatic transmissions manufactured since 1967, except the 1977 and later C-6, The Automatic Overdrive, and Jatco models PLA-A, PLA-A1, PLA-A2 transmissions.

The container identification number for the new fluid is Ford part number D7AZ-19582-A and carries a qualification number ESP-M2C138-CJ.

TYPE CJ

CAUTION: *The CJ fluid is NOT compatible with clutch friction material of other Ford transmissions and must only be used in the 1977 and later C-6, the Automatic Overdrive and Jatco models PLA-A, PLA-A1 and PLA-A2 automatic transmissions.*
Do not mix or interchange the fluids through refills or topping off as the Type F and the Type CJ fluids are not compatible.

A technical bulletin has been issued by Ford Motor Company, dated 1978, advising the compatibility of Dexron® II, series D fluid with the CJ fluid. It can be substituted or mixed, if necessary, in the 1977 and later C-6, the Automatic Overdrive and the Jatco PLA-A, PLA-A1 and PLA-A2 automatic transmissions.

NOTE: With approved internal modifications, CJ fluid can be used in the 1979 and prior C-4 automatic transmissions. Refer to the C-4 automatic transmission section for specific information and refer to the dipstick before adding fluid to late model C-4 automatic transmissions. Should the modifications be made and the fluid changed, that should be noted at the vehicle dipstick area so that the proper fluid can be installed as needed.

FLUID CONDITION

During the checking of the fluid level, the fluid condition should be inspected for color and odor. The normal color of the fluid is deep red orange-red and should not be a burned brown or black color. If the fluid color should turn to a green/brown shade at an early stage of transmission operation and have an offensive odor, but not a burned odor, the fluid condition is considered normal and not a positive sign of required maintenance or transmission failure.

With the use of absorbent white paper, wipe the dipstick and examine the stain for black, brown or metallic specks, indicating clutch, band or bushing failure, and for gum or varnish on the dipstick or bubbles in the fluid, indicating either water or anti-freeze in the fluid.

Should there be evidence of water, anti-freeze or specks of residue in the fluid, the oil pan should be removed and the sediment inspected. If the fluid is contaminated or excessive solids are found in the removed oil pan, the transmission should be disassembled, completely cleaned and overhauled. In addition to the cleaning of the transmission, the converter and transmission cooling system should be cleaned and tested.

FLUID OVERFILL PROBLEMS

When the automatic transmission is overfilled with fluid, the rotation of the internal units can cause the fluid to become aerated. This aeration of the fluid causes air to be picked up by the oil pump and causes loss of control and lubrication pressures. The fluid can also be forced from the transmission assembly through the air vent, due to the aerated condition.

FLUID UNDERFILL PROBLEMS

When the fluid is low in the transmission, slippage and loss of unit engagement can result, due to the fluid not being picked up by the pump. This condition is evident when first starting after the vehicle has been sitting and cooled to room temperature, in cold weather, making a turn or driving up a hill. This condition should be corrected promptly to avoid costly transmission repairs.

THROTTLE VALVE AND KICKDOWN CONTROL INSPECTION

Inspect the throttle valve and kickdown controls for proper operation, prior to the road test. Refer to the individual transmission section for procedures.

Throttle Valve Controls

The throttle valve can be controlled by linkage, cable or engine vacuum.

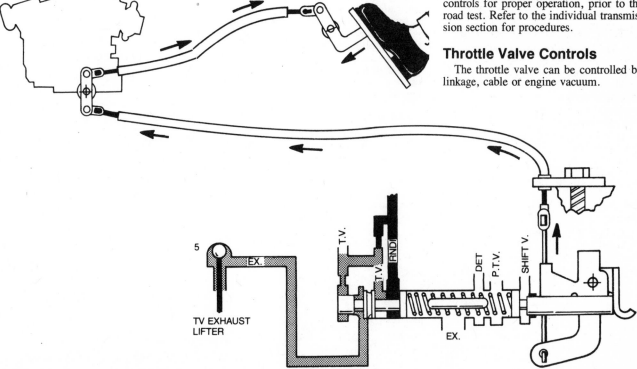

Throttle valve control using cable between the accelerator pedal, carburetor and throttle control valve

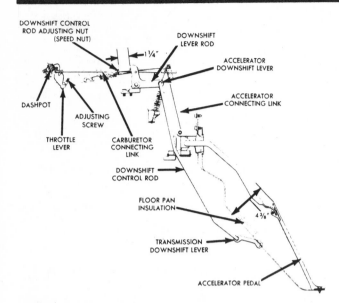

Typical throttle and kickdown linkage showing adjustment points and measurement (© Ford Motor Co.)

Manual control linkage phasing—typical (© Ford Motor Co.)

LINKAGE CONTROL

Inspect the linkage for abnormal bends, looseness at the bellcrank connections and linkage travel at the wide open throttle stop. Be sure the linkage operates without binding and returns to the closed position upon release of the accelerator.

CABLE CONTROL

Inspect the cable for sharp bends or crimps, secured retainers, freedom of cable movement throughout the full throttle position and the return to the closed throttle position without binding or sticking, and connection of the throttle return spring.

ENGINE VACUUM CONTROLS

Inspect for sufficient engine vacuum, vacuum hose condition and routing, and signs of transmission fluid in the vacuum hoses indicating a leaking vacuum diaphragm (modulator).

Kickdown Controls

The transmission kickdown is controlled by linkage, cable or electrical switches and solenoid.

LINKAGE CONTROLS

The linkage control can be a separate rod connected to and operating in relation with the carburetor throttle valves, or incorporated with the throttle linkage. Inspect for looseness, bends, binding and movement into the kickdown detent upon the movement of the throttle to the wide open stop.

NOTE: It is advisable to inspect for the wide open throttle position at the carburetor from inside the vehicle, by depressing the accelerator pedal, rather than inspecting movement of the linkage from under the hood. Carpet matting, dirt or looseness of the accelerator can prevent the opening of the throttle to operate the kickdown linkage.

CABLE CONTROLS

The kickdown cable control can be a separate cable or used with the throttle valve control cable. It operates the kick-

down valve at the wide open throttle position. Inspect for kinks and bends on bracket retention of the cable. Inspect for freedom of movement of the cable and see that the cable drops into the kickdown detent when the accelerator pedal is fully depressed and the throttle valves are fully open.

ELECTRICAL CONTROLS

The electrical kickdown controls consist of a switch, located on the accelerator linkage and a solenoid control, either mounted externally on the transmission case or mounted internally on the control valve body, in such a position as to operate the kickdown valve upon demand of the vehicle operator. Inspect the switch for proper operation which should allow electrical current to pass through, upon closing of the switch contacts by depressing the accelerator linkage. Inspect the wire connector at the transmission case or the terminals of the externally mounted solenoid for current with the switch contacts closed. With current present at the solenoid, either externally or internally mounted, a clicking noise should be heard, indicating that the solenoid is operating.

IMPORTANT: Operate the switch several times with the accelerator linkage and observe its operation. The switch contacts should return to their open position upon release of the accelerator linkage. If the switch remains closed, the 2-3 shift will not occur.

MANUAL LINKAGE CONTROL INSPECTION

The manual linkage adjustment is one of the most critical, yet the most overlooked,

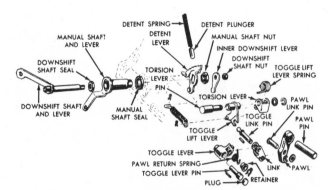

Typical manual control linkage and parking pawl assembly within the transmission (© Ford Motor Co.)

adjustment on the automatic transmission. The controlling quadrant, either steering column or console mounted, must be properly phased with the manual control valve detent. Should the manual valve be out of adjustment or position, hydraulic leakage can occur within the control valve assembly and can result in delay of unit engagement and/or slippage of the clutches and bands upon application. The partical opening of apply passages can also occur for clutches and bands not in applying sequence, and result in dragging of the individual units or bands during transmission operation.

Inspect the selector lever and quadrant, the linkage or cable control for looseness, excessive wear or binding. Inspect the engine/transmission assembly for excessive lift during engine torque application, due to loose or broken engine mounts, which can pull the manual valve out of position in the control vlave detent.

CAUTION: *The neutral start switch should be inspected for operation in Park and Neutral positions, after any adjustments are made to the manual linkage or cable.*

ROAD TEST

As noted previously in this section, be aware of the engine's performance. For example, if a vacuum diaphragm (modulator valve) is used to control the throttle pressure, an engine performing poorly cannot send the proper vacuum signals to the transmission for proper application of the throttle pressure and control pressure, in the operation of the bands and clutches. Slippages and changes in shift points can occur.

During the road test, the converter operation must be considered. Related converter malfunctions affecting the road test are as follows, with the converter operation, diagnosis and repairs discussed later in the General Information Section.

STATOR ASSEMBLY FREE WHEELS

When the stator roller clutch freewheels in both directions, the vehicle will have poor acceleration from a standstill. At speeds above approximately 45 MPH, the vehicle will act normally. A check to make on the engine is to accelerate to a high RPM in neutral. If the engine responds

properly, this is an indication that the engine is operating satisfactorily and the problem may be with the stator.

STATOR ASSEMBLY REMAINS LOCKED UP

When the stator remains locked up, the engine RPM and the vehicle speed will be restricted at higher speeds, although the vehicle will accelerate from a standstill normally. Engine overheating may be noticed and visual inspection of the converter may reveal a blue color, resulting from converter overheating.

NOTE: Refer to the converter chapter of the General Information Section for operation, diagnosis and repairs.

Clutch and Band Application

During the road test, operate the transmission in each gear position and observe the shifts for signs of any slippage, variation, spongyness or harshness. Note the speed when the upshifts and downshifts occur. If slippage and engine flare-up occur in any gear, clutch, or band or if overrunning clutch problems are indicated and depending upon the degree of wear, a major overhaul may be indicated.

The clutch and band application chart in each transmission section provides a basis for the road test analysis to determine the internal units applied or released in a specific gear ratio.

To properly analyze the automatic transmission shifts during the road test, examples of using the clutch and band application charts are given.

EXAMPLE NUMBER ONE: CHRYSLER

Using the Chrysler type automatic transmission clutch and band application chart (less lock-up) as a guide, the following conditions can be determined.

1. When moving the selector lever through the detents, the following condition could exist: the transmission slips in the D-first gear, but does not slip in the 1-first gear. The clutch and band application chart indicates the rear clutch is applied in both gear positions, but the overrunning clutch is applied in D-first gear, while the low and reverse band is applied in 1-first gear only. Therefore, we can de-

termine that the overrunning clutch is slipping. Had both forward gears slipped, the rear clutch unit would have been at fault.

2. Using this process of elimination, we can locate the malfunctioning unit in the following explanation of transmission operation.

The transmission is in the D-third gear and the transmission slips. By referring to the clutch and band application chart, we find the front and rear clutches are applied in third gear with the selector lever in the D position. To determine which clutch unit is at fault, another gear is selected which does not use one of the clutch units. In this instance, if the transmission is placed in the R (reverse) position and the transmission still slips, the front clutch unit is at fault and if the transmission does not slip, the rear clutch unit is at fault and slipping.

EXAMPLE NUMBER TWO: FORD

Using the Ford C-3 type automatic transmission clutch and band application chart as a guide, the following conditions can be determined.

1. Placing the selector lever in reverse, we find the transmission slipping. To locate the slipping unit, we find the low and reverse band is applied in both reverse and the 1 selector lever position, and the reverse and high clutch is applied in both reverse and D-third gear. If the transmission is operated D-third gear and slippage occurs, the reverse and high clutch unit is assumed to be at fault. If no slippage occurs while in D-third gear, but does occur while in 1 selector lever position, the low and reverse band is at fault, either through adjustment or extreme wear.

2. Assume the transmission is placed in D position and the vehicle begins to move through the D-first gear and shifts into the D-second gear. Upon shifting into second gear, a slippage and engine flare-up is experienced, but as the vehicle speed is increased and the shift is made into D-third gear, the slippage ceases. Referring to the clutch and band application chart, we find the forward clutch is applied in D-first, second and third gears and also applied in 2-second gear, while the intermediate band is applied in D-second gear and in 2-second gear. Since no slippage occurred while the transmission was in D-first and third gears, the forward clutch can be eliminated and since the intermediate band is applied in D-second gear, we can assume the band to

CLUTCH AND BAND APPLICATION CHART
Example 1—Chrysler Corp.—Less lock-up converter

Drive Position	P	R	N	D 1	D 2	D 3	2 1	2 2	1
Front Clutch		●				●			
Front Band					●			●	
Rear Clutch				●	●	●	●	●	●
Rear Band		●							●
Overrunning Clutch				●			●		

CLUTCH AND BAND APPLICATION CHART
Example 2—Ford C-3

Drive Position	P	R	N	D 1	D 2	D 3	2 1	2 2	1
Forward Clutch				●	●	●	●	●	
Reverse and High Clutch		●				●			
Intermediate Band					●			●	
Low and Reverse Band		●							●
One-Way Clutch				●					

CLUTCH AND BAND APPLICATION CHART
Example 3—GM-TH350

Drive Position	P/N	Drive			Super		Low	Reverse
		1	2	3	1	2		
Intermediate Clutch			●	●		●		
Direct Clutch				●				●
Forward Clutch		●	●	●	●	●	●	
Low & Reverse Clutch							●	●
Intermediate Overrunning Roller Clutch			●			●		
Low & Reverse Roller Clutch		●			●		●	
Intermediate Overrun Band						●		

be slipping due to adjustment or extreme wear. To further check for band slippage, place the selector lever in the 2-second gear position and slippage should occur under torque, since the 2-second gear has the same clutch and band applied as the D-second gear position.

EXAMPLE NUMBER THREE: GM

Using the General Motors Turbo Hydra-Matic, model 350 clutch and band application chart, another example of transmission operation diagnosis can be made to determine the internal unit that is malfunctioning.

1. By examining the clutch and band application chart and having a condition of slippage in all forward gears, but not in reverse, we find the forward clutch unit is at fault since it is engaged in all forward gears, but not in reverse.

2. If no reverse or low gears are available, but a D-third gear is present, we can assume the low and reverse clutch is defective while the direct clutch is operating properly.

Summary of Road Test

This process of elimination is used to locate the unit that is malfunctioning and to

confirm the proper operation of the transmission. Although the slipping unit can be defined, the actual cause of the malfunction cannot be determined by the band and clutch application charts. It is necessary to perform hydraulic and air pressure tests to determine if a hydraulic or mechanical component failure is the cause of the malfunction.

PRESSURE TESTS
Oil Pressure Gauge

The oil pressure gauge is the primary tool used by the diagnostician to determine the source of malfunctions of the automatic transmissions.

Oil pressure gauges are available with different rates, ranging from 0-100, 0-300 and 0-400 PSI that are used to measure the pressures in the various hydraulic circuits of the automatic transmissions. The high-rated pressure gauges (0-300, 0-400 PSI) are used to measure the control line pressures while the low-rated gauge (0-100 PSI) is used to measure the governor, lubrication and throttle pressures on certain automatic transmissions.

The gauges may be an individual unit

with a 4- to 10-foot hose attached, or may be part of a console unit with other gauges, normally engine tachometer and vacuum.

To measure the hydraulic pressures, select a gauge rated over the control pressure specifications and install the hose fitting into the control pressure line tap, located either along the side or the rear of the transmission case. Refer to the individual automatic transmission sections for the correct locations, since many transmission cases have more than one pressure tap to check pressures other than the control pressure.

The pressure gauges can be used during a road test, but care must be excerised in routing the hose from under the vehicle to avoid dragging or being entangled with objects on the roadway.

The gauge should be positioned so the dial is visible to the diagnostician near the speedometer area. If a console of gauges are used, the console is normally mounted on a door window or on the vehicle dash.

CAUTION: *During the road test, traffic safety must be exercised. It is advisable to have a helper to assist in the reading or recording of the results during the road test.*

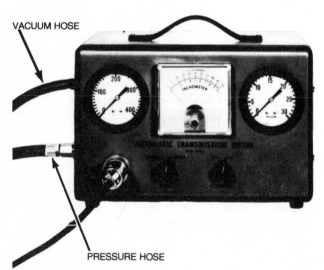

Typical automatic transmission oil pressure, vacuum and tachometer console gauge unit (© Ford Motor Co.)

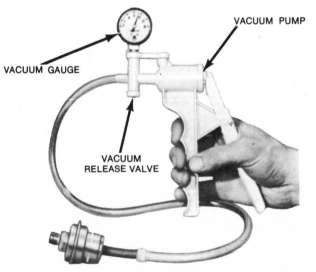

Typical vacuum pump used to test vacuum diaphragms (modulators) (© Ford Motor Co.)

Control Pressure Testing

The methods of obtaining control pressure readings vary from manufacturer to manufacturer when vacuum diaphragms (modulators) are used to modulate the pressures. Since engine vacuum is the controlling factor, the amount of vacuum allowed to act upon the diaphragm must be controlled.

Ford Motor Company recommends the use of an air bleed valve, placed in the vacuum line from the engine to the diaphragm, that can be adjusted to control the amount of vacuum to the diaphragm in relation to engine speed.

NOTE: When an air bleed valve is not used, a stall test can be done.

CAUTION: *Refer to the stall test chapter before any attempt is made to perform a stall test.*

General Motors recommends the use of a vacuum pump to control the amount of vacuum directed to the vacuum diaphragm (modulator) during the pressure test. All vacuum lines should be checked for cracks and breaks. Before the test, apply 15-20 inches of vacuum to the vacuum diaphragm and check for leakdown. The engine must be operating at a specific RPM, and the test is performed both with the vehicle at a standstill and while driving.

CAUTION: *When wheels are locked up and the engine is forced to 1000 RPM while in a specific gear range, the total test time should not exceed 2 minutes. Place the transmission in N position and increase the engine speed to 1500 RPM for approximately 1-2 minutes.*

EXAMPLE NUMBER 1: FORD

To illustrate a typical control pressure test on a typical Ford Motor Company automatic transmission, the following explanation is given concerning the accompanying chart.

NOTE: It is assumed the oil pressure gauge, the vacuum gauge, the air bleed valve and tachometer are attached to the engine/transmission assembly, while having the brakes locked and the wheels chocked. The fluid temperature should be 150°—200° F.

1. Start the engine and allow to idle.
2. Place the selector lever in the D position and adjust the air bleed valve to attain 15 inches or more of vacuum.
3. The control pressure should read 50-60 PSI in the D, 2 and 1 positions, as indicated on the chart.

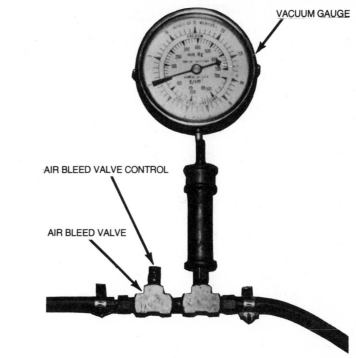

VACUUM GAUGE

AIR BLEED VALVE CONTROL

AIR BLEED VALVE

Vacuum gauge and air bleed valve assembly

4. Place the selector in the R position while maintaining 15 inches or more of vacuum. The control pressure should change to 66-77 PSI, as indicated on the chart.
5. While maintaining the 15 inches of vacuum, place the selector lever in the P and N positions. The pressure should read 50-60 PSI, as indicated on the chart.

By adjusting the air bleed valve, the manifold vacuum at engine idle can be regulated to any specified vacuum readings and the control pressure checked for being within specifications. For example, we will assume the vacuum has been regulated to 12 inches by the air bleed valve. By moving the selector lever through the various ranges, we find the control pressure should be 50-69 PSI with the selector lever in the D, 2 or 1 positions, while in the R position, the control pressure should be 66-123 PSI. Placing the selector lever in the P and N positions, the control pressures should read 50-69 PSI, as indicated on the chart.

In the "wide open throttle thru detent" column on the chart, the pressures listed for the D, 2 or 1 positions, and the R position have greatly increased. The air bleed valve

is closed and the engine is operating at normal engine vacuum.

Being positive the brakes are locked and the wheels chocked, place the selector lever in either the D, 2, or 1 position. While observing the pressure gauge, quickly depress the accelerator pedal through the detent (accelerator pedal to the floor), and release the pedal as soon as possible. Record or match the pressure reading to the reading of the chart. Repeat the procedure for each of the ranges of the selector lever, except N and P positions. The pressures should be 140-165 PSI in the forward ranges and 232-273 in the R range, as per the specifications listed on the chart.

CAUTION: *Do not use the above pressure readings when actually testing the control pressures. The above readings are only used as a guide for the chart explanation. Refer to the individual transmission specification charts for the correct readings.*

EXAMPLE NUMBER 2: GM

To illustrate a typical control pressure test on a typical General Motors Automatic Transmission, the following explanation is

CONTROL PRESSURE (IN P.S.I.) AT ZERO GOVERNOR R.P.M.
(OIL TEMPERATURE 150° – 200° F)

Range	15" & Above	14"	13"	12"	11"	10"	Wide Open Throttle Thru Detent
			Manifold Vacuum at Engine Idle Speed				
D, 2, 1	50–60	50–62	50–62	50–69	56–77	65–85	140–165
R	66–77	66–98	66–110	66–123	66–135	—	232–273
P, N	50–60	50–60	50–62	50–69	56–77	—	—

TRANSMISSION MALFUNCTION RELATED TO OIL PRESSURE

15-20″ vacuum applied to modulator							0″ vacuum to modulator	
③ Drive Brakes Applied 1000 RPM	③ Reverse Brakes Applied 1000 RPM	③ Super or Lo Brakes Applied 1000 RPM	Neutral Brakes Applied 1000 RPM	③ Drive 1000 RPM Brakes on Detent* Activated	Drive Idle	① Drive 30 MPH Closed Throttle	Drive—from 1000 to 3000 RPM Wheels free to move	Pressure Test Conditions
60-90	85-150	85-110	55-70	90-110	60-85	55-70	Pressure drop of 10 PSI or more	Normal Results Note2
							DROP	Malfunction in Control Valve Assembly
							NO DROP	Malfunction in Governor or Governor Feed System
ALL PRESSURES HIGH WITH LESS THAN 35 PSI BETWEEN PRESSURE READINGS							—	Malfunction in Detent System
ALL PRESSURES HIGH WITH MORE THAN 35 PSI BETWEEN PRESSURE READINGS							—	Malfunction in Modulator
Low					—	Low to Normal	—	Oil Leak in Feed System to the Direct Clutch
Low		Low to Normal		Low to Normal	—	Low to Normal	—	Oil Leak in Feed System to the Forward Clutch
				Low			—	Detent System

A blank space = Normal pressure
A dash (—) in space = Pressure reading has no meaning
① Coast for 30 mph—read before reaching 20 mph

② If high line pressures are experienced see "High Line Pressures" note.

③ Cable pulled or blocked thru detent position or downshift switch closed by hand

given concerning the accompanying chart. Because of the variation of the engine vacuum supply due to the addition of emission controls, the use of an external vacuum source is recommended. A hand-operated vacuum pump allows definite amounts of vacuum to be applied to the vacuum diaphragm (modulator) so that a consistent control pressure can be obtained for evaluation.

NOTE: It is assumed the oil pressure gauge, the vacuum gauge, the vacuum pump, and the tachometer are attached to the engine/transmission assembly, while having the brakes locked and the wheels chocked. The fluid temperature should be at normal operating temperature.

1. Start the engine and allow to idle.
2. Apply 15-20 inches of vacuum to the vacuum diaphragm (modulator) with the vacuum hand pump.
3. Place the selector lever in the Drive position and increase the engine speed to 1000 RPM. The pressure reading should be 60-90 PSI, as indicated on the chart.
4. Taking another example from the chart, place the selector lever in the Super or Low position and increase the engine speed to 1000 RPM. As indicated on the chart, the pressure should read 85-110 PSI.

5. Referring to the chart column to test pressures with the detent activated, the following must be performed. Place the selector lever in the Drive position and increase the engine speed to 1000 RPM. With the aid of a helper, if necessary, pull the detent cable through the detent, or if the transmission is equipped with an electrical downshift switch, close the switch by hand. The pressure reading should be 90-110 PSI, as indicated on the chart.
6. A pressure test conditions column is included in the chart to assist the diagnostician in determining possible causes of transmission malfunctions from the hydraulic system.

CAUTION: *Do not use the above pressure readings when actually testing the control pressures. The above readings are only used as a guide for the chart explanation. Refer to the individual transmission section for the correct specifications and pressure readings.*

HIGH CONTROL PRESSURE

If a condition of high control pressure exists, the general causes can be categorized as follows.
1. Vacuum leakage or low vacuum
2. Vacuum diaphragm (modulator) damaged

3. Pump pressure excessive
4. Control valve assembly
5. Throttle linkage or cable misadjusted

LOW CONTROL PRESSURE

If a condition of low control pressure exists, the general causes can be categorized as follows.
1. Transmission fluid low
2. Vacuum diaphragm (modulator) defective
3. Filter assembly blocked or air leakage
4. Oil pump defective
5. Hydraulic circuit leakage
6. Control valve body

NO CONTROL PRESSURE RISE

If a control pressure rise does not occur as the vacuum drops or the throttle valve linkage/cable is moved, the mechanical connection between the vacuum diaphragm (modulator) or throttle valve linkage/cable and the throttle valve should be inspected. Possible broken or disconnected parts are at fault.

VACUUM DIAPHRAGM (MODULATOR) TESTING

A defective vacuum diaphragm (modulator) can cause one or more of the following conditions.

A—Vacuum diaphragm (modulator) bellows bad

MODULATOR
IN QUESTION

KNOWN GOOD
MODULATOR

C℄

B—Vacuum diaphragm (modulator) bellows good

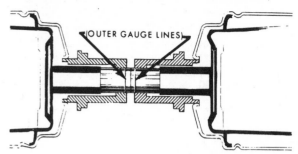

OUTER GAUGE LINES

Comparison of internal spring load with special gauge tool to check bellows

a. Engine burning transmission fluid
b. Transmission overheating
c. Harsh upshifts
d. Delayed shifts
e. Soft up and down shifts

Whenever a vacuum diaphragm (modulator) is suspected of malfunctioning, a vacuum check should be made of the vacuum supply.

1. Disconnect the vacuum line at the transmission vacuum diaphragm (modulator) connector pipe.

2. With the engine running, the vacuum gauge should show an acceptable level of vacuum for the altitude at which the test is being performed.

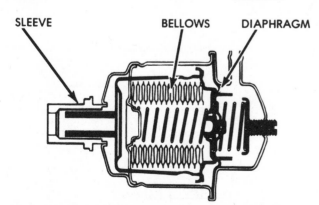

APPLY 18 INCHES
VACUUM AND HOLD

SINGLE
DIAPHRAGM

MANIFOLD
VACUUM PORT

EGR
VACUUM
PORT

CONNECT
TESTER

DUAL AND
DIAPHRAGM

Testing single and dual vacuum diaphragm (modulator) with outside vacuum source

3. If the vacuum reading is low, check for broken, split or crimped hoses and for proper engine operation.

4. If the vacuum reading is acceptable, accelerate the engine quickly. The vacuum should drop off and return immediately upon release of the accelerator. If the gauge does not register a change in the vacuum reading, indications are that the vacuum lines are plugged, restricted or connected to a reservoir supply.

5. Correct the vacuum supply as required.

When the vacuum supply is found to be sufficient, the vacuum diaphragm (modulator) must be inspected and this can be accomplished on or off the vehicle.

On the Vehicle Tests

1. Remove the vacuum line and attach a vacuum pump to the diaphragm (modulator) connector pipe.

2. Apply 18 inches of vacuum to the diaphragm (modulator). The vacuum should remain at 18 inches without leaking down.

3. If the vacuum reading drops sharply or will not remain at 18 inches, the diaphragm is leaking and the unit must be replaced.

4. If transmission fluid is present on the vacuum side of the diaphragm or in the vacuum hose, the diaphragm is leaking and the unit must be replaced.

NOTE: Gasoline or water vapors may settle on the vacuum side of the diaphragm. Do not diagnose as transmission fluid.

Off the Vehicle Tests

1. Remove the vacuum diaphragm (modulator) from the transmission.

2. Attach a vacuum pump to the vacuum diaphragm (modulator) connector pipe and apply 18 inches of vacuum.

3. The vacuum should hold at 18 inches, if the diaphragm is good and will drop to zero if the diaphragm is leaking.

4. With the control rod in the transmission side of the vacuum diaphragm (modulator), apply vacuum to the connector pipe. The rod should move inward with light finger pressure applied to the end of the rod. When the vacuum is released, the rod will move outward by pressure from the internal spring.

Internal Spring Load Comparison Test

The vacuum diaphragm (modulator) can be tested for the internal spring load, by using a known good unit of the same design and number, mounted in a special comparison gauge (example—Kent Moore part number J-24466 or equivalent).

NOTE: Different numbered special comparison gauges are available for different model transmission vacuum diaphragm (modulator) usage.

1. Install the known good vacuum diaphragm (modulator) in one end of the gauge. Place the unit to be tested into the other end of the gauge.

2. While holding the assembly horizontally, force both vacuum diaphragms (modulator) towards the center of the

SLEEVE   BELLOWS   DIAPHRAGM

Cross-section of typical vacuum diaphragm (modulator)

COMPARISON GAUGE J-24466   OUTER GAUGE LINE

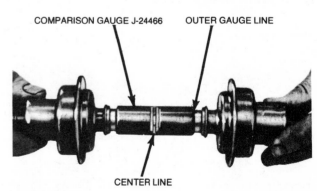

CENTER LINE

Aligning comparison gauge to the vacuum diaphragm (modulator)

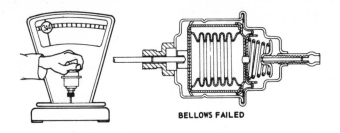

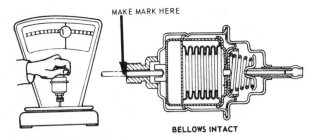

BELLOWS FAILED

MAKE MARK HERE

BELLOWS INTACT

Measuring vacuum diaphragm (modulator) internal spring tension with a weigh scale (© Ford Motor Co.)

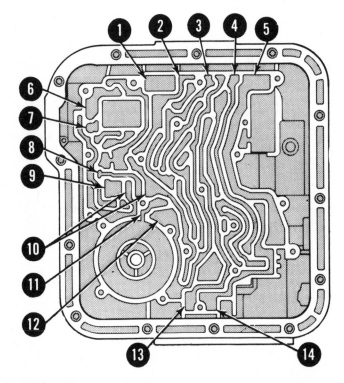

1. Pump inlet
2. Reverse-and-high
3. Forward clutch
4. Torque converter inlet
5. Pump outlet
6. Intermediate servo apply
7. Intermediate servo release

8. Main pressure to throttle valve
9. Throttle pressure return
10. Throttle pressure to main control
11. Low-and-reverse servo apply
12. Low-and-reverse servo release

13. Governor, pressure to main control
14. Main pressure to governor

Typical automatic transmission case passage identification (Ford C-3 case illustrated) (© Ford Motor Co.)

gauge, until one sleeve end touches the center line scribed on the gauge. The distance between the center line and the sleeve end on the opposite vacuum diaphragm (modulator) should not exceed $1/16$ inch.

NOTE: Certain comparison gauges may use the control rod that is used with the vacuum diaphragm (modulator) or a special rod. Follow the manufacturer's instructions.

3. Inspect the control rod for straightness and the vacuum diaphragm (modulator) stem for being concentrical to the can.

Other Methods of Measuring Spring Pressure

Other methods of measurements to determine the internal spring load can be done with the use of a known good vacuum diaphragm (modulator), ruler and a weigh scale (bathroom type).

1. Place the control rod or valve into the stem of a known good vacuum diaphragm (modulator).
2. Force the vacuum diaphragm (modulator) downward until the rod or valve begins to compress the inner spring.
3. Apply more force until the internal spring is near its compressed limit and record the pound force from the scale.
4. While holding the vacuum diaphragm (modulator) at the recorded pound force, measure the distance from the end of the stem to the scale platform and record.
5. Perform the same test on the vacuum diaphragm (modulator) in question and record the results. The distance should not exceed $1/16$ inch between the good and the tested units.

NOTE: It is most important to have the same numbered vacuum diaphragm (modulator) and to use it in the same application as the one being tested. Otherwise, different test results will occur.

AIR PRESSURE TESTS

The automatic transmission has many hidden passages and hydraulic units that are controlled by internal fluid pressures, supplied through tubes, shafts and valve movements.

The air pressure tests are used to confirm the findings of the fluid pressure tests and to further pinpoint the malfunctioning area. The air pressure test can also confirm the hydraulic unit operation after repairs have been made.

To perform the air pressure test, the control valve body must be removed from the transmission case, exposing the case passages. By referring to the individual transmission section, identify each passage before any attempt is made to proceed with the air pressure test.

CAUTION: *It is a good practice to protect the diagnostician's face and body with glasses and protective clothing and the surrounding area from the oil spray that will occur when air pressure is applied to the various passages.*

NOTE: The air pressure should be controlled to approximately 25 PSI and the air should be clean and dry.

When the passages have been identified and the air pressure applied to a designated passage, reaction can be seen, heard and felt in the various units. Should air pressure be applied to a clutch apply passage, the piston movement can be felt and a soft dull thud should be heard. Some movement of the unit assembly can be seen.

When air pressure is applied to a servo apply passage, the servo rod or arm will move and tighten the band around the drum. Upon release of the air pressure, spring tension should release the servo piston.

When air pressure is applied to the governor supply passage, a whistle, click or buzzing noise may be heard.

When failures have occurred within the transmission assembly and the air pressure tests are made, the following problems may exist.
1. No clutch piston movement
2. Hissing noise and excessive fluid spray
3. Excessive unit movement
4. No servo band apply or release

EFFECT OF ALTITUDE ON ENGINE VACUUM

Elevation in Feet	Number of Engine Cylinders		
	Four	Six	Eight
Zero to 1000	18 to 20	19 to 21	21 to 22
1000 to 2000	17 to 19	18 to 20	19 to 21
2000 to 3000	16 to 18	17 to 19	18 to 20
3000 to 4000	15 to 17	16 to 18	17 to 19
4000 to 5000	14 to 16	15 to 17	16 to 18
5000 to 6000	13 to 15	14 to 16	15 to 17

STALL SPEED TESTS

The stall speed test is performed to evaluate the condition of the transmission as well as the condition of the engine.

The stall speed is the maximum speed at which the engine can drive the torque converter impeller while the turbine is held stationary. Since the stall speed is dependent upon the engine and torque converter characteristics, it can vary with the condition of the engine as well as the condition of the automatic transmission, so it is most important to have a properly performing engine before a stall speed test is attempted, thereby eliminating the engine from any malfunction that may be present.

Because engines perform differently between high and low altitudes, the stall speeds given in specification charts are for vehicles tested at sea level and cannot be considered representative of stall speed tests performed at higher altitudes. Unless specific stall speed tests specification charts are available for use at higher altitudes, representative stall speeds can be determined by testing several vehicles known to be operating properly, averaging the results and recording the necessary specifications for future reference.

Performing the Stall Speed Tests

1. Attach a techometer to the engine and an oil pressure gauge to the transmission control pressure tap. Position the tachometer and oil pressure gauge so the operator can read the dials.
2. Mark the specified maximum engine RPM on the tachometer clear cover with a grease pencil so that the operator can immediately check the stall speed to see if it is over or under specifications.
3. Start the engine and bring to normal operating temperature.
4. Check the transmission fluid level and correct as necessary.
5. Apply the parking brake and chock the wheels.

CAUTION: *Do not allow anyone in front of or behind the vehicle during the preparation or during the stall speed test.*

6. Apply the service brakes and place the selector lever in the ''D'' position and depress the accelerator pedal to the wide open throttle position.

STALL TEST

Range	Specified Engine RPM	Record Actual Engine RPM
D		
2		
1		
R		

Results _____

MARK WITH GREASE PENCIL

TACHOMETER

Preparation for stall test (© Ford Motor Co.)

SAMPLE STALL TEST DIAGNOSIS CHART

Selector Lever Range	Specified Engine RPM	Actual Engine RPM	Control Pressure PSI	Holding Members Applied
D (Drive)				
1				
2				
R (Reverse)				

NOTE: The range identifications are to be taken from the selector quadrant of the vehicle being tested. Before stall test, fill in the specified engine RPM and the holding members applied columns from the clutch and band application and specification charts of the automatic transmission being tested.

7. Do not hold the throttle open any longer than necessary to obtain the maximum engine speed reading and never over five (5) seconds for each test. The stall will occur when no more increase in engine RPM at wide open throttle is noted.

CAUTION: *If the engine speed exceeds the maximum limits of the stall speed specifications, release the accelerator immediately as internal transmission slippage is indicated.*

8. Shift the selector lever into the Neutral position and operate the engine from 1000 to 1500 RPM for at least 30 seconds to two minutes to cool the transmission fluid.

9. If necessary, the stall speed test can be performed in other forward and reverse gear positions. Observe the transmission fluid cooling procedure between each test as outlined in step 8.

Results of the Stall Speed Tests

The stall speed RPM will indicate possible problems. If the engine RPM is high, internal transmission slippage is indicated. If the engine RPM is low, engine or converter problems can exist.

The transmission will not upshift during the stall test and by knowing what internal transmission members are applied in each test range, an indication of a unit failure can be pinpointed.

It is recommended a chart be prepared to assist the diagnostician in the determination of the stall speed results in comparison to the specified engine RPM, and should include the range and holding member applied.

LOW ENGINE RPM ON STALL SPEED TEST

The low engine RPM stall speed indicates either the engine is not performing properly or the converter stator one-way clutch is not holding. By road testing the vehicle, the determination as to the defect can be made.

If the stator is not locked by the one-way clutch, the performance of the vehicle will be poor up to approximately 30-35 MPH. If the engine is in need of repairs or adjustments, the performance of the vehicle will be poor at all speeds.

HIGH ENGINE RPM ON STALL SPEED TEST

When the engine RPM is higher than specifications, internal transmission unit slippage is indicated. By following the holding member application chart, the defective unit can be pinpointed.

It must be noted that a transmission using a one-way overrunning clutch while in the "D" position, first gear, and having a band applied in the "2" position, first gear (for vehicle braking purposes while going downhill), may have the band slipping unnoticed during the stall test, because the overrunning clutch will hold. To determine this, a road test must be performed to place the transmission in a range where the band is in use without the overrunning clutch.

NORMAL ENGINE RPM ON STALL SPEED TEST

When the engine RPM is within the specified ranges, the holding members of the transmission are considered to be operating properly.

A point of interest is if the converter oneway clutch (overrunning) is seized and locks the stator from turning either way, the engine RPM will be normal during the test, but the converter will be in reduction at all times and the vehicle will probably not exceed a speed of 50-60 miles per hour. If this condition is suspected, examine the fluid and the converter exterior for signs of overheating, since an extreme amount of heat is generated when the converter remains in constant reduction.

TRANSMISSION NOISES

During the stall speed test and the road test, the diagnostician must be alert to any abnormal noises from the transmission area or any excessive movement of the engine/transmission assembly during torque application or transmission shifting.

CAUTION: *Before attempting to diagnose automatic transmission noises, be sure the noises do not originate from the engine components, such as the water pump, alternator, air conditioner compressor, power steering or the air injection pump. Isolate these components by removing the proper drive belt and operate the engine. Do not operate the engine longer than two minutes at a time to avoid overheating.*

1. Whining or siren type noises—Can be considered normal if occurring during a stall speed test, due to the fluid flow through the converter.
2. Whining noise (continual with vehicle stationary)—If the noise increases and decreases with the engine speed, the following defects could be present.
   a. Oil level low
   b. Air leakage into pump (defective gasket, "O"-ring or porosity of a part)
   c. Pump gears damaged or worn
   d. Pump gears assembled backward
   e. Pump crescent interference
3. Buzzing noise—This type of noise is normally the result of a pressure regulator valve vibrating or a sealing ring broken or worn out and will usually come and go, depending upon engine/transmission speed.
4. Rattling noise (constant)—Usually occurring at low engine speed and resulting from the vanes stripped from the impeller or turbine face or internal interference of the converter parts.
5. Rattling noise (intermittent)—Reflects a broken flywheel or flex plate and usually occurs at low engine speed with the transmission in gear. Placing the transmission in "N" or "P" will change the rattling noise or stop it for a short time.

6. Gear noise (one gear range)—This type of noise will normally indicate a defective planetary gear unit. Upon shifting into another gear range, the noise will cease. If the noise carries over to the next gear range, but at a different pitch, defective thrust bearings or bushings are indicated.
7. Engine vibration or excessive movement—Can be caused by transmission filler or cooler lines vibrating due to broken or disconnected brackets. If excessive engine/transmission movement is noted, look for broken engine/transmission mounts.

CAUTION: *When necessary to support an engine equipped with metal safety tabs on the mounts, be sure the metal tabs are not in contact with the mount bracket after the engine/transmission assembly is again supported by the mounts. A severe vibration can result.*

8. Squeal at low vehicle speeds—Can result from a speedometer driven gear seal, a front pump seal or rear extension seal being dry.

The above list of noises can be used as a guide. Noises other than the ones listed can occur around or within the transmission assembly. A logical and common sense approach will normally result in the source of the noise being detected.

TRANSMISSION FLUID LEAKAGE

Fluid leakage can occur from various points on the automatic transmission assembly and if not located and corrected, serious transmission failure can occur.

Seepage of fluid is acceptable through normal transmission operation, but when the seepage becomes droplets, the fluid loss occurs rapidly and the fluid level is normally not corrected as promptly as it should be, therefore causing the transmission to be operated in the "low fluid level" condition.

Locating Fluid Leakage

The following outline is a suggested method of determining the area of fluid leakage.

1. Raise the vehicle and support safely.
2. Clean the area of suspected leakage with a solvent to remove traces of the old fluid.
3. If fluid leakage is suspected from the converter area, remove the flywheel lower cover and clean the accessible area with solvent and compressed air.
4. After cleaning the suspected area of leakage, a white powder can be sprayed in the area to help pinpoint the leakage.
5. An alternate method of leak detection is to use a "black" lamp. The comparison of the engine oil and the transmission fluid must be made under the "black" lamp before the leak inspection is performed, in order to identify the leaking fluid.
6. Route the exhaust gases to the outside of the building and start the engine. Allow the engine/transmission to reach normal operating temperature while observing the areas of suspected fluid leakage.

7. Check for leaks with the transmission in Low (1), D 2, S (2), or R gear ranges if leakage does not occur in the N or the D positions.

8. Observe leakage areas around the rear of the engine and the front of the transmission. The air stream produced as the vehicle is being driven can cause the leaking fluid to be carried to the rear of the transmission and appear as a leak at the wrong location.

9. As the fluid leakages are pinpointed, the proper repair procedures should be followed to correct the leakage defects.

CAUTION: *Fluid leakage from the converter housing area can be caused by engine oil leaking past the rear main bearing seal, or from oil gallery plugs, or from the power steering system. Before starting repair procedures, be sure of the exact cause of the leakage.*

TRANSMISSION NOISE DIAGNOSIS

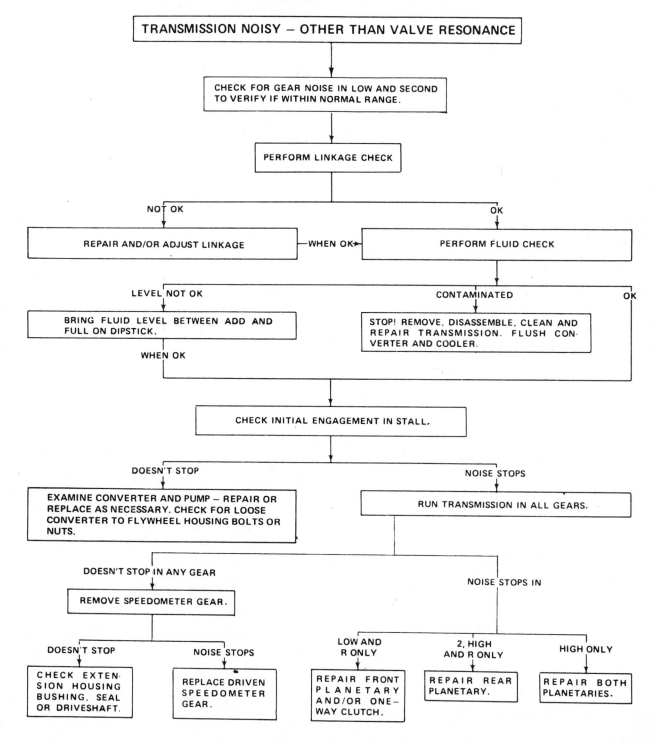

General Information

Seepage of fluid from oil pan—acceptable since so droplets of fluid have formed

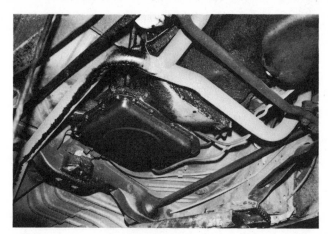

Leakage of fluid from fluid overfill, fluid forced out of fill pipe or leakage at fill pipe seat. Unacceptable and should be repaired.

Red dye is used in the automatic transmission fluids to help the diagnostician to distinguish between engine oil and transmission fluid. With the use of additives and certain types of transmission fluids, the color can sometimes become deceiving. A remedy for this situation is the use of a fluorescent dye and the "black" lamp. A mixture of oil-soluble aniline or fluorescent dyes, premixed at a rate of ½ teaspoon of dye powder to ½ pint of automatic transmission fluid, is used. With the use of the "black" lamp, determination can be made as to the location of the leak.

Possible Points of Fluid Leakage and Causes

1. Pan leakage
   a. Loose bolts
   b. Damaged gasket
   c. Hole in pan
2. Extension Housing
   a. Extension housing seal
   b. Housing to case gasket
   c. Loose bolts
   d. Speedometer driven gear seal
   e. Porous housing
3. Transmission case leakage
   a. Porous case
   b. Pressure taps loose or cross-threaded
   c. Manual shaft lip seal
   d. Band adjustment bolts and nuts
   e. Filler pipe sealing ring
   f. Cooler line fittings loose or cross-threaded
   g. Governor or modulator "O"-rings damaged
   h. Down shift or detent cable leakage
   i. Transmission case vent
4. Converter housing to transmission case
   a. Defective gasket
   b. Defective "O" or square cut seal ring
   c. Loose converter housing bolts
5. Oil pump
   a. Oil pump seal for converter hub
   b. Oil pump to case seal or gasket
   c. Oil pump bolts loose
6. Converter
   a. Converter leakage in welded area
   b. Converter hub worn
   c. Converter drain plugs loose

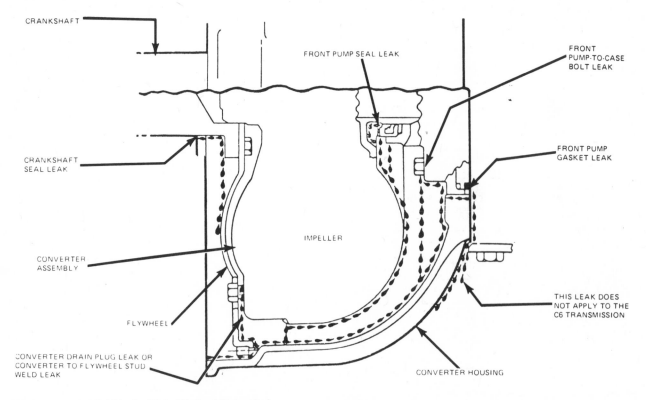

Typical converter fluid leak points (© Ford Motor Co.)

d. Converter to flywheel stud nuts loose and worn into converter or attaching bolts of a length greater than original causing hole in converter shell.

7. Coolant contaminated with transmission fluid
   a. Radiator transmission cooler leakage

8. Fluid loss, but no signs of external leakage
   a. Vacuum diaphragm (modulator valve) leaking causing fluid to be drawn into the engine

Repairing Case Porosity

Case porosity can be repaired with the transmission in the vehicle, providing the leakage area is accessible. An epoxy cement kit is available for this purpose through dealers' parts systems and parts supply houses.

1. Have the transmission at normal operating temperature and running.

2. After locating the leakage area, stop the engine and clean the leakage area with solvent. Allow to air dry.

3. Follow the manufacturer's recommendations and mix a sufficient amount of epoxy.

4. While the transmission case is still hot, cover the leaking area with the epoxy.

5. Allow the epoxy to cure for at least three hours.

6. Road test the vehicle and reinspect the repaired area for leakage.

NOTE: If the porosity leak is located in an inaccessible area on the case, the transmission must be removed for the repair. An alternate method, after obtaining the owner's permission, is to drill an access hole through the floor panel. After making the epoxy repair, seal the floor panel to avoid exhaust fume entrance into the vehicle.

Torque Converter Leakage

When indications of a converter leakage are evident, the converter should be removed from the vehicle and tested.

Special tools are available and the manufacturer's instructions should be followed to complete the test because of the safety factor involved with the use of air pressure.

General converter testing procedures are as follows.

1. Install the converter testing tool on the converter, securing it safely by following the manufacturers instructions.

2. Fill the converter with air pressure to approximately 80-90 PSI.

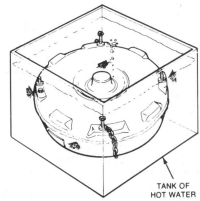

TANK OF HOT WATER

Converter air linkage points during immersion in a tank of hot water (© Ford Motor Co.)

3. Place the converter in a container of hot water and check for air bubbles from the converter, indicating a leak.

4. If leakage is indicated, perform the appropriate repairs or replace the converter assembly, if required.

5. Release the air pressure before removing the tool from the converter.

Engine Coolant Mixing with Transmission Fluid

During the transmission leakage inspection procedure, engine coolant may be found in the transmission fluid reservoir, resulting from a leakage of the transmission fluid cooler, located within the radiator assembly. Should this defect occur, the following procedure of repair should be used.

1. Remove the transmission assembly from the vehicle in the prescribed manner.

2. Completely disassemble the transmission and sub-assemblies, following the recommended procedures.

3. Thoroughly clean all transmission parts and passages with solvent and air pressure. Inspect all bushings, bearings and gear assemblies for galling or other damage.

4. Repair or replace the fluid cooler assembly and flush the cooler lines.

5. If the repair shop is not equipped with a converter flushing machine, the converter must be replaced.

NOTE: The contaminated fluid cannot be completely removed from the converter by draining.

6. Proceed with the rebuilding of the transmission assembly by replacing all rubber type seals, all composition clutch plates and all nylon washers, speedometer gears and governor gears.

NOTE: The engine coolant will affect:
1. the sealing rubbers and cause pressure leakage;
2. the composition clutch plates by separating the facing from the steel plate;
3. the nylon parts by causing them to swell.

CLEANING AND INSPECTION OF PARTS

Whenever an automatic transmission is overhauled because of burned clutches or

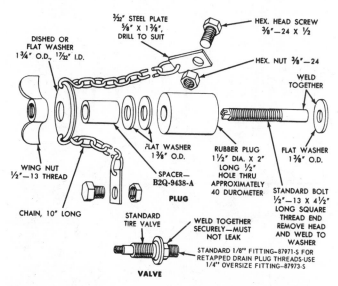

Special tool used to check converter leakage with air pressure (© Ford Motor Co.)

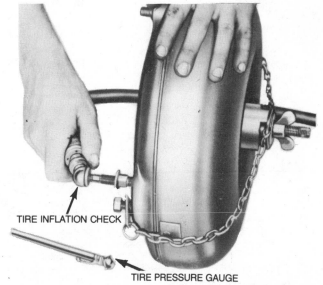

TIRE INFLATION CHECK

TIRE PRESSURE GAUGE

Filling converter with air for leakage test (© Ford Motor Co.)

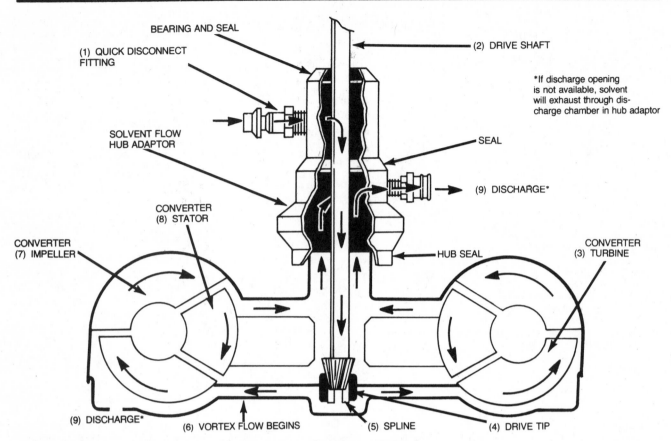

(1) QUICK DISCONNECT FITTING

BEARING AND SEAL

(2) DRIVE SHAFT

*If discharge opening is not available, solvent will exhaust through discharge chamber in hub adaptor

SOLVENT FLOW HUB ADAPTOR

SEAL

(9) DISCHARGE*

CONVERTER (8) STATOR

CONVERTER (7) IMPELLER

HUB SEAL

CONVERTER (3) TURBINE

(9) DISCHARGE* **(6) VORTEX FLOW BEGINS** **(5) SPLINE** **(4) DRIVE TIP**

Typical method of maching flushing torque converters (© General Motors Corp.)

bands, worn bushings or bearings, or coolant contamination of the fluid, the entire transmission assembly must be disassembled and cleaned to remove the contamination from the transmission parts, such as the gear train, control valve body, servos and passages.

During the cleaning of the disassembled parts, an inspection should be made of the components as to their usability during the rebuilding procedure.

Converter Assembly

The converter assembly need not be replaced because of clutch or band material contamination or sludge accumulation due to overheating of the transmission assembly, if a means of slushing the converter is available for use. When fluid contamina-

tions of bronze, steel or aluminum particles are found, thrust washer or bearing failure has occurred, the replacement of the converter should be done.

DRAINING AND FLUSHING THE CONVERTER

When the converter has become filled with contaminated fluid that can be flushed out and the converter reused, different manufacturers recommended procedures will have to be followed, as many different converter flushing machines are available to the rebuilding shops.

Many converters do not have a drain hole in the converter shell and must be drained from the converter hub. Certain vehicle manufacturers recommend drilling an ⅛ inch hole near one of flywheel attach-

ing bolt lugs and allowing the fluid to drain through the hole. After the fluid has all drained, a blind "pop" rivet is installed in the drilled hole. Various after-market suppliers have drill and tap guide kits available to drill a hole in the converter shell and install a threaded plug that can be removed for future draining.

NOTE: If the rivet leaks, the converter must be replaced, since the rivet cannot be drilled out. Part of the rivet would remain in the converter.

DRILLING THE CONVERTER

1. Drill the recommended hole in the converter shell.

NOTE: Coat the drill bit with grease to hold the metal drilling chips and install a sleeve on the drill bit to control the entrance of the bit to no more than ¼ inch.

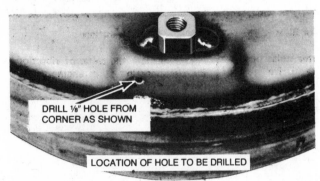

Drilling converter for draining

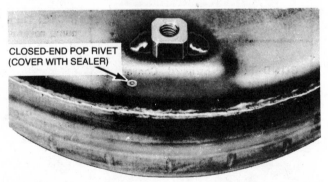

Installation of closed end "POP" rivet to close drilled hole in the converter

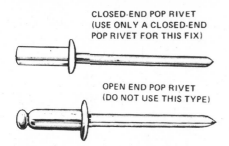

CLOSED-END POP RIVET
(USE ONLY A CLOSED-END
POP RIVET FOR THIS FIX)

OPEN END POP RIVET
(DO NOT USE THIS TYPE)

Comparison of closed end "POP" rivet and open end "POP" rivet

2. Prop the converter upright on a drain pan with the drilled hole on the bottom. Allow to drain for approximately 15 to 30 minutes.

3. Charge the converter with air pressure to force as much of the contaminated fluid from the converter as possible.

4. A flushing tool is available for use with a regular parts cleaner, the type that does general cleaning duties within the repair shop, and is equipped with a flexible hose. When using this type, the solvent should be well filtered or new.

5. Install the flushing tool plug into the converter hub with the hose connected to the hose of the parts cleaner. With the parts cleaner solvent flushing the converter for a period of 20 to 30 minutes, move the converter assembly intermittently to loosen the contamination within the converter.

6. Stop the parts cleaner and remove the special flushing tool from the converter hub. Allow the converter to drain by referring to step 2.

NOTE: It must be understood that this type of flushing will not clean the converter as well as the commercial type converter cleaner and flusher.

7. Remove any burrs from the drilled hole, apply sealer to the hole and blind "pop" rivet, install the rivet into the drilled hole and crimp the rivet. Brush sealer over the head of the rivet.

8. Pressure test the converter for leaks at 80–90 PSI. If the "pop" rivet leaks, the converter must be replaced, since the rivet

cannot be drilled out as part of it would remain inside.

9. If the threaded plug is installed in the converter shell, the tap should be coated with grease to catch the metal particles. The tool manufacturer's instructions should be followed during the installation of the threaded plug.

INSPECTION OF THE CONVERTER SHELL

The converter assembly can cause one or more of the following problems.

1. Malfunction and poor vehicle performance.

2. Noise and vibration.

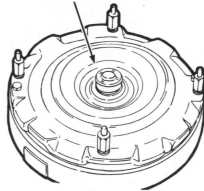

CHECK FOR DAMAGE
CLEAN UP BURRS OR NICKS

Checking converter cover for nicks and burrs (© Ford Motor Co.)

3. Fluid leakage.

4. Fluid contamination.

During the inspection of the unit, the outside shell must be inspected for damages. The defects to look for are as follows.

1. Flywheel attaching bolts or nuts missing.

2. Starter ring gear loose or damaged.

3. Flywheel or flex-plate cracked or broken.

4. Balance weights loose or missing.

5. Loose parts on or in the converter shell.

6. Scoring of the impeller hub.

NOTE: If the impeller hub is only lightly scored, clean with light crocus cloth, grit #600. If the scoring cannot be removed with light polishing or if the scoring is too deep, the converter must be replaced.

INSPECTION OF THE CONVERTER INTERNAL PARTS

Most converters are of the welded design and cannot be disassembled by the repair shops. Certain manufacturers recondition used converter assemblies and offer them for sale to the transmission rebuilding trade.

The average automatic transmission repair shop can and should inspect the converter assembly for internal wear before any attempt is made to reuse the unit after rebuilding the transmission unit. Special converter checking tools are needed and can be obtained through various tool supply channels.

CHECKING CONVERTER END PLAY (STATOR AND TURBINE)

1. Place the converter on a flat surface with the flywheel side down and the converter hub opening up.

2. Insert the special end play checking tool into the drive hub opening until the tool bottoms.

NOTE: Certain end play checking tools have a dual purpose, to check the stator and turbine end play and to check the stator one-way clutch operation. The dual purpose tool will have an expandable sleeve (collet) on the end, along with splines to engage the internal splines of the stator one-way clutch inner race.

3. Install the cover or guide plate over the converter hub and tighten the screw nut firmly to expand the split sleeve (collet) in the turbine hub.

4. Attach a dial indicator tool on the tool screw and position the indicator tip or button on the converter hub or the cover. Zero the dial indicator.

5. Lift the screw upward as far as it will go, carrying the dial indicator with it. Read

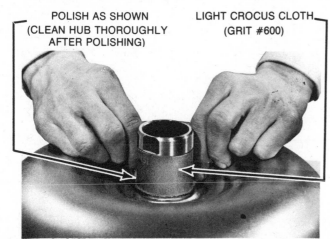

POLISH AS SHOWN
(CLEAN HUB THOROUGHLY
AFTER POLISHING)

LIGHT CROCUS CLOTH
(GRIT #600)

Removing light score marks from the converter hub with light crocus cloth—600 grit (© Ford Motor Co.)

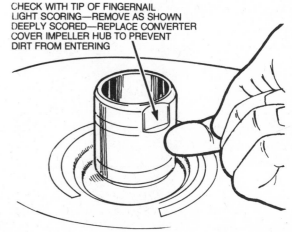

CHECK WITH TIP OF FINGERNAIL
LIGHT SCORING—REMOVE AS SHOWN
DEEPLY SCORED—REPLACE CONVERTER
COVER IMPELLER HUB TO PREVENT
DIRT FROM ENTERING

Checking converter hub for light or heavy scoring (© Ford Motor Co.)

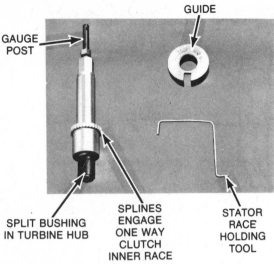

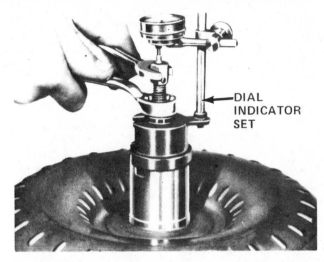

Checking converter end play with dial indicator and special tool

Typical converter checking tool set (© Ford Motor Co.)

the measurement from the indicator dial. The reading represents the converter end play.

6. Refer to the individual automatic transmission sections for the permissable converter end play.

7. Remove the tools from the converter assembly. Do not leave the split sleeve (collet) in the turbine hub.

STATOR TO IMPELLER INTERFERENCE CHECK

1. Place the transmission oil pump assembly on a flat surface with the stator splines up.

2. Carefully install the converter on oil pump and engage the stator splines.

3. Hold the pump assembly and turn the converter counterclockwise.

4. The converter should turn freely with no interference. If a slight rubbing noise is heard, this is considered normal, but if a binding or loud scraping noise is heard, the converter should be replaced.

TURBINE INTERFERENCE CHECK

1. Place the converter assembly on a flat

surface with the flywheel side down and the converter hub opening up.

2. Install the oil pump on the converter hub. Install the input shaft into the converter and engage the turbine hub splines.

3. While holding the oil pump and converter, rotate the input shaft back and forth.

4. The input shaft should turn freely with only a slight rubbing noise. If a binding or loud scraping noise is heard, the converter should be replaced.

STATOR ONE-WAY CLUTCH CHECK

Because the stator one-way clutch must hold the stator for torque multiplication at low speed and free-wheel at high speeds during the coupling phase, the stator assembly must be checked while the converter is out of the vehicle.

1. Place the converter on a flat surface with the flywheel side down and the hub opening up.

2. Install the stator race holding tool into the converter hub opening and insert the end into the groove in the stator to prevent the stator from turning.

3. Place the special tool post, without the screw and split sleeve (collet), into the converter hub opening. Engage the splines of the tool post with the splines of the stator race. Install the cover or guide plate to hold the tool post in place.

4. With a torque wrench, turn the tool post in a clockwise manner. The stator should turn freely.

5. Turn the tool post with the torque wrench in a counterclockwise rotation and the lock-up clutch should lock up with a 10 ft.-lb. pull.

6. If the lock-up clutch does not lock up, the one-way clutch is defective and the converter should be replaced.

Certain vehicle manufacturers do not recommend the use of special tools or the use of the pump cover stator shaft as a testing device for the stator one-way clutch unit. Their recommendations are to insert a finger into the converter hub opening and contact the splined inner race of the one-way clutch. An attempt should be made to rotate the stator inner race in a clockwise direction and the race should turn freely. By turning the inner race in a counterclockwise rotation, it should either lock-up or turn with great difficulty.

Checking for stator to impeller interference (© Ford Motor Co.)

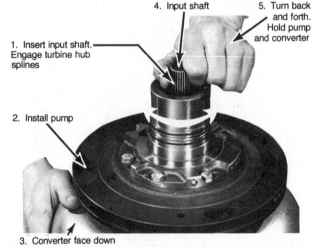

Checking for turbine interference clearance (© Ford Motor Co.)

CAUTION: *Care should be exercised to remove any metal burrs from the converter hub before placing a finger into the opening. Personal injury could result.*

Planetary Gear Set Inspection

Inspect the gear teeth surfaces for signs of scoring, chipping or wear on the thrust ends of the gears and housings. Certain transmission models will use needle bearings between the planetary gears and the pin or shaft. If the gears rotate smoothly when turned by hand, the needle bearings should be in good condition. Examine the thrust washer or bearing surfaces for nicks, burrs or scores. Light scores, nicks or burrs can be removed with crocus cloth or a fine stone. Inspect the planetary gear carriers for cracks or loose pins or shafts. When oil contamination is the result of a thrust washer or bearing failure, the metal flakes will be embedded in the surface of the gear teeth and cause noises on acceleration, depending upon the gear ratio the transmission is in and the reduction phase of the planetary gear set. Normally, replacement of the planetary gear set is required to overcome the noise factor.

Overrunning Clutch Inspection

Inspect the inner and outer races for wear or scoring. Turn the overrunning clutch by hand to be sure the unit will freewheel in one direction and lock-up in the other direction.

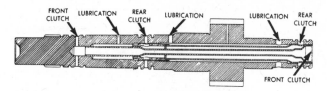

Cross-section of primary sun gear shaft with internal tubes

ROLLER TYPE OVERRUNNING CLUTCH

Inspect the rollers for a smooth, round surface, free of flat spots, chips, scoring or flaking. Inspect the springs for distortion or breakage. Inspect the roller contact surfaces in the cam notches for brinelling (surface indentations caused by impact loading, such as the lock-up of the unit).

SPRAG TYPE OVERRUNNING CLUTCH

Inspect the sprag faces for scoring, chips, flaking or being flattened. Inspect the sprag retaining springs for proper positioning on the sprag lugs, for discoloration due to excessive heat, distortion or breakage. Replace the races should they show signs of abnormal brinelling (surface indentations caused by impact loading, such as the lock-up of the unit).

REPAIRS OR REPLACEMENT OF THE OVERRUNNING CLUTCH

Numerous manufacturers have replacement parts available for the repairs to the overrunning clutch. Overrunning clutch cams, riveted to the transmission case, can be replaced with bolts replacing the rivets.

The installation instructions of the parts supplier should be followed to avoid damage to the transmission case.

CAUTION: *Upon installation of the overrunning clutch unit, be sure the installation of the roller or sprag unit is correct to lock-up in the right direction, otherwise, freewheeling will occur in the gear ratio when the unit should be locked up.*

Clutch Body, Hubs, Drums and Shaft Inspection

Inspect the transmission "Hard Parts" (common terminology for metal movable parts), for cracks, scores, worn splines and serrations, burrs, scratches and nicks. Inspect and clean the fluid passageways and check for obstructions. Inspect the shafts for abnormal wear and should the shaft be fitted with internal tubes for fluid delivery, be sure the tubes remain in their original position. Should plugs be used in the shafts to seal passages off, check for looseness or absence of the plug.

CAUTION: *Cases of tube movement have occurred, causing transmission malfunctions that are extremely difficult to diagnose and can cause loss of lubricating fluid to the rotating parts.*

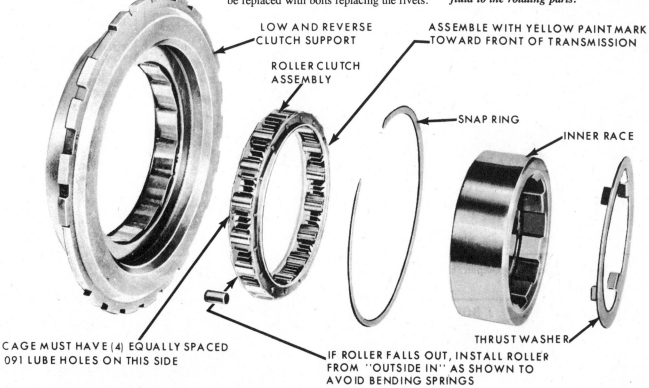

Exploded view of roller type overrunning clutch unit with installation instructions

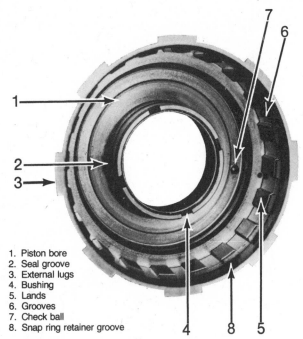

1. Piston bore
2. Seal groove
3. External lugs
4. Bushing
5. Lands
6. Grooves
7. Check ball
8. Snap ring retainer groove

Typical clutch pack housing showing areas of inspection

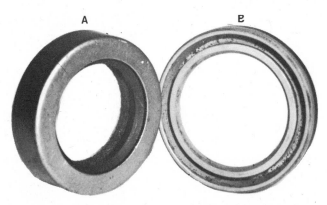

Two types of rear extension seals: (A) Single lip type with felt liner: (B) Lip type with double sealing lips

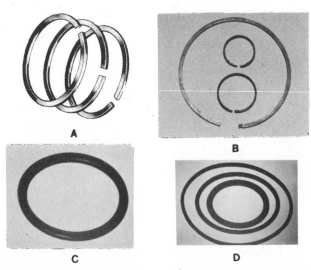

A. Butt-end type oil seal; B. Hook type oil seal; C. "O" ring rubber type seals;
D. Square lip rubber seals

Pistons, Seals and Sealing Ring Inspection

PISTONS

Inspect the piston bore and seal grooves for score marks and burrs. Inspect the check valve in the piston body, if equipped, for freedom of movement and its sealing under vacuum. (Normally, mouth vacuum is sufficient to seat the ball.) Inspect the piston springs, spring retainers and the snapring retainers for breakage, distortion or warpage, due to excessive heat.

SEALS

The seals used in the automatic transmissions are normally of three types, round "O"-ring seals, square section seals and lip type seals. It is common practice to replace all seals during the rebuilding process to avoid possible unit malfunction or fluid leakage due to the reusing of seals.

CAUTION: *During the installation of the new seals, lubricate with ATF fluid, petroleum jelly or door-ease type lubricate.*

SEALING RINGS

The sealing rings are of two types, cast iron and teflon. As with the seals, the rings should always be replaced during the rebuilding process.

The mating surfaces of the rings should be inspected for wear grooves and if present, the parts should be replaced.

Certain cast iron rings are manufactured with hook connectors at the ring ends so that a gap is not present when installed. During the installation, be certain the hooks are connected properly.

Certain Teflon® sealing rings are angled or chamfered on the ring ends to achieve a more positive seal when installed. During the installation, be sure the angles are placed together. Other Teflon rings are butted together at the gap end or are an endless circle.

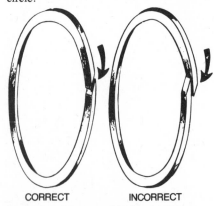

Correct and incorrect Teflon seal ring installation (© General Motors Corp.)

Bushings, Thrust Bearings and Washer Inspection

The bushings should be inspected for scoring, looseness in the bore and loss of metal from lack of lubrication or overheating.

The following guide can be used to determine the acceptability or unacceptability of the bushings.

ACCEPTABLE

Bushing A— Shows very light wear marks and discoloration and no scoring. Excellent.

Bushing B— Shows very light wear, moderate discoloration and no scoring. Very good.

Bushing C— Shows light wear and discoloration with some light scoring and pitting. Good condition.

Bushing D— Shows moderate wear and scoring, with some pitting and light discoloration. Overall bushing bearing surface is still smooth. Acceptable condition.

UNACCEPTABLE

Bushing E— Shows heavy scoring, no discoloration. The bushing bearing surface is rough and pitted.

Bushing F— Shows heavy scoring and pitting, discoloration and moderate wear. The bearing surface is rough, uneven, pitted and flaking.

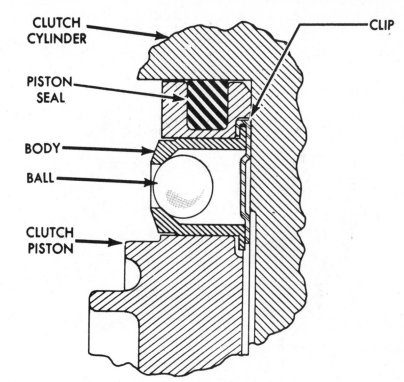

Cross-section of front clutch piston seal and check valve

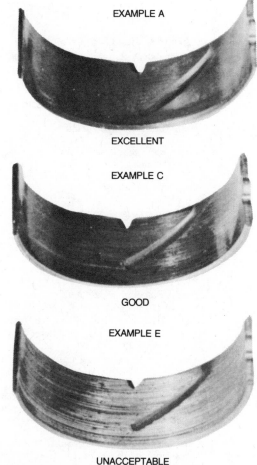

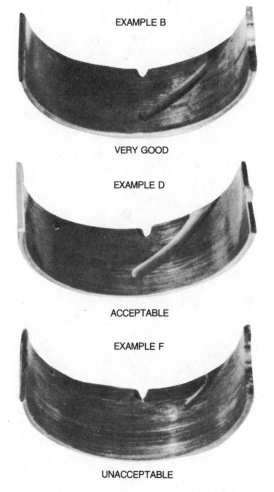

Guide for determining bearing wear

THRUST BEARINGS

Check the cages and the races of the bearings for heat discoloration, corrosion (frettage), cracks, abrasions, surface indentations (brinelling) or flaking (spalling).

Replace the necessary bearings and races as a unit. Do not replace individually.

THRUST WASHERS

Should the transmission end play be checked during the disassembly and the clearances are within specification, the indications are that the wear on the thrust washers has not been excessive. Should the visual check of the washers indicate wear,

Typical clutch plate assortment

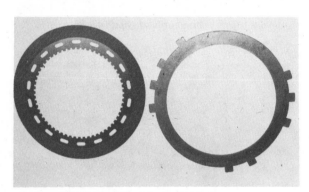

Drive (composition) and Driven (steel) plate comparison

Types of steel clutch plates used—typical

or loss of any locating lugs, the washer should be replaced.

Clutch Plate Inspection
COMPOSITION
CLUTCH PLATES

Clean the clutch plates with an approved solvent and allow to air dry.

CAUTION: *Do not use compressed air for drying purposes. The air pressure can lift the composition from the clutch plates.*

Inspect the clutch plate composition or friction material for pitting, flaking, wear, glazing, cracking, having chips of metal particles imbedded in the lining or being charred from transmission overheating.

Inspect the internal splines for missing or broken spline teeth.

Inspect the grooves on the lining, if so equipped. The grooves should be 0.005 inch deep.

NOTE: The grooves allow the fluid to escape from between the steel and the composition plates during application, to prevent a hydraulic "planing" effect, which results in slipping clutches. Poor grooving can cause shift problems and premature clutch burnout.

If any of these conditions exist, replace the composition clutch plates.

NOTE: Most rebuilders replace the clutch plates as a matter of standard operating procedure during the rebuilding operation.

STEEL PLATES

The steel plates should be inspected for defects such as surface pitting, scratches, rust, scale, gouges or delamination (separation of the metal surface due to a metal flaw). The externally splined teeth should also be inspected for burrs, chips, ragged edges or for broken or missing teeth.

The scored steel plates can cause the friction material to be simply torn from the lined clutch plates during transmission operation and cause the hydraulic circuit to become plugged.

CLUTCH PLATE DISH
OR FLATNESS

It is most important to know if the clutch plates are required to be "dished" or flat before installation into the transmission clutch unit.

The steel plates are normally used if any clutch plates are required to be dished.

NOTE: The individual automatic transmission sections will include this information as required.

The methods of measuring the plates for the required dish can be accomplished in the following manner.

Method A—The preferred method of measuring the flatness or dish of the clutch plates is with the use of a dial indicator and a flat surface plate. The clutch plates must be measured at approximately six different locations of the circle, both on the inside diameter (ID) and the outside diameter (OD). Record the results. For example,

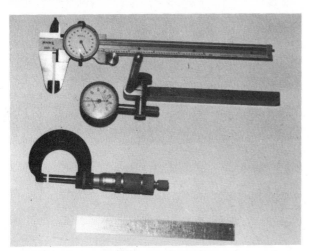

Tools used for clutch plate measurement

Measuring steel clutch plate with micrometer

should the dish specifications be 0.010 to 0.020 inch, the difference between the ID and the OD, measured at the same location of the circle, should be within the specified measurement. Should the specifications be zero dish, any difference in the readings between the ID and the OD should result in the replacement of the defective plate.

NOTE: Should one or more plates, either lined or steel need replacement, the entire clutch pack should be replaced.

Method B—An alternate method of measurement is with the use of a flat surface and a feeler gauge. By laying the clutch plate on the flat surface with the dish up, insert the feeler gauge between the inside diameter of the clutch plate and the flat surface. For example, should the dish specifications be 0.010 to 0.020 inch, a feeler gauge blade of 0.010 inch should be used as a "go" and a feeler gauge blade of 0.020 inch should be used as a "no go".

NOTE: Should one or more plates, either lined or steel need replacement, the entire clutch pack should be replaced.

CAUTION: *The clutch plates can appear to be the same from one clutch unit to another, but one clutch unit may require the use of dished plates while another unit may not. Do not interchange the clutch plates.*

MEASURING CLUTCH PLATE AND CLUTCH PACK THICKNESS

An important phase of the transmission rebuilding process is the measuring of the clutch plate thickness and the height of the clutch plate pack, before or after installation into the clutch plate retainer housing.

The thickness of the original clutch plates should be known, so that replacement plates of the proper thickness can be obtained. The individual clutch plates should be measured and remain within a 0.003 to 0.005 inch tolerance of each other and within a 0.001 inch parallelism, as checked at six equally spaced points around the plates.

Certain manufacturers require the adding or deletion of clutch plates, replacement of the pressure plates or retaining snap rings,

to bring the clutch plate pack to the proper height. Other manufacturers require the clutch pack be assembled on a flat surface and measured after a specific amount of pressure is applied to the clutch pack.

The method recommended for the measurement of the clutch plates or the clutch packs are noted in the individual automatic transmission sections.

The tools generally used to measure the clutch plates the clutch packs are the feeler gauges, dial indicators, depth gauges, measuring scales and micrometers.

Band Inspection

Inspect the band lining material for wear and bonding of the material to the band metal. Inspect the material for flaking, burn marks, glazing or abnormal wear. Should the lining be worn at the ends or center of the band material as indicated by groove wear, the band should be replaced.

Inspect the band strut pockets for breakage, cracks, looseness of welds or rivets and worn or broken strut locator pins. Inspect the band metal for distortion or signs of overheating.

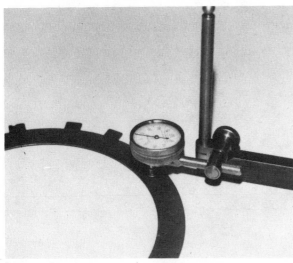

Measuring steel clutch plate with dial indicator

Measuring steel clutch plate dish—typical

37

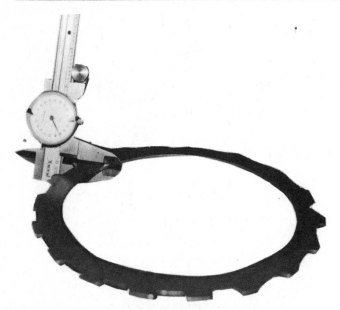

Measuring steel plate with dial caliper gauge

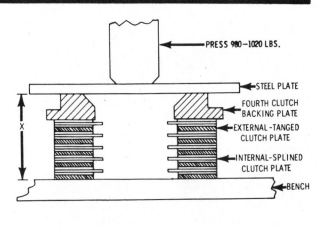

DIMENSION X	USE PISTON
0.9965-1.0205	C
1.0205-1.0465	B
1.0465-1.0725	A

Example of measuring clutch pack under a specific pressure with typical measurement guide for correct piston installation (© Detroit Diesel Allison)

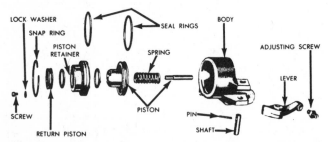

Exploded view of typical servo assembly, externally mounted within oil pan area (© Ford Motor Co.)

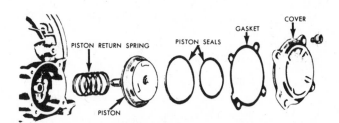

Exploded view of typical servo assembly located in the transmission case (© Ford Motor Co.)

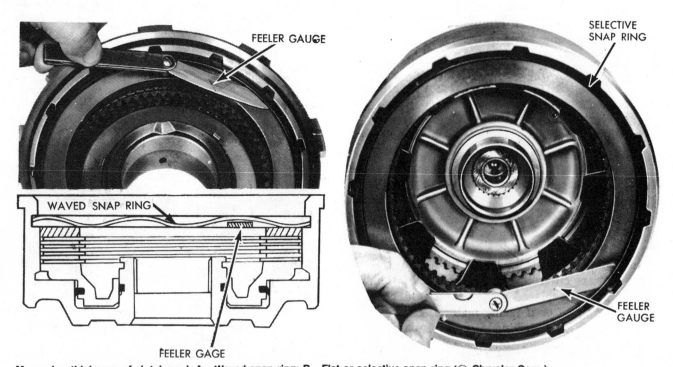

Measuring thickness of clutch pack A—Waved snap-ring; B—Flat or selective snap-ring (© Chrysler Corp.)

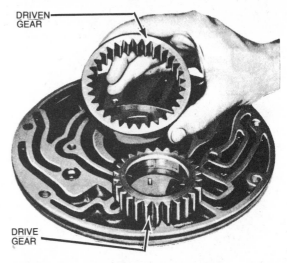

Checking oil pump body to gear face clearance—typical (© General Motors Corp.)

Identification of oil pump gears

Servo and Accumulator Inspection

Inspect the servo pistons for cracks, burrs, scores and abnormal wear. Should secondary pistons or plugs be used with the primary pistons, their movements should be free operating with no signs of abnormal wear.

Inspect the servo piston springs for distortion, weakness and fatigue. Inspect the piston bores in the servo body or in the transmission case for scores, burrs and abnormal wear. Light score marks can be removed by using crocus cloth #600 grit.

Oil Pump Inspection

Inspect the oil pump body for wear, scroes or cracks, the machine surfaces for nicks and burrs and the pump internal gears for scoring or pitting. Inspect the bushing for wear or scoring.

Should a pressure regulator valve or a pressure relief valve be incorporated in the oil pump housing, the valves, springs and bores should be inspected for any defects.

With the oil pump assembled, measure the clearance between the face of the gears and the pump body, the clearance between the inner and outer gear teeth and the clearance between the outer gear and the bore of the pump body. Refer to the specifications in the individual automatic transmission sections.

CAUTION: *Should an oil pump need replacement, install the pump assembly designed for the specific transmission that is being rebuilt. Interchanging a pump assembly on appearance assumption only can result in premature automatic transmission failure.*

Valve Body and Governor Inspection

The valve body bores do not change dimensionally with use, and if the valve body was operating satisfactorily before overhaul, it should operate properly when cleaned and reinstalled. Usually, there is no need to replace a valve body unless it is damaged through careless handling.

The equipment used for cleaning the control valve body and the governor valve assemblies can be gathered from regular kitchen or household items.

A partial list is as follows:

1. A large cake pan can be used for cleaning, washing and drying the valves and bodies.

2. A small pan can hold transmission fluid to be used to lubricate the parts before assembly.

3. A large, low-sided tray for the disassembly and assembly of the units.

4. A strainer can be used to drain the parts after washing.

NOTE: Use only *hot* water when rinsing the parts. The heat will help dry the parts and rust can be avoided.

5. To avoid mixing the valve springs, a spring holder can be fabricated from an electrical round box cover. Drill holes through the cover on its outer circle and install long screws in the holes and lock fast with nuts. Number each long screw around the circle, starting with one and counting upward.

6. An exploded view of the valve body or governor assembly is needed to correspond the springs to the holder. Number the springs in the exploded view from one upward and place the corresponding numbered spring on the spring holder.

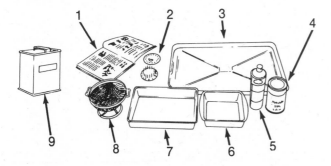

1. Shop manual
2. Spring holder
3. Large tray
4. Automatic transmission fluid
5. Carburetor and combustion cleaner
6. Small pan (8 x 8)
7. Large pan
8. Strainer
9. Mineral spirits

Typical valve body and governor cleaning arrangement (© Ford Motor Co.)

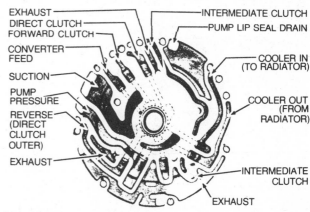

Identification of fluid passages in pump body—typical (© General Motors Corp.)

7. Another method of spring and valve order retention is to use a multi-channeled hard wood block or a multi-channeled steel block to lay the valves and springs in as they are removed from the valve body or governor body. The multi-channeled block can very easily be made by drilling ½ or ¾ inch holes, lengthwise through a block, approximately 8 to 10 inches long. Cut the block in half and two multi-channeled blocks are available.

8. Cleaners to be used are either an approved solvent, mineral spirits or carburetor/combustion area cleaner, and can be obtained at most automotive supply houses.

The valve body and governor assembly can be disassembled, cleaned in solvent and air dried with compressed air. Be sure all passages are clean and free of obstruction. Lay all valves and springs in their proper order of disassembly so as not to mix the valves or springs.

Inspect the throttle and manual linkage levers and shafts for being bent, worn or loose.

CAUTION: *Do not attempt to straighten bent levers. If necessary to weld, use only silver solder.*

Inspect all mating surfaces for scratches, nicks, burrs or warpage. Inspect all valve springs for distortion, broken or collapsed coils. Inspect the valves for scores, nicks and burrs.

CAUTION: *Do not bevel the sharp edges of the valves during the attempt to remove small nicks and scores from the valves with crocus cloth. The sharpness of the valve edges is important to prevent sticking of the valve in the bore due to foreign material lodging between the valve and the bore.*

When the bores are dry and clean, the valves should drop freely into their bores. Certain transmission manufacturers supply the spring lengths, color codes, and modification information concerning the spring usage.

CAUTION: *Under no circumstances should springs be stretched or shimmed to change pressures as possible internal transmission damage can occur.*

Transmission Housing Assembly Inspection
TRANSMISSION CASE

Inspect the transmission case for cracks, stripped threads, porosity leakage areas, nicks or burrs on the mating surfaces, and obstructions in the vent system or fluid passages. Inspect the manual linkage assembly and parking mechanism for proper operation, if not previously removed.

TRANSMISSION EXTENSION HOUSING

Inspect the transmission extension housing for cracks and defective mating surfaces. Inspect the housing rear bushing for a worn or scored condition. Should it be necessary to replace the bushing, inspect the bushing bore of the housing for being out of round.

TRANSMISSION CONVERTER HOUSING

Should the converter housing be a separate assembly, inspect for nicks and burrs on the mating surfaces. Inspect the bolt holes for cracks and signs of transmission misalignment. If the converter housing and the transmission case are one unit, inspect for cracks and strain marks on the housing area assembly, along with the above checks.

Drive Line Service

Drive line vibrations can affect the operation and longevity of the automatic transmissions and should be diagnosed during the road test and inspected during the transmission removal phase.

The drive shafts are designed for specified applications and the disregard for the correct application can result in drive shaft failure with extremely violent and hazardous consequences. A replacement shaft assembly must always be of the same design and material specifications as the original to assure proper operation.

Drive Line Operation
UNIVERSAL JOINTS

Universal joints are designed to transmit torque under various loading and driveline angle changes due to movement of the drive shaft (conventional drive) or drive axles (trans-axle assemblies) during the vehicle operation.

When the variation of angles exceeds the designed limitations of the drive line, its operational life will be greatly decreased.

A single universal joint, operating between two shafts and set at a given angle, will have the drive yoke turning at a constant speed, while the driven yoke will turn at a fluctuating speed, slowing down and speeding up, twice during one revolution of the shaft. The changing speed of the driven yoke increases as the angle between the two yoke shafts increase. Two universal joints are used to avoid the pulsation and vibration that would occur from the drive line assembly.

When two joints are placed in the drive line, one at each end of the drive shaft and the angles remain the same, the rear joint will provide the direct opposite reaction to the front joint fluctuation and cause a cancellation of the fluctuation speed, by having the rear joint slow down by the same amount that the forward joint speeds up, resulting in the driving and driven shafts turning at a constant and identical speed.

The universal joints use needle bearings in the bearing cups which are mounted between the yokes and the spider or cross, to aid in a smooth transfer of torque through the drive line angle. The bearing cups are held in the yokes by either snap-rings, plastic injection or by bolts. Truck applications may require the use of two or more driveshaft assemblies, supported in the center by a bearing unit.

SLIP JOINTS

Slip joints are used to permit the forward and rearward movement of the drive shaft during the vehicle operation. The splines of the yoke are normally lubricated internally by the transmission fluid and contained by an oil seal to avoid leakage. The splined yoke is held in alignment by the rear extension housing bushing. In truck applications, the slip joints are normally positioned between two driveshaft members.

It is most important to mark the slip joint components before removal so that they may be put back in their proper phase during the reassembly of the shaft.

"C" TYPE BEARINGS

DRIVE LUG

"C" TYPE BEARINGS

Universal joint with bolted bearing cup retainer

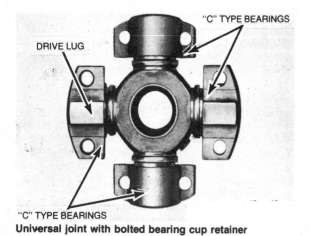

1. Trunnion
2. Seal
3. Bearings
4. Washer
5. Cap
6. Snap ring

Universal joint with external bearing cup retainer

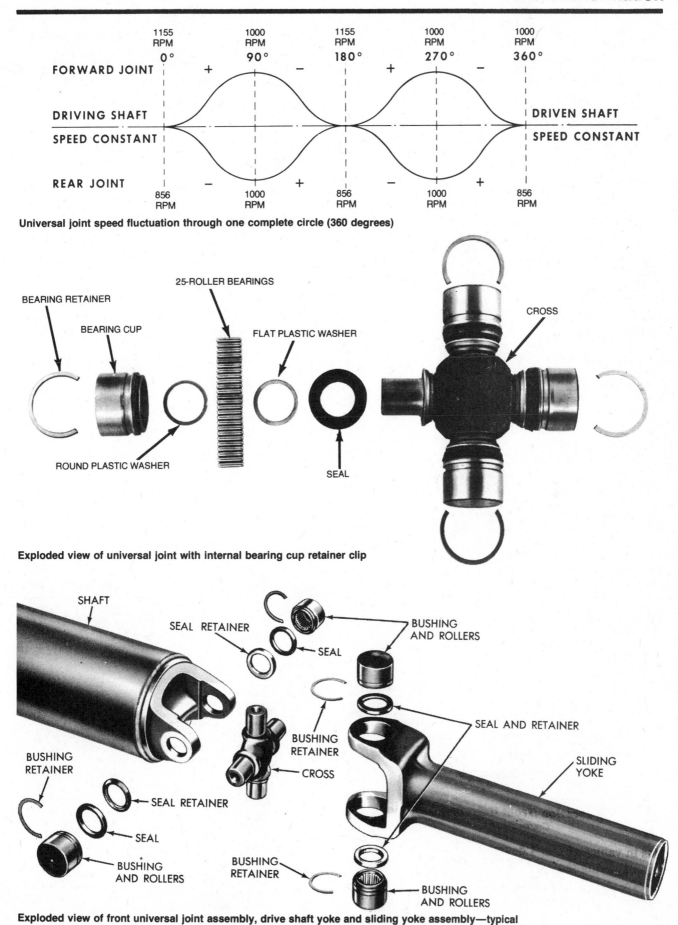

Universal joint speed fluctuation through one complete circle (360 degrees)

Exploded view of universal joint with internal bearing cup retainer clip

Exploded view of front universal joint assembly, drive shaft yoke and sliding yoke assembly—typical

41

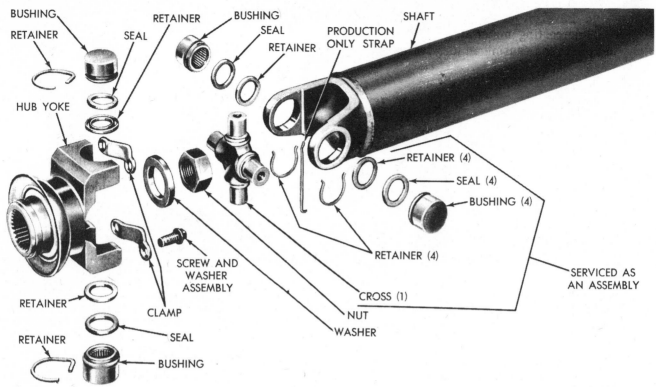

Exploded view of rear universal joint assembly, drive shaft yoke and hub yoke. U bolts can be used in place of the bolts and clamps (© Chrysler Corp.)

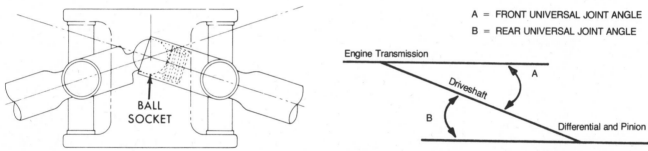

Simplified illustration of constant velocity universal joint

A = FRONT UNIVERSAL JOINT ANGLE

B = REAR UNIVERSAL JOINT ANGLE

Typical parallel type drive line illustration—Angles A and B are equal

DRIVE SHAFT DIAGNOSIS AND TESTING

Condition	Cause	Correction
Leak at front slip yoke. An occasional drop of lubricant leaking from splined yoke is normal and requires no attention.	a) Rough outside surface on splined yoke.	a) Replace seal if cut by burrs on yoke. Minor burrs can be smoothed by careful use of crocus cloth or honing with a fine stone. Replace yoke if outside surface is rough or burred badly.
	b) Defective transmission rear oil seal.	b) Replace transmission rear oil seal. Bring transmission oil up to proper level after correction.
Knock in drive line, clunking noise when car is operated under floating condition at 10 mph in high gear or neutral.	a) Worn or damaged universal joints.	a) Disassemble universal joints, inspect and replace worn or damaged parts.
	b) Slide gear hub counterbore in differential worn oversize.	b) Replace differential carrier and/or side gears as required.

DIAGNOSIS AND TESTING (CONT'D)

Condition	Cause	Correction
Ping, Snap or Click in drive line. Usually occurs on initial load application after transmission has been put into gear, either forward or reverse.	a) Loose upper or lower control arm bushing bolts. b) Loose companion flange.	a) Tighten bolts to specified torque. b) Remove companion flange, turn 180° from its original position, apply white lead to splines and reinstall. Tighten pinion nut to specified torque.
Roughness usually at low speeds, light load, 15-35 mph.	a) Improper joint angles. Usually rear joint angle is too large. b) Improperly adjusted front joint angle.	a) Check rear trim height at curb weight. a) Check rear joint angle. b) Check and adjust front joint angle by shimming transmission support.
Scraping noise.	a) Slinger, companion flange, or end yoke rubbing on rear axle carrier.	a) Straighten slinger to remove interference.
Roughness—above 35 mph felt and/or heard.	a) Tires unbalanced or worn.	a) Balance or replace as required.
Squeak	a) Lack of lubricant in universal joint. b) Bearing cup retainer on rear joint rubbing on cross.	a) Replace universal joint with service package. b) Pry retainer from cross. If one end is broken, remove retainer.
Roughness, Vibration or Body Boom at any speed. With tachometer installed in car, determine whether drive shaft is cause of complaint by driving through speed range and note the engine speed (rpm) at which vibration (roughness) is most pronounced. Then, shift transmission to a different gear range and drive car at same engine speed (rpm) at which vibration was noted before. Note the effect on the vibration. If vibration occurs at the same engine speed (rpm), regardless of transmission gear range selected, drive shaft assembly is NOT at fault, since the shaft speed (rpm) varies. If vibration decreased, or is eliminated, in a different gear range but at the same engine speed (rpm), check the possible causes:	a) Bent or dented drive shaft. b) Undercoating on drive shaft. c) Tire unbalance. (30-80 mph, not throttle conscious) d) Tight universal joints. e) Worn universal joints. f) Burrs or gouges on companion flange. g) Drive shaft or companion flange unbalance. h) Incorrect rear joint angle. The angle is usually too large when it is a factor. i) Excessive looseness at slip yoke spline. j) Drive shaft runout (50-80 mph throttle conscious)	a) Replace. b) Clean drive shaft. c) Balance or replace as required. d) Strike yokes with a hammer to free up. Overhaul joint if unable to free up or if joint feels rough when rotated by hand. e) Overhaul, replacing necessary parts. f) Rework or replace companion flange. g) Check for missing balance weights on drive shaft. Remove and reassemble propeller shaft to companion flange, 180° from original position. h) Check and correct trim height at curb weight. Measure joint angle. i) Replace necessary parts. j) Check drive shaft runout at front and rear. Should be less than specified. If above rotate shaft 180° and recheck. If still above specified, replace shaft.
Shudder on acceleration, low speed.	a) Incorrectly set front joint angle.	a) Shim under transmission support mount to decrease front joint angle.

CONSTANT VELOCITY DOUBLE CARDAN JOINT

When large angles are encountered in the driveline, the constant velocity-double Cardan joint is used to transmit the torque in a constant, vibrationless rotation by having two universal joints closely coupled by a coupling yoke, phased properly for constant velocity. A centering ball socket between the two joints maintains the relative position of the two joints and causes each of the two units to operate through one-half of the complete angle between the driveshaft and the output shaft during its revolution.

DRIVESHAFT ANGLE

During the installation of a replacement transmission, the drive shaft angle can change through mounting differences and a vibration can occur during the vehicle operation. If the repair shop does not recognize this problem, much lost time and money can result in the attempt to correct it.

To understand the angles used in the drive line, a brief explanation is given. The front universal joint angle is the angle between the engine-transmission centerline and the drive shaft, while the rear universal joint angle is the angle between the differential pinion centerline and the drive shaft. These angles cannot be altered on certain vehicles, while on others, the angles can be altered by adding or removing shims from the transmission mount area or by movement of cam bolts or shim changes on the differential housing.

Special tools are available for the specific purpose of checking the drive line angle and the manufacturer's instructions should be followed.

ALUMINUM THREAD REPAIR

Damaged or worn threads in the aluminum transmission case and valve body can be repaired with the use of the Heli-Coils® or equivalent thread restorer. This repair consists of the drilling out of the stripped or worn threads, tapping the hole with a special tap and installing a Heli-Coil or equivalent insert into the tapped hole, which will bring the hole back to its original dimension and thread size.

The Heli-Coil or equivalent tools with inserts are available from most automotive parts jobbers.

Some thread drag may occur when using the Heli-Coil or equivalent tool, when attempting to install a screw into a repaired hole. A torque reading should be taken of the thread drag with an inch-pound torque wrench and added to the required torque reading before securing a specific bolt.

INCH SYSTEM AND METRIC FASTENERS

Metric bolt sizes and thread pitches are more commonly used for all fasteners on the automatic transmissions now being manufactured. The metric bolt sizes and thread dimensions are very close to the dimensions of the similar inch system fasteners and for this reason, replacement fasteners must have the same measurement and strength as those removed.

Do not attempt to interchange metric fasteners for inch system fasteners. Mismatched and incorrect fasteners can result in damage to the transmission unit through malfunction, breakage or possible personal injury.

Care should be exercised to re-use the fasteners in the same locations as removed, whenever possible.

SHIFTING MODIFICATIONS

To overcome slow or overlapping shifts of the automatic transmission and to give firmer and quicker shifts, various control valve body modification kits are available for the transmission rebuilding trade. The modification kits make it possible for the recalibration of the shift points by replacement of valves, springs, spacer plates and gaskets.

This phase of automatic transmission modifications should be by a repairman, skilled in valve body operation and overhaul. If this type of repair is performed by a novice, internal transmission damage could occur.

Special Tools
SPECIAL TOOLS AND WORK AREA

There are an unlimited number of special tools and accessories available to the transmission rebuilder to lessen the time and effort required in performing the diagnosing and overhaul of the automatic transmission. Specific tools are necessary during the disassembly and assembly of each transmission and its subassemblies. Certain tools can be fabricated, but it becomes the responsibility of the rebuilder to judge their worth as a substitute for the proper tool.

The commercial labor saving tools range from puller sets, bushing and seal installer sets, compression tools and presses (both mechanically and hydraulically operated), holding fixtures, oil pump aligning tools (a necessity on most automatic transmissions) to work bench arrangements, degreaser tanks, steam cleaners, converter flushing machines, transmission jacks and lifts, to name a few. For specific information concerning the various tools, a parts and tool supplier should be consulted.

In addition to the special tools, a complete tool chest with the necessary hand tools should be available to the repairman.

Work Area

The size of the work area depends upon the space available within the service shop to perform the rebuilding operation by having the necessary benches, tools, cleaners and lifts arranged to provide the most logical and efficient approach to the removal, disassembly, assembly and installation of the automatic transmission. Regardless of the manner in which the work area is arranged, it should be well lighted, ventilated and clean.

Precautions to Observe When Handling Solvents

All solvents are toxic or irritating to the skin to some degree. The amount of toxicity or irritation normally depends upon the skin exposure to the solvent. It is a good practice to avoid skin contact with solvent by using rubber gloves and parts drainers when cleaning the transmission parts.

CAUTION: *Do not, under any circumstances, wash grease from hands or arms by dipping into the solvent tank and air drying with compressed air. Blood poison can result.*

TOWING

Proper towing is important for the safe and reliable transfer of an inoperative vehicle from one point to another.

The basic towing instructions and procedures described are general in nature and the towing procedures may not apply to the same make, model or vehicle throughout the model years without procedure changes, modification to equipment or the use of auxiliary equipment designed for a specific purpose.

It is important to minimize the risk of personal injury, to avoid damage to the towed vehicle or to render it unsafe while being towed. There are many conceivable methods of towing with possible hazardous consequences for each method. Therefore, it is the responsibility of the tow truck operator to determine the correct connection of the towing apparatus in a safe and secure manner.

Towing manuals are available from varied sources, explaining the vehicle manufacturer's recommended lifting procedures, towing speeds and towing distances. The operating instructions for the towing truck should be understood and followed by the operator.

The disabled vehicle should never be pushed or pulled on a highway because of safety reasons.

GENERAL MOTORS BODY IDENTIFICATION

Because of the many mechanical and body similarities between General Motors vehicles, the corporation has begun using Body Identification codes, rather than model names, when referring to the same chassis design used by different divisions.

The accompanying Body Identification Chart will assist the repair shop in verifying the type of vehicle being serviced, and in obtaining replacement parts. Many mechanical items are interchangeable between divisions.

The Transmission Coding Chart can also be helpful in determining transmission type by General Motors' new coding system.

GENERAL MOTORS BODY IDENTIFICATION

A BODY—Buick Century, Regal; Chevrolet Malibu, Monte Carlo; Olds Cutlass; Pontiac LeMans, Grand AM.

B BODY—Buick LeSabre, LeSabre Estate, Electra Estate; Chevrolet Caprice, Impala; Olds Delta 88, Custom Cruiser; Pontiac Catalina, Bonneville.

C BODY—Buick Electra (exc. Electra Estate); Olds Ninety-Eight; Cadillac Fleetwood (exc. 4dr sedan), DeVille.

D BODY—Cadillac Fleetwood 4dr sedan, Limousine.

E BODY—Buick Riviera; Cadillac Eldorado; Olds Toronado.

F BODY—Chevrolet Camaro; Pontiac Firebird.

G BODY—Pontiac Grand Prix.

H BODY—Buick Skyhawk; Chevrolet Monza, Vega; Olds Starfire; Pontiac Astre, Sunbird.

K BODY—Cadillac Seville.

T BODY—Buick Opel; Chevrolet Chevette.

X BODY (pre-1980)—Buick Apollo, Skylark; Chevrolet Nova; Olds Omega; Pontiac Phoenix.

X BODY (1980)—Buick Skylark; Chevrolet Citation; Olds Omega; Pontiac Phoenix.

Y BODY—Chevrolet Corvette.

Z BODY—Cadillac Commercial Chassis.

AUTOMATIC TRANSMISSION IDENTIFICATION

The need to identify automatic transmissions occurs when transmission units are obtained; for example, through bulk buying for overhaul and storage as replacement units. To assist the repairman in coupling the proper transmission to the vehicle/engine combination, a listing is given of automatic transmission codes and their corresponding vehicle model and engine usage. The torque converter identification is difficult, as most replacement converters are rebuilt and distributed to suppliers for resale. The transmission model and serial number should be considered when converter replacement is required.

The following listings contain the transmission models or codes for the most commonly used automatic transmissions. Should a model or code be needed for a nonlisted vehicle or transmission, refer to the individual transmission section and, if available, the model or code will be noted along with the model and code location.

CHRYSLER CORPORATION AUTOMATIC TRANSMISSION LISTING

Year	Assembly No.	Engine Cu. In.	Type Trans.	Other Information
			Chrysler Corp. Cars	
1974	3681841	198/225	A-904	Standard (with alum. carrier)
	3743421	198/225	A-904	Standard (with steel carrier)
	3681842	225	A-904	Export (with alum. carrier)
	3743422	225	A-904	Export (with steel carrier)
	3681843	318	A-904-1	Standard (with alum. carrier)
	3743423	318	A-904-1	Standard (with steel carrier)
	3681844	360	A-999	Standard
	3681861	225	A-727	Police/taxi/truck
	3681862	318/360	A-727	Standard
	3681863	360 4 bbl	A-727	Hi-performance
	3681864	400	A-727	Standard
	3681865	400 4 bbl	A-727	Standard
	3681866	440 (A-120)	A-727	Standard
	3681867	440 (A-134)	A-727	Hi-performance
1975	3743421	198/225	A-904	Standard
	3743476	198/225	A-904	Standard (without console shift)
	3743422	225	A-904	Export
	3743423	318	A-904-1	Standard
	3743479	318	A-904-1	Standard (with A-999 type front clutch retainer)
	3681861	225	A-727	Police/taxi/truck
	3681862	318/360	A-727	Standard
	3681863	360 4 bbl	A-727	Hi-performance
	3681864	400	A-727	Standard
	3681865	400 4 bbl	A-727	Hi-performance
	3681866	440	A-727	Standard
	3681867	440	A-727	Hi-performance

CHRYSLER CORPORATION
AUTOMATIC TRANSMISSION LISTING

Year	Assembly No.	Engine Cu. In.	Type Trans.	Other Information
			Chrysler Corp. Cars	
1976	4028462	198/225	A-904	Standard
	4028463	198/225	A-904	Standard (without console shift)
	4028466	225	A-904	Export
	4028464	318	A-904-1	Standard
	4028468	318	A-904-1	Standard (with A-999 type front clutch retainer)
	4028465	360	A-999	Standard
	4028405	225	A-727	Police/taxi/truck
	4028406	318/360	A-727	Standard
	4028407	360 4 bbl	A-727	Hi-performance
	4028412	400	A-727	Standard
	4028418	400 4 bbl	A-727	Hi-performance
	4028424	440	A-727	Standard
	4028425	440	A-727	Hi-performance
1977	4028462	198/225	A-904	Standard
	4028463	198/225	A-904	Standard (without console shift)
	4028804	225	A-904	Export
	4028464	318	A-904-1	Standard
	4028468	318	A-904-1	Standard (with A-999 type front clutch retainer)
	4028465	360	A-999	Standard
	4028405	225	A-727	Police/taxi/truck
	4028406	318/360	A-727	Standard
	4028407	360 4 bbl	A-727	Hi-performance
	4028412	400	A-727	Standard
	4028418	400 4 bbl	A-727	Hi-performance
	4028424	440	A-727	Standard
	4028425	440	A-727	Hi-performance
	4028808	360 2 bbl	A-727	Standard with studs
1978	4028811	225	A-904	Standard (with lockup & with console shift)
	4028812	225	A-904	Standard (with lockup)
	4028813	225	A-904	Heavy duty (non-lockup)
	4028876	225	A-904	Standard (non-lockup)
	4028814	318	A-904-1	Standard (with lockup)
	4028877	318	A-904-1	Standard (non-lockup)
	4028815	360	A-999	Standard (with lockup)
	4058111	360	A-999	B-body wagon, C-body (with lockup)
	4028878	360	A-999	Standard (non-lockup)
	4028822	318/360	A-727	Standard (non-lockup)
	4028823	360	A-727	Hi-performance (non-lockup)
	4028824	400	A-727	Standard (with lockup)
	4058114	400	A-727	Standard (non-lockup)
	4028825	400	A-727	Hi-performance (non-lockup)
	4028826	440	A-727	Standard (non-lockup)
	4028827	440	A-727	Hi-performance (non-lockup)
	5212256	1.7 liter	A-404	3.50 axle ratio
	5222187	1.7 liter	A-404	3.76 axle ratio

CHRYSLER CORPORATION
AUTOMATIC TRANSMISSION LISTING

Year	Assembly No.	Engine Cu. In.	Type Trans.	Other Information
Chrysler Corp. Cars				
1979	4130411	225	A-904	Standard (with lockup & console shift)
	4130413	225	A-904	Standard (with lockup)
	4130641	225	A-904	Standard (non-lockup)
	4130432	225	A-904	Heavy duty (with lockup)
	4130721	225	A-904	Heavy duty (non-lockup)
	4130414	318	A-998	Standard (with lockup)
	4130415	360	A-999	Standard (with lockup)
	4130645	360	A-999	Calif. & EFM (with lockup)
	4130431	318/360	A-727	Standard (with lockup)
	4130433	360	A-727	Hi-performance (with lockup)
	4130705	360	A-727	Hi-performance (non-lockup)
	5224023	1.7 liter	A-404	3.48 axle ratio
	5224196	1.7 liter	A-404	3.67 axle ratio
1980	4130951	225	A-904	—
	4130952	225	A-904	—
	4130953	225	A-904	—
	4130954	—	A-904	—
	4130955	318	A-904	—
	4130956	318	A-904	—
	4130957	318	A-904	—
	4130958	—	A-904	—
	4130975	—	A-727	—
	4130976	360	A-727	—
	5224442	1.7 liter	A-404	3.48 axle
American Motors Cars				
1974	3222774	232	A-904	Standard (with alum. carrier)
	3222775	258	A-904	Standard (with alum. carrier)
	3222776	309	A-904-1	Standard (with alum. carrier)
	3222777	360 2 & 4 bbl (304)	A-727	Standard (304 heavy duty)
	3222778	401 (360)	A-727	Standard (360 heavy duty)
	3222779	401	A-727	Heavy duty
	3223759	232	A-904	Standard (with steel carrier)
	3223760	258	A-904	Standard (with steel carrier)
	3223761	309	A-904-1	Standard (with steel carrier)
1975	3223759	232	A-904	Standard
	3223760	258	A-904	Standard
	3223761	304	A-904-1	Standard
	3222777	360	A-727	Standard (304 heavy duty)
	3222779	401	A-727	Standard (360 heavy duty)
1976	3228968	232	A-904	Standard
	3228969	258	A-904	Standard
	3228970	304	A-904-1	Standard
	3228832	360 2 bbl	A-727	Standard (258-304 heavy duty)
	3228644	360 4 bbl	A-727	Standard (360 heavy duty)
	3228645	401	A-727	Standard (360 4 bbl heavy duty)

CHRYSLER CORPORATION
AUTOMATIC TRANSMISSION LISTING

Year	Assembly No.	Engine Cu. In.	Type Trans.	Other Information
		American Motors Cars		
1977	3228968	232	A-904	Standard
	3230823	258	A-904	Standard
	3228970	304	A-904-1	Standard
	3228832	360 2 bbl	A-727	Standard (258-304 heavy duty)
	3228644	360 4 bbl	A-727	Standard (360 heavy duty)
	3228645	401	A-727	Standard (360 4 bbl heavy duty)
	3229281	2.0 liter	A-904	Standard
1978	3232864	2.0 liter	A-904	Standard
	3232861	232	A-904	Standard
	3232862	258	A-904	Standard
	3232863	304	A-998	Standard
	3232866	360 2 bbl	A-727	Standard (258-304 heavy duty)
	3232865	360 4 bbl	A-727	Standard (360-2 heavy duty)
1979	3234754	2.0 liter	A-904	Standard (non-lockup)
	3234750	232	A-904	Standard (non-lockup)
	3234751	258	A-904	Standard (non-lockup)
	3234752	304	A-998	Standard (with lockup)
1980	3235770	2.0 liter	A-904	—
	3237269	6 cyl.	A-904	Less 4-wheel drive
	3236581	6 cyl.	A-904	With 4-wheel drive
		Dodge Trucks		
1974	3736627	318/360	A-727	Long extension, heavy duty
	3736625	440	A-727	Long extension, heavy duty
	3736139	318/360	A-727	Medium extension
	3736140	225	A-727	Long extension
	3720902	318/360	A-727	Long extension
	3736143	440	A-727	Medium extension
	3736144	225	A-727	Short extension
	3736145	318/360	A-727	Short extension
	3736146	225	A-727	Long extension, heavy duty
	3736147	318/360	A-727	Long extension, heavy duty
	3736148	440	A-727	Long extension, heavy duty
	3736149	440	A-727	Short extension
1975	3893759	318/360	A-727	Long extension, heavy duty (rear timing)
	3893763	440	A-727	Long extension, heavy duty (rear timing)
	3893761	318/360	A-727	Medium extension (rear timing)
	3893755	318/360	A-727	Long extension (rear timing)
	3893765	440	A-727	Medium extension (rear timing)
	3736144	225	A-727	Short extension
	3736145	318/360	A-727	Short extension
	3736146	225	A-727	Long extension, heavy duty
	3736147	318/360	A-727	Long extension, heavy duty
	3736149	440	A-727	Short extension
	3820696	225	A-727	4-wheel drive
	3820697	318/360	A-727	4-wheel drive
	3820698	440	A-727	4-wheel drive

CHRYSLER CORPORATION
AUTOMATIC TRANSMISSION LISTING

Year	Assembly No.	Engine Cu. In.	Type Trans.	Other Information
			Dodge Trucks	
1976	3898855	225	A-727	4-wheel drive
	3898863	225	A-727	Short extension
	3898857	318/360	A-727	4-wheel drive
	3898864	318/360	A-727	Long extension, heavy duty
	3898865	318/360	A-727	Long extension (rear timing)
	3898866	318/360	A-727	Long extension
	3898867	318/360	A-727	Sort extension
	3898868	360	A-727	Medium extension (rear timing)
	3897784	400/440	A-727	Long extension (rear timing)
	3897785	400/440	A-727	Short extension (rear timing)
	3898856	400/440	A-727	4-wheel drive
	3898869	400/440	A-727	Long extension, heavy duty (rear timing)
	3898870	440	A-727	Medium extension (rear timing)
1977	3898855	225	A-727	4-wheel drive
	3898863	225	A-727	Short extension
	3898857	318/360	A-727	4-wheel drive
	3898864	318/360	A-727	Long heavy duty extension
	3898865	318/360	A-727	Long extension (rear timing)
	3898866	318/360	A-727	Long extension
	4036786	318/360	A-727	Short extension (rear timing)
	3898868	360	A-727	Medium extension (rear timing)
	3897784	400/440	A-727	Long extension (rear timing)
	3897785	400/440	A-727	Short extension (rear timing)
	3898856	400/440	A-727	4-wheel drive
	3898869	400/440	A-727	Long extension, heavy duty (rear timing)
	3898870	400	A-727	Medium extension (rear timing)
1978	4039530	225	A-727	Short extension
	4058301	225	A-727	Long extension
	4039541	225	A-727	4-wheel drive
	4039531	318/360	A-727	Short extension (rear timing)
	4058302	318/360	A-727	Long extension
	4058303	318/360	A-727	Long extension (rear timing)
	4039536	318/360	A-727	Long extension heavy duty
	4039542	318/360	A-727	4-wheel drive
	4039533	360	A-727	Medium extension (rear timing)
	4039532	400/440	A-727	Short extension (rear timing)
	4039540	400/440	A-727	Long extension (rear timing)
	4039539	400/440	A-727	Long extension heavy duty (rear timing)
	4039543	400/440	A-727	4-wheel drive
	4039534	440	A-727	Medium extension (rear timing)
	4089070	MMC	A-727	Short extension (diesel)
	4089071	MMC	A-727	Long extension (diesel)
	4089072	MMC	A-727	4-wheel drive (diesel)
1979	4058323	225	A-727	Short extension (non-lockup)
	4058311	225	A-727	Long extension (with lockup)
	4058327	225	A-727	Long extension (non-lockup)

CHRYSLER CORPORATION
AUTOMATIC TRANSMISSION LISTING

Year	Assembly No.	Engine Cu. In.	Type Trans.	Other Information
			Dodge Trucks	
1979	4058333	225	A-727	4-wheel drive (non-lockup)
	4058318	318/360	A-727	Short extension (non-lockup)
	4058314	318	A-727	Long extension (with lockup)
	4058328	318/360	A-727	Long extension (non-lockup)
	4058316	318	A-727	Long extension, heavy duty (with lock-up)
	4058332	318/360	A-727	Long extension, heavy duty (non-lock-up)
	4058334	318/360	A-727	4-wheel drive (non-lockup)
	4058319	360	A-727	Medium extension (non-lockup)
	4130705	360	A-727	Hi performance long extension (non-lockup)
	4058324	440	A-727	Short extension (non-lockup, rear timing)
	4058326	440	A-727	Medium extension (non-lockup, rear timing)
	4058329	440	A-727	Long extension (non-lockup, rear timing)
	4058325	Diesel	A-727	Short extension (non-lockup)
	4058331	Diesel	A-727	Long extension (non-lockup)
	4058335	Diesel	A-727	4-wheel drive (non-lockup)
			American Motors General Corp.	
1976	5902621	AMG 232	A-727	Short extension
1977	5982294	AMG 232	A-727	Short extension (same as 5902621 except 4-disc front clutch)
1978	5989527	258	A-727	Short extension
1979	5902848	258	A-727	Short extension (non-lockup)
			International Harvester Trucks	
1974	451874-C91	258	A-727	4 × 4
	451875-C91	258	A-727	Short extension
	451876-C91	304/345	A-727	4 × 4
	451877-C91	304/345	A-727	Short extension
	444495-C91	304/345	A-727	Long extension, heavy duty extension
	451878-C91	392	A-727	Short extension
	451879-C91	392	A-727	Medium extension
	444496-C91	392	A-727	Long extension, heavy duty
	451880-C91	400	A-727	Short extension
	444498-C91	400	A-727	Long extension, heavy duty
1975	451876	304/345		4-wheel drive
	451877	304/345		Short extension
	444495	304/345		Long extension, heavy duty
	451878	392		Short extension
	451879	392		Medium extension
	444496	392		Long extension, heavy duty
1976-77	472293	304/345	A-727	4-wheel drive
1978	492448-C91	304/345	A-727	Scout
	492449-C91	Diesel	A-727	Scout
1979	492448-C91	304/345	A-727	Scout (non-lockup)

FORD MOTOR COMPANY
AUTOMATIC TRANSMISSION LISTING

Year	Vehicle Model	Engine CC/CID	Transmission Code
		Ford C-3	
1974	Mustang II	2300/140	74DT-BKB
	Pinto	2300/140	74DT-BNA
	Bobcat	2300/140	74DT-BNA
1975	Mustang II	2300/140	75DT-AA, -AD, -AE
	Pinto	2300/140	75DT-BB, -BC, -BD, -BE
		2800/171	75DT-EA, -EB, -FA
	Bobcat	2300/140	75DT-BB, -BC, -BD, -BE
		2800/171	75DT-FA, -EA, -EB
	Capri	2300/140	75DT-CA
		2800/171	75DT-DA
1976	Mustang II	2300/140	76DT-CA, -CB
	Pinto	2300/140	76DT-AA, -AB
		2800/171	76DT-EA, -EB, -FA, -FB
	Bobcat	2300/140	76DT-AA, -AB
		2800/171	76DT-EA, -FA, -EB, -FB
	Capri	2300/140	76DT-HA, -ACA
		2800/171	76DT-JA, -ADA
1977	Mustang II	2300/140	77DT-BA, -CA, -DA, -EA
		2800/171	77DT-FA, -GA
	Pinto	2300/140	77DT-BA, -CA
		2800/171	77DT-FA, -GA
	Bobcat	2300/140	77DT-BA, -CA
		2800/171	77DT-FA, -GA
	Capri II	2300/140	77DT-AFA
		2800/171	77DT-AGA
1978	Mustang II	2300/140	78DT-ACA, -ADA
		2800/171	78DT-AGA
	Pinto	2300/140	78DT-ACA, -ADA
		2800/171	78DT-AGA
	Fairmont	2300/140	78DT-SA
		3300/200	78DT-BAA, -BCA, -BJA
	Bobcat	2300/140	78DT-ACA, -ADA
		2800/171	78DT-AGA
	Capri II	2300/140	—
		2800/171	—
	Zephyr	2300/140	78DT-SA
		3300/200	78DT-BAA, -BCA, -BJA
1979	Mustang	2300/140	79DT-EA, -GA
		2800/171	79DT-HA
	Pinto	2300/140	79DT-AA
		2800/171	79DT-DA
	Fairmont	2300/140	79DT-KA, -MA
		3300/200	79DT-NA
	Bobcat	2300/140	79DT-AA
		2800/140	79DT-DA

FORD MOTOR COMPANY
AUTOMATIC TRANSMISSION LISTING

Year	Vehicle Model	Engine CC/CID	Transmission Code
		Ford C-3	
1979	Capri	2300/140	79DT-EA, -GA
		2800/171	79DT-HA
	Zephyr	2300/140	79DT-KA, -MA
		3300/200	79DT-NA
1980	Mustang	2300/140	80DT-CA, -CDA
		3300/200	80DT-EA
	Pinto	2300/140	80DT-AA
	Fairmont	2300/140	80DT-CA, -HA, -CDA
		3300/200	80DT-EA, -LA
	Bobcat	2300/140	80DT-AA
	Capri	2300/140	80DT-CA, -CDA
		3300/200	80DT-EA
	Zephyr	2300/140	80DT-CA, -HA, -CDA
		3300/200	80DT-EA, -LA
		Ford C-4	
1974	Ford	351	PEF-A5
	Torino	302	PEE-V7, -M7, -M8
		351	PEF-D6, -E6, -E7
	Mustang	140	PEJ-H
		171	PEJ-E, -F
	Pinto	122	PEJ-B6
		140	PEJ-G, -G1, -C, -C1
	Maverick	200	PEB-D7, -H6
		250	PEE-AK5, -AM5
		302	PEE-AH5, -BH, -AL5, -BJ
	Bronco	302	PEA-AL1, -BG
	Montego	302	PEE-V7
		351	PEF-D6
	Cougar	351	PEF-D6, -E6, -E7
	Capri	122	PED-E1, -E2, -BB, -BB1
		171	PED-L1, -Y1, -BC, -BC1
	Bobcat	122	PEJ-B6
		140	PEJ-C, -G
	Comet	200	PEB-D7, -H6
		250	PEE-AK5, -AM5
		302	PEE-AH5, -AL5, -BH, -BJ
	Econoline	240	PEA-AG3, -AH3, -R7, -S7
		300	PEA-AR3, -AT3
		302	PEA-AJ4, -AV2, -AK4, -BF, -AW2, -AY2, -T8. -U8, -BL, -AZ2
	F Series Trucks	240	PEA-AN1, -J8
		300	PEA-BB2
		302	PEA-BH, -F9, -BJ, -K9
1975	Torino	351W	PEF-D7, -T, -F8, -U
		351M	PEF-S, -U, -W, -AA, -AB

FORD MOTOR COMPANY
AUTOMATIC TRANSMISSION LISTING

Year	Vehicle Model	Engine CC/CID	Transmission Code
		Ford C-4	
1975	Mustang	171	PEJ-J1, -E1
		302	PEE-CC, -BY
	Granada	250	PEE-BT, -AM6, -BU, -CH, -CK
		302	PEE-CA, -CA1, -BV, -AL6, -AL7, -CB, -BJ1, -BW, -CG
		351	PEF-Y, -Q, -Z, -R
	Pinto	140	PEJ-G2
		171	PEJ-J1, -K, -L
	Maverick	200	PEB-D8, -D9, -H7, -H8
		250	PEE-AK6, -AM6, -CH, -CJ
		302	PEE-AH6, -AH7, -BH1, -AL6, -AL7, -DJ1
	Bronco	302	PEA-AL2, -BG1
	Montego	351W	PEF-D7, -T
		351M	PEF-AA, -S
	Cougar	351W	PEF-D7, -E8, -U
		351M	PEF-S, -AA, -AB
	Monarch	250	PEE-BT, -CK, -AM6, -BU, -CH
		302	PEE-CA, -CA1, -BV, -AL6, -AL7, -BJ
		351	PEF-Y, -Q, -Z, -R
	Bobcat	140	PEJ-G2
		171	PEJ-J1
	Comet	200	PEB-D8, -D9, -H7, -H8
		250	PEE-AK6, -CJ, -AM6, -CH
		302	PEE-AH6, -AH7, -BH1, -AL6, -AL7, -BJ1
	Econoline	300	PEA-AH4, -BU
	F Series Trucks	300	PEA-AH2, -J9, -BW, -BA3, -BB3
		302	PEA-F10, -K10
1976	Torino	351W	PEF-D8, -E9
		351M	PEF-S, -U, -W, -AA1, -AB1, -AG, -AH
	Mustang	171	PEJ-M1, -N
		302	PEE-CC
	Granada	250	PEE-CH1, -CK1, -DB, -DD
		302	PEE-BV1, -CA2, -CB, -CE, -CA3, -CE1, -DE, -DF
		351W	PEF-AE, -AF
	Pinto	171	PEJ-M1, -N
	Maverick	200	PEB-N, -P
		250	PEE-CH1, -CJ1, -DB, -DC
		302	PEE-AH7, -AL7, -CS, -CT
	Bronco	302	PEA-AL2
	Montego	351M	PEF-S, -AA1, -AG
		351W	PEF-D8
	Cougar	351M	PEF-S, -V, -AA1, -AB1, -AG
		351W	PEF-D8, -E9
	Monarch	250	PEE-CH1, -CK1, -DB, -DD
		302	PEE-BV1, -CB, -CA2, -CE, -CA3, -CE1, -DE, -DF
		351W	PEF-AE, -AF

FORD MOTOR COMPANY
AUTOMATIC TRANSMISSION LISTING

Year	Vehicle Model	Engine CC/CID	Transmission Code
		Ford C-4	
1976	Bobcat	140	PEJ-G3
		171	PEJ-N, -M1
	Comet	200	PEB-N, -P
		250	PEE-CH1, -CJ, -DB, -DC
		302	PEE-AH7, -AL7, -CS, -CT
	Econoline	300	PEA-AH4, -CB, -BU, -BP, -CA
	F Series Trucks	300	PEA-BB3, -BB4, -AH2, -J9, -BW, -CC, -BY, -BZ, -BA3
		302	PEA-F10, -K10
1977	Ford	302	PEA-CH
		351W	PEF-AP, -AP1
	LTD II	302	PEE-M9, -V8
		351M	PEF-AA2, -AA3, -AB2, -AB3
		351W	PEF-K, -K1
	Mustang	140	PEJ-H
		171	PEJ-M2
		302	PEE-BY, -CC1
	Granada	250	PEE-BU, -CK2, -CD1, -CP1
		302	PEE-CE2, -CB2, -DN, -CW
		351W	PEF-AE1, -AE2, -AF1, -AF2
	Pinto	140	PEJ-P
		171	PEJ-L, -M2
	Maverick	200	PEB-N1
		250	PEE-CJ2
		302	PEE-CS1, -CZ
	Thunderbird	302	PEE-M9, -V8
		351M	PEF-AA2, -AB2
		351W	PEF-K, -K1
	Bronco	302	PEA-AL4, -BG
	Cougar	302	PEE-V8
		351M	PEF-AA2, -AA3, -AB, -AB2, -AB3
		351W	PEF-K, -K1
	Monarch	250	PEE-BU, -CDI, -CK2, -CPI
		302	PEE-CB2, -CE2, -CW, -DN
		351W	PEF-AE1, -AE2, -AF1, -AF2
	Bobcat	140	PEJ-P
		171	PEJ-L, -M2
	Comet	200	PEB-N
		250	PEE-CJ2
		302	PEE-CS1, -CZ
	Versailles	302	PEE-DH, -DH1, -DJ, -DJ1, -DV, -DW
		351W	PEE-AM, -AN
	Econoline	300	PEA-BPI, -BUI, -CG
	F Series Trucks	300	PEA-K12, -BA4, -CE, -CF
		302	PEΛ-CE, -CF, -CJ, -CK
1978	Ford	302	PEA-CH2, -CH3, -CL, -CL1
		351W	PEF-AP3, -AP4

FORD MOTOR COMPANY
AUTOMATIC TRANSMISSION LISTING

Year	Vehicle Model	Engine CID	Transmission Code
		Ford C-4	
1978	LTD II	302	PEE-M12, -M13, -M11, -V11, -V12, -V10, -DY
		351W	PEF-K3, -K4
		351M	PEF-AA5, -AA6, -AB5, -AB6
	Mustang	140	PEJ-P2, -P3
		171	PEJ-M4, -M5
		302	PEE-BY2, -BY3, -CC3, -CC4
	Granada	250	PEE-CD3, -CD4, -CP3, -CP4
		302	PEE-CA6, -CA7, -CE4, -CE5, -CW2, -CW3
	Pinto	140	PEJ-P2, -P3
		171	PEJ-M4, -M5
	Fairmont	200	PEB-N3, -N4, -P1, -P2, -R, -R1, -S, -S1, -ED, -ED1, -EDI
		302	PEE-CS3, -CS4, -CS5, -CT1, -CT2, -DA, -DA1
	Thunderbird	302	PEE-M12, -M13, -M11, -V11, -V12, -V10
		351W	PEF-K3, -K4
		351M	PEF-AA5, -AA6, -AB5, -AB6
	Cougar	302	PEE-M12, -M13, -M11, -V11, -V12, -V10, -DY, -DY1
		351W	PEF-K3, -K4
		351M	PEF-AA5, -AB5, -AB6
	Monarch	250	PEE-CD3, -CD4, -CP3, -CP4
		302	PEE-CA6, -CA7, -CE4, -CE5, -CW2, -CW3
	Bobcat	140	PEJ-P2, -P3
		171	PEJ-M4, -M5
	Zephyr	200	PEB-N3, -N4, -P1, -P2, -R, -R1, -S, -S1
		302	PEE-CS3, -CS4, -CS5, -CT1, -CT2, -CT3, -DA, -DA1, -ED, -ED1
	Versailles	302	PEE-DH3, -DH4, -DJ3, -DJ4, -EF, -EF1, -EG, -EG1, -EH, -EJ
	F Series Trucks	300	PEA-CE3, -CE4, -CM, -CM1
		302	PEA-CE3, -CE4, -CM, -CM1
1979	Ford	302	PEE-DZ1, -DZ, -EA, -EA1, -EM1, -EM, -FB, -FB1, -FC, -FE
		351	PEF-AT, -AT1, -AU, -AU1, -AZ, -AZ1
	LTD II	302	PEE-M14, -V13
		351M	PEF-AA7, -AB7
	Mustang	171	PEJ-S, -SI, -W, -WI
		302	PEE-BY4
	Granada	250	PEE-CD5, -CP5
		302	PEE-CA8, -CE6, -CW4
	Pinto	140	PEJ-Y
		171	PEJ-M6, -V
	Fairmont	200	PEB-N5, -P3, -R2, -S2, -T
		302	PEE-CS6, -CT4, -DA2, -ED2, -ES, -ET, -EU, -EV, -FD
	Thunderbird	302	PEE-M14, -V13
		351M	PEF-AA7, -AB7
	Mercury	302	PEE-DZ, -DZ1, -EA, -EA1, -EM, -EM1, -FB, -FB1, -FC, -FE

FORD MOTOR COMPANY
AUTOMATIC TRANSMISSION LISTING

Year	Vehicle Model	Engine CID	Transmission Code
		Ford C-4	
1979	Mercury	351	PEF-AT, -AT1, -AU, -AU1, -AZ, -AZ1
	Cougar	302	PEE-M14, -V13
		351M	PEF-AA7, -AB7
	Monarch	250	PEE-CD5, -CP5
		302	PEE-CA8, -CE6, -CW4
	Capri	171	PEJ-S, -S1, -W, -W1
		302	PEE-BY4
	Bobcat	140	PEJ-Y
		171	PEJ-M6, -V
	Zephyr	200	PEB-N5, -P3, -R2, -S2, -T
		302	PEE-CS6, -CT4, -DA2, -ED2, -ES, -ET, -EU, -EV, -FD
	Versailles	302	PEE-DH5, -DJ5, -EW, -EW1, -EY, -EY1, -FA
	F Series Trucks	302	PEA-CE5
1980	Ford	302	PEE-DZ3, -EA3, -EM3, -FC1, -FE1
	Mustang	200	PEB-P4
		140	PEJ-AC
		225	PEM-B, -E
	Granada	302	PEE-CW5, -FP, -FR
		250	PEL-A, -B
		255	PEM-J, -K
	Pinto	140	PEJ-Z
	Fairmont	200	PEB-N6, -P4, -S3, -T1, -U
		140	PEJ-AC, -AD
		255	PEM-B, -C, -D, -E, -G, -H
	Thunderbird	302	PEE-FL, -FN
		255	PEM-D, -L
	Mercury	302	PEE-DZ3, -EA3, -EM3, -FC1, -FE1
	Cougar	302	PEE-FL, -FN
		255	PEM-D, -L
	Monarch	302	PEE-CW5, -FP, -FR
		250	PEL-A, -B
		255	PEM-J, -K
	Capri	200	PEM-P4
		140	PEJ-AC
		255	PEM-B, -E
	Bobcat	140	PEJ-Z
	Zephyr	200	PEB-N6, -P4, -S3, -T1, -U
		140	PEJ-AC, -AD
		255	PEM-B, -C, -D, -E, -G, -H
	Versailles	302	PEE-EY2, -FV
		Ford C-6	
1974	Ford	400	PJA-C5
		460	PJC-H6, -PJD-G3

FORD MOTOR COMPANY
AUTOMATIC TRANSMISSION LISTING

Year	Vehicle Model	Engine CID	Transmission Code
			Ford C-6
1974	Torino	351	PGA-AW2, -AY2, -AY3, -AF3, -AG4
		351W	PGA-AZ1
		400	PJA-G3, -H3, -H4
		460	PJC-J2, -PJD-R, -S, -S1
	Thunderbird	460	PJD-E4
	Mercury	400	PJA-C5
		460	PJC-H6, -PJD-G3
	Montego	351	PGA-AW2, -AF3
		400	PJA-G3
		460	PJC-J2, -PJD-R
	Lincoln	460	PJD-F4
	Cougar	351	PGA-AW2, -AF3, -AY2, -AY3, -AG3, -AG4
		400	PJA-G3, -H3, -H4
		460	PJD-R, -S, -S1
	Mark IV	460	PJD-E4
	F Series Trucks	360	PGB-BH, -AC9, -BJ, -AW1, -BF, -AY1,-BB1,-BC1
		390	PGB-BF, -BK, -BB1, -BM, -BC1, -BN
		460	PJD-K1, -L1, -J1
1975	Ford	400	PJA-C6, -C7, -P1
		460	PJA-P1, -PJC-H7, -H8, -H9, -K1, -K2, -PJD-G4, -G5, -P, -P1, -P2
	Torino	351M	PJA-V, -V1, -W, -W1
		400	PJA-G4, -G5, -R1, -H5, -H6, -S1
		460	PJC-J3, -L, -PJD-R1, -R2, -T1, -T2, -S2, -S3, -V1, -V2
	Thunderbird	460	PJD-E5, -E6, -M1, -M2
	Mercury	400	PJA-C6, -C7, -P1
		460	PJC-H7, -H8, -H9, -K, -K1, -K2, -PJD-G4, -G5, -P, -P1, -P2
	Montego	351M	PJA-U, -V1
		400	PJA-G4, -G5, -R1
		460	PJC-J3, -L, -PJD-R1, -R2, -T1, -T2
	Lincoln	460	PJD-F5, -F6, -N, -N1, -N2
	Cougar	351M	PJA-V, -V1, -W, -W1
		400	PJA-G4, -G5, -R1, -H5, -H6, -S1
		460	PJD-RL, -R2, -T1, -T2, -S2, -S3, -V1, -V2
	Mark IV	460	PJD-E5, -E6, -M, -M1, -M2
	Econoline	300	PGD-A, -A1, -C, -C1
		351	PGD-G, -H, -T, -V, -N, -R
		460	PJD-W, -Z
	F Series Trucks	360	PGB-AY3, -BB2, -BB3, -BC2, -BC3, -BC4
		390	PGB-CJ, -BM1, -BM2, -BN1, -BN2, -BN3
		460	PJD-L3, -L4, -L5, -L6, -J3, -J4, -J5, -J6
1976	Ford	400	PJA-C7, -C8, -C9, -C10, -C11, -P, -P1, -P2, -P3, -P4
		460	PJC-H9, -H10, -H11, -H12, PJD-G5, -G6, -G7, -G8, -AH, -AH1

FORD MOTOR COMPANY
AUTOMATIC TRANSMISSION LISTING

Year	Vehicle Model	Engine CID	Transmission Code
			Ford C-6
1976	Torino	351M	PJA-V3, -V4, -V5, -W3, -W4, -W5
		400	PJA-G5, -G6, -G7, -G8, -G9, G10, -H6, -H7, -H8, -H9, -H10, -H11, -R, -R1, -R2, -R3, -S, -S1, -S2, -S3
		460	PJD-R2, -R3, -R4, -R5, -R6, -S5, -S6, -AK1, -AJ1, -S3, -S4, -AJ, -AK, PJC-J5, -J3, -J4
	Thunderbird	460	PJD-E6, -E7, -AF, -AF1, -E8, -E9, -E10
	Mercury	400	PJA-C7, -C8, -C9, -C10, -C11, -C12, -P, -P1, -P2, -P3, -P4
		460	PJC-H9, -H10, -H11, -H12, PJD-G5, -G6, -G7, -G8, -AH1, -AH
	Montego	351M	PJA-V5
		400	PJA-G5, -G6, -G7, -R, -R1, -G8, -G9, -G10, -R2, -R3
		460	PJD-R2, -R3, -AJ, -R4, -R5, -R6, -AJ1, PJC-J3, -J4
	Lincoln	460	PJD-F6, -F7, -AG, -F8, -F9, -F10, -AG1
	Cougar	351M	PJA-V3, -V4, -W3, -W4, -V5, -W5
		400	PJA-G5, -G6, -G7, -R, -R1, -S, -S1, -G8, -G9, -G10, -H9, -H10, -H11, -R2, -R3, -S2, -S3
		460	PJD-R2, -R3, -S3, -S4, -AJ, -AK, -R4, -R5, -R6, -S5, -S6, -AJ1, -AK1
	Mark IV	460	PJD-E8, -E9, -E10, -AF1, -E6, -E7, -AF
	Econoline	300	PGD-A1, -A2, -C1, -C2, -A3, -A4, -C3, -C4, -C5, -Z1, -Z2
		351	PGD-G, -G1, -H, -H1, -Y, -Y1, -V1, -V2, -N1, -N2, -R1, -R2, -G2, -G3, -H2, -N3, -N4, -N5, -R3, -R4, -R5, -V3, -V4, -V5, -Y2, -Y3
		460	PJD-W, -W1, -Z, -Z1
	F Series Trucks	360	PGB-AC12, -AC13, -AC14, -CH1, -CH2, -BJ2, -BJ3, -BJ4, -CL, -CL1, -CL2, -CH3, -CH4, -CH5, -AY3, -AY4, -CK, -CK1, -BB3, -BB4, -BC4, -BC5, -AC15, -AC16, -AC17, -BJ5, -BJ6, -CH6, -CH7, -CH8, -CL3, -CL4, -CL5, -CK2, -CK3, -CK4, -BC6, -BC7, -BC8, -AY5, -AY6, -AY7, -BB5, -BB6, -BB7,
		390	PGB-BK3, -CJ2, -CJ3, -CJ4, -BM2, -BM3, -BN3, -BN4, -CJ5, -CJ6, CJ7, -BN5, -BN6, -BN7, -BM5
		460	PJD-L5, -L6, -J5, -J6, -J7, -J8, -L7, -L8, -L9
1977	Ford	351M	PJA-BU, -BU1
		400	PJA-BV, -BV1, -C13, -C14, -C15
		460	PJD-G10, -G11, -G12, -AH2, -AH3, -AP, -AP1, -AR, -AR1, PJC-H13, -H14
	LTD II	351M	PJA-AU, -AU1
		400	PJA-G11, -G12, -G13, -R4, -R5, -R6, -S4, S5, -S6, -H12, -H13, -H14, -BJ, -BJ1
	Thunderbird	400	PJA-R4, -R5, -R6, -S4, -S5, -S6, -H12, -H13, -H14, -BJ1, -BJ

FORD MOTOR COMPANY
AUTOMATIC TRANSMISSION LISTING

Year	Vehicle Model	Engine CID	Transmission Code
		Ford C-6	
1977	Lincoln	400	PJA-BL, -BL1, -BL2, -CA, -CA1
		460	PJD-F12, -F13, -F14, -AG2, -AG3, -AN, -AN1
	Mercury	351M	PJA-BU, -BU1
		400	PJA-C13, -C14, -C15, -BH, -BH1, -BV, -BV1
		460	PJC-H13, -H14, -M, -M1, PJD-G10, -G11, -G12, -AH2, -AH3, -AP, -AP1, -AR, -AR1
	Cougar	351M	PJA-AT, -AU, -AU1
		400	PJA-G11, -G12, -G13, -H12, -H13, -H14, -R4 -R5, -R6, -S4, -S5, -S6, -BJ, -BJ1
	Mark V	400	PJA-BK, -BK1, -BK2, -BW, -BW1
		460	PJD-E12, -E13, -E14, -AF2, -AF3, -AM, -AM1
	Econoline	300	PGD-C6, -C7, -Z3, -Z4, -AL
		351	PGD-N6, -N7, -N8, -R6, -R7, -R8, -R9, -Y4, -Y5, -V6, -V7, -V8, -V9, -AA, -AA1, -AC1, -AC2
		460	PJD-W5, -W6, -Z5, -Z6
	F Series Trucks	351	PJA-Z, -Z1, -Z2, -AB, -AB1, -AE, -AE1, -AF, -AF1, -AJ, -AJ1, -AK, -AN, -AN1, -BS, -BS1, -BY
		400	PJA-AC, -AC1, -AC2, -AG, -AG1, -AG2, -AH, -AH1, -AK, -AK1, -AM, -AP, -AP1, -BT, -BT1, -BZ
		460	PJD-J9, -J10, -L10, -L11, -AL, -AL1
1978	Ford	351M	PJA-BU3, -BU4
		400	PJA-C17, -C18, -BH3, -BH4, -PJC-H16, -H17, -PJD-AP3, -AP4
	LTD II	351	PJA-BJ3, -BJ4
		400	PJA-BF3, -BF4, -G15, -G16, -H16, -H17
	Thunderbird	351	PJA-BF3, -BF4
		400	PJA-BF3, -BF4, -G15, -G16, -H17, -H16
	Bronco	351	PJA-AJ3, -AJ4, -BY2, -BY3, -CD, -CD1, -CE, -CE1
		400	PJA-AL3, -BZ2, -BZ3
	Mercury	351M	PJA-BU3, -BU4
		400	PJA-C17, -C18, -BH3, -BH4
		460	PJC-M3, -M4, PJD-AP3, -AP4
	Lincoln	400	PJA-CA3, -CA4, -CC, -CC1, -CZ
		460	PJD-AN3, -AN4
	Cougar	351	PJA-BJ3, -BJ4
		400	PJA-BJ3, -BJ4, -G15, -G16, -H17, -H16
	Mark V	400	PJA-BW3, -BW4, -LY, -CB, -CB1, PJD-AM3, -AM4
	F Series Trucks	300	PGD-AN, -AN1, -AN2, -AP, -AP1, -AP2, -AR, -AR1, -AR2, -AU, -AU1, -AU2, -AV, -AV1, -AV2, -AW, -AW1, -AW2, -BC, -BC1, -BC2, -BD, -BD1, -BD2
		302	PGD-AS, -AS1, -AS2, -AT, -AT1, -AT2, -AY, -AY1, -AY2, -AZ, -AZ1, -AZ2, -BE, -BE1, -BE2, -BF, -BF1, -BF2
		351	PJA-CE, -CE1, -CD, -CD1, -AB3, -AB4, -AE3, -AE4, -AF3, -AF4, -BS3, -BS4 -Z3, -Z4, -CH, -CH1, -CJ, -CJ1, -CN, -CN1, -CS, -CS1, -CT, -CT1, -AJ3, -AJ4, -BY2, -BY3, -BY4

FORD MOTOR COMPANY
AUTOMATIC TRANSMISSION LISTING

Year	Vehicle Model	Engine CID	Transmission Code
		Ford C-6	
1978	F Series Trucks	400	PJA-AG4, -AG5, -AH3, -AH4, -BT3, -BT4, -AL3, -AL4, -BZ2, -BZ3
		460	PJD-AL3, -AL4, -J12, -J13, -L14, -L13
	Econoline	300	PGD-A8, -A9, -A10, -C9, -C10, -C11, -Z6, -Z7, -Z8, -AD, -AD1, -AD2, -AE, -AE1, -AE2, -AH, -AH1, -AH2, -AL2, -AL3, -AL4, -AK, -AK1, -AK2, -AM, -AM1, -AM2
		351	PGD-AA3, -AA4, -AA5, -AC4, -AC5, -AC6, -N10, -N11, -N12, -R11, -R12, -R13, -Y7, -Y8, -Y9, -AF, -AF1, -AF2, -AF3, -AG, -AG1, -AG2, -AG3, -CB, -CB1, -CC, -CC1
		460	PJD-W8, -W9, -Z8, -Z9
1979	Ford	351	PGD-BH, -BZ, -CF, -CH, -CR, -CT, -CU
	LTD II	351M	PJA-DK
	Thunderbird	351M	PJA-DK
	Bronco	351	PJA-AJ6, -AJ7
		400	PJA-AL7, -BZ7, -DH, -DJ
	Mercury	351	PGD-BH, -BZ, -CF, -CH, -CR, -CT, -CU
	Lincoln	400	PJA-CA6, -DB, -DD
	Cougar	351M	PJA-DK
	Mark V	400	PJA-BW6, -DA, -DC
	Econoline	300	PGD-A11, -A12, -C12, -C13, -Z9, -Z10, -AL5, -AL6, -AC8, -AC9, -N14, -N15, -R16, -R15, -Y11, -Y12
		302	PGD-AD4, -AD5, -AE4, -AE5, -BN, -BN1, -BS, -BS1
		351	PGD-AC8, -AC9, -N14, -N15, -R15, -R16, -Y11, -Y12, -CL, -CL1, -CM, -CM1, -CN, -CN1, -CP, -CP1
		460	PJD-W11, -W12, -Z11, -Z12
	F Series Trucks	300	PGD-AW4, -AW5, -AW6, -BC4, -BC5, -BC6, -BD4, -BD5, -BD6,
		302	PEA-CE5, -PGD-BE4; -BE5, -BF4, -BF5, -JC, -JC1
		351	PJA-AE6, -AE7, -AJ6, -AJ7, -BY7, -BY8, -CJ3, -CJ4, -Z6, -Z7
		400	PJA-AH6, -AH7, -AG7, -AG8, -AL7, -AL8, -BZ7, -BZ8, -BT6, -BT7, -DE, -DE1, -DF, -DF1, -DG., -DG1, -DH, -DH1, -DJ, -DJ1
		460	PJD-J15, -J16, -L16, -L17
		Ford FMX	
1974	Ford	351	PHB-AA1, -L8
		400	PHB-AC1
	Torino	302	PHA-J5
		351	PHB-Z5
	Mercury	351	PHB-AA1
		400	PHB-AC1

FORD MOTOR COMPANY
AUTOMATIC TRANSMISSION LISTING

Year	Vehicle Model	Engine CID	Transmission Code
		Ford FMX	
1974	Montego	302	HHA-J3
		351	PHB-Z5
	Cougar	351	PHB-Z5
	F Series Trucks	300	PHC-E, -D, -G, -F
		302	PHC-B2, -B1
1975	Ford	351M	PHB-AN, -AH, -AR
		400	PHB-AC2, -AC3, -AK
	Torino	351W	PHB-Z6
		351M	PHB-AP, -AJ, -AS
		400	PHB-AL, -AL1, -AM
	Mercury	351M	PHB-AN, -AR
		400	PHB-AC3, -AC2, -AK
	Montego	351W	PHB-Z, -Z6
		351M	PHB-AP, -AJ, -AS
		400	PHB-AL, -AL1, -AM
	Cougar	351W	PHB-Z
		351M	PHB-AP, -AJ, -AS
		400	PHB-AL, -AM
	F Series Trucks	300	PHC-B2
		302	PHC-D1
1976	Ford	351M	PHB-AR, -AR1
		400	PHB-AC3
	Torino	351M	PHB-AP1, -AT
		351W	PHB-Z7
		400	PHB-AL1
	Mercury	351M	PHB-AR, -AR1
		400	PHB-AC3
	Montego	351W	PHB-Z7
		351M	PHB-AT, -AP1
		400	PHB-AL1
	Cougar	351M	PHB-AP1, -AT
	F Series Trucks	300	PHC-B2
		302	PHC-D1
1977	Ford	351M	PHB-AW, -AW1
		400	PHB-AW, -AW1, -AC4
	LTD II	351M	PHB-AT1, -AT2, -AY, -AL2, -AL3
		400	PHB-AL2, -AL3
	Thunderbird	351W	PHB-Z8
		351M	PHB-AT1, -AT2, -AY, -AL2, -AL3
		400	PHB-AL2, -AL3
	Mercury	351M	PHB-AW, -AW1
		400	PHB-AW, -AW1, -AC4
	Cougar	351W	PHB-Z8
		351M	PHB-AT1, -AY, -AL2, -AL3
		400	PHB-AL2, -AL3
	F Series Trucks	300	PHC-B3

FORD MOTOR COMPANY
AUTOMATIC TRANSMISSION LISTING

Year	Vehicle Model	Engine CID	Transmission Code
		Ford FMX	
1978	Ford	351M	PHB-BB
		400	PHB-BC, -BC1
	LTD II	302	PHB-BG
		351W	PHB-Z9
		351M	PHB-AT2
		400	PHB-BD, -BE
	Thunderbird	302	PHB-VG
		351W	PHB-Z9
		351M	PHB-AT2
		400	PHB-BD, -BE
	Mercury	351M	PHB-BB
		400	PHB-BC, -BC2
	Cougar	302	PHB-BG
		351W	PHB-Z9
		351M	PHB-AT2
		400	PHB-BD, -BE
1979	Ford	302	PHB-BH
	LTD II	302	PHB-BG
		351M	PHB-AT2, -AT3
		351W	PHB-Z9
	Thunderbird	302	PHB-BG
		351W	PHB-Z9
		351M	PHB-AT3
	Mercury	302	PHB-BH
	Cougar	302	PHB-BG
		351M	PHB-AT2, -AT3
		351W	PHB-Z9
1980	Ford	302	PHB-BH2, -BH3
		351W	PHB-BK, -BK1, -BP, -BP1, -BT, -BT1, -BU, -BU1
	Mercury	302	PHB-BH2, -BH3
		351W	PHB-BK, -BK1, -BP, -BP1, -BT, -BT1, -BU, -BU1
		Ford 4 Speed Overdrive	
1980	Mercury	302	PKA-W
		351W	PKA-C, -R, -T, -Z
	Mark VI	302	PKA-M
		351W	PKA-D, -U
	Cougar, Thunderbird	302	PKA-Y
		Ford Jatco	
1977	Granada, Monarch	6-250	PLA-A
1978	Granada, Monarch	6-250	PLA-A1
1979	Granada, Monarch	6-250	PLA-A2
1980	Granada, Monarch	6-250	PLA-A2

GENERAL MOTORS
AUTOMATIC TRANSMISSION LISTING

Year	Vehicle Model	Engine	Transmission Code
	Turbo Hydra-Matic 425		
1974–78	Eldorado, exc. E.F.I.	—	AJ
1975–78	Eldorado, w/E.F.I.	—	AK
1974–78	Toronado	—	OJ
1974–77	Toronado	—	OM
	Turbo Hydra-Matic 325		
1979	Riviera	—	AJ
		V6-231	BJ
		—	OH, OK
		V8-350	OJ
		—	OL, OM
1980	Riviera	V6-231	BJ
		V8-350	OJ
1979	Eldorado	—	AJ, BJ
		—	OJ, OK
	Eldorado, Seville	—	AJ
	Toronado	—	AJ, OG, OH
		V8-350 diesel	OJ
		V8-350 diesel	OK
		—	OL, OM
1980	Toronado	V8-307	OH
		V8-350 diesel	OJ
	Turbo Hydra-Matic 125 Transaxle		
1980	Citation	—	CU, PZ
	Omega	—	CU, PZ
	Phoenix	—	CU, PZ
	Skylark	—	CU, PZ
	Turbo Hydra-Matic 180		
1977–78	Chevette	4-98	ND
1979	Chevette	4-98	NS
	Turbo Hydra-Matic 200		
1976	**BUICK**		
	Skyhawk	V6-231	BH
	Skylark	V6-231	BZ
		V6-260	OZ
1977	Skyhawk, exc. Calif.	V6-231	BH
	Skylark, Calif.	V6-231	BL
	Skylark, exc. Calif.	V6-231	BZ
	LeSabre	V8-350	OS
		V8-301	PZ
1978	Century S.W., LeSabre	V6-231	5BZ
	LeSabre	V8-301	5PZ
	Century, exc. S.W.	V8-305	5CO
	LeSabre	V8-350	5BA, 5OS
1979	Century	V8-231	6BZ
		V8-301 2 661 carb.	6PG
		V8-301 4 661 carb.	6PH
		V8-305	6CR

GENERAL MOTORS
AUTOMATIC TRANSMISSION LISTING

Year	Vehicle Model	Engine	Transmission Code
		Turbo Hydra-Matic 200	
1980	Century, Regal	V8-301	PW
	LeSabre	V8-301	PW
	Century, Regal	V8-305	CC
	LeSabre, eng. Code X	V8-350	BA
	LeSabre, Electra	V8-350 diesel	OT
1978–79	**CADILLAC**		
	Seville	V8-350 diesel	AS, AX
	Fleetwood, DeVille	V8-350	AS
1976	**CHEVROLET**		
	Nova	V8-305	CE
	Chevette, exc. Calif.	4-97	CN
	Nova	6-250	CQ
	Chevette, Calif.	4-98	CU
	Chevette	4-85	CU
1977	Nova	V8-305	CE
	Monza	V8-305	CD, CK
	Chevette, wo/AC	4-98	CN
	Chevrolet, full size	V8-305	CO, CR
	Chevette	4-85	CU
	Chevrolet, full size	V8-350	CY
	Chevette, w/AC	4-98	CZ
1978	Monza	4-151	PA
	El Camino, Malibu, Monte Carlo	V6-200	CS, CW
		V6-231	BZ
		V8-305	CO
	Chevrolet, full size	V8-305	CO
1979	Chevette	4-98	CN
	Monza	4-51	PA, PC
	El Camino, Malibu, Monte Carlo	V6-200	CS
		V6-231	BZ
		V8-267	CA
		V8-305	CR
	Chevrolet, full size	V8-305	CU
		V8-350	CY
1980	Chevette	L4-98	CN
	Monza	L4-151	PA, PC
	Impala, Caprice	V6-229	CK
		V8-305	CC
1976	**OLDSMOBILE**		
	Omega	6-250	CQ
	Omega	V8-260	CZ
1977	Starfire, exc. Calif.	V6-231	BH
	Omega, Calif.	V6-231	BL
	Omega, exc. Calif.	V6-231	BZ
	Starfire	V8-305	CD
	Full size	V8-350	OS
	Omega	V8-260	OZ

GENERAL MOTORS
AUTOMATIC TRANSMISSION LISTING

Year	Vehicle Model	Engine	Transmission Code
	Turbo Hydra-Matic 200		
1978	Starfire	4-151	PY
	Cutlass	V6-231	BZ
	Cutlass, exc. high alt.	V8-260	OW
	Cutlass, high alt.	V8-260	OR
	Cutlass	V8-305 2 bbl carb.	CO
		V8-305 4 bbl carb.	CR
	88 & 98, exc. S.W. & high alt.	V8-350 diesel	OT
	88 & 98, high alt. exc. S.W.	V8-850 diesel	OX
1979	Starfire	4-151	PB, PY
	Cutlass	V6-231	BZ
	Cutlass, exc. high alt.	V8-260	OW
	Cutlass, high alt.	V8-260	OR
	Cutlass, exc. high alt.	V8-260 diesel	OZ
	Cutlass	V8-305	CR
	88 & 98, exc. S.W. & high alt.	V8-350 diesel	OT
	88 & 98, high alt. exc. S.W.	V8-350	OX
1980	Starfire	L4-151	PY, PB
	Cutlass	V6-231	BZ
		V8-260	OZ
		V8-350 diesel	OT
	88 Custom Cruiser, eng. code N	V8-350	OT
	88 Custom Cruiser, eng. code R	V8-350	OS
1977	**PONTIAC**		
	Sunbird, exc. Calif.	V6-231	BH
	Ventura, Calif.	V6-231	BL
	Sunbird, Calif.	V6-231	BU
	Ventura, exc. Calif.	V6-231	BZ
	Sunbird	4-151	PY
	Ventura, exc. Calif.	4-151	PY
	Full size, exc. Calif.	V8-301	PZ
	Full size	V8-350	OS
1978	Phoenix, Sunbird	4-151	PY
	LeMans S.W., full size	V6-231	BZ
	Grand LeMans, Grand Prix	V8-301 2 bbl carb.	PG
	Grand AM & LeMans, exc. S.W.	V8-301 2 bbl carb.	PG
	Grand LeMans, Grand Prix	V8-301 4 bbl carb.	PH
	Grand AM & LeMans, exc. S.W.	V8-301 4 bbl carb.	PH
	Grand AM, Grand LeMans	V8-305 2 bbl carb.	CO
	Grand Prix & LeMans, exc. S.W.	V8-305 2 bbl carb.	CO
	Grand AM, Grand LeMans	V8-305 4 bbl carb.	CR
	Grand Prix & LeMans, exc. S.W.	V8-305 4 bbl carb.	CR
1979	Sunbird	4-151	6PB, 6PY
	LeMans Sta. Wag.	V6-231	6BZ
	Grand AM, Grand LeMans	V8-301 2 bbl carb.	6PH, 6PL
	Grand Prix & LeMans, exc. S.W.	V8-301 2 bbl carb.	6PH, 6PL
	Grand AM, Grand LeMans	V8-301 4 bbl carb.	6PG
	Grand Prix & LeMans, exc. S.W.	V8-301 4 bbl carb.	6PG

GENERAL MOTORS
AUTOMATIC TRANSMISSION LISTING

Year	Vehicle Model	Engine	Transmission Code
Turbo Hydra-Matic 200			
1979	Grand AM, Grand LeMans	V8-305	6CC, 6CR
	Grand Prix & LeMans, exc. S.W.	V8-305	6CC, 6CR
1980	Sunbird	L4-151	PB, PY
	Grand Prix, Grand AM, LeMans	V8-260	PG
	Catalina, Bonneville	V8-260	PG
	Firebird	V8-301	PD
Turbo Hydra-Matic 250, 350, 375			
1974	**BUICK**		
	Apollo	6-250	JE
	Century	V8-350	KA
	LeSabre	V8-350	KL
1975	Skyhawk	V6-231	KD, KC
	Century	V6-231	KB
	Apollo	6-250	JE
	Apollo, Skylark	V-260	LA
	Apollo, Skylark	V8-350	KA
	Century	V8-350	KE
	LeSabre	V8-350	KL
1976	Skyhawk	V6-231	KD
	Skylark	V6-231	KC
	Century	V6-231	KB
	LeSabre, Estate Wag.	V6-231	KM
	Skylark, exc. Calif.	V6-260	LA
	Skylark, Calif.	V8-260	NC
	Skylark	V8-350	KA
	Century	V8-350	KE
	LeSabre, Estate Wag.	V8-350	KL
	LeSabre, Electra, Estate Wag.	V8-350	KK
1977	Skylark	V6-231	KK, KS, MX
	Century, LeSabre	V6-231	KE, KW
	Skylark	V8-305	KJ
		V8-350	KX
	Century	V8-350	KA, KH, LE
	LeSabre, Estate Wag.	V8-350	LC
	Electra, Riviera	V8-350	LC
		V8-350	KC
	Riviera	V8-350	LT
	Century	V8-403	LM, LS
	LeSabre, Estate Wag.	V8-403	LA
	Electra, Riviera	V8-403	LA
1978	Century	V6-196	5KD
	Skyhawk	V6-231	5KA
		V6-231	5KL
	Skylark, Century Sta. Wag.	V6-231	5KE
	LeSabre	V6-231	5KE

GENERAL MOTORS
AUTOMATIC TRANSMISSION LISTING

Year	Vehicle Model	Engine	Transmission Code
	Turbo Hydra-Matic 250, 350, 375		
1978	Century, LeSabre	V6-231	5KJ
	LeSabre, Est. Wag.	V8-301	5MO
	Skylark, Century Sta. Wag.	V8-305	5KC
	Skylark, Century, LeSabre	V8-305	5JC
	Skylark, Century Sta. Wag.	V8-350	5JD
	LeSabre	V8-350	5JD
	LeSabre, Estate Wag.	V8-350	5KH, 5LA, 5LH
	LeSabre, Estate Wag.	V8-403	5LC, 5LE, 5LK
1979	Century	V-196	6KD
	Skyhawk	V6-231	6KA, 6KL, 6KX
	Century	V6-231	6KE
	Century, LeSabre	V6-231	6KJ
	LeSabre	V6-231	6KC
	Century	V8-301	6MA, 6MP
	LeSabre	V8-301	6MD
	Skylark, Century	V8-305	6JC
	Skylark	V8-305	6TA
	Skylark, Century	V8-350	6JD, 6TR
	LeSabre, Estate Wag.	V8-350	6KH, 6LA, 6LH
	LeSabre, Estate Wag.	V8-403	6LC, 6LE, 6LK
1980	Century & Regal, eng. code A	V6-231	KC
	Century & Regal, eng. code C	V6-231	KJ
	Century Limited, eng. code A	V6-231	KH
	LeSabre, eng. code A	V6-231	KH
	LeSabre, eng. code C	V6-231	KJ
	Skyhawk	V6-231	KA
	LeSabre, Electra	V6-252	KD
	Century, Regal	V8-301	TB
	LeSabre	V8-301	WB, TB
	Century, Regal	V8-305	JE, LJ
	LeSabre, eng. code R	V8-350	LA
	LeSabre, eng. code N	V8-350	WC
	LeSabre, eng. code X	V8-350	TV
1974	**CHEVROLET**		
	Chevelle	V8-400 2 bbl carb.	AW
	Camaro, exc. Z-28	V8-350	AW
	Chevelle, Nova	V8-350	FB
	Chevelle	V8-400 4 bbl carb.	FB
	Chevrolet, full size	V8-400 4 bbl carb.	FH, FD
		V8-400 2 bbl carb.	FH
		V8-400 4 bbl carb.	FW
	Camaro, Chevelle, Nova	6-250	TT
	Nova	6-250	TZ
1975	Monte Carlo	V8-350	JA
	Chevrolet, full size	V8-350	YC, YD
	Nova	V8-262	DD
	THM 250	6 cyl.	TZ
	Nova	V8-350 4 bbl carb.	YA

GENERAL MOTORS
AUTOMATIC TRANSMISSION LISTING

Year	Vehicle Model	Engine	Transmission Code
	Turbo Hydra-Matic 250, 350, 375		
1975	Chevelle	6-250	DE
		V8-350	HB
		V8-400 4 bbl carb.	YB
	Monza, THM 250	V8-262	FU
	Vega, THM 250	4-140	DJ
1976	Monza, Vega	4-140	DJ
	Chevelle	6-250	TA
	Nova	6-250	JB
	Camaro, Nova	6-250	TK
	Monza	V8-262	FU
	Camaro, Nova	V8-305	XE
	Chevelle	V8-305	TH
	Camaro, Nova	V8-350	TE
	Chevelle, Monte Carlo	V8-350	XX
	Corvette	V8-350	XH
	Chevelle & Monte Carlo, exc. Calif.	V8-400	XB
	Chevelle & Monte Carlo, Calif.	V8-400	XA
	Chevelle & Monte Carlo, Calif.	V8-400	XX
	Chevrolet Sta. Wag., exc. Calif.	V8-400	DG
	Chevrolet Sta. Wag., Calif.	V8-400	TB
	Chevrolet, w/2.73 rear axle, Calif.	V8-400	TC
	Chevrolet, w/3.08 rear axle	V8-400	TD
1977	Monza & Vega, exc. Calif.	4-140	AP
	Monza & Vega, Calif.	4-140	AO
	Chevelle & Monte Carlo, exc. Calif.	6-250	AD
	Chevrolet, full size, exc. Calif.	6-250	AD
	Nova & Camaro, exc. Calif.	6-250	WK
	Nova & Camaro, Calif.	6-250	AN
	Chevelle, Monte Carlo	V8-305	AF
	Chevrolet, full size	V8-305	AF
	Nova, Camaro	V8-305	AG
	Chevelle, Monte Carlo	V8-350	AH
	Chevrolet, full size	V8-350	AH
	Nova & Camaro, exc. Z-28	V8-350	AJ
	Z-28, Corvette	V8-350	AM
1978	Monza	V6-196	5WC
	Malibu, Monte Carlo	V6-200	5TE
	Monza, exc. high alt.	V6-231	5KK
	Monza, exc. Calif.	V6-231	5KL
	Malibu, Monte Carlo	V6-231	5KE
	Camaro & Chevrolet, exc. Calif.	6-250	WK
	Camaro & Chevrolet, Calif.	6-250	AN
	Nova	6-250	SWL
	Camaro, Chevrolet	V8-305	AG
	Malibu, Monte Carlo, Nova	V8-305	AG
	Camaro, exc. Z-28, Chevrolet	V8-350	AJ
	Malibu, Monte Carlo, Nova	V8-350	AJ

GENERAL MOTORS
AUTOMATIC TRANSMISSION LISTING

Year	Vehicle Model	Engine	Transmission Code
	Turbo Hydra-Matic 250, 350, 375		
1978	Corvette	V8-350	TL, WB
	Camaro	V8-350	WB
	Nova	V8-350	AH, WZ
1979	Monza	V6-196	6KK
	Malibu, Monte Carlo	V6-200	6WA
	Monza	V6-231	6KA, 6KL
	Malibu, Monte Carlo	V6-231	6KE
	Camaro & Chevrolet, exc. Calif.	6-250	6TP
	Camaro & Chevrolet, Calif.	6-250	6TT
	Nova, exc. Calif.	6-250	6TC
	Nova, Calif.	6-250	6TS
	Malibu, Monte Carlo	V8-267	6WG
	Camaro, Chevrolet, Malibu	V8-305	6TA
	Monte Carlo, Nova	V8-305	6TA
	Monza	V8-305	6WD
	Camaro exc. Z-28, Chevrolet	V8-350	6TR
	Malibu Sta. Wag., Nova	V8-350	6TR
	Camaro, Z-28	V8-350	6WF
	Corvette	V8-350	TB, WB
1980	Malibu, Monte Carlo	V6-229	JA
		V6-231	KC
	Monte Carlo, eng. code C	V6-231	KJ
	Impala, Caprice	V6-231	KH
	Camaro	V6-231	KF
	Monza	V6-231	KA
	Malibu, Monte Carlo	V8-267	JJ
	Impala, Caprice	V8-267	WH, TZ
	Camaro	V8-267	JJ
	Malibu, Monte Carlo	V8-305	JE, WD
	Impala, Caprice	V8-305	JE, WD, WL
	Camaro, exc. Z-28	V8-305	JK
	Camaro Z-28	V8-305	JD
	Impala, Caprice	V8-350 diesel	WC
	Camaro	V8-350	JL
1974	**OLDSMOBILE**		
	Omega	6-250	JE
		V8-350	LA
	Cutlass	V8-350	LC
	Cutlass Wag.	V8-350	LE
1975	Starfire	V6-231	KD, KX
	Omega	6-250	JB
	Cutlass	6-250	JE
	Omega	V8-260	LA, NC
	Cutlass	V6-260	LH
	Omega	V8-350	KA
	Cutlass, Wagon	V8-350	LC, LE

GENERAL MOTORS
AUTOMATIC TRANSMISSION LISTING

Year	Vehicle Model	Engine	Transmission Code
	Turbo Hydra-Matic 250, 350, 375		
1976	Starfire	V6-231	KD
	Omega	6-250	JB
	Cutlass	6-250	JE
	Omega, exc. Calif.	V8-260	LA
	Omega, Calif.	V8-260	NC
	Cutlass	V8-260	LH
	Omega	V8-350	KA
	Cutlass, exc. Calif.	V8-350	LB
	Cutlass, Calif.	V8-350	LC
	Intermediate Wagons	V8-350	LE
	88, exc. Calif.	V8-350	LL
1977	Starfire, exc. Calif.	4-140	AP
	Starfire	4-140	AO
	Omega	V6-231	KK
	Cutlass, 88	V6-231	KE
	Cutlass	V8-260	LH
	Omega	V8-305	KJ
		V8-350	KX
	Omega, high alt.	V8-350	LK
	Omega, Calif.	V8-350	LC
	Cutlass, exc. high alt. & Calif.	V8-350	LX
	Cutlass, exc. high alt.	V8-350	LE
	Cutlass, high alt.	V8-350	LD
	88, exc. high alt.	V8-350	LC
	88 Wagon	V8-350	LZ
	88 Wagon, high alt.	V8-350	LK
	98	V8-350	LC
	98, exc. high alt.	V8-350	LT
	98, high alt.	V8-350	LK
	Cutlass, exc. high alt.	V8-403	LM
	Cutlass, high alt.	V8-403	LP
	88, exc. high alt.	V8-403	LA
	88, high alt.	V8-403	LJ
	98	V8-403	LA
	98, high alt.	V8-403	LS, LJ
1978	Starfire, w/2.56 rear axle	V6-231	KA
	Starfire, w/2.93 rear axle	V6-231	KL
	Omega	V6-231	KC
	Cutlass, 88	V6-231	KE
	Cutlass, 88	V8-260	LD
	Starfire	V8-305	WC
	Omega, Cutlass, 88	V8-305	JC
	Omega, Cutlass	V8-350	JD
	88, exc. high alt.	V8-350	LA
	88 high alt.	V8-350	LH
	88 Sta. Wag.	V8-350	LJ
	88, exc. high alt. w/3.23 rear axle	V8-403	LK

GENERAL MOTORS
AUTOMATIC TRANSMISSION LISTING

Year	Vehicle Model	Engine	Transmission Code
	Turbo Hydra-Matic 250, 350, 375		
1978	88, exc. high alt, less 3.23 rear axle	V8-403	LC
	88, high alt.	V8-403	LE
1979	Starfire, w/2.56 rear axle	V6-231	KA
	Starfire, w/2.93 rear axle	V6-231	KL
	Omega, Calif.	V6-231	KX
	Omega, Cutlass	V6-231	KE
	88	V6-231	KC
	Cutlass, 88	V8-260	LD
	88	V8-301	MA
	Starfire	V8-305	WC
	Omega, Cutlass	V8-305	JC
		V8-350	JD
	Cutlass & 88, exc. high alt.	V8-350	LA
	88, exc. Sta. Wag. & high alt.	V8-350	LH
	88 Sta. Wag., exc. Calif.	V8-350	LH
	Cutlass Sta. Wag., exc. high alt.	V8-350 diesel	LS
	88 & 98, exc. Sta. Wag.	V8-350 diesel	LJ
	88 Sta. Wag., exc. high alt. w/3.23 rear axle	V8-403	LJ
	88 Sta. Wag., exc. high alt.	V8-403	LK
	88 Sta. Wag., exc. high alt. w/2.41 or 3.08 rear axle	V8-403	LC
	88 Sta. Wag., exc. Calif. & high alt.	V8-403	LW
	88 Sta. Wag., high alt.	V8-403	LE
1980	Cutlass	V6-231	KC
	Delta 88, Custom Cruiser	V6-231	KH
	Starfire	V6-231	KA
	Cutlass	V6-260	LC, LD
		V8-305	JE, JK
	Delta 88, Custom Cruiser	V8-307	WA, TT
	Cutlass	V8-350	LJ
	Delta 88, Custom Cruiser	V8-350	TY, WC, LA
1974	**PONTIAC**		
	Ventura	6-250	TZ
	Firebird, LeMans	6-250	JE
	Firebird, LeMans, w/2.73 or 2.93 rear axle	V8-350 2 bbl carb.	MA
	Firebird, LeMans, w/3.08 rear axle	V8-350 2 or 4 bbl carb.	ME
1975	Ventura, Firebird	6-250	JB
	LeMans	6-250	JE
	Ventura	V8-260	LA
	Ventura	V8-350	KA
	Firebird	V8-350	MA
	LeMans	V8-350	ME
	Firebird	V8-400 2 bbl carb.	MF
	Firebird	V8-400 4 bbl carb.	MG
	Astre, THM 250	4-140	DJ
1976	Astre, Sunbird	4-140	DJ
	Sunbird	V6-231	KD

GENERAL MOTORS
AUTOMATIC TRANSMISSION LISTING

Year	Vehicle Model	Engine	Transmission Code
	Turbo Hydra-Matic 250, 350, 375		
1976	Firebird, Ventura	6-250	JB
	LeMans	6-250	JE
	Ventura, exc. Calif.	V8-260	LA
	Ventura, Calif.	V8-260	NC
	LeMans	V8-260	LH
	Firebird	V8-350	MA
	LeMans	V8-350 2 bbl carb.	ME
		V8-350 4 bbl carb.	MM, MB
	Ventura	V8-350	KA
	Firebird	V8-400	MG
1977	Astre & Sunbird, exc. Calif. & high alt.	4-140	AO
	Astre & Sunbird, Calif. & high alt.	4-140	AP
	Astre, Sunbird	V6-231	KD
	Astre & Sunbird, 2.56 rear axle	V6-231	KL
	Astre & Sunbird, 2.93 rear axle	V6-231	KP
	Firebird	V6-231	MC
	LeMans	V6-231	JB
	LeMans, Catalina, Bonneville	V6-231	MD
	Phoenix, Ventura, Firebird	V8-301	MX
	LeMans, Catalina, Bonneville	V8-301	MM
	Phoenix, Ventura	V8-350	AJ
	Firebird	V8-350	MA, MJ
	LeMans, Catalina, Bonneville	V8-350	ME
	LeMans, Grand Prix	V8-350	LE
		V8-350 alt. perf. pkg.	LD
	Catalina, Bonneville	V8-350	ML
	Catalina, Bonneville, Ventura	V8-350	LC
	Firebird	V8-400	MG
	LeMans, Catalina, Bonneville	V8-400	MR
	Catalina, Bonneville	V8-400	MS, MP
	Firebird	V8-403	MZ
	LeMans, Grand Prix	V8-403	LM
	Catalina & Bonneville, exc. Police	V8-403	LA
1978	Sunbird	V6-231	KA
	Sunbird, exc. Calif.	V6-231	KL
	Phoenix, Firebird	V6-231	KC
	LeMans, exc. Sta. Wag.	V6-231	KE
	Grand AM, Grand LeMans	V6-231	KE
	Grand Prix, Catalina, Bonneville	V6-231	KE
	LeMans, exc. Sta. Wag.	V8-301	MP
	Grand Prix, Catalina, Bonneville	V8-301	MP
	Catalina, Bonneville	V8-301	MD, MH
	Phoenix, Firebird	V8-305	JC
	LeMans, Grand AM, Grand Prix	V8-305	JC
	Phoenix, Firebird & LeMans Sta. Wag.	V8-350	JD
	Catalina, Bonneville	V8-350	KD
	Catalina & Bonneville, Calif.	V8-350	LA
	Catalina & Bonneville, high alt.	V8-350	LH

GENERAL MOTORS
AUTOMATIC TRANSMISSION LISTING

Year	Vehicle Model	Engine	Transmission Code
	Turbo Hydra-Matic 250, 350, 375		
1978	Firebird, exc. Trans Am	V8-400	MC
	Firebird, Trans Am	V8-400	MK
	Catalina, Bonneville	V8-400	ME, MJ
	Firebird	V8-403	LP
	Catalina, Bonneville	V8-403	LC
		V8-403	LE
1979	Sunbird	V6-231	6KA
	Sunbird, exc. Calif.	V6-231	6KL
	Phoenix, Firebird	V6-231	6KX
	LeMans, Grand LeMans	V6-231	6KE
	Grand Am, Grand Prix	V6-231	6KE
	Catalina, Bonneville	V6-231	6KC
	Firebird	V8-301	6ME, 6MJ
	LeMans Sta. Wag.	V8-301	6MA, 6MP
	Catalina, Bonneville	V8-301	6MA, 6MP
	Sunbird	V8-305	6WD
	Phoenix, Firebird & LeMans Sta. Wag.	V8-305	6JC
	Phoenix	V8-305	6TA
	Phoenix, Firebird & LeMans Sta. Wag.	V8-350	6JD
	Phoenix	V8-350	6TR
	Catalina & Bonneville, exc. Calif. & high alt.	V8-350	6KH
	Catalina & Bonneville, Calif. & high alt.	V8-350	6LA
	Catalina & Bonneville, Calif. & high alt.	V8-350	6LH
	Firebird	V8-403	6LM, 6LP
	Catalina, Bonneville	V8-403	6LC, 6LE, 6LK
1980	Grand Prix, LeMans, Grand Am	V6-231	KC, KH
	Catalina, Bonneville	V6-231	KH
	Sunbird	V6-231	KA
	Firebird	V6-231	KF
		V8-260	MC
	Grand Prix, LeMans, Grand Am	V8-301	TB, MD
	Catalina, Bonneville	V8-301	TB, WB
	Firebird	V8-301	MJ, MT
	Grand Prix, LeMans, Grand Am	V8-305	JE
	Firebird	V8-305	JK, JD
	Catalina & Bonneville, eng. code N	V8-350	WC
	Catalina & Bonneville, eng. code R	V8-350	TY, LA
	Catalina & Bonneville, eng. code X	V8-350	TV
	Turbo Hydra-Matic 375, 400, 475		
1974	**BUICK**		
	Century Regal	V8-455, less stage 1	BS
		V8-455, stage 1	BB

GENERAL MOTORS
AUTOMATIC TRANSMISSION LISTING

Year	Vehicle Model	Engine	Transmission Code
	Turbo Hydra-Matic 375, 400, 475		
1974	All others, high perf.	V8-455	BT
	All others, exc. high perf.	V8-455	BC
1975	LeSabre	V8-350	BK
		V8-400	OC
		V8-455	BC, BT
1976	LeSabre	V8-455	BC
		V8-455	BT
1977	LeSabre	V8-350	BB
		V8-350	OB
		V8-403	OC
1978	Electra Riviera	V8-350	5BB
		V8-350	5OB
		V8-403	5OD
1979	Electra	V8-350	6BB
		V8-350	6OB
1980	Electra	V8-403	6OC, 6OD
	Electra, eng. code X	V8-350	BB
1974–75	**CADILLAC**		
	All models	—	AA
1976	All models Cadillac, exc. Seville	E.F.I.	AB
	Seville	E.F.I.	AC
1977	Fleetwood & DeVille, exc. E.F.I. & high alt.	—	AD
	Fleetwood & DeVille, E.F.I. exc. Calif.	—	AB
	Fleetwood & DeVille, Calif. Em. exc. E.F.I.	—	AE
	Fleetwood & DeVille, exc. E.F.I., high alt. & Calif.	—	AA
	Fleetwood & DeVille, high alt.	—	AL
	Seville, 2.56 axle	—	AC
	Seville, 3.08 axle	—	AH
1978	Brougham & DeVille, exc. E.F.I., Calif. & high alt., 2.28 axle	—	5AE
	Brougham & DeVille, 2.73 axle, exc. E.F.I., Calif. & high alt.	—	5AD
	Brougham & DeVille, E.F.I. exc. Calif. & high alt.	—	5AB
	Brougham & DeVille, Calif.	—	5AE
	Brougham & DeVille, high alt.	—	5AL
	Seville, 2.56 axle, exc. high alt.	—	5AC
	Seville, 3.08 axle, Calif.	—	5AH
	Seville, high alt.	—	5AT
1979	Brougham & DeVille, 2.28 axle, exc. E.F.I., Calif. & high alt.	—	AE
	Brougham & DeVille, 2.73 axle, exc. E.F.I., Calif. & high alt.	—	AD
	Brougham & DeVille, E.F.I. exc. Calif.	—	AB

GENERAL MOTORS
AUTOMATIC TRANSMISSION LISTING

Year	Vehicle Model	Engine	Transmission Code
	Turbo Hydra-Matic 375, 400, 475		
1979	Brougham & DeVille, 2.28 axle, Calif. & high alt.	—	AE
	Brougham & DeVille, 2.73 axle, Calif. & high alt.	—	AD
	Brougham, DeVille	diesel	AZ
	Seville, 2.24 axle	—	AA
	Seville, 2.56 axle	exc. diesel	AC
		diesel	AX
	Seville, 3.08 axle, exc. high alt.	—	AH
	Seville, 3.08 axle, high alt.	—	AS
1980	Fleetwood & DeVille, Federal	V8-368	AE
	Fleetwood & DeVille, Calif.	V8-368	AB
	Fleetwood & DeVille, high alt.	V8-368	AD
1974	**CHEVROLET**		
	Full size	V8-400	CA
	Chevrolet	V8-400 2 bbl carb.	CB
		V8-400	CG
	Corvette	V8-350 190 H.P.	CK
	Chevrolet	V8-454	CR
	Corvette	V8-454	CS
		V8-350 250 H.P.	CZ
1975	Police	V8-350	CR
	Corvette, exc. sp. high perf.	V8-350	CK
	Corvette, sp. high perf.	V8-350	CZ
	Chev., with 3.08, 3.42 axles, exc. Calif.	V8-400	CA
	Chev., with 2.56, 2.73 axles, exc. Calif.	V8-400	CB
	Chevelle	V8-454	CF
	Chevrolet	V8-454	CR
1976	Corvette	V8-350	AM, CZ
	Chevrolet	V8-400	CA
		V8-454	CR
1977	Corvette	V8-350	CB
1974	**OLDSMOBILE**		
	88	V8-350	OA
	Cutlass, exc. Calif.	V8-455	OD
	Cutlass, Calif.	V8-455	OW
	Cutlass	V8-455 275 H.P.	OX
	88 & 98, single exh., exc. Calif.	V8-455	OR, OL
	88 & 98, dual exhaust	V8-455	OK
	Custom Cruiser, exc. Calif.	V8-455	OK
	Custom Cruiser, Calif.	V8-455	OE
1975	88, exc. Calif.	V8-350	OA
	88, Calif.	V8-350	OB
	Custom Cruiser	V8-400	OC

GENERAL MOTORS
AUTOMATIC TRANSMISSION LISTING

Year	Vehicle Model	Engine	Transmission Code
	Turbo Hydra-Matic 375, 400, 475		
1975	98	V8-400	OF
	Cutlass, exc. Calif.	V8-455	OD
	Cutlass, Calif.	V8-455	OW
1976	88, Calif.	V8-350	OB
	Cutlass, exc. Calif.	V8-455	OD
	Cutlass, Calif.	V8-455	OW
	Custom Cruiser	V8-455	OK
	88 & 98, exc. Calif.	V8-455	OR
	98, exc. Calif.	V8-455	OC
	88 & 98, Calif.	V8-455	OL
1977	Cutlass Wag., 98	V8-350	OB
	Cutlass Wag.	V8-403	OC
	98	V8-403	OD
	Custom Cruiser, exc. Calif.	V8-455	OK
	Custom Cruiser, Calif.	V8-455	OE
	88 & 98, exc. Calif.	V8-455	OR
	88 & 98, Calif.	V8-455	OL
1978–79	98	V8-350	OB
	98, 3.08 & 3.23 axle	V8-403	OC
	98, 2.41 & 2.56 axle	V8-403	OD
1979	Exc. Calif., 2.41 axle	V8-403	OF
1980	98	V8-307	OA
	98	V8-350	OB
1974	**PONTIAC**		
	Pontiac Sta. Wag.	V8-400 2 bbl carb.	PA
	Pontiac	V8-400 4 bbl carb.	PB
	Pontiac	V8-455 4 bbl carb.	PC
	Pontiac, exc. Calif.	V8-400 2 bbl carb.	PD
	Pontiac, Calif.	V8-400 2 bbl carb.	PF
	LeMans & Firebird, exc. Calif.	V8-400 4 bbl carb.	PG
	LeMans & Firebird, Calif.	V8-400 4 bbl carb.	PW
	Firebird	V8-455 S.D.	PQ
	LeMans, Grand Prix	V8-455 4 bbl carb.	PR
	LeMans & Firebird, exc. Calif.	V8-400 2 bbl carb.	PT
	LeMans & Firebird, Calif.	V8-400 2 bbl carb.	PL
	Grand Prix	V8-400 4 bbl carb.	PX
	Firebird	V8-455 4 bbl carb.	PZ
1975	Pontiac	V8-400 2 bbl carb.	PD
	LeMans, Grand Prix	V8-400 2 bbl carb.	PT
	Pontiac, exc. Calif.	V8-400 2 bbl carb.	PB
	Pontiac, Calif.	V8-400 4 bbl carb.	PF
	LeMans & Grand Prix, exc. Calif.	V8-400 4 bbl carb.	PX
	LeMans & Grand Prix, Calif.	V8-400 4 bbl carb.	PG

GENERAL MOTORS
AUTOMATIC TRANSMISSION LISTING

Year	Vehicle Model	Engine	Transmission Code
	Turbo Hydra-Matic 375, 400, 475		
1975	Pontiac, exc. Calif.	V8-455	PC
	Pontiac, Calif.	V8-455	PA
	LeMans & Grand Prix, exc. Calif.	V8-455	PR
	LeMans & Grand Prix, Calif.	V8-455	PL
1976	Grand Prix	V8-350	PS
	LeMans, Grand Prix	V8-400 2 bbl carb.	PT
	Pontiac	V8-400 2 bbl carb.	PD
	LeMans, Grand Prix	V8-400 4 bbl carb.	PX
	Pontiac	V8-400 4 bbl carb.	PB
	Pontiac, Police	V8-400	PA
	LeMans	V8-455	PL, PR
	Pontiac	V8-455	PH, PC
1977	LeMans, Grand Prix	V8-301	PA
	LeMans Sta. Wag.	V8-350 4 bbl carb.	OB
	Grand Prix	V8-350	PB
	LeMans, Grand Prix	V8-400	PD
	LeMans Sta. Wag.	V8-400	PC
	Grand Prix	V8-400	PC
	LeMans Sta. Wag.	V8-403	OC

AMC TORQUE COMMAND   **CHRYSLER TORQUEFLITE**

TROUBLESHOOTING
AMC TORQUE COMMAND   CHRYSLER TORQUEFLIGHT

THE PROBLEM	ITEMS TO CHECK	
	IN THE CAR	OUT OF CAR
HARSH N TO D OR N TO R SHIFT	Oil pressure (check). Kickdown band. Low-reverse band. Improper engine idle. Servo linkage. Valve body assembly. Manual valve lever.	Front kickdown clutch. Rear clutch.
DELAYED SHIFT—N TO D	Oil level. Oil pressure (check). Valve body assembly. Accumulator. Perform air pressure check. Manual valve lever.	Front pump and/or sleeve. Front kickdown clutch.
RUNAWAY ON UPSHIFT & 3-2 KICKDOWN	Oil level. Control linkage. Oil pressure check. Kickdown band. Kickdown servo or linkage. Valve body assembly. Accumulator. Perform air pressure check.	Rear clutch.
HARSH UPSHIFT & 3-2 KICKDOWN	Control linkage. Oil pressure check. Kickdown band. Kickdown servo or linkage. Valve body assembly. Accumulator. Manual valve lever.	Rear clutch.
NO UPSHIFT	Oil level. Control linkage. Oil pressure check. Kickdown band. Kickdown servo or linkage. Valve body assembly. Accumulator. Perform air pressure check. Governor.	Rear clutch.
NO KICKDOWN ON NORMAL DOWNSHIFT	Oil level. Control linkage. Gear shift cable. Oil pressure check. Kickdown band. Kickdown servo or linkage. Valve body assembly. Accumulator. Perform air pressure check. Governor.	Overrunning clutch.
ERRATIC SHIFTS	Oil level. Control linkage. Gear shift cable. Oil pressure check. Improper engine idle. Regulator valve and/or spring. Output shaft bushing. Strainer. Vadve body assembly. Perform air pressure check. Governor. Manual valve lever.	Front pump and/or sleeve.
SLIPS IN FORWARD DRIVE POSITIONS	Oil level. Oil pressure check. Valve body assembly. Accumulator. Perform air pressure check.	Front kickdown clutch. Rear clutch. Overrunning clutch.
SLIPS IN REVERSE ONLY	Oil pressure (check). Low-reverse band. Servo linkage. Valve body assembly. Perform air pressure check.	
SLIPS IN ALL POSITIONS	Oil level. Oil pressure (check). Regulator valve and/or spring. Valve body assembly. Perform air pressure check.	Converter.
NO DRIVE IN ANY POSITION	Oil level. Oil pressure (check). Regulator valve and/or spring. Strainer. Valve body assembly. Perform air pressure check. Manual valve lever.	Front pump and/or sleeve. Front kickdown clutch.
NO DRIVE IN FORWARD POSITION	Oil pressure (check). Kickdown band. Regulator valve and/or spring. Kickdown servo or linkage. Valve body assembly. Accumulator. Perform air pressure check.	Front kickdown clutch. Rear clutch. Overrunning clutch.
NO DRIVE IN REVERSE	Oil pressure (check). Low-reverse band. Servo linkage. Perform air pressure check. Governor. Valve body assembly.	Rear clutch.
DRIVES IN NEUTRAL	Gear shift cable. Valve body assembly. Manual valve lever.	Front kickdown clutch.
DRAGS OR LOCKS	Kickdown band. Low-reverse band. Kickdown servo or linkage. Servo linkage.	Front kickdown clutch. Rear clutch. Planetary. Overrunning clutch.
NOISES	Oil level. Regulator valve and/or spring. Converter control valve. Output shaft bushing. Governor. Valve body assembly. Manual valve lever.	Front kickdown clutch.
HARD TO FILL OR FLUID BLOWS OUT	Oil level. Regulator valve and/or spring. Converter control valve. Breather clogged. Cooler or lines. Strainer. Valve body assembly.	
TRANSMISSION OVERHEATS	Oil level. Kickdown band. Low-reverse band. Regulator valve and/or spring. Converter control valve. Cooler or lines. Governor. Valve body assembly.	Front kickdown clutch. Rear clutch. Converter.

AMC TORQUE COMMAND   CHRYSLER TORQUEFLIGHT

AMC TORQUE COMMAND CHRYSLER TORQUEFLIGHT

Identification

The transmission identification pad is located on the left side of the transmission oil pan flange. The chart below relates the part number shown on this pad to the engine and vehicle application.

Applications

Model #	Year & Engines
Application AMC	
904	6 Cyl 232 CID
998	6–258 V8 304 Cid V8
727	V8–360–401–CID
V8	
Application Chrysler	
A 904	6 Cyl Exc Police and Taxi
A 727 RG or	
Aa 727 A	Police and Taxi
A 904 LA	318 CID V8
A 727 A	340–360 CID V8
A 7273	400–426 Hemi-440 CID V8

In-Car Testing Procedures
Air Pressure Tests

The front clutch, rear clutch, kickdown servo and low and reverse servo may be checked with air pressure, after the valve body assembly has been removed.

To make air pressure tests, proceed as follows:

CHILTON CAUTION: *Compressed air must be free of dirt and moisture. Use pressure of 30–100 psi.*

FRONT CLUTCH

Apply air pressure to the front clutch apply passage and listen for a dull thud. This will indicate operation of the front clutch. Hold the air pressure at this point for a few seconds and check for excessive oil leaks.

NOTE: If a dull thud cannot be heard in the clutch, place finger tips on clutch housing and again apply air pressure. Movement of piston can be felt as clutch is applied.

REAR CLUTCH

Apply air pressure to the rear clutch apply passage and proceed in an identical manner as that described in the previous paragraph.

KICKDOWN SERVO

Air pressure applied to the kickdown servo apply passage should tighten the front band. Spring tension should be sufficient to release the band.

LOW & REVERSE SERVO

Direct air pressure into the low and reverse servo apply passage. Response of the servo will result in a tightening of the rear band. Spring tension should be enough to release the band.

If clutches and servos operate properly, no upshift or erratic shift conditions existing, trouble exists in the control valve body assembly.

GOVERNOR

Governor troubles can usually be found during a road or pressure test.

Hydraulic Control Pressure Checks

LINE PRESSURE & FRONT SERVO RELEASE PRESSURE
NOTE: These pressure checks must be made in the D position with the rear wheels free to turn. The transmission fluid must be at operating temperature (150°–200°F).

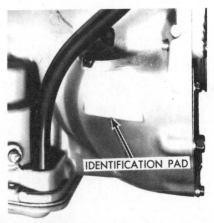

Transmission I.D. pad

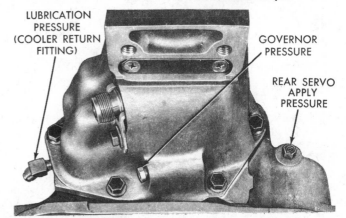

Pressure test locations

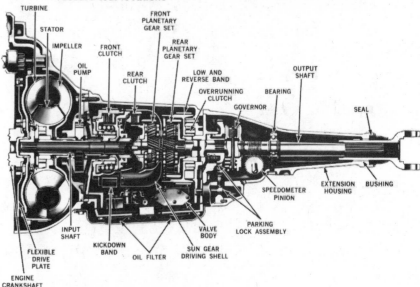

Cross-section view of transmission

AMC TORQUE COMMAND   CHRYSLER TORQUEFLITE

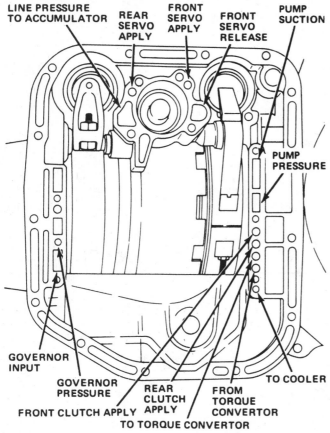

Transmission case channels

1 Install an engine tachometer, then, raise the car on a hoist and locate the tachometer so it can be read from under the car.
2 Connect two 0–100 psi pressure gauges to pressure takeoff points at side of the accumulator and at the front servo release.
3 With the selector in D position, increase engine speed gradually until the transmission shifts into High. Reduce engine speed slowly to 1000 rpm. The line pressure must be 54–60 psi with front servo release having no more than a 3 psi drop.
4 Disconnect throttle linkage from transmission throttle lever and move throttle lever gradually to full throttle position. Line pressure must rise to maximum of 90–96 psi just before or

at kickdown into low gear. Front servo release pressure must follow line pressure up to kickdown point and should not be more than 3 psi below line pressure. If pressure is not 54–60 psi at 1000 rpm, adjust line pressure.

If line pressure is not as above, adjust the pressure as outlined below under the heading: Hydraulic Control Pressure Adjustments—Line Pressure.

If front servo release pressures are less than specified, and line pressures are within limits, there is excessive leakage in the front clutch and/or front servo circuits.

REAR SERVO APPLY PRESSURE
1 Connect a 0–300 psi pressure gauge, to the apply pressure take-off point at the rear servo.
2 With the control in the R position, and the engine running at 1600 rpm, the reverse servo apply pressure should be 230–300 psi.

GOVERNOR PRESSURE
1 Connect a 0–100 psi gauge to the governor pressure take-off point. This location is at the lower left rear corner of the extension mounting flange.
2 Governor pressure should fall within limits in chart and should return to 0–1½ psi when car is stopped. If it is above this, transmission will not downshift.
Pressure should change smoothly with car speeds.

If governor pressures are incorrect at the prescribed speeds, the governor valve and/or weights are probably sticking.

LUBRICATION PRESSURES
A lubrication pressure check should be made when line pressure and front servo release pressures are checked.
1 Install a T fitting between the cooler return line fitting and the fitting hole in the transmission case at the rear left

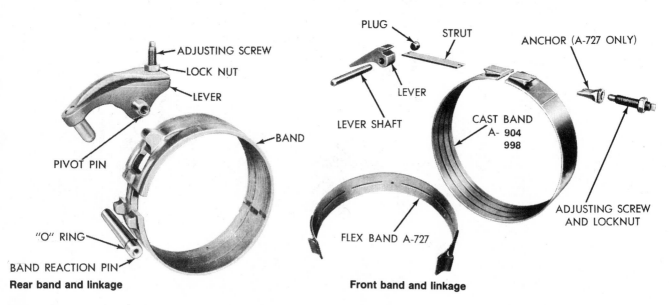

Rear band and linkage   **Front band and linkage**

AMC TORQUE COMMAND   CHRYSLER TORQUEFLITE

side of the transmission. Connect a 0–100 psi pressure gauge to the T-fitting.

2 At 1000 engine rpm, with throttle closed and transmission in High, lubrication pressure should be 5–15 psi. Lubrication pressure will approximately double as throttle is opened to maximum line pressure.

In-Car Service Procedures

FRONT BAND ADJUSTMENT

The front band adjusting screw is located on the left side of the transmission case just above the manual value and throttle control levers.

1 Loosen adjusting screw locknut and back off five turns.
2 Be sure adjusting screw turns freely in case.
3 Tighten adjusting screw to 72 inch-pounds
4 Back off adjusting screw as follows:
  a. Model 904—two turns
  b. Models 998 and 727—two and one-half tuens
5 Hold adjusting screw in position and tighten locknut to 35 foot-pounds.

REAR BAND ADJUSTMENT

1 Remove adjusting screw locknut.
2 Tighten adjusting screw to 41 inch-pounds
3 Back off adjusting screw seven turns.
4 Hold adjusting screw in position and install nut.
5 Tighten locknut to 35 foot-pounds torque.

On Models 998–727, adjust band as follows:

1 Loosen locknut. Back off locknut five turns.
2 Tighten adjusting screw to 72 inch-pounds torque.
3 Back off adjusting screw as follows:
  a. Model 998—four turns
  b. Model 727—two turns
4 Hold adjusting screw in position and tighten locknut to 35 foot-pounds torque.

FLUID LEAKS

Some leaks that can normally be corrected without transmission removal are:

1 Transmission output shaft oil seal.
2 Extension housing gasket.
3 Speedometer pinion seal and cable seal.
4 Oil filler tube seal.
5 Oil pan gasket and drain plug.
6 Gearshift control cable seal.
7 Throttle shaft seal.
8 Neutral starting switch seal.
9 Oil cooler line fittings and pressure take-off plugs.

Oil found inside the converter housing should be positively identified as transmission oil before diagnosing the need for any major transmission work.

HYDRAULIC CONTROL PRESSURE ADJUSTMENTS

LINE PRESSURE

An incorrect throttle pressure setting will cause incorrect line pressure even though line pressure adjustment is correct. Always inspect and correct throttle pressure adjustment before adjusting line pressure.

NOTE: Before adjusting line pressure, measure distance between manual valve (valve in 1-low position) and line pressure adjusting screw. This measurement must be 1⅞ in. Correct by loosening spring retainer screws and repositioning spring retainer. The regulator valve may cock and hang up in its bore if spring retainer is out of position.

If line pressure is not correct remove valve body assembly to adjust. The correct adjustment is 1-5/16 in. measured from valve body to inner edge of adjusting nut. Vary adjustment slightly to obtain specified line pressure.

One complete turn of the adjusting screw (Allen head) changes closed throttle line pressure about 1.66 psi. Turning the screw counterclockwise increases pressure, clockwise decreases pressure.

THROTTLE PRESSURE

Because throttle pressures cannot be

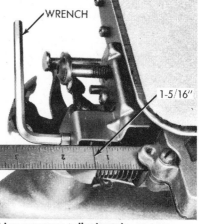

Line pressure adjustment

checked, exact adjustments should be checked and made correct whenever the valve body is disturbed.

1 Remove the valve body assembly, as outlined in a succeeding coverage entitled, Valve Body Assembly and Accumulator Piston.
2 Loosen throttle lever stop screw locknut and back off the screw about five turns.
3 Insert gauge pin between the throttle lever cam and the kickdown valve.
4 Push on the tool and compress the kickdown valve against its spring, so that the throttle valve is completely bottomed inside the valve body.
5 As the spring is being compressed, finger tighten the throttle lever stop screw against the throttle lever tang, with the lever cam touching the tool and the throttle valve bottomed. (Be sure the adjustment is made with the spring fully compressed and the valve bottomed in the valve body.)
6 Remove the tool and secure the stop screw locknut.

Speedometer Pinion R&R

Rear axle gear ratio and tire size determine pinion gear size.

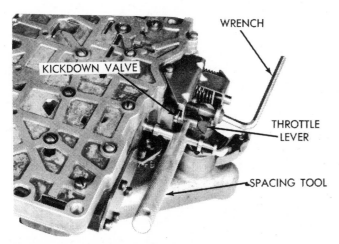

Throttle pressure adjustment

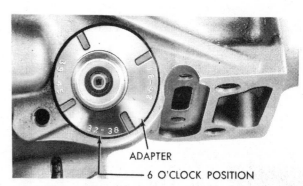

Speedometer pinion and adapter

Component Removal and Overhaul

AMC TORQUE COMMAND   CHRYSLER TORQUEFLITE

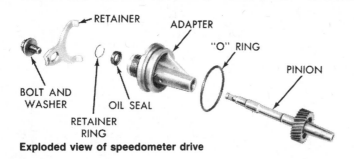

Exploded view of speedometer drive

1 Remove bolt and retainer securing speedometer pinion adapter in extension housing.
2 With cable housing connected, carefully work adapter and pinion out of extension housing.
3 If transmission fluid is found in cable housing, replace seal in adapter. Start seal and retainer ring in adapter, then push them into adapter until tool bottoms.

CHILTON CAUTION: *Before installing pinion and adapter assembly make sure adapter flange and mating area on extension housing are perfectly clean. Dirt or sand will cause misalignment and speedometer pinion gear noise.*

4 Note number of gear teeth and install pinion gear into adapter.
5 Rotate pinion gear and adapter assembly so that number on adapter corresponding to number of teeth on gear is in six o'clock position as assembly is installed.

6 Install retainer and bolt with retainer tangs in adapter positioning slots. Tap adapter firmly into extension housing and tighten retainer bolt to 100 in./lbs.

Output Shaft Oil Seal R&R

1 Mark parts for reassembly. Disconnect driveshaft at rear universal joint. Carefully pull shaft yoke out of transmission extension housing. Be careful not to scratch or nick ground surface of sliding spline yoke.
2 Remove extension housing yoke seal by gently tapping out around circumference of seal with slide hammer.
3 To install new seal, place seal in opening of extension housing and drive it into housing with suitable drift.
4 Carefully guide front universal joint yoke into extension housing and onto the mainshaft splines. Align marks made at removal and connect driveshaft to pinion shaft yoke.

Extension Housing

REMOVAL

1 Mark parts for reassembly. Disconnect driveshaft at rear universal joint. Carefully pull shaft out of extension housing.
2 Remove speedometer pinion and adapter assembly. Drain approximately two quarts of fluid from transmission.
3 Remove bolts securing extension housing to crossmember. Raise transmission slightly with service jack and remove center crossmember and support assembly.
4 Remove extension housing to transmission bolts. On console shifts, remove two bolts securing gearshift torque shaft lower bracket to extension housing. Swing bracket out of way.

NOTE: Gearshift lever must be in 1-Low position so that parking lock control rod can be engaged or disengaged with parking lock sprag.

5 Remove two screws, plate, and gasket from bottom of extension housing mounting pad.
6 Spread large snap ring from output shaft bearing with Tool C-3301 or equivalent.
7 With snap-ring spread as far as possible tap extension housing gently off output shaft bearing.
8 Carefully pull extension housing rearward to bring parking lock control rod knob past parking sprag and remove housing.
   To install:
9 Install snap ring in front groove on output shaft.
10 Install bearing on shaft with its outer race ring groove toward front. Press or tap bearing tight against front snap ring.
11 Install rear snap ring. If so equipped, slide the yoke seal front stop ring onto the shaft. Install the seal with lips toward the rear, then install the rear stop ring on the shaft.
12 Place new extension housing gasket on transmission case.
13 Place output shaft bearing retaining snap-ring in extension housing. Spread ring as far as possible, then carefully tap extension housing into place. Make sure snap-ring is fully seated in bearing groove.
14 Install and torque extension housing bolts to specification.
15 Install gasket, plate, and two screws on bottom of extension housing mounting pad.
16 Install speedometer pinion and adapter assembly.

Governor R&R

1 Remove extension housing.

NOTE: Remove output shaft support bearing if so equipped.

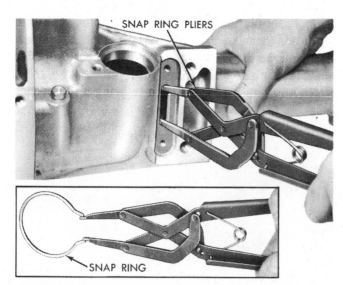

Removing/installing extension housing

AMC TORQUE COMMAND   CHRYSLER TORQUEFLITE

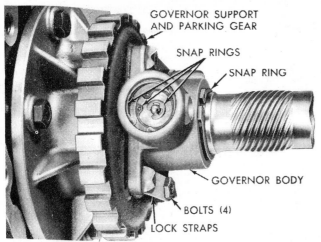

Governer shaft & weight snap rings

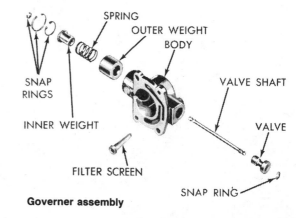

Governer assembly

2 With a screwdriver, carefully pry the snap-ring from the weight end of governor valve shaft. Slide the valve and shaft assembly out of the governor housing.

3 Remove the large snap-ring from the weight end of the governor housing and lift out the governor weight assembly.

4 Remove snap-ring from inside governor weight, remove inner weight and spring from the outer weight.

5 Remove the snap-ring from behind the governor housing, then slide the governor housing and support assembly from the output shaft. If necessary, remove the four screws and separate the governor housing from the support.

6 The primary cause of governor operating trouble is sticking of the valve or weights, brought about by dirt or rough surfaces. Thoroughly clean and blow dry all of the governor parts. Remove any burrs or rough bearing surfaces with crocus cloth and clean again. If all moving parts are clean and operating freely, the governor may be reassembled.

7 Assemble the governor housing to the support, then finger tighten the screws. Be sure the oil passage of the governor housing aligns with the passage in the support.

8 Align the master spline of the support with the master spline on the output shaft, slide the assembly into place. Install the snap-ring behind the governor housing. Torque bolts to 100 in./lbs.

9 Assemble the governor weights and spring, and fasten with snap-ring inside of large governor weight. Place the weight assembly in the governor housing shafts retaining snap-ring. Install output shaft support bearing if so equipped.

10 Place the governor valve on the valve shaft, insert the assembly into the housing and through the governor weights. Install the shaft retaining

snap-ring. Install output shaft support bearing if so equipped.

11 Install the extension housing. Connect the driveshaft.

Governor & Support R&R

1 Remove the snap-ring from the weight end of the governor valve shaft. Slide the valve and shaft assembly from the governor housing.

2 Remove the snap-ring from behind the governor housing, then slide the governor housing and support from the output shaft.

To install:

3 Place the governor and support on the output shaft. Position it so that the governor valve shaft hole aligns with the hole in the output shaft, then slide

the assembly into place. Install snap-ring behind the governor housing.

Torque housing-to-support screws to specification.

4 Place the governor valve on the valve shaft, insert the assembly into the housing and through the governor weights. Install the valve shaft retaining snap-ring.

Valve Body Assembly & Accumulator Piston R&R

1 Raise vehicle on hoist and loosen oil pan bolts, tap pan to break it loose, allowing fluid to drain.

2 Remove pan and gasket.

3 Disconnect throttle and gearshift linkage from levers on transmission.

GOVERNOR PRESSURE SPECIFICATIONS

Axle ratio	Speed (mph)	Press. limit (psi)	Axle ratio	Speed (mph)	Press. limit (psi)
CHRYSLER			**ALL MODELS**		
2.76	20–22	15	**8–318 ENGINE**		
	48–57	50	2.71	20–22	15
	72–79	75		56–64	50
3.23	17–19	15		80–88	75
	41–49	50			
	66–73	75	**8–340 ENGINE**		
			3.23	17–18	15
DODGE DART/PLYMOUTH				49–53	50
VALIANT				66–71	75
6–198 & 6–225 ENGINES					
2.76	19–20	15	**8–360 ENGINE**		
	49–55	50	2.71	20–22	15
	72–79	75		54–63	50
				78–86	75
HIGH PERFORMANCE VEHICLES					
DODGE & PLYMOUTH			**8–383 & 8–400 ENGINES**		
8–383 & 8–400 ENGINES			2.76	21–22	15
3.23	18–19	15		55–64	50
	53–58	50		80–87	75
	71–77	75			
			8–440 ENGINE		
8–440HP ENGINE			2.94	20–22	15
3.23	18–19	15		54–62	50
	53–58	50		78–85	75
	71–77	75			

Component Removal and Overhaul

GOVERNOR PRESSURE SPECIFICATIONS A.M.C.

MODEL	GREMLIN		
ENGINE—	6	6	V8
CID	232	258	304
AXLE GEAR	2.73	2.73	2.87
15 p.s.i. at mph	19-23	19-23	17-21
50 p.s.i. at mph	46-54	46-54	49-57
75 p.s.i. at mph	64-73	64-73	69-77

MODEL	HORNET			
ENGINE—	6	6	V8	V8
CID	232	258	304	360
AXLE GEAR	2.73	2.73	2.87	2.87
15 p.s.i. at mph	20-24	20-24	17-21	17-21
50 p.s.i. at mph	47-55	47-55	49-57	49-57
75 p.s.i. at mph	66-74	66-74	69-77	69-77

MODEL	JAVELIN				
ENGINE	6	6	V8	V8	V8
CID	232	258	304	360	401
AXLE GEAR	3.08	3.08	2.87	2.87	2.87
15 p.s.i. at mph	17-21	17-21	17-21	17-21	17-21
50 p.s.i. at mph	42-50	42-50	49-57	50-58	49-57
75 p.s.i. at mph	59-66	59-66	69-78	70-79	69-78

MODEL	MATADOR				
ENGINE	6	6	V8	V8	V8
CID	232	258	304	360	401
AXLE GEAR	3.15	3.15	3.15	3.15	3.15
15 p.s.i. at mph	17-21	18-22	16-20	17-21	16-20
50 p.s.i. at mph	42-49	44-52	45-53	49-57	45-53
75 p.s.i. at mph	58-66	62-70	64-72	68-77	64-72

MODEL	AMBASSADOR		
ENGINE	V8	V8	V8
CID	304	360	401
AXLE GEAR	3.15	3.15	3.15
15 p.s.i. at mph	16-20	17-21	16-20
50 p.s.i. at mph	46-54	49-57	46-54
75 p.s.i. at mph	65-73	68-77	65-73

Loosen clamp bolts and remove levers.

4 Remove E clip securing parking lock rod to valve body manual lever.

5 Remove backup light and neutral start switch.

6 Place drain pan under transmission, remove ten hex head valve body to transmission case bolts. Hold valve body in position while removing bolts.

7 While lowering valve body down out of transmission case, disconnect parking lock rod from lever. To remove rod pull it forward out of case. If necessary rotate driveshaft to align parking gear and sprag to permit knob on end of control rod to pass sprag.

8 Withdraw accumulator piston from transmission case. Inspect piston for scoring, and rings for wear or breakage.

9 If valve body manual lever shaft seal requires replacement, drive it out of case with punch.

10 Drive new seal into case with 15/16 in. socket and hammer.
    To install:

11 If parking lock rod was removed, insert it through opening in rear of case with knob positioned against plug and sprag. Move front end of rod toward center of transmission while exerting rearward pressure on rod to force it past sprag. Rotate driveshaft if necessary.

12 Install accumulator piston in transmission case. On 440-4v. models, (1974) with dual exhausts, the spring is longer and must be installed before the piston.

13 Place accumulator spring on valve body.

14 Place valve body manual lever in low position. Lift valve body into its approximate position, connect parking lock rod to manual lever and secure with E clip. Place valve body in case, install retaining bolts finger tight.

15 With neutral start switch installed, place manual lever in neutral position. Shift valve body if necessary to center neutral finger over neutral switch plunger. Snug bolts down evenly. Torque to 100 in./lbs.

16 Install gearshift lever and tighten clamp bolt. Check lever shaft for binding by moving lever through all detents. If lever binds, loosen valve body bolts and re-align.

17 Make sure throttle shaft seal is in place, install flat washer and lever and tighten clamp bolt. Connect throttle and gearshift linkage and adjust as required.

18 Install oil pan using new gasket. Add transmission fluid to proper level.

Component Removal

The following reconditioning data covers the removal, disassembly, inspection, repair, assembly and installation procedures for each sub-assembly in detail.

NOTE: Should any transmission component fail, the converter should be thoroughly flushed to insure the removal of fine particles that may cause damage to the reconditioned transmission.

OIL PAN REMOVAL

1 Secure transmission in a repair stand.

2 Unscrew attaching bolts and remove oil pan and gasket.

VALVE BODY REMOVAL

1 Loosen clamp bolts and remove throttle and gearshift levers from transmission.

2 Remove backup light and neutral start switch.

3 Remove ten hex head valve body to transmission bolts. Remove E clip securing parking lock rod to valve body manual lever.

4 While lifting valve body upward out of transmission case, disconnect parking lock rod from lever.

ACCUMULATOR PISTON AND SPRING REMOVAL

1 Lift the spring from the accumulator piston and withdraw the piston from the case. Spring will be on the small end of the piston on 440 dual exhaust engines.

CHECKING DRIVE TRAIN END-PLAY

1 Mount dial indicator on the transmission bell housing with the plunger seated against the end of the input shaft.

2 Move the input shaft in and out to obtain the end-play reading.

3 Record this reading for future use when assembling the transmission. Correct end-play is found at the front of the section.

GOVERNOR AND SUPPORT REMOVAL

1 Remove the snap-ring from the weight end of the governor valve shaft. Slide the valve and shaft assembly from the governor housing.

2 Remove the snap-ring from behind the governor housing, then slide the governor housing and support from the output shaft.

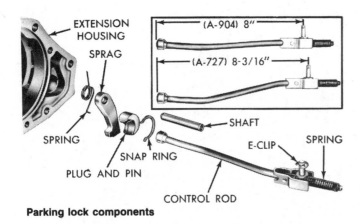

Parking lock components

AMC TORQUE COMMAND   CHRYSLER TORQUEFLITE

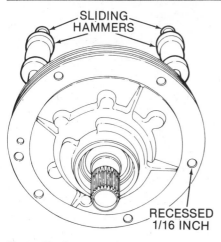

Removing front pump

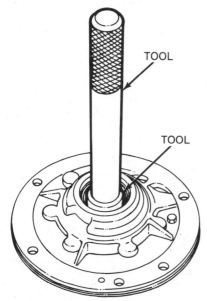

Pump bushing removal

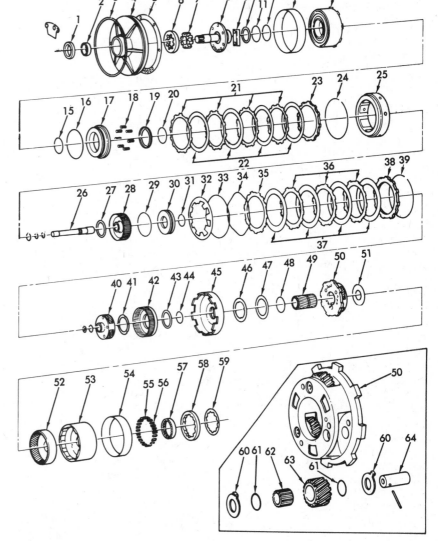

Torqueflite gear train

FRONT OIL PUMP AND REACTION SHAFT SUPPORT REMOVAL

1 Tighten the front band adjusting screw until the front band is tight on the front clutch retainer. This prevents the front clutch from coming out with the pump and damaging the clutches.
2 Remove the front pump housing retaining bolts.
3 Attach slide hammers to the pump housing flange, using the three and nine o'clock hole locations.
4 Bump out evenly, with the tool, to withdraw the oil pump and reaction shaft support assembly from the case.

FRONT BAND AND FRONT CLUTCH REMOVAL

1 Loosen the front band adjuster, remove the band strut and slide the band from the case.
2 Slide the front clutch assembly from the case.

1. Oil seal
2. Pump housing bushing
3. O-ring
4. Oil pump housing
5. Oil pump gasket
6. Oil pump rotor pakage
7. Oil pump rotor package
8. Reaction shaft w/support
9. Baffle
10. Thrust washer
11. Seal ring
12. Seal ring
13. Kickdown band
14. Front clutch reatainer
15. Inner seal
16. Outer seal
17. Front clutch piston
18. Front spring
19. Retainer
20. Inner ring
21. Front clutch plate
22. Front clutch disc

23. Front clutch pressure plate
24. Front clutch outer ring
25. Rear clutch retainer
26. Input shaft
27. Thrust washer
28. Rear piston retainer
29. Outer seal
30. Rear clutch piston
31. Inner seal
32. Rear spring
33. Spacer
34. Wave spring
35. Front clutch pressure plate
36. Rear clutch plate
37. Rear clutch disc
38. Rear clutch pressure plate
39. Snap-ring
40. Front pinion carrier
41. Thrust washer
42. Front annulus gear
43. Thrust washer
44. Sun gear lock-ring

45. Sun gear shell
46. Sun gear driving plate
47. Thrust washer
48. Lock-ring
49. Sun gear
50. Rear pinion carrier
51. Thrust plate
52. Rear annulus gear
53. Reverse drum
54. Reverse band
55. Springs
56. Roller
57. Race
58. Clutch cam
59. Spring retainer
60. Washer
61. Ring
62. Roller
63. Pinion gear
64. Shaft

Component Removal and Overhaul

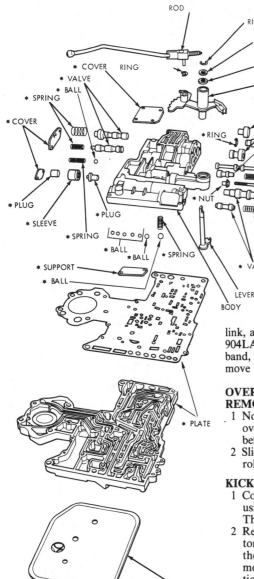

Exploded view of Torqueflite valve body

INPUT SHAFT AND REAR CLUTCH REMOVAL

1 Grasp the input shaft and slide the shaft and rear clutch assembly out of the case.

NOTE: Don't lose the thrust washer located between the rear end of the input shaft and the front end of the output shaft.

PLANETARY GEAR ASSEMBLIES, SUN GEAR AND DRIVING SHELL REMOVAL

1 While hand-supporting the output shaft and driving shell, carefully slide the assembly forward and out of the case.

REAR BAND AND LOW-REVERSE DRUM REMOVAL

Remove low-reverse drum, loosen rear band adjuster, remove band strut and link, and remove band from case. On A-904LA transmissions with double wrap band, loosen band adjusting screw and remove band and low-reverse drum.

OVERRUNNING CLUTCH REMOVAL

1 Notice the established position of the overrunning clutch rollers and springs before disassembly.

2 Slide out the clutch hub and remove rollers and springs.

KICKDOWN SERVO REMOVAL

1 Compress kickdown servo spring using engine valve spring compressor. Then remove snap ring.

2 Remove the rod guide, spring and piston rod from the case. Don't damage the piston rod or guide during removal. Tag the springs for identification.

3 Withdraw piston from the transmission case.

4 If equipped, disassemble the "Controlled Load" servo piston by removing the small snap-ring and washer from the piston. Remove the spring and piston rod.

LOW AND REVERSE SERVO REMOVAL

1 Using a suitable tool, depress the piston spring retainer and remove the snap ring.

2 Remove the spring retainer, spring, servo piston and plug assembly from the case.

FLUSHING THE TORQUE CONVERTER

1 The torque converter must be removed in order to flush it.

2 Place the converter in a horizontal position and pour two quarts of new clean solvent into the converter through the impeller hub.

3 Turn and shake the converter, swirling the solvent through the internal parts. Turn the turbine and stator with the input and reaction shafts to dislodge foreign material.

4 Position the converter in its normal position with the drain plug at its lowest point. Drain the solvent. At the same time, rotate the turbine and stator and shake the converter, to prevent dirt particles from settling.

5 Repeat the flushing operation at least once, until the solvent or kerosene is clear.

6 After flushing, shake and rotate the converter several times, with the drain plug removed, to drain out any residual solvent or dirt.

7 Flush any remaining solvent or dirt with two quarts of new transmission fluid. Tighten the drain plug.

8 Flush and blow out the oil cooler and lines.

Before removing any of the transmission sub-assemblies, thoroughly clean the exterior of the unit, preferably by steam. When disassembling, each part should be washed in a suitable solvent, and either set aside to drain or dried with compressed air. Do not wipe with shop towels. All of the transmission parts require extremely careful handling to avoid nicks and other damage to the accurately machined surfaces.

Component Overhaul

The following procedures cover the disassembly, inspection, repair and assembly of each sub-assembly as removed from the transmission.

The use of crocus cloth is permissible but not encouraged as extreme care must be used to avoid rounding off sharp edges of valves. The edge portion of valve body and valves is very important to proper functioning.

NOTE: Use all new seals and gaskets, and coat each part with automatic transmission fluid, type A, suffix A, during assembly.

VALVE BODY DISASSEMBLY—

1 Place the valve body on a clean repair stand. Never place any part of the valve body or transfer plate in a vise, since distortion can cause sticking valves and excessive leakage.

2 Remove the three screws from the fluid filter and remove the filter.

3 Remove the transfer plate retaining screws and 2 of the spring retainer mounting screws.

4 Lift off the transfer plate assembly. Separate the stiffener and separator plate for cleaning.

5 Remove the seven balls and spring from the valve body. Tag all springs as they are removed, for identification.

6 Turn the valve body over and remove the shuttle valve cover plate.

7 Remove the governor plug end plate and slide out the shuttle valve, throttle

AMC TORQUE COMMAND   CHRYSLER TORQUEFLITE

valve and spring, 1–2 shift valve governor plug and the 2–3 shift valve governor plug.

8 Remove the shuttle valve E-clip and remove the shuttle valve. If equipped, also remove the secondary spring and guides which are held by the clip.

9 Hold the spring retainer firmly against the spring and remove the last screw from the valve body.

10 Remove the spring retainer, line pressure adjusting screw (do not disturb the setting), and the line pressure and torque converter regulator springs.

11 Slide the torque converter and line pressure valves out of their bores.

12 Remove the E-clip and washer from the throttle lever shaft. Remove any burrs from the shaft, and, while holding the manual lever detent ball and spring in their bore, slide the manual lever from the throttle shaft. Remove detent ball and spring.

13 Slide the manual valve from its bore.

14 Remove the throttle lever stop screw assembly from the valve body and slide out the kickdown detent, kickdown valve, throttle valve spring and throttle valve.

15 Remove the line pressure regulator valve end plate and slide out the regulator valve sleeve, line pressure plug and throttle pressure plug. If equipped, remove end plate and downshift housing assembly.

16 Remove the throttle plug housing and slide the throttle plug out. If equipped, remove retainer, limit valve, and spring.

17 Remove the shift valve springs and slide both shift valves from their bores.

CLEANING AND INSPECTION

Inspect all components for scores, loose or bent levers, burrs and warping. Don't straighten bent levers; renew them. Loose levers may be silver soldered at the shaft. Burrs and minor nicks may be carefully removed with crocus cloth. Check for valve body warpage or distortion with a surface plate (plate glass will do) and a feeler gauge. Do not attempt to service a distorted plate or valve body, since this is a very critical area. Check all springs for distortion or fatigue. Check valves for scores and freedom of movement in the bores, they should fall of their own weight, in and out of the bore.

ASSEMBLY

1 Slide the shift valves and springs into the proper valve body bores.

2 If so equipped, assemble the downshifts housing. Insert the limit valve and spring into the housing. Slide the spring retainer into the groove.

3 Insert the throttle plug into the housing bore and install the housing on the valve body. Torque the screws to 28 in./lbs.

4 Install the throttle pressure plug, line

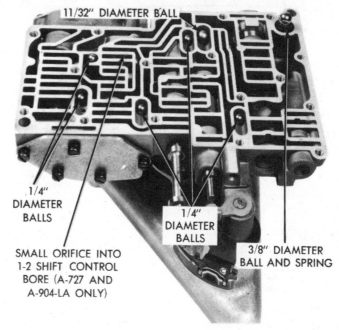

Ball check location

pressure plug and sleeve, fastening the end plate. Torque to 28 in./lbs.

5 Install the throttle valve, throttle valve spring, kickdown valve, kickdown detent and throttle lever stop screw with the locknut (do not adjust yet).

6 Slide the manual valve into its bore.

7 Install the throttle lever and shaft on the valve body. Insert the detent ball and spring into its bore in the valve body. Depress the ball and spring and slide the manual lever over the throttle shaft. Be sure that it engages the manual valve and detent ball. Install the seal, retaining washer and E-clip on the throttle shaft.

8 Insert the torque converter control valve and spring into the valve body.

9 Insert the line pressure regulator and spring into the valve body.

10 Install the line pressure adjusting screw assembly and spring retainer on the springs temporarily with one screw.

11 Place the 1–2 and 2–3 shift valve governor plugs in their proper bores.

12 Install the shuttle valve, spring, and shuttle valve throttle plug. If so equipped, install the secondary spring with two guides and clip on the other end.

13 Install the governor plug end plate and torque the screws to 28 in./lbs.

14 On those valve bodies not having a secondary spring, install the E-clip on the end of the shuttle valve.

15 Install the shuttle valve cover plate and torque screws to 28 in./lbs.

16 Install the spring and seven balls in the valve body. The seven include: five ¼ in. balls, one ⅜ in. diameter ball in the corner and one 11/32 in. diameter ball in the large chamber.

17 Place the separator plate on the transfer plate. Make sure all bolt holes are aligned and torque the two transfer plate screws and two stiffener plate screws to 28 in./lbs.

18 Place the transfer plate assembly on the valve body. Align the spring loaded ball as the 17 shorter screws are installed. Start at the center and work outward, tightening the screws to 35 in./lbs.

19 Install the oil filter and torque to 35 in./lbs.

20 Check spring engagement with the tang and adjusting nut. Install the remaining spring retainer screws. Check alignment and torque to 28 in./lbs.

21 After valve body has been serviced and completely assembled, adjust throttle and line pressures. If pressures were satisfactory prior to disassembly, use the original settings.

ACCUMULATOR PISTON AND SPRING

Inspect both seal rings for wear and freedom in the piston grooves. Check the piston for scores, burrs, nicks and wear. Check the piston bore for corresponding damage and check piston spring for distortion and fatigue. Replace parts as required.

GOVERNOR

1 Carefully remove the snap-ring from weight end of governor valve shaft and pull out valve and shaft. Remove the large snap-ring from the weight end of governor housing and lift out the governor weight assembly.

2 Remove the snap-ring from inside the governor weight, remove the inner weight and spring from the outer weight.

Component Removal and Overhaul

NOTE: Thoroughly clean all parts in a suitable and clean solvent. Check for damage and free movement before assembly.

3 If lugs on support gear are damaged, remove four bolts and separate support from governor body.

ASSEMBLY

1 If support was separated from governor body, assemble and tighten bolts finger-tight. Make sure the oil passage of the body aligns with passage in the support. Position support and governor on output shaft so that the valve shaft hole in the governor aligns with the hole in the output shaft. Install a snap-ring behind the governor body and tighten the bolts to 100 in./lbs.
2 Assemble the governor weights and spring, then secure with snap-ring inside large governor weight.
3 Place the weight assembly in the governor housing and install snap-ring.

FRONT PUMP AND REACTION SHAFT SUPPORT

1 Remove bolts from rear side of reaction shaft support and lift support from the pump.
2 Dye-spot the face of the inner and outer rotors so they may be reinstalled in their original relationship, then remove the rotors.
3 Remove the rubber seal ring from front pump body flange.
4 Drive out the oil seal with blunt punch.

INSPECTION

Clean and inspect interlocking seal rings on the reaction shaft support for wear or broken interlocks, be sure they turn freely in their grooves. Check all machined surfaces of pump body and reaction shaft support for scuff marks and burrs. Inspect pump rotors for scores and pits. With rotors clean and installed into the pump body, apply a straightedge across the face of the rotors and pump body. With a feeler gauge, check straightedge to rotor face clearance. Clearance should be 0.005 to 0.010 inch.

ASSEMBLY

1 Place reaction shaft support in assembling tool and place it on the bench, with the support hub resting on the bench. Screw two pilot studs, or satisfactory substitutes, into threaded holes of reaction shaft support flange.
2 Assemble rotors with dye marks aligned, place rotors in center of the support. The two driving lugs inside rotor must be next to the face of the reaction shaft support.
3 Lower pump body over pilot studs, insert an aligning shaft through pump body and engage pump inner rotor. Turn the rotors, with the tool, to enter them into pump body. With the pump body firmly against the reaction shaft

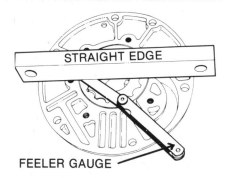

Pump rotor end clearance 0.005 to 0.010 inch

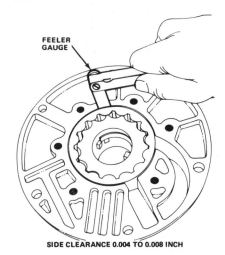

SIDE CLEARANCE 0.004 TO 0.008 INCH

Side clearance 0.004 to 0.008 inch

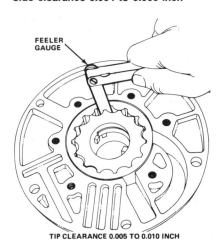

TIP CLEARANCE 0.005 TO 0.010 INCH

Tip clearance 0.005 to 0.010 inch

support, tighten ring squeezer or clamping tool securely.
4 Invert the front pump and reaction shaft support assembly, with the clamping tool intact. If so equipped, position the vent baffle over the vent opening. Install support to pump body bolts. Remove clamping tool, pilot studs and rotor aligning studs.
5 Insert new oil seal into opening of front oil pump housing (with lip of

seal facing inward). Drive seal into housing.

FRONT CLUTCH

DISASSEMBLY

1 With screwdriver or pick, remove large snap-ring, which holds the pressure plate in the clutch piston retainer. Lift pressure plate and clutch plates out of the retainer.
2 Install compressor, or similar tool, over piston spring retainer. Compress spring and remove snap-ring, then slowly release tool until the spring retainer is free of the hub. Remove the compressor, retainer and spring.
3 Turn the clutch retainer upside down and bump on a wooden block to remove the piston. Remove seal rings from the piston and clutch retainer hub.

INSPECTION

Inspect clutch discs for evidence of burning, glazing and flaking. A general method of determining clutch plate breakdown is to scratch the lined surface of the plate with a finger nail. If material collects under the nail, replace all driving discs. Check driving splines for wear or burrs. Inspect steel plates and pressure plate surfaces for discoloration, scuffing or damaged driving lugs. Replace if necessary.

Check steel plate lug grooves in clutch retainer for smooth surfaces. Plate travel must be free. Inspect band contacting surface of clutch retainer, being sure the ball moves freely. Check seal ring surfaces in clutch retainer for scratches or nicks,

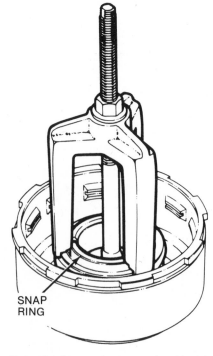

SNAP RING

Front clutch snap ring removal and installation

AMC TORQUE COMMAND   CHRYSLER TORQUEFLITE

light annular scratches will not interfere with the sealing of neoprene rings.

Inspect inside bore of piston for score marks. If light marks exist, polish with crocus cloth. Check seal ring grooves for nicks and burrs. Inspect neoprene seal rings for deterioration, wear and hardness. Check piston spring, retainer, and snap-ring for distortion and fatigue.

ASSEMBLY

1 Lubricate and install inner seal ring onto hub of clutch retainer. Be sure that lip of seal faces down and is properly seated in the groove.
2 Lubricate and install outer seal ring onto clutch piston, with lip of seal toward the bottom of the clutch retainer. Place piston assembly in retainer and, with a twisting motion, seat the piston in the bottom of the retainer.
3 Place spring on the piston hub and position spring retainer and snap-ring on spring. Compress spring with tool, or suitable ring compressor, and seat snap-ring in the hub groove. Remove compressor.
4 Lubricate all clutch plates, then, install a steel plate, followed by a lined plate, until all plates are installed. Install the pressure plate and snap-ring. Be sure the snap-ring is correctly seated.
5 With front clutch assembled, insert a feeler gauge between the pressure plate and snap-ring. The clearance should be to specification. If not, install a snap-ring of proper thickness.

REAR CLUTCH—

DISASSEMBLY

1 With a small screwdriver or pick, remove the large snap-ring that secures the pressure plate in the clutch piston retainer. Lift the pressure plate, clutch plates, and inner pressure plate from the retainer.
2 Carefully pry one end of wave spring out of its groove in clutch retainer and remove wave spring, spacer ring (if so equipped) and clutch piston spring.
3 Remove compressor tool and piston spring. Turn clutch retainer assembly upside down and bump on a wood block to remove the piston. Remove seal rings from the piston.
4 If necessary, remove the snap-ring and press input shaft from the piston retainer.

INSPECTION

Inspect driving discs for indication of damage; handle as previously outlined under front clutch inspection.

ASSEMBLY

1 If removed, press input shaft into the piston retainer and install the snap-ring.
2 Lubricate, then install inner and outer seal rings onto the clutch piston. Be sure the seal lips face toward the head of the clutch retainer and seals are properly seated in the piston grooves.
3 Place piston assembly in retainer and, with a twisting motion, seat piston in bottom of retainer.
4 On the A-727 models, place the clutch retainer over the piston retainer and support so that the clutch retainer remains in place. Place the clutch piston spring and spacer ring on top of the piston in clutch retainer. Make sure the spring and spacer ring are placed in the retainer recess. Start one end of the wave spring in the retainer groove. Progressively push or tap the spring into place making sure it is fully seated in its groove.
5 Install inner pressure plate into clutch

retainer, with raised portion of plate resting on the spring.
6 Lubricate all clutch plates, then install one lined plate, followed by a steel plate, until all plates are installed. Install outer pressure plate and snap-ring.
7 With rear clutch completely assembled, insert a feeler gauge between the pressure plate and snap-ring. The clearance should be to specification. If not, install snap-ring of proper thickness to obtain the required clearance.

NOTE: Rear clutch plate clearance is very important to obtaining satisfactory clutch performance. Clearance is influenced by the use of various thickness outer snap-rings.

PLANETARY GEAR TRAIN—

DISASSEMBLY

Refer to illustrations for assembly and disassembly of these units.

1 Remove thrust washer from forward end of output shaft.
2 Remove snap-ring from forward end of output shaft, then, slide front planetary assembly from the shaft.
3 On A-904 transmissions, remove snap ring and thrust washer from forward hub of planetary gear assembly. Slide front annulus gear and support off planetary gear set. Remove thrust washer from rear side of planetary gear set. If necessary, remove snap ring from front of annulus gear to separate support from annulus gear. On A-727 transmissions, slide front annulus gear off planetary gear set. Remove thrust washer from rear side of planetary gear set.
4 Slide the sun gear, driving shell, and rear planetary assembly, with low and reverse drum, from the output shaft.
5 Remove sun gear and driving shell from the rear planetary assembly. On

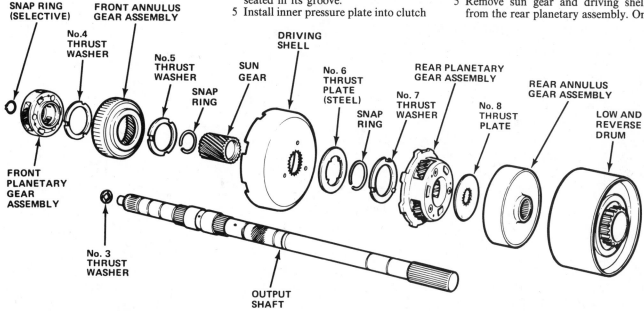

SNAP RING (SELECTIVE)   FRONT ANNULUS GEAR ASSEMBLY   No.4 THRUST WASHER   No.5 THRUST WASHER   SNAP RING   SUN GEAR   DRIVING SHELL   No. 6 THRUST PLATE (STEEL)   SNAP RING   No. 7 THRUST WASHER   REAR PLANETARY GEAR ASSEMBLY   No. 8 THRUST PLATE   REAR ANNULUS GEAR ASSEMBLY   LOW AND REVERSE DRUM   FRONT PLANETARY GEAR ASSEMBLY   No. 3 THRUST WASHER   OUTPUT SHAFT

Planetary assembly-model 727

Component Removal and Overhaul

AMC TORQUE COMMAND   CHRYSLER TORQUEFLITE

A-727 transmissions, remove thrust washer from inside driving shell. On all transmissions, remove snap-ring and steel washer from sun gear (rear side of driving shell). Slide sun gear out of driving shell, then remove snap-ring and steel washer from opposite end of sun gear, if necessary.

6 Remove thrust washer from forward side of rear planetary assembly. Remove snap-ring from front side of low and reverse drum, then slide rear planetary assembly out of drum. If necessary, remove snap-ring from rear of annulus gear in order to separate the support from the annulus gear.

INSPECTION

Inspect output shaft bearing surfaces for burrs or other damage. Light scratches or burrs may be polished out with crocus cloth or a fine stone. Check speedometer drive gear for damage, and make sure all oil passages are clear.

Check bushings in the sun gear for wear or scores. Replace sun gear assembly if bushings show wear or other damage. Inspect all thrust washers for wear and scores. Replace if necessary. Check lockrings for distortion and fatigue. Inspect annulus gear and driving gear teeth for damage. Inspect planetary gear carrier for cracks and the pinions for broken or worn gear teeth.

ASSEMBLY—A-904

1 Locate the rear annulus gear support

in the annulus gear and install snap-ring.

2 Position the rear planetary gear assembly in the rear annulus gear, then slide the assembly into low and reverse drum. Put thrust washer on the front side of the planetary gear assembly.

3 Insert output shaft into the rear opening of the drum. Carefully work the shaft through the annulus gear support and planetary gear assembly. Make sure the shaft splines are fully engaged in the splines of the annulus gear support.

4 Install steel washer and snap-ring onto the shortest end of the sun gear. Insert sun gear through front side of the driving shell, then install rear steel washer and snap-ring. (The longer end of the sun gear must be toward the rear, extending from the driving shell).

5 Carefully slide the driving shell and sun gear assembly onto the output shaft, engaging the sun gear teeth with the planetary pinion teeth.

6 Place front annulus gear support in annulus gear and install snap-ring.

7 Position front planetary gear assembly in front annulus gear, place thrust washer over planetary gear assembly hub and install snap-ring. Position thrust washer on rear side of planetary gear assembly.

8 Carefully work the front planetary and annulus gear assembly onto the output shaft, meshing the planetary pinions with the sun gear teeth.

9 With all components properly assembled, install snap-ring onto the front end of the output shaft. Clearance can be adjusted through the use of various thickness snap-rings, listed at front of section.

ASSEMBLY—A-727

1 Install rear annulus gear on output shaft. Apply thin coat of grease on thrust plate, place it on shaft, and in annulus gear making sure teeth are over shaft splines.

2 Position rear planetary gear assembly in rear annulus gear. Place thrust washer on front side of planetary gear assembly.

3 Install snap-ring in front groove of sun gear (long end of gear). Insert sun gear through front side of driving shell. Install rear steel washer and snap-ring.

4 Carefully slide driving shell and sun gear assembly on output shaft, engaging sun gear teeth with rear planetary pinion teeth. Place thrust washer inside front driving shell.

5 Place thrust washer on rear hub of front planetary gear set. Slide assembly into front annulus gear.

6 Carefully work front planetary and annulus gear assembly on output shaft, meshing planetary pinions with sun gear teeth.

7 With all components properly positioned, install selective snap-ring on front end of output shaft. Measure end-play of assembly. Adjust end-play with selective snap rings.

OVERRUNNING CLUTCH INSPECTION

Inspect clutch rollers for smooth round surfaces, they must be free of flat spots, chipped edges and flaking. Inspect roller contacting surfaces on both cam and race for pock marks and roller wear-marks. Check springs for distortion and fatigue and inspect low and reverse drum thrust. On A-727 transmissions, inspect cam set screw for tightness. If loose, tighten and restake the case.

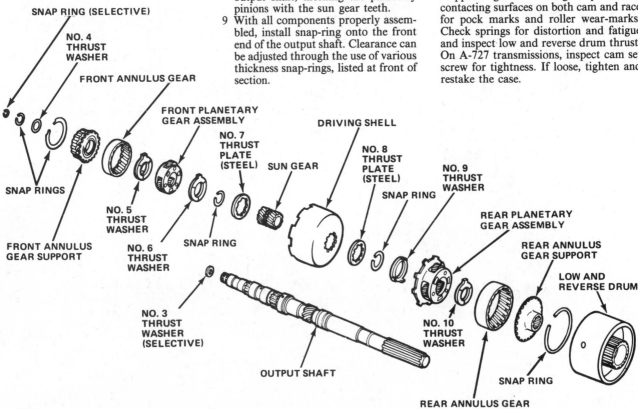

Planetary assembly-models 904–998

AMC TORQUE COMMAND   CHRYSLER TORQUEFLITE

KICKDOWN SERVO AND BAND INSPECTION

Inspect piston and guide seal rings for wear, and be sure of their freedom in grooves. It is not necessary to remove seal rings, unless circumstances warrant. Inspect piston for scores, burrs or other damage. Check fit of guide on piston rod. Check piston for distortion and fatigue. Inspect band lining for wear and fit of lining material to the metal band. This lining is grooved; if grooves are not still visible at the ends or any part of the band, replace the band. Inspect band for distortion or cracked ends.

LOW AND REVERSE SERVO AND BAND

DISASSEMBLY

Remove snap-ring from piston and remove the piston plug and spring.

INSPECTION

Inspect neoprene seal ring for damage, rot, or hardness. Check piston and piston plug for nicks, burrs, scores and wear. The piston plug must operate freely in the piston. Check the piston bore in the case for scores or other damage. Examine springs for distortion and fatigue.

Check band lining for wear and the fit of the lining to the metal band. This lining has a grooved surface; if the grooves are worn away at the ends or at any part of the band, replace the band. Inspect the band for distortion or cracked ends.

ASSEMBLY

Lubricate and insert the piston plug and spring into the piston, and secure with the snap-ring.

Component Installation

The following assembly procedures include the installation of sub-assemblies into the transmission case and adjustment of the drive train end-play. Do not use force to assemble any of the mating parts. Always use new gaskets during the assembly operations.

NOTE: Use only automatic transmission fluid, type A, suffix A, or fluid of equivalent chemical structure, to lubricate automatic transmission parts during, or after, assembly.

OVERRUNNING CLUTCH

With transmission case in upright position, insert clutch race inside cam. Install overrunning clutch rollers and springs as shown in figure.

LOW AND REVERSE SERVO AND BAND

1 Carefully work servo piston assembly into the case with a twisting motion. Place spring, retainer and snap-ring over the piston.
2 Using a valve spring compressor, compress the spring and install the snap-ring.

3 Position rear band in the case, install the short strut, then connect the long lever and strut to the band. Screw in band adjuster just enough to hold struts in place. Install low-reverse drum. On A-727 transmissions, be sure long link and anchor assembly is installed to provide running clearance for low-reverse drum.

LOW-REVERSE BAND A-904-LA V8 ONLY

This transmission has a double wrap band supported at two points by a band reaction pin in case and acted upon at one point by a servo lever adjusting screw.

1 Push band reaction pin with new O-ring into case flush with gasket surface.

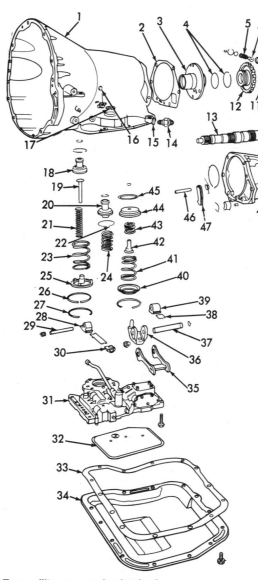

Torqueflite case and valve body

1. Transmission case
2. Gasket
3. Governor Support
4. Seal rings
5. Governor weight springs
6. Inner weights
7. Outer weights
8. Output shaft bearing
9. Governor valve shaft
10. Governor valve
11. Governor repair kit
12. Governor support
13. Output shaft
14. Neutral safety switch
15. Washer
16. Band adjusting screw
17. Seal
18. Kickdown piston
19. Pitson rod
20. Accumulator piston
21. Spring
22. Large ring
23. Spring
24. Accumulator piston spring
25. Kickdown guide
26. Seal ring
27. Snap-ring
28. Kickdown band lever
29. Kickdown lever shaft
30. Kickdown band anchor
31. Valve body and plate
32. Filter
33. Pan gasket
34. Oil pan
35. Reverse band link and anchor
36. Reverse band lever and stem
37. Reverse lever shaft
38. Reverse band lever srtut
39. Lever with short stem
40. Reverse servo retainer
41. Reverse servo piston spring
42. Reverse servo plug
43. Reverse servo cushion spring
44. Reverse piston
45. Reverse servo piston seal
46. Park sprag shaft
47. Park lock sprag
48. Speedometer pinion
49. Speedometer outer seal
50. Speedometer adaptor
51. Inner seal
52. Extension housing
53. Extension houing bushing
54. Extension housing seal

Component Removal and Overhaul

2 Place band in case resting two lugs against band reaction pin.

3 Install low-reverse drum into overrunning clutch and band.

4 Install operating lever with pivot pin flush in case and adjusting screw touching center lug on band.

KICKDOWN SERVO

1 If equipped with a controlled load servo piston, sub-assemble the unit as follows: grease the O ring and install on the piston rod; install the piston rod into the servo piston; install the spring, flat washer and snap-ring.

2 Carefully insert servo piston into case bore. Install piston rod, two springs and guide. Depress guide and install snap-ring.

NOTE: A-904 transmissions and maximum performance A-727 transmissions use only one small spring.

PLANETARY GEAR ASSEMBLIES, SUN GEAR, DRIVING SHELL, LOW AND REVERSE DRUM

1 While supporting the assembly in the case, insert the output shaft through the rear support. Carefully work the assembly rearward, engaging the carrier lugs with low-reverse drum slots.

CHILTON CAUTION: *Be careful not to damage the ground surfaces of the output shaft during installation.*

2 Apply a daub of grease to the selective thrust washer and install washer on the front end of the output shaft.

NOTE: If the drive train end-play was not within specifications when checked (Checking Drive Train End-Play), replace the thrust washer with one of proper thickness.

FRONT AND REAR CLUTCH

1 The following method may be used to support the transmission; cut a 3½" hole in a bench, small drum or box, strong enough to support the transmission; file notches at the edge so the output shaft support will lie flat; insert the output shaft into the hole and support the transmission upright.

2 On A-904 transmissions, apply a coat of grease to the selective thrust washer and install the washer on the front end of the output shaft. If drive train end-play was not correct before disassembly, replace the thrust washer with one of suitable thickness.

3 On A-727 transmissions, apply a coat of grease to the input shaft to output shaft thrust washer and install the thrust washer on the front end of the output shaft.

4 Align the front clutch plate inner splines, and place the assembly in position on the rear clutch. Be sure the clutch plate splines are fully engaged.

5 Align the rear clutch plate inner splines and lower the two clutch assemblies in to the case.

6 Carefully, work the clutch assemblies in a circular motion to engage the rear clutch splines over the front annulus gear splines. Make sure the front clutch drive lugs are fully engaged in the driving shell.

FRONT BAND

1 Slide the band over the front clutch assembly.

2 Install band strut, screw in the adjuster just enough to hold the band in place.

FRONT OIL PUMP AND REACTION SHAFT SUPPORT

1 On A-904 transmissions, install the thrust washer on the reaction shaft support hub. On A-727 transmissions, if drive train end-play was not within specifications, replace the thrust washer on the reaction shaft support hub, with one of proper thickness (see specifications).

2 Screw (two) pilot studs into front pump opening in the case.

3 Place a new rubber seal ring in groove

Specifications

Type—three speed with torque converter	A-904	A-727
Torque converter diameter		
standard	10¾ in.	11¾ in.
high performance	—	10¾ in.
Converter oil capacity		
std.	17 pts	19 pts
Cooling method	Water	Water
Lubrication pump	Rotor type	Rotor type
Number of clutch plates and discs		
Pump Clearances		
End Clearance	0.0015-0.003	0.0015-0.003
Outer rotor tip to inner rotor tip measured diametrically	0.005-0.010	0.005-0.010
Planetary Assembly end play		
All	0.006-0.033	0.010-0.037
Gear train end play		
All	0.030-0.089	0.036-0.084
Clutch Plate clearance		
front 3 disc	0.077-0.122	0.076-0.123
front 4 disc	0.089-0.144	0.088-0.145
front 5 disc	—	0.090-0.122
rear 3 disc	0.026-0.055	—
rear 4 disc	0.032-0.055	—
rear 4 disc	—	0.025-0.045
Snap Rings, Front and Rear Clutches		
Rear (selective)	0.060-0.062	0.060-0.062
	0.068-0.070	0.074-0.076
	0.076-0.078	0.088-0.090
	—	0.106-0.108
Output Shaft (forward end)	0.040-0.044	0.048-0.052
	0.048-0.052	0.055-0.059
	0.059-0.065	0.062-0.066
Thrust Washers		
Reaction shaft support to front clutch retainer	#1 0.061-0.063	#1 Selective
(Natural)		0.061-0.063
(Red)		0.084-0.086
(Yellow)		0.102-0.104
Front clutch to Rear clutch (input shaft to output shaft)	#2 0.061-0.063	#2 0.061-0.063
	#3 Selective	#3 0.062-0.064
(Natural)	0.052-0.054	
(Red)	0.068-0.070	
(Black)	0.083-0.085	
Front annulus support to front carrier	#4 0.121-0.125	#4 0.060-0.062
Front carrier to driving shell thrust plate	#5 0.048-0.050	
Front annulus support to driving shell		#5 0.060-0.062
Driving shell thrust plate	#6 0.034-0.036	#6 0.034-0.036
	#7 0.034-0.036	
Rear carrier to driving shell	#8 0.048-0.050	#7 0.060-0.062
Rear annulus thrust plate		#8 0.034-0.036

AMC TORQUE COMMAND   CHRYSLER TORQUEFLITE

on outer flange of pump. Be sure the seal ring is not twisted. Two small holes in the handle of the tool are vertical. This is to locate the inner rotor so the converter impeller shaft will engage the inner rotor lugs during installation.

4 Install the assembly into the case, tap lightly with a soft mallet if necessary. Install four bolts, remove pilot studs, install remaining bolts and pull down evenly, then, torque the bolts to specification.

5 Rotate the pump rotors until the

GOVERNOR

1 Place the governor and support on the output shaft. Position it so that the governor valve shaft hole aligns with

the hole in the output shaft, then slide the assembly into place. Install snap-ring behind the governor housing. Torque housing-to-support screws to specification.

2 Place the governor valve on the valve shaft, insert the assembly into the housing and through the governor weights. Install the valve shaft retaining snap-ring.

EXTENSION HOUSING

1 Install snap ring in front groove on output shaft.

2 Install bearing on shaft with its outer race ring groove toward front. Press or tap bearing tight against front snap ring.

3 Install rear snap ring. If so equipped,

slide the yoke seal front stop ring onto the shaft. Install the seal with lips toward the rear, then install the rear stop ring on the shaft.

4 Place new extension housing gasket on transmission case.

5 Place output shaft bearing retaining snap-ring in extension housing. Spread ring as far as possible, then carefully tap extension housing into place. Make sure snap-ring is fully seated in bearing groove.

6 Install and torque extension housing bolts to specification.

7 Install gasket, plate, and two screws on bottom of extension housing mounting pad.

8 Install speedometer pinion and adapter assembly.

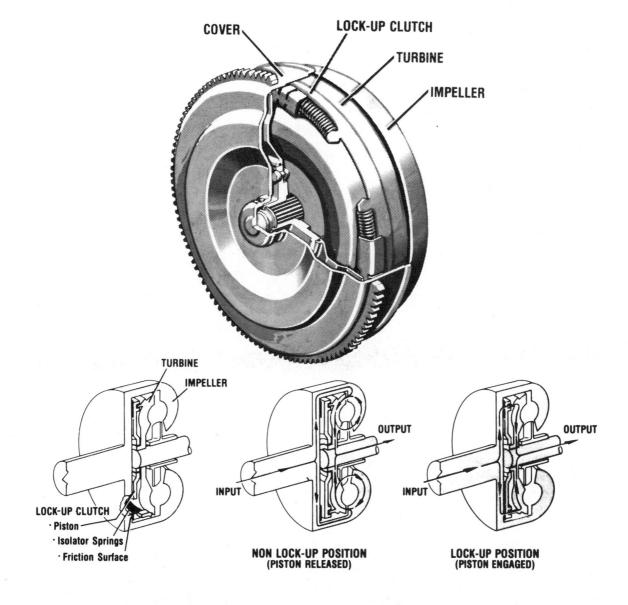

The lock-up clutch type torque converter in some 1978 and later Chrysler Corporation vehicles engages automatically at 28 mph providing mechanical rather than hydraulic power transmission

CHRYSLER A-404 TORQUEFLITE AUTOMATIC TRANSAXLE

TROUBLESHOOTING CHRYSLER A-404 TORQUEFLITE AUTOMATIC TRANSAXLE

POSSIBLE CAUSE		HARSH ENGAGEMENT FROM NEUTRAL TO D OR R	DELAYED ENGAGEMENT FROM NEUTRAL TO D OR R	RUNAWAY UPSHIFT	NO UPSHIFT	3-2 KICKDOWN RUNAWAY	NO KICKDOWN OR NORMAL DOWNSHIFT	SHIFTS ERRATIC	SLIPS IN FORWARD DRIVE POSITIONS	SLIPS IN REVERSE ONLY	SLIPS IN ALL POSITIONS	NO DRIVE IN ANY POSITION	NO DRIVE IN FORWARD DRIVE POSITIONS	NO DRIVE IN REVERSE	DRIVES IN NEUTRAL	DRAGS OR LOCKS	GRATING, SCRAPING GROWLING NOISE	BUZZING NOISE	HARD TO FILL, OIL BLOWS OUT FILLER TUBE	TRANSMISSION OVERHEATS	HARSH UPSHIFT	DELAYED UPSHIFT
Overrunning clutch inner race damaged.	34																	×				
Overrunning clutch worn, broken or seized.	33								×				×			×	×					
Planetary gear sets broken or seized.	32											×	×	×		×	×					
Rear clutch dragging.	31														×							
Worn or faulty rear clutch.	30	×	×						×				×	×	×							
Insufficient clutch plate clearance.	29														×					×		
Faulty cooling system.	28																			×		
Kickdown band adjustment too tight.	27															×				×		
Hydraulic pressure too high.	26	×																			×	
Breather clogged.	25																		×			
High fluid level.	24																		×			
Worn or faulty front clutch.	23		×	×	×	×		×		×				×								×
Kickdown servo band or linkage malfunction.	22			×	×	×	×	×														×
Governor malfunction.	21				×		×	×														×
Work or broken reaction shaft support seal rings.	20		×	×	×	×		×		×				×								×
Governor support seal rings broken or worn.	19				×			×														×
Output shaft bearing and/or bushing damaged.	18																×					
Overrunning clutch not holding.	17								×				×									
Kickdown band out of adjustment.	16					×											×				×	×
Incorrect throttle linkage adjustment.	15			×	×	×	×	×	×												×	×
Engine idle speed too low.	14		×																			
Aerated fluid.	13		×	×		×		×	×	×	×							×	×			
Worn or broken input shaft seal rings.	12		×						×		×		×									
Faulty oil pump.	11		×					×	×	×	×	×								×		
Oil filter clogged.	10		×	×				×	×		×	×							×			
Incorrect gearshift control linkage adjustment.	9		×		×			×	×	×				×	×					×		
Low fluid level.	8		×	×	×	×		×	×	×	×	×	×					×		×		
Low-reverse servo, band or linkage malfunction.	7		×							×				×								
Valve body malfunction or leakage.	6	×	×	×	×	×	×	×	×	×	×	×	×	×	×			×				
Low-reverse band out of adjustment.	5									×				×		×	×					
Hydraulic pressures too low.	4		×	×	×	×		×	×	×	×	×	×	×						×	×	
Engine idle speed too high.	3	×																		×		
Stuck switch valve.	1																			×		

CONDITION

CHRYSLER A-404 TORQUEFLITE AUTOMATIC TRANSAXLE

Chrysler A-404 Torqueflite Automatic Transaxle

The transaxle combines a torque converter, fully automatic three speed transmission, final drive gearing and differential into a single assembly. The unit is a metric design.

Application

Horizon and Omni

In Car Testing Procedures

Hydraulic Pressure Tests

NOTE: For hydraulic pressure tests, install a tachometer, raise the vehicle to allow the front wheels to turn and disconnect the throttle and shift cable from transmission levers so they can be controlled from outside the vehicle. Connect 150 psi gauges to required ports. A 300 psi gauge is needed for the "reverse" pressure test at rear servo.

TEST ONE

This tests pump output, pressure regulation, and condition of rear clutch and rear servo hydraulic circuits.
1 Attach gauges to "line" and "rear servo" ports, place selector lever on transaxle all the way forward ("1" position) and operate the engine at 1000 rpm.
2 Read pressures on both gauges as throttle lever on transaxle is moved from full forward to full rearward positions.
3 Line pressure should read 60 to 66 psi with throttle lever forward and gradually increase to 97 to 103 psi as lever is moved rearward. Rear servo pressure should be within 3 psi of line pressure.

TEST TWO

This tests pump output, pressure regulation, and condition of rear clutch and lubrication hydraulic circuits.
1 Connect gauge to "line" pressure port, and "tee" into lower cooler line fitting to read lubrication pressure. Place transaxle selector lever one detent rearward from full forward ("2" position), and operate engine at 1000 rpm.
2 Read pressures on both gauges as throttle lever on transaxle is moved from full forward to full rearward positions.
3 Line pressure should be 60 to 66 psi with throttle lever forward and gradually increase to 97 to 103 psi as the lever is moved rearward. Lubrication pressure should be 10 to 25 psi with lever forward and 10 to 35 psi with lever rearward.

TEST THREE

This tests pump output, pressure regulation and condition of rear clutch, front clutch and hydraulic circuits.
1 Connect gauges to "line" and "front servo release" ports. Place transaxle selector lever two detents rearward from full forward position ("D" position), and operate engine at 1600 rpm.
2 Read pressure on both gauges as throttle lever on transaxle is moved from full forward to full rearward positions.
3 Line pressure should be 60 to 66 psi with throttle lever forward and gradually increase as lever is moved rearward. Front servo release is pressurized only in direct drive and should be within 3 psi of line pressure up to downshift point.

TEST FOUR

This tests pump output, pressure regulation and condition of front clutch and rear servo hydraulic circuits. It also tests for leakage into rear servo due to case porosity.
1 Connect a 300 psi gauge to "rear servo apply" port. Place transaxle selector lever in "R" position (four detents rearward from full forward), and operate engine at 1600 rpm.
2 Rear servo pressure should be 176 to 180 psi with throttle lever forward and gradually increase to 270 to 280 psi as the lever is moved rearward. Move transaxle selector lever to "D" position, and check that rear servo pressure drops to zero.

TEST RESULT INDICATIONS

1 If proper line pressure, minimum to maximum, is found in any one test, the pump and pressure regulator are working properly.

2 Low pressure in "D," "1," and "2," but correct pressure in "R," indicates rear clutch circuit leakage.
3 Low pressure in "D" and "R," but correct pressure in "1," indicates front clutch circuit leakage.
4 Low pressure in "R" and "1," but correct pressure in "2," indicates rear servo circuit leakage.
5 Low line pressure in all positions indicates a defective pump, clogged filter or a stuck pressure regulator valve.

GOVERNOR PRESSURE

Test only if transaxle shifts at wrong vehicle speeds when throttle cable is correctly adjusted.
1 Connect a 0–150 psi gauge to governor pressure take off point located at the lower right side of the case below differential cover.
2 Operate transaxle in third gear to read pressure and compare speeds shown in chart.
3 If governor pressures are incorrect at the given vehicle speeds, the governor valve and/or weights are probably sticking. The governor pressure should respond smoothly to changes in mph and should return to 0 to 3 psi when vehicle is stopped. Pressure over 3 psi at standstill will prevent downshifting.

THROTTLE PRESSURE

1 No gauge port is provided for throttle pressure. Incorrect throttle pressure should only be suspected if part throttle upshift speeds are either delayed or occur too early. Engine runaway on upshift or downshift can indicate a too low throttle pressure setting.
2 Verify that throttle cable adjustment is correct before adjusting throttle pressure.

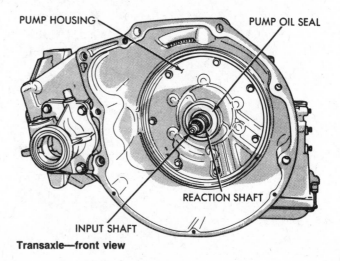

Transaxle—front view

PUMP HOUSING
PUMP OIL SEAL
REACTION SHAFT
INPUT SHAFT

CHRYSLER A-404 TORQUEFLITE AUTOMATIC TRANSAXLE

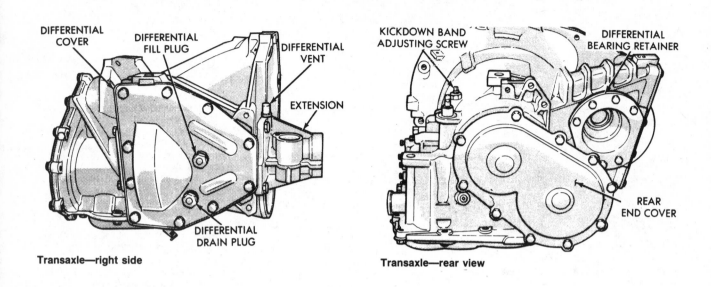

Transaxle—right side

Transaxle—rear view

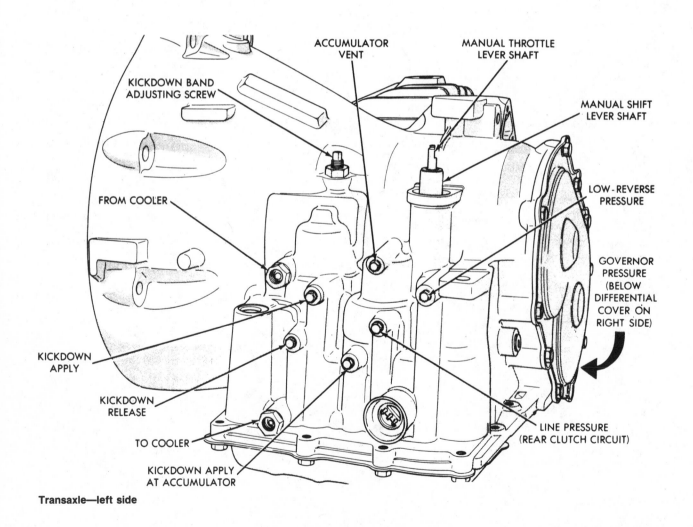

Transaxle—left side

CHRYSLER A-404 TORQUEFLITE AUTOMATIC TRANSAXLE

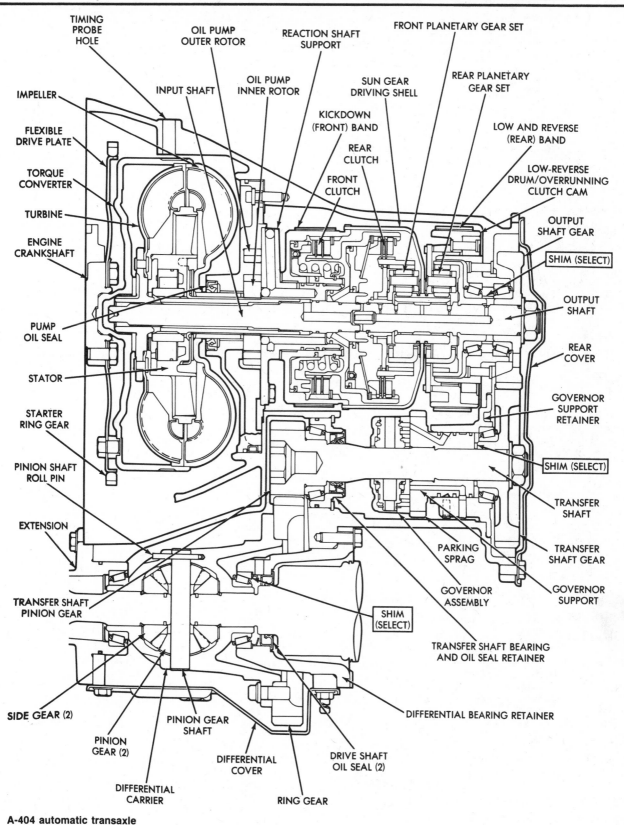

TIMING PROBE HOLE

OIL PUMP OUTER ROTOR

REACTION SHAFT SUPPORT

FRONT PLANETARY GEAR SET

IMPELLER

INPUT SHAFT

OIL PUMP INNER ROTOR

SUN GEAR DRIVING SHELL

REAR PLANETARY GEAR SET

FLEXIBLE DRIVE PLATE

KICKDOWN (FRONT) BAND

LOW AND REVERSE (REAR) BAND

TORQUE CONVERTER

REAR CLUTCH

FRONT CLUTCH

LOW-REVERSE DRUM/OVERRUNNING CLUTCH CAM

TURBINE

OUTPUT SHAFT GEAR

ENGINE CRANKSHAFT

SHIM (SELECT)

OUTPUT SHAFT

PUMP OIL SEAL

REAR COVER

STATOR

GOVERNOR SUPPORT RETAINER

STARTER RING GEAR

SHIM (SELECT)

PINION SHAFT ROLL PIN

TRANSFER SHAFT

EXTENSION

TRANSFER SHAFT GEAR

TRANSFER SHAFT PINION GEAR

PARKING SPRAG

GOVERNOR ASSEMBLY

GOVERNOR SUPPORT

SIDE GEAR (2)

SHIM (SELECT)

TRANSFER SHAFT BEARING AND OIL SEAL RETAINER

DIFFERENTIAL BEARING RETAINER

PINION GEAR (2)

PINION GEAR SHAFT

DIFFERENTIAL CARRIER

DIFFERENTIAL COVER

DRIVE SHAFT OIL SEAL (2)

RING GEAR

A-404 automatic transaxle

CHRYSLER A-404 TORQUEFLITE AUTOMATIC TRANSAXLE

AUTOMATIC SHIFT SPEEDS AND GOVERNOR PRESSURE CHART
(APPROXIMATE MILES PER HOUR)

Carline	Federal MZ	California MZ
Engine (Litre)	1.7	1.7
Axle Ratio	3.48	3.74
Throttle Minimum		
1-2 Upshift	8-15	7-14
2-3 Upshift	11-21	10-19
3-1 Downshift	8-15	7-14
Throttle Wide Open		
1-2 Upshift	32-44	38-44
2-3 Upshift	52-65	48-60
Kickdown Limit		
3-2 WOT Downshift	48-61	44-57
3-2 Part Throttle Downshift	35-50	32-46
3-1 WOT Downshift	30-57	28-34
Governor Pressure*		
15 psi	21-23	19-21
40 psi	35-41	32-38
60 psi	50-56	46-52

*Governor pressure should be from zero to 3 psi at stand still or downshift may not occur.

NOTE: Changes in tire size will cause shift points to occur at corresponding higher or lower vehicle speeds.

Clutch and Servo Air Pressure Tests

The front and rear clutches, kickdown servo and low-reverse servo may be tested by applying air pressure after the valve body has been removed.

NOTE: Compressed air must be free of dirt and moisture. Use a pressure of 30 psi.

CLUTCHES

1 Apply air pressure to each individual clutch "apply" passage and listen for a dull thud which indicates that the clutch is operating. If the dull thud cannot be heard in the clutches, place finger tips on clutch housing to feel piston movement.
2 Inspect each clutch system for oil leaks while air pressure is applied.

SERVOS

1 Apply air pressure to each individual servo "apply" passage. Operation of kickdown servo is indicated by a tightening of the front band. Operation of the low-reverse servo is indicated by a tightening of the rear band. Spring tension on servo pistons should release the bands.

In-Car Service and Adjustment Procedures

Band Adjustment

KICKDOWN BAND (FRONT)

1 The adjusting screw is located on the left side (top front) of the transmission case. Loosen lock nut, and back off nut approximately five turns. Test adjusting screw for free turning in the transmission case.
2 Using wrench, Tool C-3380-A with adaptor C-3705, tighten band adjusting screw 47 to 50 inch-pounds. If the adaptor is not used, tighten adjusting screw to 72 inch-pounds.

LOW-REVERSE BAND (REAR)

1 This band is not adjustable, but lining inspection is necessary to determine the need for replacement.
2 Band is within service limits if the grooves in the lining are visible and groove depth at any point is no less than 0.2 mm. (.080 inch). At the same time the band end gap as installed on the drum and 100 pounds force applied to the ends must not be less than 0.5 mm.

Hydraulic Control Pressure Adjustments

LINE PRESSURE

An incorrect throttle pressure setting will cause incorrect line pressure readings even though line pressure adjustment is correct. Always inspect and correct throttle pressure adjustment before adjusting the line pressure.

1 The approximate adjustment is 1 5/16 inches measured from valve body to inner edge of adjusting nut. The adjustment can be varied to obtain specified line pressure.
2 One complete turn of the adjusting screw changes closed throttle line pressure approximately 1 2/3 psi. Turning counter-clockwise increases pressure, and clockwise decreases pressure.

THROTTLE PRESSURE

Throttle pressures cannot be tested accurately. The adjustment should be measured if a malfunction is evident.

1 Insert gauge pin of Tool C-3763 between the throttle lever cam and kickdown valve. By pushing in on tool, compress kickdown valve against its spring so throttle valve is completely bottomed inside the valve body.
2 As force is being exerted to compress spring, turn throttle lever stop screw with adaptor C-4553 with handle until head of screw touches the throttle lever cam touching tool and the throttle valve bottomed. Be sure adjustment is made with spring fully compressed and valve bottomed in the valve body.

Speedometer Pinion Gear

1 Remove the bolt securing the pinion adapter in the extension housing, and, with the cable connected, carefully work the adapter and pinion out of the extension housing. Remove the retainer and remove the retainer from the adapter.
2 If the cable housing contains transmission fluid, install a new speedometer pinion and seal assembly. If transmission fluid is found to be leaking between the cable and the adapter, replace the small "O" ring on the cable.
3 Before installing, make sure adapter flange and its mating areas on the extension housing are clean. Dirt will cause misalignment resulting in speedometer pinion gear damage.

CHRYSLER A-404 TORQUEFLITE AUTOMATIC TRANSAXLE

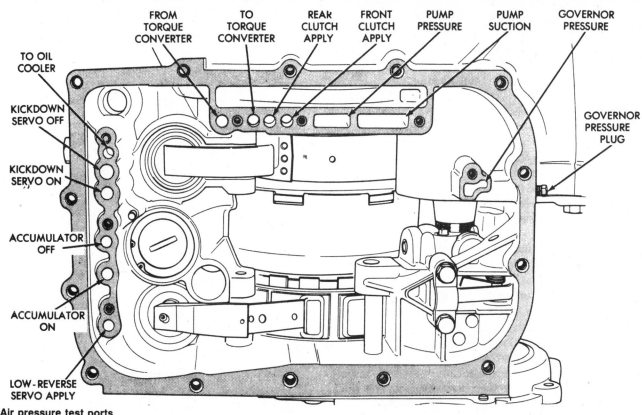

Air pressure test ports

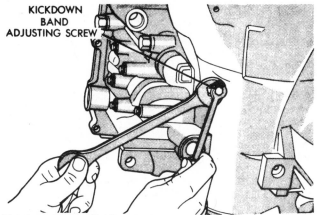

Kickdown band adjusting screw

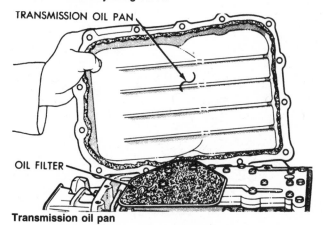

Transmission oil pan

Removing parking rod "E" clips

Out Of Vehicle Service Procedures

Valve Body Removal

1 Remove the oil pan and filter.
2 Remove parking rod "E" clip, and remove the rod.
3 Remove valve body attaching bolts, and remove it with governor tubes attached.

CHRYSLER A-404 TORQUEFLITE AUTOMATIC TRANSAXLE

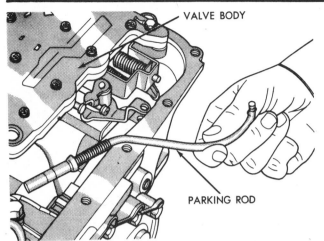

Parking rod after retaining "E" clips have been removed

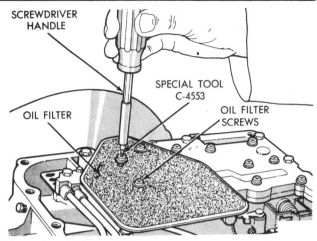

Removing oil filter screen

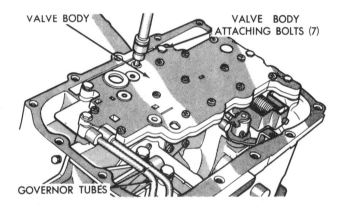

Removing valve body attaching bolts

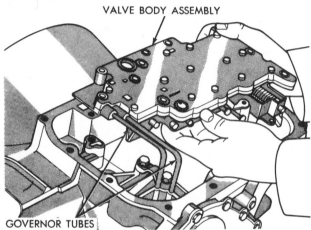

Removing valve body with governor tubes attached

Valve Body Recondition

1 Remove detent spring.
2 Remove valve body screws, and separate valve body and transfer and steel plates.
3 Remove throttle shaft "E" clip.
4 Remove manual valve lever assembly.
5 Remove valve lever assembly.
6 Remove the manual valve.
7 Remove pressure regulator and adjusting screw bracket.
8 Remove governor plugs.
9 Remove pressure regulator valve plugs.
10 Remove shift valves and shuttle valves.
11 Reverse procedure to assemble.

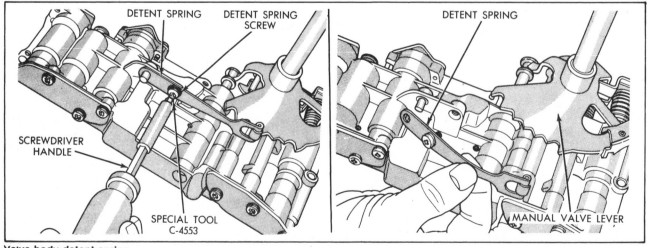

Valve body detent spring

CHRYSLER A-404 TORQUEFLITE AUTOMATIC TRANSAXLE

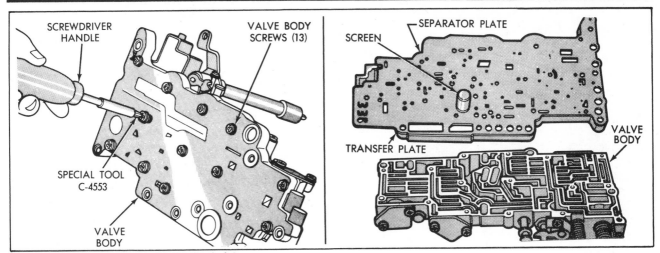

Valve body screws and transfer and steel plates

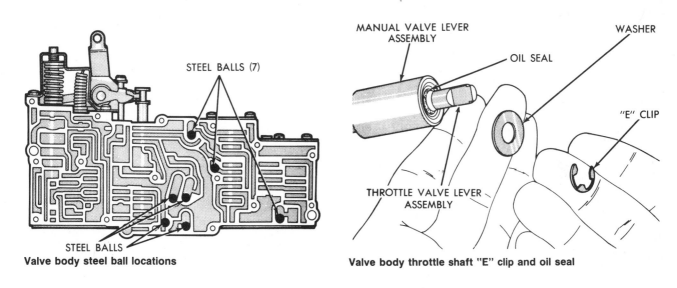

Valve body steel ball locations

Valve body throttle shaft "E" clip and oil seal

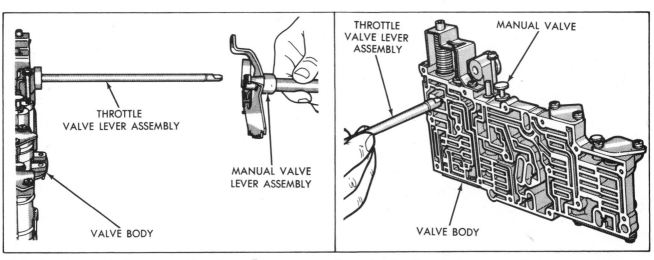

Removing manual valve lever assembly

Removing throttle valve lever assembly from valve body

Component Removal and Overhaul

CHRYSLER A-404 TORQUEFLITE AUTOMATIC TRANSAXLE

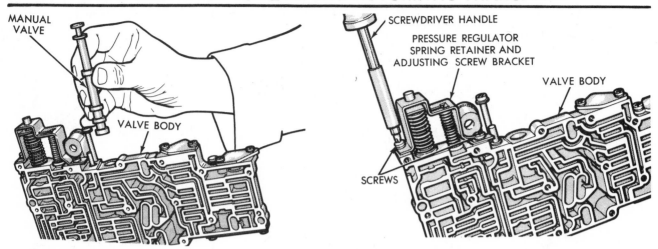

Removing manual valve from valve body

Removing pressure regulator and adjusting screw bracket

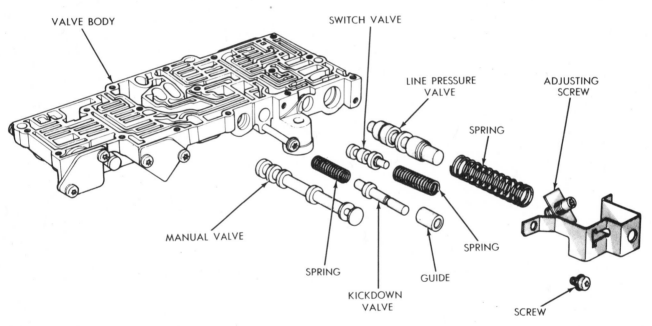

Valve body pressure regulators and manual controls

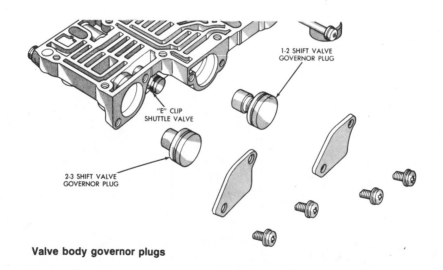

Valve body governor plugs

CHRYSLER A-404 TORQUEFLITE AUTOMATIC TRANSAXLE

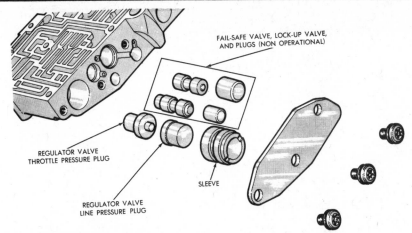

Valve body pressure regulator valve plugs

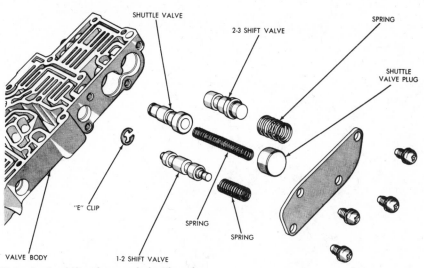

Valve body shift valves and shuttle valve

Pump Oil Seal

The pump oil seal can be replaced without removing pump and reaction shaft support assembly from the transmission case.

1 Remove the seal using Tool C-3981.
2 Install a new seal with lip side facing inward. Drive the seal into the housing until it bottoms.

Input Shaft End Play

Measuring input shaft end play before disassembly will usually indicate when a thrust washer change is required. The thrust washer is located between the input and output shafts.

1 Attach a dial indicator with its plunger seated against end of input shaft. Move shaft in and out for end play measurement reading.
2 Record indicator reading for assembly reference.

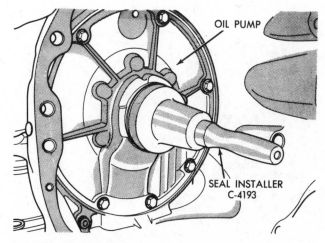

Installing oil pump seal

CHRYSLER A-404 TORQUEFLITE AUTOMATIC TRANSAXLE

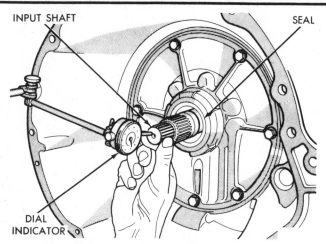

Measuring input shaft end play

Transaxle Disassembly

1 Tighten kickdown band adjusting screw, and remove oil pump attaching bolts.
2 Remove oil pump with Tool C-3752.
3 Loosen kickdown band adjusting screw, and remove the kickdown band and strut.
4 Remove the front clutch assembly.
5 Remove no. 2 thrust washer and the rear clutch assembly.
6 Remove no. 3 thrust washer and front planetary gear snap ring, and remove the front planetary gear assembly.
7 Remove no. 6 thrust washer and sun gear driving shell.
8 Remove no. 9 thrust washer and rear planetary gear assembly.
9 Remove no. 10 thrust washer and overrunning clutch cam assembly.
10 Remove overrunning clutch rollers and springs.
11 Remove the low-reverse band and strut, and remove the no. 11 thrust washer.
12 Reverse this procedure for assembly.

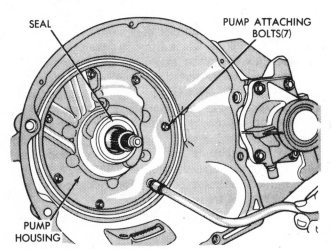

Oil pump attaching bolts

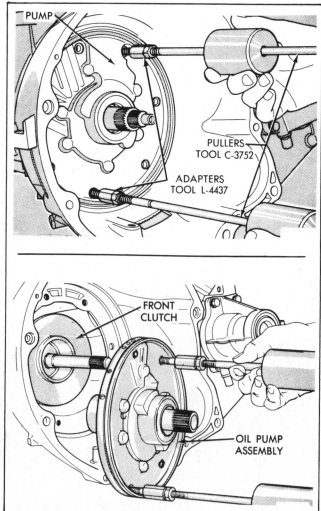

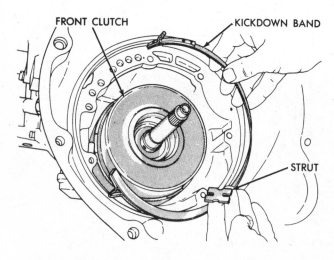

Kickdown band and strut

Removing oil pump with special tools

CHRYSLER A-404 TORQUEFLITE AUTOMATIC TRANSAXLE

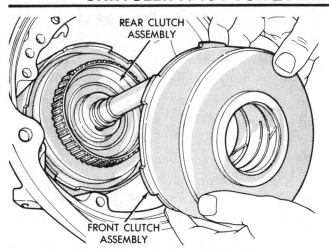

Removing front clutch

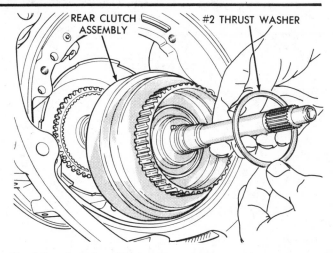

No. 2 thrust washer and rear clutch assembly

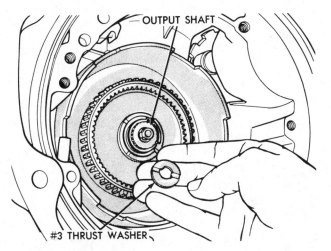

Thrust washer no. 3

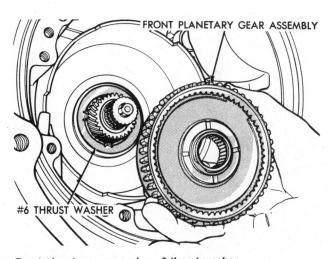

Front planetary gear and no. 6 thrust washer

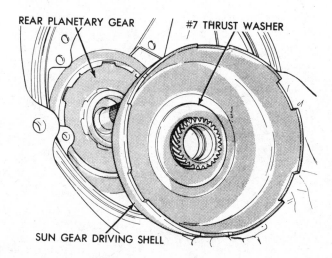

Sun gear driving shell and no. 7 thrust washer

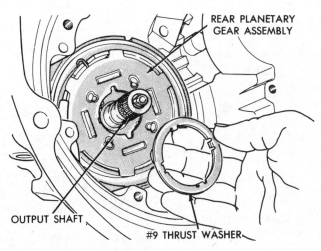

No. 9 thrust washer and rear planetary gear assembly

Component Removal and Overhaul

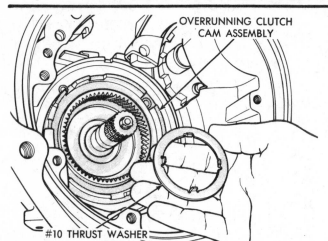

No. 10 thrust washer and overrunning clutch cam assembly

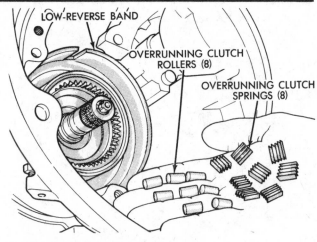

Overrunning clutch rollers and springs

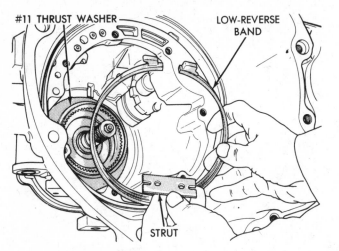

Low-reverse band and strut

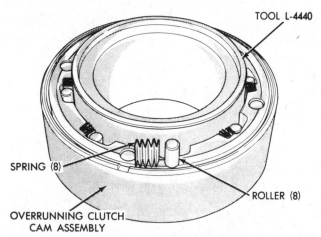

Use tool L-4440 to retain rollers and springs when installing overrunning clutch

Oil Pump Recondition

1 Remove the reaction shaft support, measure the pump clearance, and remove the inner and outer rotor.

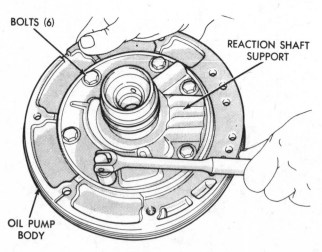

Oil pump reaction shaft removal

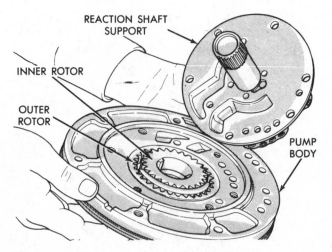

Oil pump reaction shaft support and rotors

CHRYSLER A-404 TORQUEFLITE AUTOMATIC TRANSAXLE

Front Clutch

1 Remove waved snap ring, thick steel plate and discs and plates.
2 Remove the return spring snap ring, and remove the return spring and piston.

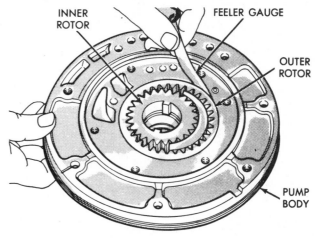

Measuring pump clearance

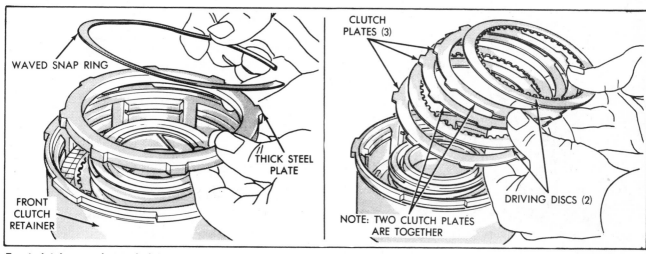

Front clutch snap ring and plates

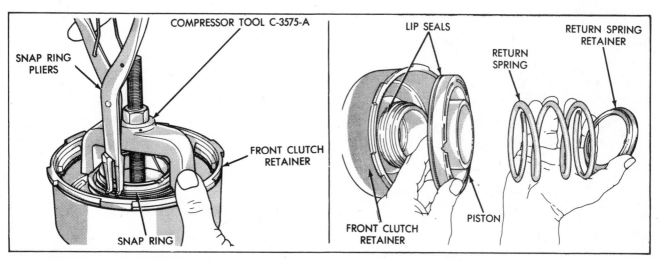

Front clutch return spring and piston

CHRYSLER A-404 TORQUEFLITE AUTOMATIC TRANSAXLE

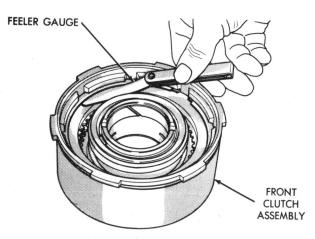

Measuring front clutch plate clearance

Rear Clutch

1 Remove the outer snap ring and then the discs and plates.
2 Remove the piston spring snap ring, and remove the piston and spring.
3 Remove the input shaft snap ring.

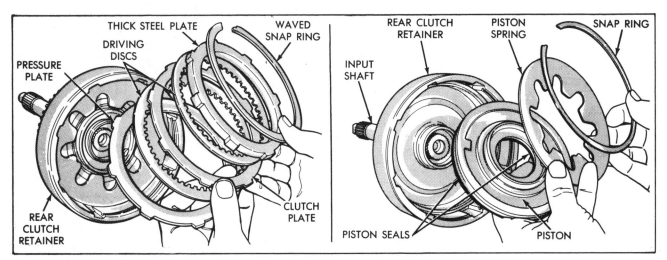

Rear clutch disassembled

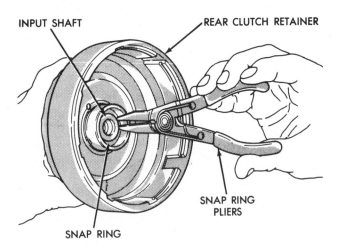

Rear clutch input shaft snap ring

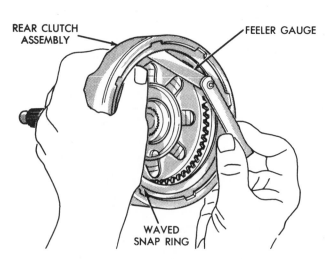

Measuring rear clutch plate clearance

CHRYSLER A-404 TORQUEFLITE AUTOMATIC TRANSAXLE

Front Planetary and Annulus Gear

1 Remove the front planetary gear snap ring and no. 4 thrust washer.
2 Remove front planetary gear annulus gear and no. 5 thrust washer.
3 Remove front snap ring from annulus gear, and remove the gear support.
4 Remove the front annulus gear support rear snap ring.

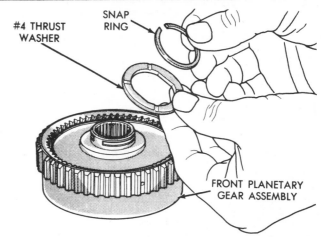

Front planetary gear snap ring and no. 4 thrust washer

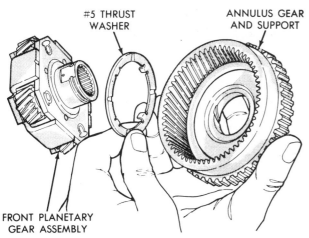

Front planetary annulus gear and no. 5 thrust washer

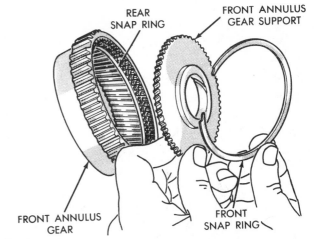

Front annulus gear support and snap ring

Low-Reverse Servo

1 Remove the low-reverse servo retainer snap ring, and remove the retainer, return spring and servo assembly.

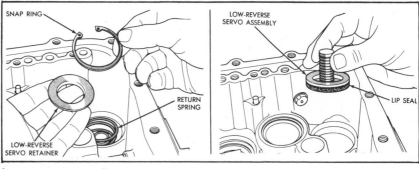

Low-reverse servo disassembly

Accumulator

1 Remove the accumulator snap ring, and remove the accumulator plug and accumulator spring and piston.

Accumulator disassembly

CHRYSLER A-404 TORQUEFLITE AUTOMATIC TRANSAXLE

Kickdown Servo

1 Remove the servo snap ring and rod guide.
2 Remove the spring and piston rod, and remove the kickdown piston.

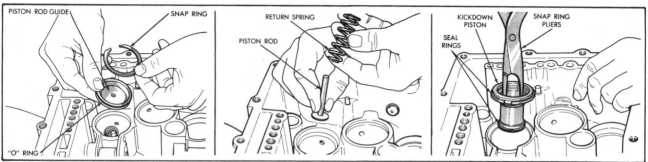

Kickdown servo disassembly

Transfer Shaft

1 Remove the rear cover, and use Tool L-4434 to remove the transfer shaft gear retaining nut.
2 Remove transfer shaft gear using Tool L-4407. Also remove shim.
3 Remove transfer shaft gear bearing cone using Tool L-4406-1.
4 Remove the governor support re-

tainer, and use Tool L-4517 to remove the governor support retainer bearing cup.
5 Remove the low-reverse band anchor pin and the governor assembly.
6 Remove the transfer shaft retainer snap ring, and use Tool L-4512 to remove the transfer shaft and retainer assembly.

7 Use Tool C-293-52 to remove the transfer shaft bearing cone.
8 Remove the oil seal from transfer shaft seal retainer.
9 Remove transfer shaft retainer bearing cup with Tool L-4518.
10 Reverse this procedure to assemble.

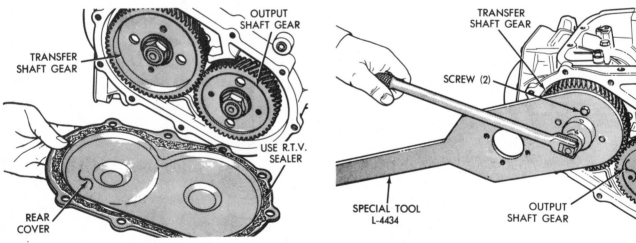

Rear cover removal

Removing transfer shaft gear retaining nut with tool L-4434

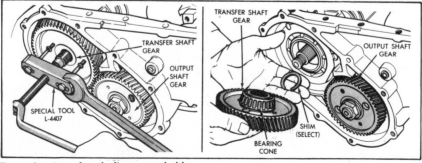

Removing transfer shaft gear and shim

CHRYSLER A-404 TORQUEFLITE AUTOMATIC TRANSAXLE

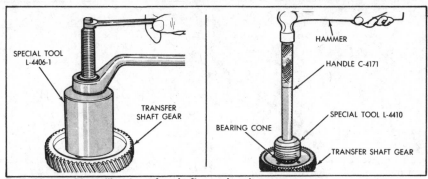

Removing and installing transfer shaft gear bearing cone

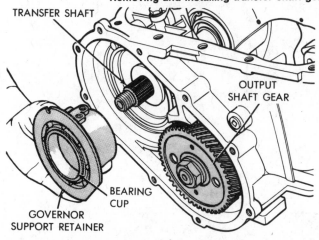

Removing governor support retainer

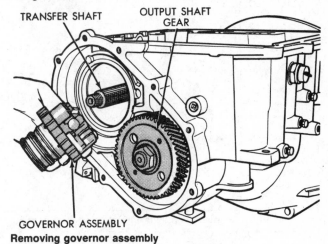

Removing governor assembly

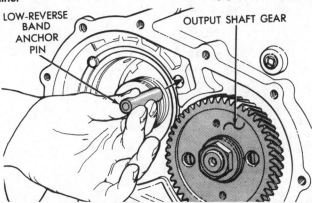

Removing and installing governor support retainer bearing cup

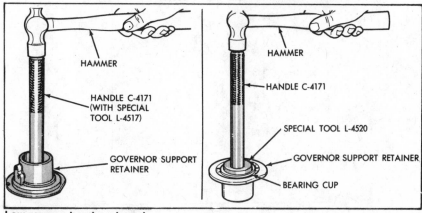

Low-reverse band anchor pin

Component Removal and Overhaul

CHRYSLER A-404 TORQUEFLITE AUTOMATIC TRANSAXLE

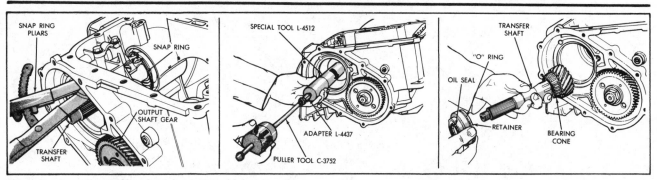

Removing transfer shaft and retainer assembly

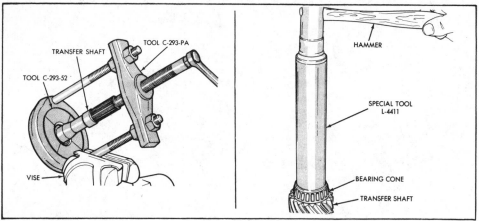

Removing and installing transfer shaft bearing cone

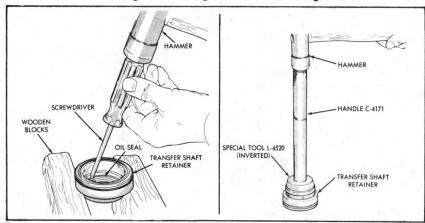

Transfer shaft retainer oil seal replacement

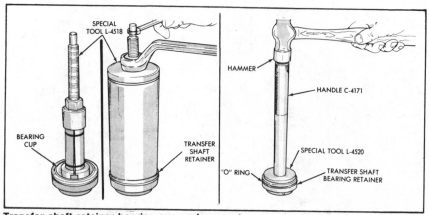

Transfer shaft retainer bearing cup replacement

CHRYSLER A-404 TORQUEFLITE AUTOMATIC TRANSAXLE

Parking Pawl

1 Remove parking pawl retainer, and remove the parking pawl, return spring and pivot shaft.

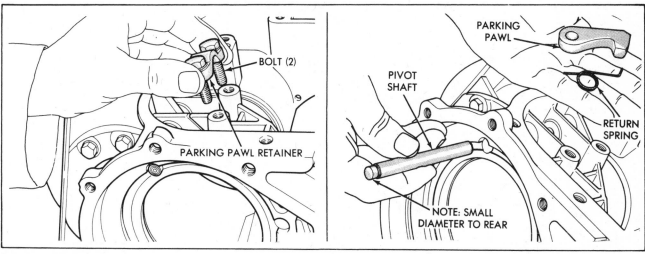

Parking pawl disassembly

Output Shaft

NOTE: Remove transfer shaft prior to output shaft removal.

1 Remove output shaft retaining nut with Tool 4-4424, and remove the output shaft gear and shim.

2 Remove the output shaft gear bearing cone with Tool L-4406-1.
3 Remove output shaft and annulus gear assembly.
4 Use a press to remove output shaft from rear planetary annulus gear.
5 Remove rear planetary annulus gear

bearing cone with Tool L-4406-1.
6 Remove output shaft gear bearing cup using Tool L-4518.
7 Remove rear planetary annulus gear bearing cup using Tool L-4517.
8 Reverse this procedure to assemble.

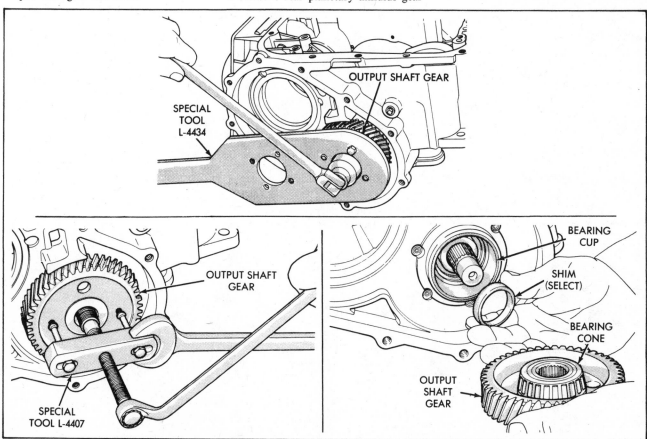

Removing output shaft retaining nut, gear and shim

Component Removal and Overhaul

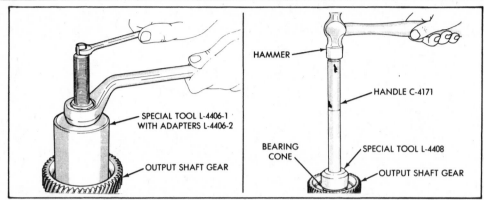

Output shaft gear bearing cone replacement

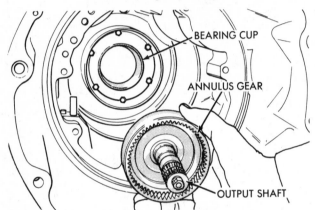

Removing output shaft and annulus gear assembly

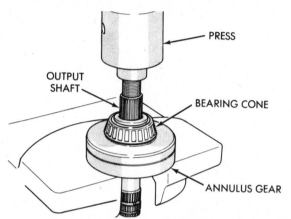

Separating output shaft from rear planetary annulus gear

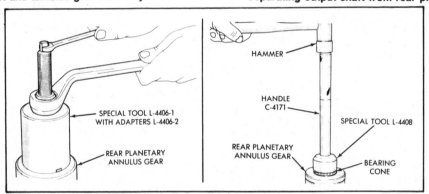

Rear planetary annulus gear bearing cone replacement

Output shaft gear bearing cup replacement

CHRYSLER A-404 TORQUEFLITE AUTOMATIC TRANSAXLE

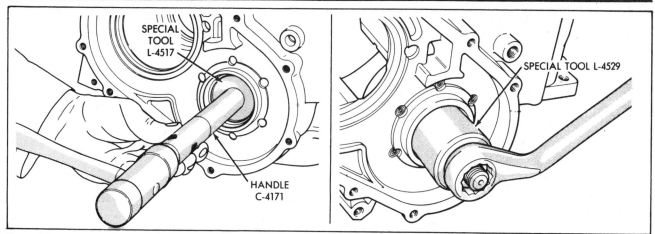

Rear planetary annulus gear bearing cup replacement

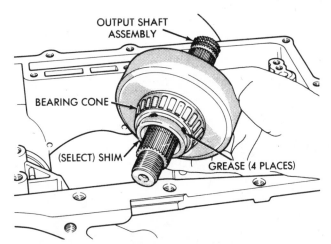

Install output shaft assembly using grease to hold shim in place on rear planetary annulus gear

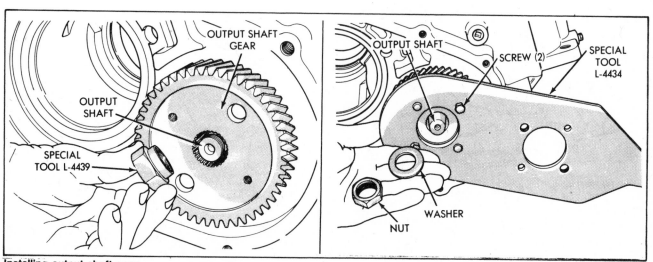

Installing output shaft gear

Component Removal and Overhaul

Differential

NOTE: Remove transfer shaft prior to output shaft removal.

1 Remove the extension housing oil seal.
2 Remove the differential cover.
3 Remove the differential bearing retainer bolts and remove the retainer using Tool L-4435.
4 Remove the extension housing bolts, and remove the housing using Tool L-4435.
5 Remove the differential assembly and extension housing.
6 Remove the differential bearing cones with Tool L-4406-1.
7 Remove the pinion shaft roll pin using an easy-out.
8 Remove the pinion shaft.
9 Remove pinion gears, side gears and four thrust washers by rotating pinion gears to opening in differential case.
10 Remove ring gear attaching bolts.
11 Remove differential bearing retainer oil seal.
12 Remove bearing cup from differential bearing retainer using Tool L-4518.
13 Reverse this procedure to assemble.

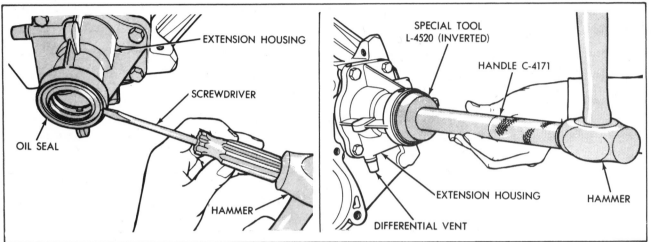

Replacing extension housing oil seal

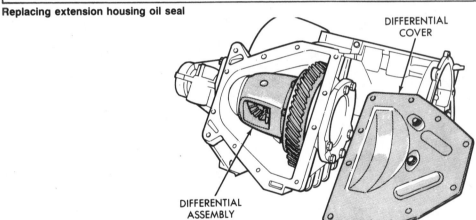

Differential cover removed

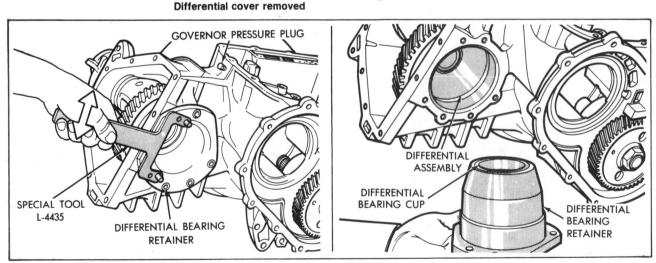

Removing differential bearing retainer

CHRYSLER A-404 TORQUEFLITE AUTOMATIC TRANSAXLE

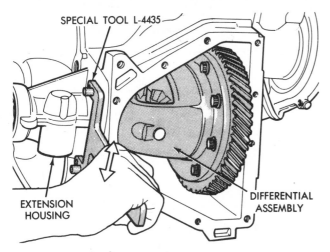

Extension housing removal using tool L-4435

Removing differential assembly and extension housing

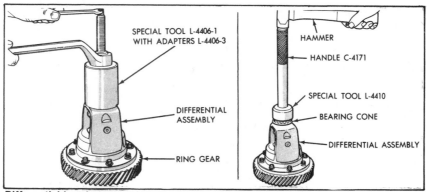

Differential bearing cone removal and installation

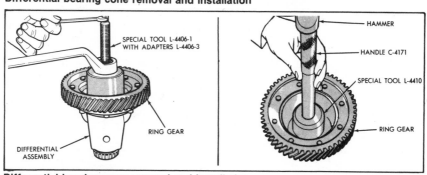

Differential bearing cone removal and installation

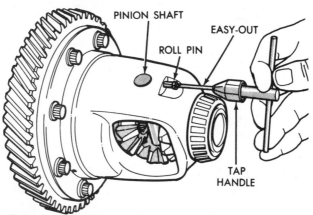

Pinion shaft roll pin removal

Component Removal and Overhaul

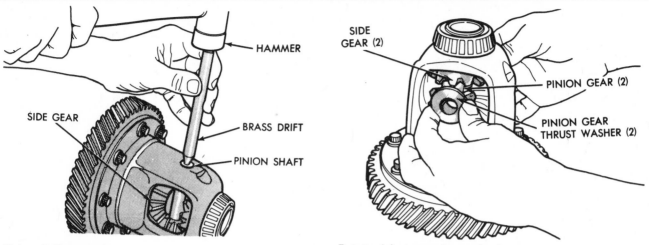

HAMMER

BRASS DRIFT

PINION SHAFT

SIDE GEAR

Pinion shaft removal

SIDE GEAR (2)

PINION GEAR (2)

PINION GEAR THRUST WASHER (2)

Rotate pinion gears to opening to remove

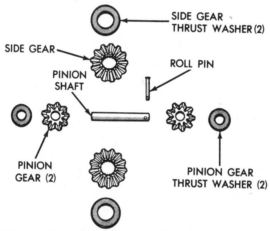

SIDE GEAR THRUST WASHER (2)

SIDE GEAR

PINION SHAFT

ROLL PIN

PINION GEAR (2)

PINION GEAR THRUST WASHER (2)

Differential gears, thrust washers and pinion shaft

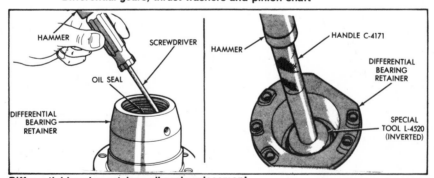

HAMMER

SCREWDRIVER

OIL SEAL

DIFFERENTIAL BEARING RETAINER

HANDLE C-4171

HAMMER

DIFFERENTIAL BEARING RETAINER

SPECIAL TOOL L-4520 (INVERTED)

Differential bearing retainer oil seal replacement

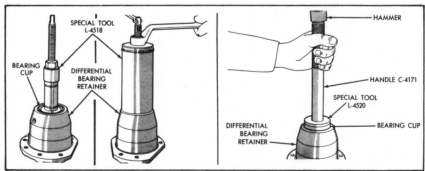

SPECIAL TOOL L-4518

BEARING CUP

DIFFERENTIAL BEARING RETAINER

HAMMER

HANDLE C-4171

SPECIAL TOOL L-4520

DIFFERENTIAL BEARING RETAINER

BEARING CUP

Removal and installation of bearing cup—differential bearing retainer

CHRYSLER A-404 TORQUEFLITE AUTOMATIC TRANSAXLE

A-404 TORQUEFLITE AUTOMATIC TRANSAXLE SPECIFICATIONS

	MM	INCH
Torque Converter Diameter	241	9.48
Pump Clearances		
Outer Gear to Pocket	0.045-0.141	.0018-.0056
Outer Gear I.D. to Crescent	0.150-0.306	.0059-.012
Outer Gear Side Clearance	0.025-0.050	.001-.002
Inner Gear O.D. to Crescent	.160-.316	.0063-.0124
Inner Gear Side Clearance	.150-.306	.0059-.012
End Play		
Input Shaft	0.76-2.69	.030-.106
Front Clutch Retainer	0.76-2.69	.030-.106
Front Carrier	0.89-1.45	.007-.057
Front Annulus Gear	0.09-0.50	.0035-.020
Planet Pinion	0.15-0.59	.006-.023
Reverse Drum	0.76-3.36	.030-.132
Clutch Clearance and Selective Snap Rings		
Front Clutch (Non-adjustable)		
Measured From Reaction Plate to Farthest Wave	1.7-2.7	.067-.106
Rear Clutch	0.40-0.94	.016-.037
Rear Clutch Selective Snap Rings (2)	1.53-1.58	.060-.062
	2.01-2.06	.079-.081

A-404 AUTOMATIC TRANSAXLE TORQUE SPECIFICATIONS

Bell Housing Cover Bolt	105 in/lbs
Flex Plate-to-Crank Bolt	50 ft/lbs
Flex Plate-to-Torque Converter Bolt	40 ft/lbs
Transmission-to-Cylinder Block Screw	70 ft/lbs
Lower Bell Housing Cover Screw	105 in/lbs
Manual Control lever Screw	105 in/lbs
Speedometer-to-Extension Screw	105 in/lbs
Cooler Tube-to-Transmission Nut	150 in/lbs
Cooler Hose-to-Radiator Connector	110 in/lbs
Starter-to-Transmission Bell Housing Bolt	40 ft/lbs
Throttle Cable-to-Transmission Case	105 in/lbs
Throttle Lever-to-Transmission Case	105 in/lbs
Manual Cable-to-Transmission Case	250 in/lbs
Front Engine Mount Bolt	40 ft/lbs
Left Engine Mount Bolt	40 ft/lbs
Cooler Line Connector	250 in/lbs
Pressure Check Plug	45 in/lbs
Neutral Safety Switch	25 ft/lbs
Ring Gear Screw	70 ft/lbs
Extension-to-Case Bolt	250 in/lbs
Differential Bearing Retainer-to-Case	250 in/lbs
Differential Cover-to-Case Screw	165 in/lbs
Differential Cover Plug	24 ft/lbs
Output Shaft Nut	200 ft/lbs
Transfer Shaft Nut	200 ft/lbs
Governor-to-Support Bolt	60 in/lbs
Governor Counterweight Screw	250 in/lbs
Rear Cover-to-Case Screw	165 in/lbs
Reverse Band Shaft Plug	60 in/lbs
Reaction Shaft Assembly Bolt	250 in/lbs
Pump-to-Case Bolt	275 in/lbs
Kickdown Band Adjustment Lock Nut	35 ft/lbs
Sprag Retainer-to-Transfer Case Bolt	250 in/lbs
Valve Body Screw	40 in/lbs
Transfer Plate Screw	40 in/lbs
Filter Screw	40 in/lbs
Transfer Plate-to-Case Screw	105 in/lbs
Oil Pan-to-Case Screw	165 in/lbs

FORD C3

TROUBLESHOOTING
FORD C3

THE PROBLEM	ITEMS TO CHECK	
	IN THE CAR	OUT OF CAR
ROUGH INITIAL ENGAGEMENT IN D1 or D2	Engine idle speed. Vacuum modulator or lines. Control pressure. Pressure regulator. Valve body. Intermediate band.	Forward clutch.
1-2 or 2-3 SHIFT POINTS INCORRECT or ERRATIC	Fluid level. Vacuum modulator or lines. Downshift linkage—inner lever position. Manual linkage. Governor. Control pressure. Valve body. Perform air pressure checks.	
ROUGH 1-2 SHIFTS	Vacuum modulator or lines. Intermediate servo. Make control pressure. Pressure regulator.	
ROUGH 2-3 SHIFTS	Vacuum modulator or lines. Intermediate band. Control pressure. Pressure regulator. Intermediate band. Valve body. Perform air pressure checks.	Reverse—high clutch. Reverse—high clutch piston air bleed valve.
DRAGGED OUT 1-2 SHIFTS	Fluid level. Vacuum modulator or lines. Intermediate servo. Control pressure. Intermediate band. Valve body. Pressure regulator. Perform air pressure checks.	Hydraulic system leakage.
ENGINE OVERSPEEDS ON 2-3 SHIFT	Manual linkage. Fluid level. Vacuum modulator or lines. Intermediate servo. Control pressure. Valve body. Pressure regulator. Intermediate band.	Reverse—high clutch. Reverse—high clutch piston air bleed valve.
NO 1-2 or 2-3 UPSHIFT	Manual linkage. Downshift linkage—inner lever position. Vacuum modulator or lines. Governor. Control pressure. Valve body. Intermediate band. Intermediate servo.	Reverse—high clutch. Hydraulic system leakage.
NO 3-1 SHIFT IN D1 or 3-2 SHIFT IN D2	Governor. Valve body.	
NO FORCED DOWNSHIFT	Downshift linkage—inner lever position. Valve body. Vacuum modulator or lines.	
RUNAWAY ENGINE ON FORCED DOWNSHIFT	Control pressure. Intermediate servo. Intermediate band. Pressure regulator. Valve body. Vacuum modulator or lines.	Forward clutch.
ROUGH 3-2 or 3-1 SHIFT AT CLOSED THROTTLE	Engine idle speed. Vacuum modulator or lines. Intermediate servo. Valve body. Pressure regulator.	
SHIFTS 1-3 IN D1 & D2	Intermediate band. Intermediate servo. Vacuum modulator or lines. Valve body. Governor. Perform air pressure checks.	
NO ENGINE BRAKING IN FIRST GEAR—MANUAL LOW	Manual linkage. Reverse servo. Governor. Perform air pressure checks.	
CREEPS EXCESSIVELY	Engine idle speed. Vehicle brakes.	
SLIPS or CHATTERS IN FIRST GEAR, D1	Fluid level. Vacuum modulator or lines. Control pressure. Pressure regulator. Valve body.	Forward clutch. Hydraulic system leakage. Planetary one-way clutch.
SLIPS or CHATTERS IN SECOND GEAR	Fluid level. Vacuum modulator or lines. Intermediate servo. Intermediate band. Control pressure. Pressure regulator. Valve body. Perform air pressure checks.	Forward clutch. Hydraulic system leakage.

FORD C3

Ford C3

Identification

The identification tag is located under the left side extension housing bolt. The C3 transmission is coded V on the vehicle certification label. The illustrations below show the transmission identification tag and the code interpretation.

Front Band Adjustment

1 Remove the downshift rod from the transmission.
2 Clean all dirt from the band adjusting screw. Remove and discard the locknut.
3 Install a new locknut, tighten the adjusting screw to the present torque with tool kit T71P-77370-A or 10

ft./lbs. Back off adjusting screw exactly 1½ turns.
4 Torque the locknut to 35–45 ft./lbs.
5 Install downshift rod.

Governor R&R

1 Remove the small snap-ring from the guide rod or connecting pin and remove the rod and valve.

FORD C3

TROUBLESHOOTING FORD C3

| THE PROBLEM | ITEMS TO CHECK | |
	IN THE CAR	OUT OF CAR
SLIPS or CHATTERS IN R	Fluid level. Vacuum modulator or lines. Control pressure. Reverse servo. Pressure regulator. Valve body. Perform air pressure checks.	Reverse—high clutch. Hydraulic system leakage. Reverse—high clutch piston air bleed valve.
NO DRIVE IN D1	Fluid level. Manual linkage. Control pressure. Valve body. Perform air pressure checks.	Planetary one-way clutch.
NO DRIVE IN D2	Fluid level. Manual linkage. Control pressure. Intermediate servo. Valve body. Perform air pressure checks.	Hydraulic system leakage. Planetary one-way clutch.
NO DRIVE IN L	Fluid level. Manual linkage. Control pressure. Valve body. Intermediate servo. Perform air pressure checks.	Hydraulic system leakage. Planetary one-way clutch.
NO DRIVE IN R	Fluid level. Manual linkage. Control pressure. Valve body. Intermediate servo. Perform air pressure checks.	Reverse—high clutch. Hydraulic system leakage. Reverse—high clutch piston air bleed valve.
NO DRIVE IN ANY SELECTOR POSITION	Fluid level. Manual linkage. Control pressure. Pressure regulator. Valve body. Perform air pressure checks.	Hydraulic system leakage. Front pump.
LOCKUP IN D1		Reverse—high clutch. Parking brake linkage. Hydraulic system leakage.
LOCKUP IN D2	Reverse servo.	Reverse—high clutch. Parking brake linkage. Hydraulic system leakage. Planetary one-way clutch.
LOCKUP IN L	Intermediate band. Intermediate servo.	Reverse—high clutch. Parking brake linkage. Hydraulic system leakage.
LOCKUP IN R	Intermediate band. Intermediate servo.	Forward clutch. Parking brake linkage. Hydraulic system leakage.
PARKING LOCK BINDS or DOES NOT HOLD	Manual linkage.	Parking brake linkage.
TRANSMISSION OVERHEATS	Oil cooler and/or connections. Pressure regulator. Vacuum modulator or lines. Control pressure.	Front oil pump seal.
MAXIMUM SPEED TOO LOW, POOR ACCELERATION	Engine performance. Vehicle brakes.	Front oil pump seal.
TRANSMISSION NOISY IN N & P	Fluid level. Pressure regulator.	Front pump. Planetary assembly.
TRANSMISSION NOISY IN ANY DRIVE POSITION	Fluid level. Pressure regulator.	Planetary assembly. Forward clutch. Front pump. Planetary one-way clutch.
FLUID LEAKS	Fluid level. Converter drain plug. Oil pan and/or fillet tube gaskets or seals. Oil cooler and/or connections. Manual or downshift lever shaft seal. Pipe plug, side of case. Extension housing-to-case gasket or washers. Extension housing rear oil seal. Vacuum modulator or lines. Reverse servo. Intermediate servo. Speedometer driven gear adaptor seal.	Engine rear oil seal. Front oil pump seal. Front pump-to-case seal or gasket.
CAR MOVES FORWARD IN N	Manual linkage.	Forward clutch.

2 Unscrew the housing from the hub.
3 Move the governor housing toward the rear and the governor hub toward the front of the output shaft to disassemble.
4 Remove the snap-ring from the housing and remove the weights.
5 Remove the snap-ring from the outer weight and remove the spring and the inner weight.
6 Clean all parts and replace any that are worn or damaged. If necessary, remove the three steel seals and carefully install new ones.
7 Assemble the spring, the inner weight,

the outer weight and the snap-ring.
8 Secure these in the housing with a snap-ring.
9 Assemble the governor housing to the output shaft from the rear, and the hub to the output shaft from the front.
10 Assemble the two together.
11 Install the valve and guide rod or connecting pin and fasten with the small snap-ring.

Control Valve Body R&R

1 Remove the separator plate bolts and carefully lift off the separator plate.

2 Remove the five ball valves and both relief valves with springs.
3 Remove the retaining plates; dowels, plugs and valves with springs. Do not mix the springs. They can be identified by color.
4 Carefully clean the oil channels and parts by blowing through with compressed air.
5 Inspect all parts for burring, unevenness and gum deposits. As necessary, replace parts.
6 Lubricate all parts with transmission fluid, then install the valves, springs, plugs and pins.

FORD C3

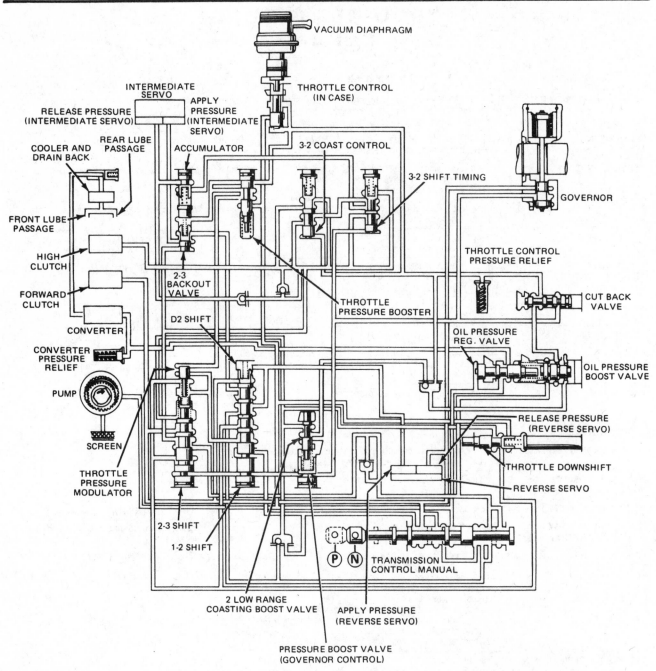

VACUUM DIAPHRAGM

INTERMEDIATE SERVO

RELEASE PRESSURE (INTERMEDIATE SERVO)

APPLY PRESSURE (INTERMEDIATE SERVO)

THROTTLE CONTROL (IN CASE)

REAR LUBE PASSAGE

COOLER AND DRAIN BACK

ACCUMULATOR

3-2 COAST CONTROL

3-2 SHIFT TIMING

GOVERNOR

FRONT LUBE PASSAGE

THROTTLE CONTROL PRESSURE RELIEF

HIGH CLUTCH

2-3 BACKOUT VALVE

THROTTLE PRESSURE BOOSTER

CUT BACK VALVE

FORWARD CLUTCH

D2 SHIFT

OIL PRESSURE REG. VALVE

CONVERTER

OIL PRESSURE BOOST VALVE

CONVERTER PRESSURE RELIEF

PUMP

SCREEN

RELEASE PRESSURE (REVERSE SERVO)

THROTTLE DOWNSHIFT

REVERSE SERVO

THROTTLE PRESSURE MODULATOR

2-3 SHIFT

1-2 SHIFT

P   N   TRANSMISSION CONTROL MANUAL

2 LOW RANGE COASTING BOOST VALVE

APPLY PRESSURE (REVERSE SERVO)

PRESSURE BOOST VALVE (GOVERNOR CONTROL)

C3 hydraulic schematic

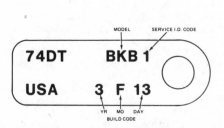

| 74DT | BKB 1 |
| USA | 3 F 13 |

MODEL — SERVICE I.D. CODE

YR MO DAY
BUILD CODE

Ford C3 transmission identification tag
(Mustang with 2300cc. engine)

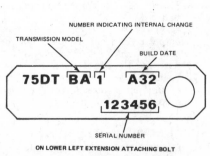

NUMBER INDICATING INTERNAL CHANGE

TRANSMISSION MODEL

BUILD DATE

| 75DT | BA 1 | A32 |
| | 123456 | |

SERIAL NUMBER

ON LOWER LEFT EXTENSION ATTACHING BOLT

Ford C3 transmission identification tag

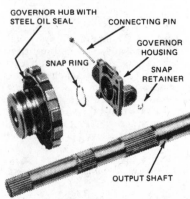

GOVERNOR HUB WITH STEEL OIL SEAL

CONNECTING PIN

GOVERNOR HOUSING

SNAP RING

SNAP RETAINER

OUTPUT SHAFT

Governor components

FORD C3

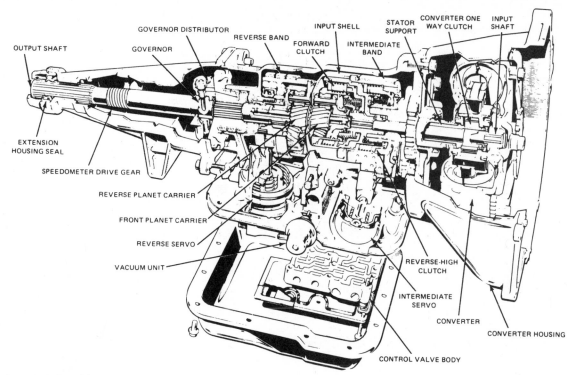

Cutaway view C3 transmission

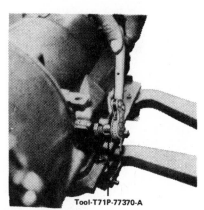

Tool-T71P-77370-A

Band adjustment

7 Using a new gasket, install the separator plate.

Extension Housing R&R

1 Scribe driveshaft rear yoke and companion flange to assist reassembly. Disconnect and remove driveshaft.
2 Support the transmission.
3 Disconnect speedometer cable at the extension housing.
4 Remove the bolts which hold rear support to the crossmember.
5 Jack up transmission slightly, and remove the rear support.
6 Loosen extension housing bolts, and allow the transmission fluid to drain. Now, remove the bolts; remove the housing.
7 In installation, reverse the above procedures, and: Make sure, when in-

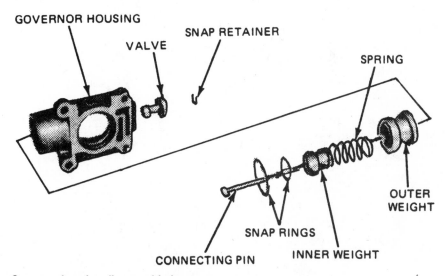

Governer housing-disassembled

stalling the extension housing, that the operating rod parking notch is correctly seated before installing bolts. Torque bolts to 27–39 ft./lbs. Check housing for leakage under operation.

Overhaul

Disassembly
1 Remove the torque converter.
2 Withdraw the input shaft.
3 Remove the oil pan.
4 Remove the oil filter screen, gasket and three spacers (if so equipped).
5 Remove the interlock spring.

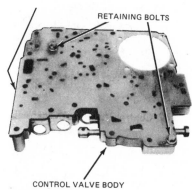

CONTROL VALVE BODY
Separator plate

FORD C3

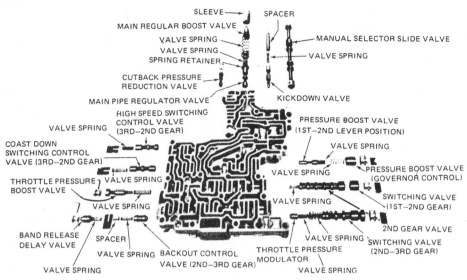

Control valve body—exploded view

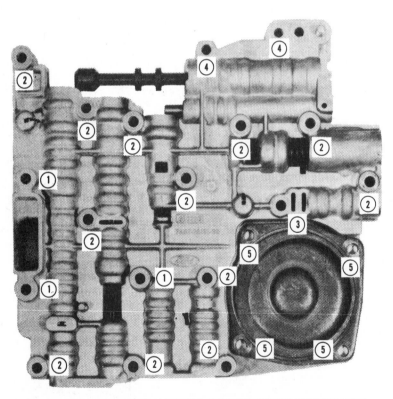

OPERATING PARKING ROD
MUST SEAT CORRECTLY

Ford C3 extension housing installation

6 Remove the rear servo cover and the gasket. (The force of the spring will raise the cover.) Take out the servo piston with the spring.

7 Remove the bolts from the control valve body. While easing the valve body out of the transmission, unlock and detach the selector lever connecting link. Remove the valve body and gasket.

8 Unbolt the converter housing and remove the housing and hydraulic pump as an assembly. Remove the No. 1 thrust washer and the gasket.

9 Holding Tool T74P-77248-A with a spanner, remove the hydraulic pump oil seal.

10 Remove the hydraulic pump from the converter housing and remove the steel plate (behind oil seal) with the O-ring.

Transmission End Play Check

Perform this check at this stage of transmission disassembly.

1 Fit the hydraulic pump with the existing No. 1 thrust washer into the case.

Position	Bolt Size Metric	Length In Inches	Quantity	Torque
1	M6 x 45	1.772	3	7-9 Ft-Lb
2	M6 x 40	1.578	12	7-9 Ft-Lb
3	M6 x 35	1.378	1	7-9 Ft-Lb
4	M6 x 30	1.141	2	7-9 Ft-Lb
5	M6 x 20	.787	4	7-10 Ft-Lb

Control valve body specifications

FORD C3

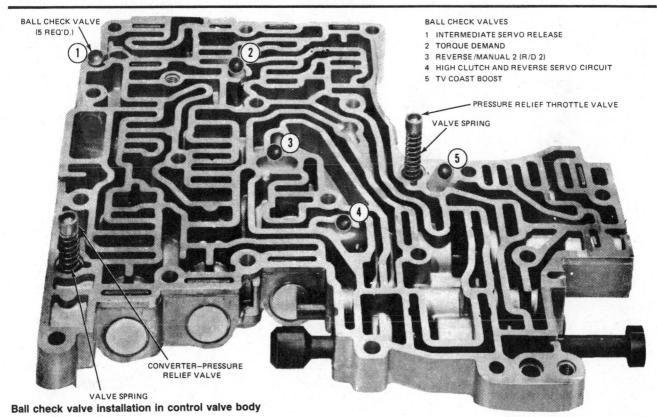

Ball check valve installation in control valve body

Be sure the assembly is correctly engaged.

2 Mount the dial gauge block on the hydraulic pump with the plunger resting on the transmission housing. Set the dial gauge on zero.

3 Swing the gauge around so the plunger contacts the hydraulic pump.

4 Check the reading on the dial. This reading is the amount of end play. End play should be from 0.007 to 0.-032. If the reading exceeds these limits, replace the No. 1 thrust washer. Refer to the specifications for the sizes available.

Following the end play check, remove the hydraulic pump and thrust washer again. Mark the installed position of the hydraulic pump gears in

relation to one another and remove them.

Front Brake and Front Servo Assemblies

Removal

1 Loosen the locknut, back out the adjustment screws and remove the struts.

2 Remove the front brake and front assembly including thrust washer No. 8.

Transmission assembly washers Nos. 5, 8, and 9 are identical. Knowing this should prevent confusion when reviewing the remaining steps of disassembly and assembly of the transmission.

3 Press inward slightly on the front

servo cover to permit removal of the snap ring.

4 Carefully force out the servo piston with compressed air.

Case and Extension Housing Parts

Removal

1 Remove the transmission extension housing bolts and slide off the extension housing. Remove the gasket.

2 Remove the return spring and the parking pawl.

3 Remove the large snap ring from the planet gear carrier (rear) and remove the planet gear carrier with thrust washer No. 5 from the transmission

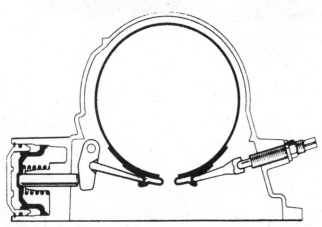

Brake band and servo piston

Using compressed air to remove servo piston

Component Removal and Overhaul

FORD C3

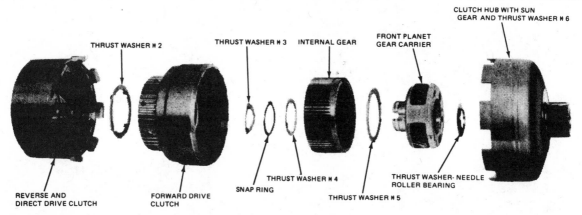

THRUST WASHER N 2   THRUST WASHER N 3   INTERNAL GEAR   FRONT PLANET GEAR CARRIER   CLUTCH HUB WITH SUN GEAR AND THRUST WASHER N 6

REVERSE AND DIRECT DRIVE CLUTCH   FORWARD DRIVE CLUTCH   SNAP RING   THRUST WASHER N 4   THRUST WASHER N 5   THRUST WASHER- NEEDLE ROLLER BEARING

Forward gear train component assembly

case. Remove the small snap ring from the output shaft.
4 Remove the output shaft and the governor with thrust washer No. 11.
5 Remove the internal gear, the No. 10 thrust washer and the intermediate brake drum.
6 Remove the rear band assembly.
7 If necessary, due to wear or damage, remove the one-way clutch inner race.
8 Remove the vacuum diaphragm.
9 Remove the neutral switch. **Use a thin-walled socket. An open-end wrench will crush the switch.**
10 To change the shift lever oil seal, press inward on the downshift lever. Remove the O-ring.
11 Remove the shift lever roll pin from the case.
12 Remove the shift lever nut (outside) and remove the parking pawl actuating rod.
13 Remove the selector lever from the outside and the downshift lever shaft from inside the case.
14 Remove the shift lever oil seal with a screwdriver.
15 Install a new seal.
16 Install the downshift lever shaft inside the case and the shift lever outside.
17 Install the nuts and the parking pawl actuating rod.
18 Install the roll pin, a new O-ring and the downshift lever.

Front Assembly

Disassembly
1 Remove the input shell and sun gear.
2 Remove the planet gear carrier with the internal gear and thrust washer No. 5.
3 If necessary, remove the sun gear from the input shell after removing the retainer. Replace thrust washer No. 7 if it is damaged.
4 Remove the forward drive clutch and thrust washer No. 2.

Reverse and Direct Drive Clutch

Disassembly
1 Remove the pressure plate retainer ring and remove the plate pack.

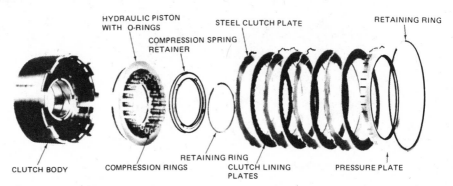

HYDRAULIC PISTON WITH O-RINGS   STEEL CLUTCH PLATE   RETAINING RING   COMPRESSION SPRING RETAINER   CLUTCH BODY   COMPRESSION RINGS   RETAINING RING   CLUTCH LINING PLATES   PRESSURE PLATE

Reverse and direct clutch disassembled

Removal of hydraulic piston

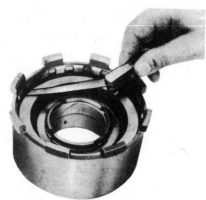

Clearance check between pressure plate and retaining ring

2 Compress the compression springs using Tool T65L-77515-A, remove the retainer and carefully release the pressure on the springs.
3 Remove the compression springs and the compression spring retainer.
4 Turn the clutch body over and, using compressed air, carefully direct pressure into the ports to force out the piston.
5 Remove the hydraulic pump and remove the O-rings from the piston and clutch body.

Assembly
1 Inspect the steel clutch plates and clutch lining plates for wear, damage or effects of overheating. Replace the entire set if necessary. If new plates are used immerse them in transmission fluid for 30 minutes before assembly.
2 Install new O-rings on the piston and clutch body.
3 Carefully install the clutch piston. Use Tool T74P-77404-A to protect the inner seal.
4 Install the compression springs (20 required) and the spring retainer.
5 Compress the springs using Tool T65L-77515-A and install the retaining ring. Release the load on the springs and remove the tool.
6 Install the clutch plates in the correct order and secure with the retaining ring.

FORD C3

7 Use a feeler gage to check the clearance between the retaining ring and the pressure plate under a 22 pound load. The clearance should be to specifications. If the clearance is not within specification, install a different, suitable retaining ring. Refer to the specifications at the end of this Part for the clearance specification and the available retaining ring sizes.

Forward Clutch
Disassembly

Plate and piston removal procedures for the forward speed clutch assembly are similar to those for the reverse and direct drive clutch.

1 Remove the retaining ring and remove the clutch plates and corrugated washer.
2 Compress the springs with Tool T65L-77515-A being careful not to damage the piston.
3 Remove the retainer and carefully release the load on the springs.
4 Remove the springs and spring retainer.
5 Carefully remove the piston using air pressure.
6 Remove the hydraulic pump.
7 Remove the O-rings from the piston and clutch body.

Assembly

1 Inspect the steel clutch plates and the clutch lining plates for wear, damage or effects of overheating. If it is necessary to replace the entire set, immerse the new set of clutch plates in oil for 30 minutes before assembly.
2 Use protective tooling to prevent damage to inner and outer seals. Install new O-rings on the piston, apply Vaseline to the rings and to the shoulder at the clutch stub, and install the piston carefully.
3 Install the compression springs and the spring retainer.
4 Use Tool T65L-77515-A to compress the springs, and install the retaining ring.
5 Install the disc spring washer and the clutch plates in the correct order and position the retaining snap ring.
6 Check the clearance between the retaining ring and the support plate following the procedure described for the Reverse and Direct Drive Clutch.
7 Install new steel seals on the clutch stub.

Internal Gear and Planet Gear Assembly
Disassembly

1 Remove the snap ring, the planet gear carrier internal gear and thrust washer No. 4.
2 Separate the planet gear carrier from the internal gear and remove thrust washer No. 5.

Assembly

1 Insert the planet gear carrier with thrust washer No. 5 in the internal

gear. Position thrust washer No. 5 again, and secure with a new snap ring (the internal gear must be free of the planet gear carrier).

2 **The needle roller bearing (No. 6 Washer) can only be replaced complete, with the planet gear carrier. If the thrust washer needle roller bearing is removed, the washer must be positioned with the collar pointing toward the rear.**

Front Assembly
Assembly

1 Place the reverse and direct drive clutch assembly vertically on a bench.
2 Position thrust washer No. 2 and the forward gear clutch assembly.
3 Secure thrust washer No. 3 with Vaseline or oil on the planet gear carrier.
4 Install the internal gears and planet gear assembly.
5 Assemble the input shell with the sun gear to the planet gear carrier, then to

the body of the reverse and direct drive clutch.

One-Way Clutch
Disassembly and Assembly

1 Remove the snap ring.
2 Lift out the cage with the springs and bearing rollers as a unit.
3 Install the cage with springs.
4 Insert the bearing rollers one by one, using a suitable screwdriver and install the snap ring.

Governor

1 Remove the small snap ring from the guide rod or connecting pin and remove the rod and valve.
2 Unscrew the housing from the hub.
3 Move the governor housing toward the rear and the governor hub toward the front of the output shaft to disassemble.
4 Remove the snap ring from the housing and remove the weights.

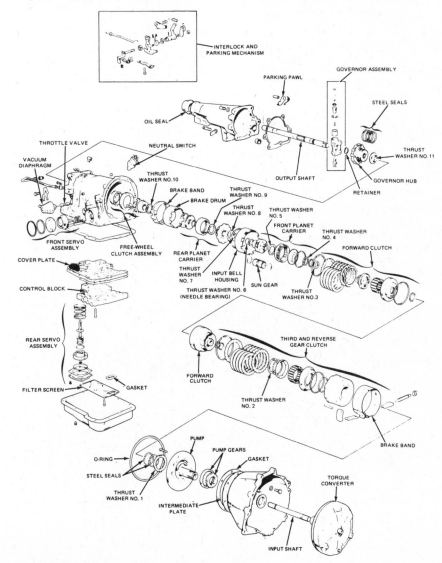

C3 exploded view

Component Removal and Overhaul

5 Remove the snap ring from the outer weight and remove the spring and the inner weight.

6 Clean all parts and replace any that are worn or damaged. **If necessary, remove the three steel seals and carefully install new ones.**

7 Assemble the spring, the inner weight, the outer weight and the snap ring.

8 Secure these in the housing with a snap ring.

9 Assemble the governor housing to the output shaft from the rear, and the hub to the output shaft from the front. Assemble the two together.

10 Assemble the two together.

11 Install the valve and guide rod or connecting pin and fasten with the small snap ring.

Control Valve Body

1 Remove the separator plate bolts and carefully lift off the separator plate.

2 Remove the five ball valves and both relief valves with springs.

3 Remove the retaining plates; dowels, plugs and valves with springs. Do not mix the springs. They can be identified by color.

4 Carefully clean the oil channels and parts by blowing through with compressed air.

5 Inspect all parts for burring, unevenness and gum deposits. If necessary replace parts.

6 Lubricate all parts with transmission fluid, then install the valves, springs, plugs and pins.

7 Using a new gasket, install the separator plate.

Transmission

Assembly

Thoroughly clean all moving parts and lubricate them with the specified automatic transmission fluid before assembly.

1 Press in a new hydraulic pump oil seal.

2 Install the vacuum diaphragm.

3 If the one-way clutch inner race was removed, install it using Tool T74P-77193-A.

4 Position thrust washer No. 11 in the case and install the output and governor assembly. **Be careful not to damage the steel oil seals.**

5 Position the rear brake band in the housing in correct relation to the guide pilots.

6 Position thrust washer No. 10 in the case and install the rear brake drum using the clutch replacing guide Tool T74P-77193-A. Then remove the guide and install the internal gear. Secure with a snap ring.

7 Position thrust washer No. 9 to the back of the planet system carrier with light grease or Vaseline. Install the planet system carrier and secure to the rear brake drum with the snap ring.

8 Position thrust washer No. 8 to the planet gear carrier with Vaseline and install the front assembly.

9 Install the spiral spring on the front servo piston assembly. Position the piston and cover and secure with the snap ring. Replace the servo piston or O-rings at this time, if necessary.

10 Install the brake band and struts starting with the one at the servo piston lever.

11 **Being careful not to damage the oil seals, turn the transmission on the assembly stand so the output shaft points downward.** Install the hydraulic pump with thrust washer No. 1.

12 Recheck the end play as described under Disassembly, and if necessary replace the thrust washer. **This step is very important.**

13 Remove the hydraulic pump. Install the inside and outside pump gears making sure that the small gear has the I.D. pump drive flat recess facing upward, and that the large gear has the chamfer facing downward.

14 Position the steel plate on the hydraulic pump in the exact position required; then install the complete assembly to the converter housing.

15 Tighten the bolts finger tight. Align the pump using Tool T74P-77103-A. **This tool must be used and the sleeve gauging surfaces must be in good condition. Otherwise the pump will not be aligned correctly with the converter housing; then seal leakage or bushing failure will result.** To use the tool, select the arbor with the smallest I.D. that will fit completely over the pump shaft. Assemble the common handle to the selected arbor and slide the tool down over the shaft until it bottoms against the pump. The outside diameter of the tool arbor will then automatically center the pump in the converter housing. Torque the bolts to specification and remove the tool.

16 Insert the input shaft into the pump and install the converter into the pump gears. Rotate the converter to check for free movement, then remove the converter and input shaft.

17 Position the selected thrust washer No. 1 to the pump housing with Vaseline. Install a new O-ring. Carefully install the converter housing with the pump using a new gasket. **Do not damage the steel oil seals.** Install the bolts and torque to specification.

18 Adjust the brake band.

19 Install the parking pawl and its return spring in the extension housing and preload.

20 Using a new gasket, install the extension housing. Be sure to correctly seat the operating parking rod in the extension guide cup. Torque the bolts to specification.

21 Remove the extension housing oil seal.

22 Install a new extension housing oil seal.

23 Using a new gasket, position the control valve body, and fit and secure the connecting rod or link to the manual

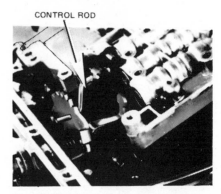

CONTROL ROD

Ford C3 control valve body installation

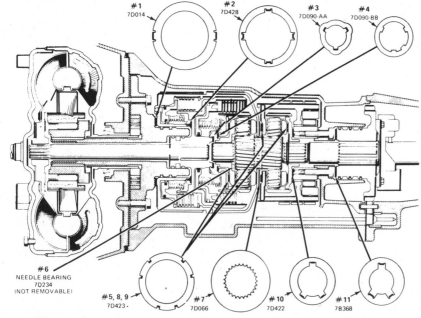

Indentification and position of thrust washers

FORD C3

valve. Install the retaining bolts. Because the bolts are of different lengths, make sure each bolt head bottoms on the control body. Attach the interlock spring to the control housing.

24 Before proceeding further be sure the inner downshift lever is seated between the stop and the downshift valve.

25 Assemble the servo piston rod, servo piston and spring. (Use available servo piston rod.)

26 Install the rear servo piston assembly and spring into the rear servo bore. Make certain the piston rod is correctly seated in the reverse band apply end.

27 Install tool T74P-77190-A with a new servo cover gasket and tighten the three attaching bolts. As the servo cover to case bolts are not long enough to attach adjusting tool to case, use three M6×30 main control to case attaching bolts.

28 Torque the servo tool adjusting screw to 36 in./lbs.

29 Install dial indicator on transmission case and position indicator on one of the three servo piston pads that is accessible through the cutout on the tool. Set dial indicator at zero.

30 Back out the servo tool adjusting screw until the servo piston bottoms out on the tool. Record the distance the servo piston travelled.

31 If piston travel is between .120 and .220 inch, it is within specification. If piston travel is greater than .220 inch, use the next longer rod. If piston travel is less than .120 inch, use the next shorter rod.

32 Using the above procedure, check the piston travel with the new selected rod (if required) to make sure that the piston travel is .120 to .220 inch.

33 Remove servo adjusting tool and install servo cover and gasket. Torque the 4 attaching bolts to 7 to 10 ft./lbs.

34 Install the oil pan with a new gasket.

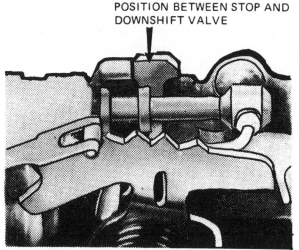

POSITION BETWEEN STOP AND DOWNSHIFT VALVE

Correct seating of inner downshift lever

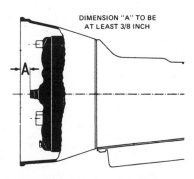

DIMENSION "A" TO BE AT LEAST 3/8 INCH

Converter hub to housing flange clearance

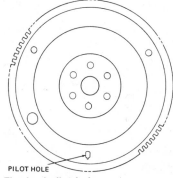

PILOT HOLE

Flywheel pilot hole

Torque the bolts to specification in two steps. Install the neutral switch.

35 Install the input shaft and the torque converter.

36 Install the transmission in the vehicle using the procedures described under Removal and Installation of Transmission in this Part.

FORD C3 TORQUE SPECIFICATIONS

Items	Ft/lbs	Items	Ft/lbs
Converter housing-to-case	27-39	Nut—manual lever, inner	30-40
Extension housing-to-case	27-39	Neutral switch-to-case	12-15
Flywheel-to-converter housing	27-30	Front band adjusting locknut	35-45
Main control-to-case	7-9	Vacuum diaphragm retaining clip-to-case	1-2
Plate-to-valve body	7-9	Oil cooler line by-pass tube-to-connector	7-10
Oil pan-to-case	12-17	Connector-to-case	10-15
Governor-to-collector body	7-10	Drain plug—converter	20-30
Converter housing-to-engine	22-27	Flywheel-to-crankshaft.	48-53
Nut—downshift lever, outer	7-11		

TROUBLESHOOTING
FORD C4-C4S

	ITEMS TO CHECK	
THE PROBLEM	**IN CAR**	**OUT OF CAR**
ROUGH INITIAL ENGAGEMENT IN D1 or D2	Engine idle speed. Vacuum modulator or lines. Control pressure. Pressure regulator. Valve body. Intermediate band.	Forward clutch.
1-2 or 2-3 SHIFT POINTS INCORRECT or ERRATIC	Fluid level. Vacuum modulator or lines. Manual linkage. Governor. Control pressure. Valve body. Perform air pressure checks.	
ROUGH 1-2 SHIFTS	Vacuum modulator or lines. Intermediate servo. Intermediate band. Control pressure. Valve body. Pressure regulator.	
ROUGH 2-3 SHIFTS	Vacuum modulator or lines. Intermediate servo. Control pressure. Pressure regulator. Intermediate band. Valve body. Perform air pressure checks.	Reverse—high clutch. Reverse—high clutch piston air bleed valve.
DRAGGED OUT 1-2 SHIFT	Fluid level. Vacuum modulator or lines. Intermediate servo. Control pressure. Intermediate band. Valve body. Pressure regulator. Perform air pressure checks.	Hydraulic system leakage.
ENGINE OVERSPEEDS ON 2-3 SHIFT	Manual linkage. Fluid level. Vacuum modulator or lines. Intermediate servo. Control pressure. Valve body. Pressure regulator. Intermediate band.	Reverse—high clutch. Reverse—high clutch piston air bleed valve.
NO 1-2 or 2-3 UPSHIFT	Manual linkage. Downshift linkage—inner lever position. Vacuum modulator or lines. Governor. Control pressure. Valve body. Pressure regulator. Intermediate band. Intermediate servo.	Reverse—high clutch. Hydraulic system leakage.
NO 3-1 SHIFT IN D1 or 3-2 SHIFT IN D2	Governor. Valve body.	
NO FORCED DOWNSHIFT	Downshift linkage—inner lever position. Vacuum modulator or lines.	
RUNAWAY ENGINE or FORCED DOWNSHIFT	Control pressure. Intermediate servo. Intermediate band. Pressure regulator. Valve body. Vacuum modulator or lines.	Hydraulic system leakage.
ROUGH 3-2 or 3-1 SHIFT AT CLOSED THROTTLE	Engine idle speed. Vacuum modulator or lines. Intermediate servo. Valve body. Pressure regulator.	
SHIFTS 1-3 IN D1 AND D2	Intermediate band. Intermediate servo. Vacuum modulator or lines. Valve body. Governor. Perform air pressure checks.	
NO ENGINE BRAKING IN FIRST GEAR—MANUAL LOW	Manual linkage. Reverse band. Reverse servo. Valve body. Governor. Perform air pressure checks.	
CREEPS EXCESSIVELY	Engine idle speed. Vehicle brakes.	
SLIPS or CHATTERS IN 1st GEAR, D1	Fluid level. Vacuum modulator or lines. Control pressure. Pressure regulator. Valve body.	Forward clutch. Hydraulic system leakage. Planetary one-way clutch.
SLIPS or CHATTERS IN 2nd GEAR	Fluid level. Vacuum modulator or lines. Intermediate servo. Intermediate band. Control pressure. Pressure regulator. Valve body. Perform air pressure checks.	Forward clutch. Hydraulic system leakage.

FORD C4-C4S

In-Car Testing Procedures

Prior to performing any of the tests or adjustments described below, engine idle speed and anti-stall dashpot adjustment, manual linkage adjustment, and transmission fluid level must be checked (see appropriate car section). If fluid level is excessively low, check for leakage as described below.

Control Pressure Check
C4 AUTOMATIC TRANSMISSION

When the vacuum diaphragm unit operates properly and the downshift linkage is adjusted correctly, all transmission shifts (automatic and kickdown) should occur within the specified road speed limits. If these shifts do not occur within the limits or if the transmission slips during a

shift point, perform the following procedure to locate the problem:

1 Connect the Automatic Transmission Tester as follows:
  a. Tachometer cable to engine.
  b. Vacuum gauge hose to the transmission vacuum diaphragm unit.
  c. Pressure hose to proper port in transmission.

FORD C4—C4S

TROUBLESHOOTING
FORD C4-C4S

THE PROBLEM	ITEMS TO CHECK	
	IN CAR	OUT OF CAR
SLIPS or CHATTERS IN R	Fluid level. Vacuum modulator or lines. Reverse band. Reverse servo. Pressure regulator. Valve body. Intermediate servo.	Reverse—high clutch. Hydraulic system leakage. Reverse—high clutch piston air bleed valve.
NO DRIVE IN D1	Fluid level. Manual linkage. Control pressure. Valve body. Perform air pressure checks.	Planetary one-way clutch.
NO DRIVE IN D2	Fluid level. Manual linkage. Control pressure. Intermediate servo. Valve body. Perform air pressure checks.	Hydraulic system leakage. Planetary one-way clutch.
NO DRIVE IN L	Fluid level. Manual linkage. Control pressure. Valve body. Reverse servo. Intermediate servo.	Hydraulic system leakage. Planetary one-way clutch.
NO DRIVE IN R	Fluid level. Manual linkage. Reverse band. Control pressure. Reverse servo. Valve body. Perform air pressure checks.	
NO DRIVE IN ANY SELECTOR POSITION	Fluid level. Manual linkage. Control pressure. Pressure regulator. Valve body. Perform air pressure checks.	Hydraulic system leakage. Front pump.
LOCKUP IN D1		Reverse—high clutch. Parking brake linkage. Hydraulic system leakage.
LOCKUP IN D2	Reverse band. Reverse servo.	Reverse—high clutch. Parking brake linkage. Hydraulic system leakage. Planetary one-way clutch.
LOCKUP IN L	Intermediate band. Intermediate servo.	Reverse—high clutch. Parking brake linkage. Planetary one-way clutch. Hydraulic system leakage.
LOCKUP IN R	Intermediate band. Intermediate servo.	Forward clutch. Parking brake linkage. Hydraulic system leakage.
PARKING LOCK BINDS or DOES NOT HOLD	Manual linkage.	Parking brake linkage.
TRANSMISSION OVERHEATS	Oil cooler and/or connections. Pressure regulator. Vacuum modulator or lines. Control pressure.	Front oil pump seal.
MAXIMUM SPEED TOO LOW, POOR ACCELERATION	Engine performance. Vehicle brakes.	Front oil pump seal.
TRANSMISSION NOISY IN N & P	Fluid level. Pressure regulator.	Front pump. Planetary assembly.
TRANSMISSION NOISY IN ANY DRIVE POSITION	Fluid level. Pressure regulator.	Planetary assembly. Forward clutch. Front pump. Planetary one-way clutch.
FLUID LEAKS	Fluid level. Converter drain plug. Oil pan and/or filler tube gaskets/seals. Oil cooler and/or connections. Manual or downshift lever shaft seal. Pipe plug, side of case. Extension housing real oil seal. Vacuum modulator or lines. Reverse servo. Intermediate servo. Speedometer driven gear adaptor seal.	Engine rear oil seal. Front oil pump seal. Front pump-to-case seal or gasket.
CAR MOVES FORWARD IN N	Manual linkage.	Forward clutch.

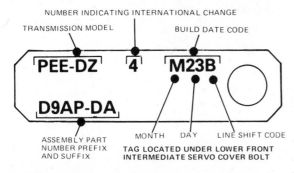

NUMBER INDICATING INTERNATIONAL CHANGE
TRANSMISSION MODEL
BUILD DATE CODE

PEE-DZ 4 M23B

D9AP-DA

ASSEMBLY PART NUMBER PREFIX AND SUFFIX
MONTH DAY LINE SHIFT CODE
TAG LOCATED UNDER LOWER FRONT INTERMEDIATE SERVO COVER BOLT

Refer to this transmission identification tag when ordering parts for a specific transmission

2 Apply the parking brake and start the engine. On a car equipped with a vacuum brake release, disconnect the vacuum line or use the service brakes since the parking brake will release automatically when the transmission is put in any Drive position.

3 Check the transmission diaphragm unit for leaks.

4 Check control pressure in all selector lever positions at specified manifold vacuum (see specifications). Record readings and compare to specifications.

Component Removal and Overhaul

FORD C4—C4S

FORD C4 & C4S
CHECKS & ADJUSTMENTS

Operation	Specification
Transmission end play	0.008-0.042 inch (selective thrust washers available)
Turbine and stator end play	Model PEB, PEG, PEJ—new or rebuilt 0.044 max. Used 0.060 max.
	Model PEE, PEA, PEF—New or rebuilt 0.023 max. Used 0.040 max.
Intermediate band adjustment	Remove and discard lock nut. Adjust screw to 10 ft./lbs. torque, then back off 1¾ turns. Install new lock nut and torque to specification.
Low-reverse band adjustment	Remove and discard lock nut. Adjust screw to 10 ft./lbs. torque, then back off 3 turns. Install new lock nut and torque to specification.

C4S Semi-Automatic Transmissions

If the shifts do not occur within the proper road speeds or the transmission slips during a shift point, perform the following procedure to find the possible trouble:

1 Connect the tachometer of the automatic transmission tester to the engine.
2 Attach the pressure gauge to the control pressure outlet at the transmission.
3 Firmly apply the parking brake and start the engine.
4 Check control pressure in all selector lever positions at specified manifold vacuum (see specifications). Record readings and compare to specifications.

Vacuum Diaphragm Unit Check

1 Remove the vacuum diaphragm unit from the transmission after disconnecting the vacuum hose.
2 Adjust a vacuum pump until the vacuum gauge shows 18 in. Hg with the vacuum hose blocked.
3 Connect the vacuum hose to the vacuum diaphragm unit and note the reading on the vacuum gauge. If the

Ford C4 & C4S
Control Pressure at Zero Output Shaft Speed

Engine Speed	Throttle	Manifold Vac. Ins. Hg.	Range	P.S.I.
As required	As required	12 and above	P, N, D	55-86
			2, 1	55-122
			R	55-197
As required	As required	10	P, N, D	98-110
Stall	Thru detent	Below 1.0	D, 2, 1	143-164
			R	239-272

Ford C4 & C4S
Control Pressure at Zero Output Shaft Speed

Engine Speed	Throttle	Manifold Vac. Ins. Hg.	Range	P.S.I.
As required	As required	15 and Above	P, N, D	52-85
			2, 1	52-115
			R	52-180
As required	As required	10	P, N, D	96-110
Stall	Thru detent	Below 1.0	D, 2, 1	143-160
			R	230-260

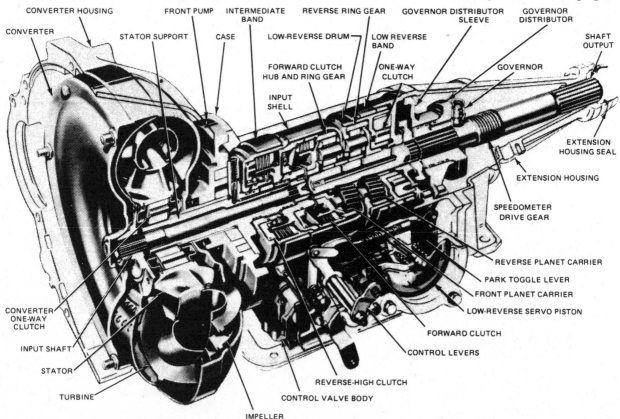

Cutaway view Ford C-4 automatic transmission

FORD C4—C4S

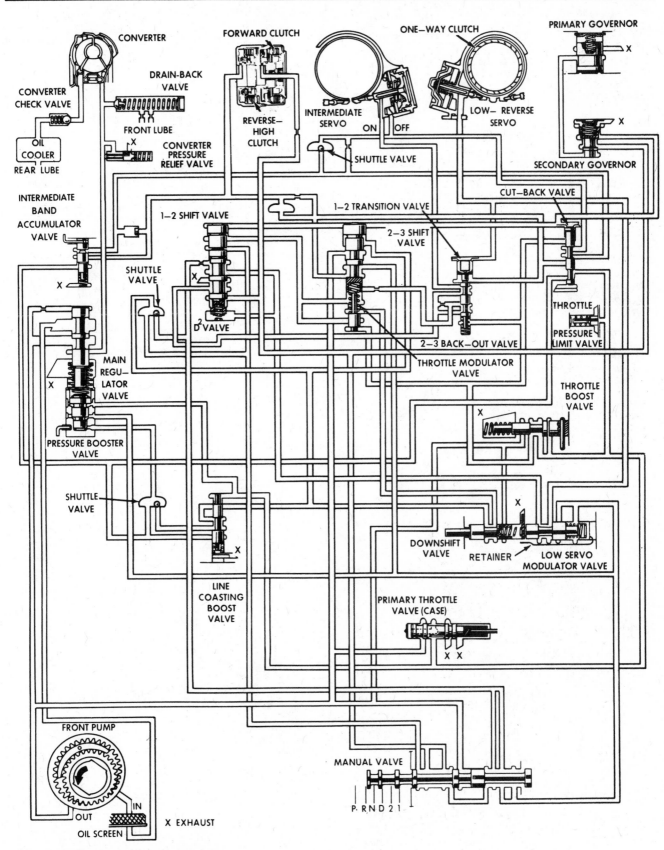

Hydraulic schematic typical C-4 and C-4S

FORD C4—C4S

reading is 18 in. Hg, the vacuum diaphragm unit is good. While removing the vacuum hose from the vacuum diaphragm unit, hold a finger over the end of the control rod. As the vacuum is released, the internal spring of the vacuum diaphragm unit will push the control rod out.

Air Pressure Checks

If the car will not move in one or more ranges, or, if it shifts erratically, the items at fault can be determined by using air pressure at the indicated passages.

Drain the transmission and remove the oil pan and the control valve assembly.

Front Clutch

Apply air pressure to the front clutch input passage. A dull thud can be heard when the clutch piston moves.

Governor

Remove the governor inspection cover from the extension housing. Apply air to

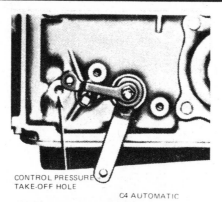

CONTROL PRESSURE TAKE-OFF HOLE

C4 AUTOMATIC

Typical vacuum diaphragm and control pressure connecting point on C4 and FMX transmissions

the front clutch input passage. Listen for a sharp click and watch to see if the governor valve snaps inward.

Rear Clutch

Apply air to the rear clutch passage and listen for the dull thud that will indicate that the rear clutch piston has moved.

Front Servo

Apply air pressure to the front servo apply tube and note if front band tightens. Shift the air to the front servo release tube, which is next to the apply tube, and watch band release.

Rear Servo

Apply air pressure to the rear servo apply passage. The rear band should tighten around the drum.

Conclusions

If the operation of the servos and clutches is normal with air pressure, the no-drive condition is due to the control valve and pressure regulator valve assemblies, which should be disassembled, cleaned and inspected.

If operation of the clutches is not normal; that is, if both clutches apply from one passage or if one fails to move, the aluminum sleeve (bushing) in the output shaft is out of position or badly worn.

Use air pressure to check the passages in the sleeve and shaft, and also check the passages in the primary sun gear shaft.

If the passages in the two shafts and the sleeve are clean, remove the clutch assemblies, clean and inspect the parts.

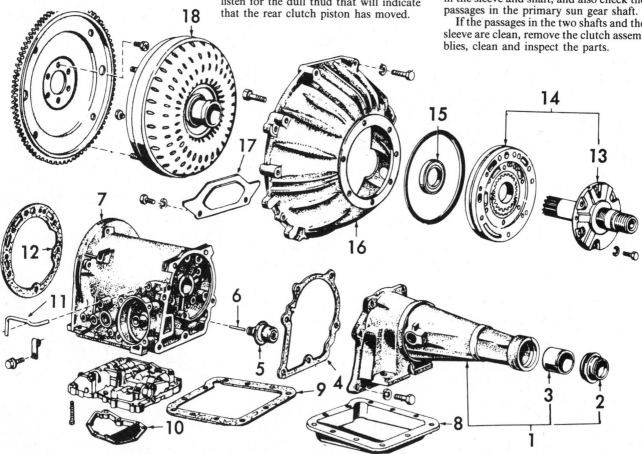

C-4 Case and parts identification

1. REAR EXTENSION CASE
2. REAR OIL SEAL
3. BUSHING
4. GASKET
5. VACUUM DIAPHRAGM
6. THROTTLE VALVE ROD
7. TRANSMISSION CASE
8. OIL PAN
9. OIL PAN GASKET
10. OIL SCREEN
11. BREATHER TUBE
12. GASKET
13. STATOR SUPPORT
14. FRONT OIL PUMP
15. FRONT PUMP OIL SEAL
16. CONVERTER HOUSING
17. HOUSING COVER
18. TORQUE CONVERTER

FORD C4—C4S

Erratic operation can also be caused by loose valve body screws. When reinstalling the valve body, tighten the control valve body screws to specifications.

Transmission Fluid Leakage Checks

Make the following checks if a leakage is suspected from the transmission case:

1 Clean all dirt and grease from the transmission case.
2 Inspect the speedometer cable connection at the extension housing of the transmission. If fluid is leaking, replace the rubber seal.
3 Inspect the oil pan gasket and attaching bolts for leaks. Torque all bolts and recheck for signs of leakage. If necessary, replace the gasket.
4 Check filler tube connection at the transmission for signs of leakage. If tube is leaking, tighten the connection to stop the leak. If necessary, disconnect the filler tube, and replace the O-ring.
5 Inspect all fluid lines between the transmission and the cooler core in the lower radiator tank. Replace any lines or fittings that appear to be worn or damaged. Tighten all fittings to the proper torque.
6 Inspect the engine coolant for signs of transmission fluid in the radiator. If there is transmission fluid in the engine coolant, the oil cooler core is probably leaking.

The oil cooler core may be tested further by disconnecting all lines to it and applying a 50–75 psi air pressure through the fittings. Remove the radiator cap to relieve any pressure buildup outside the cooler core. If air bubbles appear in the coolant or if the cooler core will not hold pressure, the oil cooler core is leaking and must be replaced.

7 Inspect the openings in the case where the downshift control lever shaft and the manual lever shaft enter. If necessary, replace the defective seal.
8 Inspect all plugs or cable connections in the transmission for signs of leakage. Tighten any loose plugs or connectors to the proper torque.
9 Remove the lower cover from the front of the bellhousing and inspect the converter drainplugs for signs of leakage. If there is a leak around the drainplugs, loosen the plug, coat the threads with a sealing compound and tighten the plug to the proper torque.

NOTE: Fluid leaks from around the converter drainplug may be caused by engine oil leaking past the rear main bearing or from the oil gallery plugs. To determine the exact cause of the leak before beginning repair procedures, an oil-soluble aniline or fluorescent dye may be added to the transmission fluid to find the source of the leak and whether the transmission is leaking. If a fluorescent dye is used, a black light must be used to detect the dye.

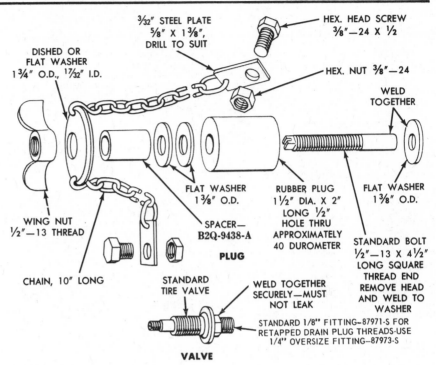

3/32" STEEL PLATE 5/8" X 1 3/8", DRILL TO SUIT

DISHED OR FLAT WASHER 1 3/4" O.D., 17/32" I.D.

HEX. HEAD SCREW 3/8"—24 X 1/2

HEX. NUT 3/8"—24

WELD TOGETHER

FLAT WASHER 1 3/8" O.D.

RUBBER PLUG 1 1/2" DIA. X 2" LONG 1/2" HOLE THRU APPROXIMATELY 40 DUROMETER

FLAT WASHER 1 3/8" O.D.

SPACER— B2Q-9438-A

PLUG

WING NUT 1/2"—13 THREAD

CHAIN, 10" LONG

STANDARD BOLT 1/2"—13 X 4 1/2" LONG SQUARE THREAD END REMOVE HEAD AND WELD TO WASHER

STANDARD TIRE VALVE

WELD TOGETHER SECURELY—MUST NOT LEAK

STANDARD 1/8" FITTING–87971-S FOR RETAPPED DRAIN PLUG THREADS-USE 1/4" OVERSIZE FITTING–87973-S

VALVE

Converter leak checking tool

Tire Inflation Chuck

Tire Pressure Gauge

Converter leak checking tool installation

If further converter checks are necessary, remove the transmission from the car and the converter from the transmission. To further check the converter for leaks, assemble and install a converter leak checking tool and fill the converter to 20 psi air pressure. Then, place the converter in a tank of water and watch for air bubbles. If no air bubbles are seen, the converter is not leaking.

In-Car Service Procedures

The following adjustments and repairs may be performed without removing the entire transmission from the car. Some of these procedures will require the use of special tools and instruments.

Throttle Linkage Adjustment

ALL FORD MID-SIZE V8

1 Disconnect downshift return spring.
2 Hold the carburetor throttle lever in wide open position, (against stop), and hold the transmission linkage in full downshift position against its internal stop.
3 Turn the carburetor downshift lever adjustment screw to within .010–.080 in. of the contacting pickup surface of the throttle lever.
4 Connect return spring.

ALL 6 CYL. MID-SIZE CARS

1 Make sure idle speed is correct and that the throttle lever is against the hot idle speed adjusting screw when the engine is hot.
2 Disconnect the throttle return spring, and remove the trunnion and cable at the bellcrank.
3 Hold the throttle wide open, and hold the transmission in full downshift position.
4 Adjust the trunnion at the bellcrank so the shaft ball stud and cable ball stud receiver are in alignment. Then, give the trunnion one additional turn so as to lengthen it.
5 Reconnect linkage and throttle return spring.

FORD PINTO
FORD MUSTANG II
MERCURY BOBCAT

1 Disconnect the downshift lever return spring.
2 Hold the throttle shaft lever in the

Component Removal and Overhaul

FORD C4—C4S

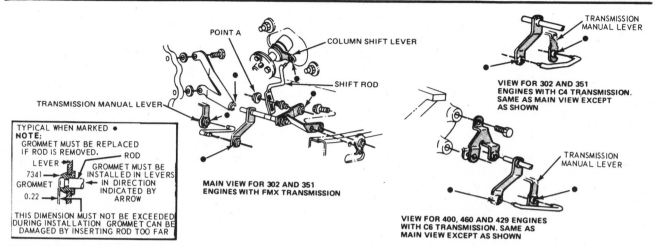

NOTE:
TYPICAL WHEN MARKED ●
GROMMET MUST BE REPLACED
IF ROD IS REMOVED.
ROD
LEVER
7341
GROMMET
0.22
GROMMET MUST BE
INSTALLED IN LEVERS
IN DIRECTION
INDICATED BY
ARROW
THIS DIMENSION MUST NOT BE EXCEEDED
DURING INSTALLATION. GROMMET CAN BE
DAMAGED BY INSERTING ROD TOO FAR

Manual linkage (column shift) Ford, Mercury full-size cars

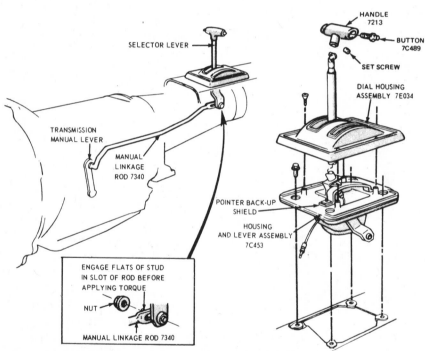

Floor shift linkage (Ford Pinto & Mustang)

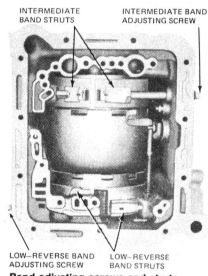

Band adjusting screws and struts

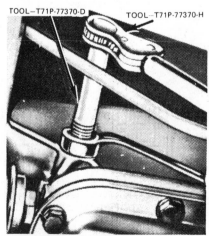

Adjusting low reverse band (C4)

wide open position. Hold the downshift rod against the through detent stop. Adjust the downshift screw to obtain 0.050–0.070 in. clearance between the screw tip and the throttle shaft lever tab.

3 Connect the downshift lever return spring.

Band Adjustments

Only the intermediate and low-reverse bands are adjustable.

INTERMEDIATE BAND

1 Clean all dirt from the adjusting screw area. Remove and discard the locknut.
2 Install new locknut. Tighten the adjusting screw to a torque of 10 ft./lb.

3 Back off screw exactly 1¾ turns.
4 Torque the locknut to 30 ft./lb.

LOW REVERSE BAND

1 Clean all dirt from adjusting screw area.
2 Remove and discard locknut. On Pinto and Bobcat it will be necessary to use a ¾ × 1½″ long socket to remove the locknut.
3 Install a new locknut. Torque the adjusting screw to 10 ft./lb. Back off screw 3 full turns.
4 Torque the locknut to 30 ft./lb.

Control Valve R&R

ALL MODELS

1 Perform the preliminary operations indicated above.
2 Shift the transmission to Park position and remove the two bolts holding the manual detent spring to the control valve body and case.
3 Remove all the valve body-to-case attaching bolts. Hold the manual valve in place and remove the valve body

from the case. If the manual valve is not held in place, it could be bent or damaged.

4 Refer to the Component Disassembly and repair section for control valve body repair procedures.

5 Thoroughly clean the old gasket material from the case and remove the nylon shipping plug from the oil filler tube hole. This nylon plug is installed before shipment and should be discarded when the transmission oil pan is removed.

6 Be sure the transmission is in the Park position (manual detent lever is in P detent position). Install the valve body in the case. Position the inner downshift lever between the downshift lever stop and the downshift valve. Be sure the two lands on the end of the manual valve engage the actuating pin on the manual detent lever.

7 Install seven valve body attaching bolts but do not tighten them.

8 Place the detent spring on the lower valve body and install the spring-to-case bolt finger tight.

9 While holding the detent spring roller in the center of the manual detent lever, install the detent spring-to-lower valve body bolt and tighten it to 80–120 in./lbs. torque.

10 Tighten the remainder of the control valve body attaching bolts to specifications.

11 Put a new gasket on the oil pan, install pan in place, and install and tighten all the pan attaching bolts to the proper torque.

12 If the filler tube was removed, reinstall it and tighten securely. If necessary, replace the oil seal around the filler tube to prevent leakage.

13 Lower the car and fill the transmission with fluid.

CHILTON CAUTION: *Always check neutral safety switch adjustment after adjusting shift linkage.*

Intermediate Servo R&R

1 Raise the car and remove the four servo cover attaching bolts (right-hand side of case). Remove the cover and identification tag *(do not lose tag).*

NOTE: To gain access to the servo on some models, the crossmember must be removed.

2 Remove the gasket, piston, and piston return spring.

3 Install the piston return spring in the case. Place a new gasket on the cover. Install new seals on the piston, lubricate the seals with transmission fluid, and install the piston in the servo cover. Install the piston and cover in the transmission case, using two 5/16–18 × 1¼ bolts 180 degrees apart to align the cover against the case.

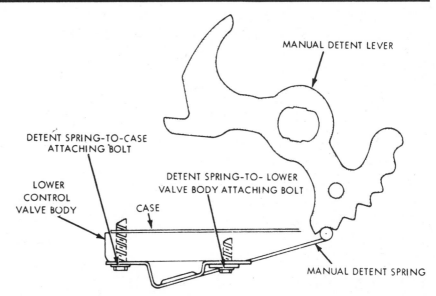

Control valve body detent spring installed

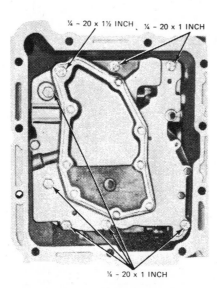

¼ – 20 x 1½ INCH   ¼ – 20 x 1 INCH

¼ – 20 x 1 INCH

Typical control valve body-to-case attaching bolts (C4)

NOTE: On 8-cylinder 1974 Comets and Mavericks, hold the servo cover in position with a pry bar between the floor pan and the cover while installing the retaining bolts and the identification tag.

4 Install the transmission identification tag and two attaching bolts. Remove the two 1¼ bolts and install the other two cover attaching bolts. Tighten all cover attaching bolts to the proper torque.

5 If removed, position the crossmember and install the attaching bolts, tightening them to the proper torque.

6 Adjust the intermediate band. If the intermediate band cannot be adjusted correctly, remove the control valve

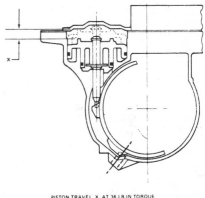

PISTON TRAVEL X AT 36 LB IN TORQUE MUST BE BETWEEN 120 AND 220 IN

Piston travel

body and see that the struts are installed correctly. Adjust the struts and reinstall the control valve body. Install the oil pan with a new gasket.

7 Lower the car and fill the transmission.

Low-Reverse Servo Piston R&R

1 Raise the car on a hoist.

2 Loosen the reverse band adjusting screw locknut and tighten the adjusting screw to 10 ft./lbs. torque. This operation will hold the band strut against the case and prevent it from falling when the reverse servo piston is removed.

3 Remove the four servo cover attaching bolts and remove the servo cover and seal from the case.

4 Remove the servo piston from the case. *The piston and piston seal are bonded together and must be replaced together.*

NOTE: To remove the piston from the shaft for replacement on a Pinto or Mustang II, insert a small screwdriver

Component Removal and Overhaul

through the hole in the shaft to hold the shaft, and remove the retaining nut. Install a new piston, and torque the nut to specifications.

5 Install the servo piston assembly into the case. Place a new cover seal on the cover and position it on the case using two 5/16–18 bolts, 1¼ in. long, 180° apart. Install two cover attaching bolts with the identification tag.
6 Remove the two positioning bolts and install the other two cover bolts. Tighten all the cover attaching bolts to the proper torque.
7 Adjust the low-reverse band. If the low-reverse band cannot be adjusted properly, remove the control valve body and check the alignment of the band struts. Reinstall the valve body and the oil pan with a new gasket.
8 Lower the car and fill the transmission.

Extension Housing Bushing & Rear Seal R&R

1 Disconnect the drive shaft from the transmission.
2 If only the rear seal needs replacing, carefully remove it with a tapered chisel or use a slide hammer. Be careful not to damage the spline seal with the bushing remover.
3 Install the new bushing.
4 Before installing a new rear seal, inspect the sealing surface of the universal joint yoke for scores. If the universal joint yoke is scored, replace the yoke.
5 Inspect the housing counterbore for burrs and remove them with crocus cloth if necessary.
6 Install the new rear seal into the housing. The seal should be firmly seated in the housing. Coat the inside diameter of the fiber portion of the seal with chassis lubricant.
7 Coat the front universal joint spline with chassis lubricant and install the drive shaft.

Extension Housing R&R

1 Raise the car on a hoist.

2 Remove the drive shaft. Place a transmission jack under the transmission for support.
3 Remove the speedometer cable from the extension housing.
4 Remove the extension housing-to-crossmember mount attaching bolts. Raise the transmission and remove the mounting pad between the extension housing and the crossmember.

NOTE: On 8-cylinder Comets and Mavericks it is necessary to disconnect the exhaust pipe from the exhaust manifolds and remove the crossmember in order to remove the crossmember in order to remove the engine rear support from the extension housing.

5 Loosen the extension housing attaching bolts to drain the transmission fluid.
6 Remove the six extension housing attaching bolts and remove the extension housing.
7 To install the extension housing, reverse the above removal instructions. Install a new extension housing gasket. When the extension housing has been installed and all parts have been secured, lower the car and fill the transmission with the correct amount of fluid. Check for fluid leaks around the extension housing area.

Governor R&R (C4 Automatic Only)

1 After removing the extension housing according to the instructions above, remove the governor housing-to-governor distributor attaching bolts. Remove the governor housing from the distributor.
2 Refer to the Component Disassembly and Repair section for instructions on repairing the governor assembly.
3 Install the governor housing on the governor distributor and tighten the attaching bolts to the proper torque.
4 Install the extension housing with a new gasket according to the instructions above.
5 When the extension housing has been installed and all bolts have been tight-

ened to the proper torque, lower the car and fill the transmission with fluid to the proper level. Check around the extension housing area for leaks.

Overhaul

Disassembly

1 Remove the converter from the transmission front pump and converter housing.
2 Remove the transmission vacuum unit, and vacuum unit control rod.
3 Remove the primary throttle valve from the opening at the rear of the case.
4 Remove the two extension housing-to-case bolts and mount the transmission in the holding fixture.
5 Remove the transmission pan attaching bolts, pan and gasket.
6 Remove the control valve body attaching bolts. Remove the control valve body from the case.
7 Loosen the intermediate band adjusting screw and remove the intermediate band struts from the case. Loosen the low-reverse band adjusting screw and remove the low-reverse band struts.

Transmission End Play Check

1 To keep the output shaft in alignment during the end play check, install the extension housing oil seal replacer tool or a front universal joint yoke in the extension housing.
2 Remove one of the converter housing-to-case attaching bolts and mount the dial indicator.
3 The input shaft is a loose part and has to be properly engaged with the spline of the forward clutch hub during the end play checking procedure. Move the input shaft and gear train toward the rear of the transmission case.
4 With the dial indicator contacting the end of the input shaft set the indicator at zero.
5 Insert a screwdriver behind the input shell. Move the input shell and the front part of the gear train forward.
6 Record the dial indicator reading. The end play should be 0.008 to 0.042 inch. If the end play is not within limits, the selective thrust washers must be replaced as required. The selective thrust washers (No. 1 and 2) can be replaced individually to obtain the specified end play.
7 Remove the dial indicator and remove the input shaft from the front pump stator support.

Removal of Case and Extension Housing Parts

1 Rotate the holding fixture to put the transmission in a vertical position with the converter housing up.
2 Remove the five counverter housing-to-case attaching bolts. Remove the converter housing from the transmission case.

SELECTIVE THRUST WASHERS
(FOR END-PLAY CORRECTION)

THRUST WASHER NO. 1		THRUST WASHER NO. 2	
0.053-0.0575	Red		
0.070-0.0745	Green	No. Stamped Washer	Metal Thrust Washer
0.087-0.0915	Natural		
		2	0.058-0.056
		3	0.075-0.073
		SPACER	0.036-0.032
		(This is a selective spacer used with washer 2 or 3. When used, install next to stator support.)	

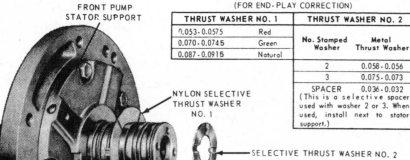

FRONT PUMP STATOR SUPPORT

NYLON SELECTIVE THRUST WASHER NO. 1

SELECTIVE THRUST WASHER NO. 2

Selective thrust washer locations and identification

FORD C4—C4S

3 Remove the seven front pump attaching bolts. Insert a screwdriver behind the input shell and move the input shell forward until the front pump seal is above the edge of the case. Remove the front pump and gasket from the case. If the selective thrust washer No. 1 did not come out with the front pump, remove it from the top of the reverse-high clutch.

4 Remove the intermediate and low-reverse band adjusting screws from the case. Rotate the intermediate band to align the band ends with the clearance hole in the case. Remove the intermediate band from the case. **If the intermediate band is to be re-used, it must not be cleaned in a vapor de-**

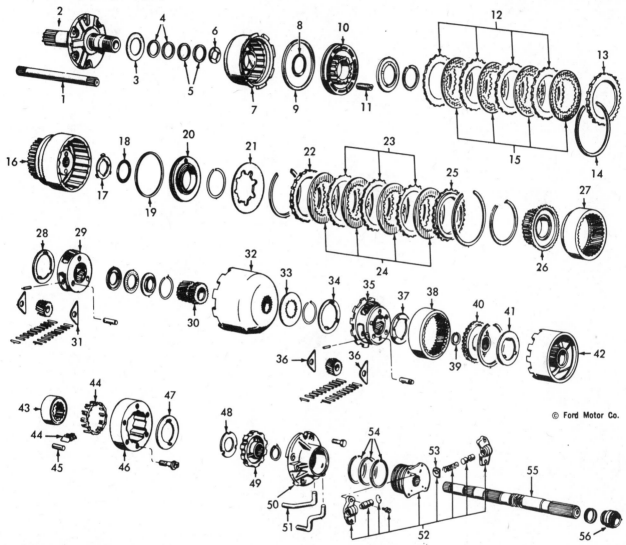

© Ford Motor Co.

Gear train parts identification

1. INPUT SHAFT
2. STATOR SUPPORT
3. THRUST WASHER
4. OIL SEAL
5. OIL SEAL
6. THRUST WASHER KIT
7. FRONT BRAKE DRUM
8. OIL SEAL
9. CLUTCH PISTON OIL SEAL
10. CLUTCH PISTON
11. CLUTCH PISTON SPRING
12. CLUTCH PLATE—DRIVEN
13. CLUTCH PRESSURE PLATE
14. SNAP RING
15. CLUTCH PLATE—DRIVE
16. FORWARD CLUTCH CYLINDER
17. THRUST WASHER
18. OIL SEAL
19. OIL SEAL, CLUTCH PISTON

20. FORWARD CLUTCH PISTON
21. PISTON DISC SPRING
22. CLUTCH PRESSURE PLATE
23. CLUTCH PLATE—DRIVEN
24. CLUTCH PLATE—DRIVE
25. CLUTCH PRESSURE PLATE
26. CLUTCH HUB
27. INTERNAL GEAR
28. THRUST WASHER
29. FOREWARD DRUM AND PLANETARY ASSEMBLY
30. SUN GEAR
31. THRUST WASHER
32. INPUT SHELL
33. THRUST WASHER
34. THRUST WASHER
35. REVERSE PLANET ASSEMBLY
36. THRUST WASHER
37. THRUST WASHER

38. INTERNAL GEAR
39. RING
40. OUTPUT SHAFT HUB
41. THRUST WASHER
42. REVERSE BREAK DRUM
43. INNER RACE
44. CLUTCH SPRING RETAINER
45. ROLLER
46. OUTER RACE
47. THRUST WASHER
48. THRUST WASHER
49. PARKING GEAR
50. OIL DISTRIBUTOR SLEEVE
51. FRONT OIL TUBE
52. GOVERNOR ASSEMBLY
53. SPRING RETAINER
54. OIL SEALS
55. OUTPUT SHAFT
56. OIL SEAL

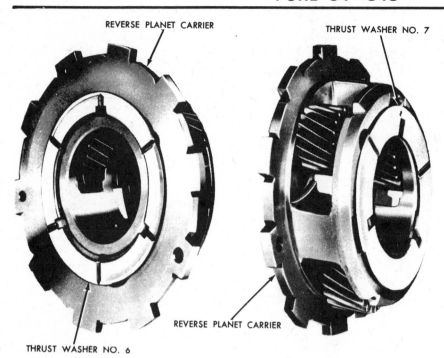

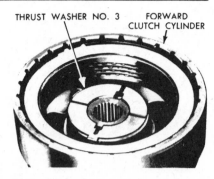

#3 thrust washer location

6 & 7 thrust washer location

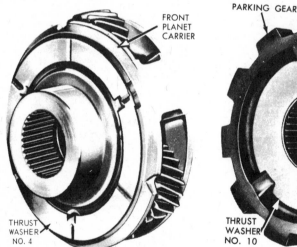

#4 and 10 thrust washer locations

greaser or with a detergent solution. Wipe it clean with a lint-free cloth.

5   Using a screwdriver between the input shell and rear planet carrier, lift the input shell upward and remove the forward part of the gear train as an assembly.

6   Place the forward part of the gear train in the holding fixture.

7   With the gear train in the holding fixture, remove the reverse-high clutch and drum from the forward clutch.

8   If thrust washer No. 2 did not come out with the front pump, remove the thrust washer from the forward clutch cylinder. If a selective spacer was used, remove the spacer. Remove the forward clutch from the forward clutch hub and ring gear.

9   If thrust washer No. 3 did not come out with the forward clutch, remove the thrust washer from the forward clutch hub.

10   Remove the forward clutch hub and ring gear from the front planet carrier.

11   Remove thrust washer No. 4 and the front planet carrier from the input shell.

12   Remove the input shell, sun gear and thrust washer No. 5 from the holding fixture.

13   From inside the transmission case remove thrust washer No. 6 from the top of the reverse planet carrier.

14   Remove the reverse planet carrier and thrust washer No. 7 from the reverse ring gear and hub.

15   Move the output shaft forward and remove the reverse ring gear hub-to-output shaft snap ring.

16   Remove the reverse ring gear and hub from the output shaft. Remove thrust washer No. 8 from the low and reverse drum.

17   Remove the low-reverse band from the case.

18   Remove the low-reverse drum from the one-way clutch inner race.

19   Rotate the one-way clutch inner race clockwise, and remove inner race from the outer race.

20   Remove the 12 one-way clutch rollers, springs and the spring retainer from the outer race. **Do not lose or damage any of the 12 springs or rollers. The outer race of the one-way clutch can-**

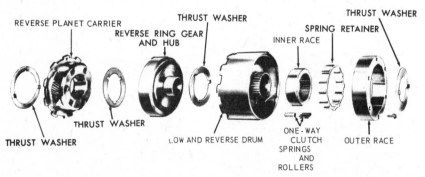

Lower part of gear train

FORD C4—C4S

not be removed from the case until the extension housing, output shaft and governor distributor sleeve are removed.

21 Remove the transmission from the holding fixture. Position the transmission on the bench in a vertical position, with the extension housing up. Remove the four extension housing-to-case attaching bolts. Remove the extension housing and gasket from the case.

22 Pull outward on the output shaft and remove shaft and governor distributor assembly from the governor distributor sleeve.

23 Remove the governor distributor snap ring from the output shaft. Remove the governor distributor from the output shaft.

24 Remove the four distributor sleeve-to-case attaching bolts. Remove the distributor sleeve from the case. Do not bend or distort the fluid tubes as the tubes are removed from the case with the distributor sleeve.

25 Remove the parking pawl return spring, pawl, and pawl retaining pin from the case.

26 Remove the parking gear and thrust washer No. 10 from the case.

27 Remove the six one-way clutch outer race-to-case attaching bolts. As the bolts are removed, hold the outer race located inside the case in position. Remove the outer race and thrust washer No. 9 from the case.

28 If the transmission case bushing is worn or damaged, replace it.

Assembly

When assembling the transmission sub-assemblies, make sure that each thrust washer is installed in the correct position. Lubricate internal parts with clean transmission fluid. Vaseline should be used to hold the thrust washers in their proper location. If the end play is not within specifications after the transmission is assembled, either the wrong selective thrust washers (No. 1 and 2) were used, or a thrust washer came out of position during the transmission assembly operation.

1 Install thrust washer No. 9 inside the transmission case.

2 Place the one-way clutch outer race inside the case. From the back of the case install the six outer race-to-case attaching bolts. Torque the bolts according to Specifications in this Part.

3 Place the transmission case in a vertical position with the back face of the case upward. Install the parking pawl retaining pin in the case. Install the parking pawl on the case retaining pin, and the parking pawl return spring.

4 Install thrust washer No. 10 on the parking gear. Place the gear and thrust washer on the back face of the case.

5 Place the two distributor tubes in the governor distributor sleeve. Install the

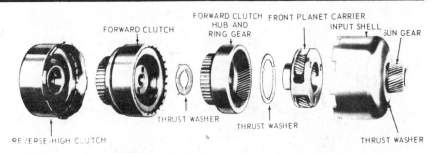

Forward part of gear train

distributor sleeve on the case. As the distributor sleeve is installed, insert the tubes in the two holes in the case and insert the parking pawl retaining pin in the alignment hole in the distributor sleeve.

6 Install the four governor distributor sleeve-to-case attaching bolts and torque the bolts to specification.

7 Install the governor distributor assembly on the output shaft. Install the distributor snap ring.

8 Check the rings in the governor distributor, making sure the rings are fully inserted in the ring grooves and rotate freely. Install the output shaft and governor distributor assembly in the distributor sleeve.

9 Place a new extension housing gasket on the case. Install the extension housing, vacuum tube clip and the extension housing-to-case attaching bolts. Torque the bolts to specification.

10 Place the transmission in the holding fixture. Rotate the holding fixture to put the transmission in a vertical position with the front pump mounting face up. Make sure thrust washer No. 9 is still located at the bottom of the transmission.

11 Install the one-way clutch spring retainer into the outer race. Install the inner race inside of the spring retainer. **Be sure the face with the step is installed toward the rear of the case, mating with the thrust washer.**

12 Install the twelve springs between the inner and outer race. Starting at the

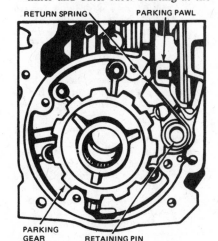

Parking pawl and gear

back of the transmission case, install the twelve one-way clutch rollers by slightly compressing each spring and positioning the roller between the spring and the spring retainer. Rotate the inner race clockwise, after assembly, to center the rollers and springs.

13 Install the low and reverse drum. The splines of the drum have to engage with the splines of the one-way clutch inner race. Check the one-way clutch operation by rotating the low and reverse drum. The drum should rotate clockwise but not rotate counterclockwise.

14 Install thrust washer No. 8 on top of the low and reverse drum. Install the low-reverse band in the case, with the end of the band for the small strut toward the low-reverse servo.

15 Install the reverse ring gear and hub on the output shaft. Move the output shaft forward and install the reverse ring gear hub-to-output shaft snap ring.

16 Place thrust washers No. 6 and 7 on the reverse planet carrier.

17 Install the planet carrier in the reverse ring gear and engage the tabs of the carrier with the slots in the low-reverse drum.

18 On the bench, install the forward clutch in the reverse-high clutch. Rotate the units to mesh the reverse-high clutch plates with the splines of the forward clutch.

19 Using the recorded dial indicator readings obtained during transmission disassembly, determine which No. 2 steel-backed thrust washer is required and proceed as follows:

a. Position the stator support vertically on the work bench and install the correct No. 2 thrust washer, or combination of washer and spacer, as required to bring the end play within limits of 0.008 to 0.042 inch.

b. Install the reverse-high clutch and the forward clutch on the stator support.

c. Invert the complete unit making sure that the intermediate brake drum bushing is seated on the forward clutch mating surface.

d. Select the thickest fiber washer (No. 1) that can be inserted between the stator support and the intermediate brake drum thrust surfaces and still maintain a slight

FORD C4—C4S

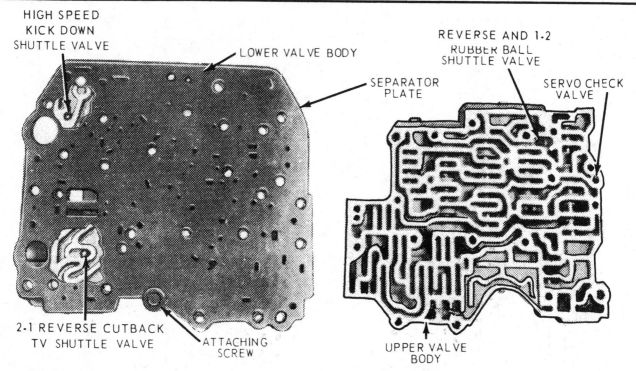

HIGH SPEED KICK DOWN SHUTTLE VALVE

LOWER VALVE BODY

REVERSE AND 1-2 RUBBER BALL SHUTTLE VALVE

SEPARATOR PLATE

SERVO CHECK VALVE

2-1 REVERSE CUTBACK TV SHUTTLE VALVE

ATTACHING SCREW

UPPER VALVE BODY

Separating upper and lower valve bodies

clearance. **Do not select a washer that must be forced between the stator support and intermediate brake drum.**

e. Remove the intermediate brake drum and forward clutch unit from the stator support.

f. Install the selected thrust washers (No. 1 and 2) on the front pump stator support using enough vaseline to hold the thrust washers in position during the front pump installation.

20 Install thrust washer No. 3 on forward clutch.

21 Install the forward clutch hub and ring in the forward clutch. Rotate the units to mesh the forward clutch plates with the splines on the forward clutch hub.

22 Install thrust washer No. 4 on the front planet carrier. Install the front planet carrier into the forward clutch hub and ring gear. **Check the forward thrust bearing race inside the planet carrier for proper location against the thrust bearing. Make sure the race is centered for alignment with the sun gear on the input shell.**

23 Install the input shell and sun gear on the gear train. Rotate the input shell to engage the drive lugs of the reverse-high clutch. **If the drive lugs do not engage, the outer race inside the forward planet carrier is not centered to engage the end of the sun gear inside the input shell.** Center the thrust bearing race and install the input shell.

24 Hold the gear train together and install the forward part of the gear train assembly in the case.

The input shell sun gear must mesh with the reverse pinion gears. The front planet carrier internal splines must mesh with the splines on the output shaft.

25 **A new band should be soaked in transmission fluid for fifteen minutes before it is installed.** Install the intermediate band through the front of the case.

26 Install a new front pump gasket on the case. Line up the bolt holes in the gasket with the holes in the case.

27 Install the front pump stator support into the reverse-high clutch. Align the pump-to-case attaching bolt holes. Install the front pump-to-case attaching bolts and torque them to specification as listed at the end of this Part.

28 Install the input shaft. **Be sure the short splined end is installed toward the rear of the transmission.** Rotate the holding fixture to place the transmission in a horizontal position. Check the transmission end play. If the end play is not within specification, either the wrong selective thrust washers were used, or one of the 10 thrust washers is not properly positioned.

29 Remove the dial indicator used for checking the end play and install the one front pump-to-case attaching bolt. Torque the bolt to specification.

30 Place the converter housing on the transmission case. Install the five converter housing-to-case attaching bolts. Torque the bolts to specification.

31 Install the intermediate and low-reverse band adjusting screws in the case. Install the struts for each band.

32 Adjust the intermediate and low-reverse band.

33 Install a universal joint yoke on the output shaft. Rotate the input and output shafts in both directions to check for free rotation of the gear train.

34 Install the control valve body. Refer to the Removal and Installation Section of this Part for the control valve body installation procedures.

35 Place a new pan gasket on the case and install the pan and pan-to-case attaching bolts. Torque the attaching bolts to specification.

36 Remove the transmission from the holding fixture. Install the two extension housing-to-case attaching bolts. Torque the bolts to specification.

37 Install the primary throttle valve in the transmission case.

38 Install the vacuum unit and control rod in the case.

39 Make sure the input shaft is properly installed in the front pump stator support and gear train. The short splined end of the shaft should be installed toward the rear of the transmission. Install the converter in the front pump and the converter housing.

Control Valve Body

Disassembly

When the main control is disassembled and the valve body-to-screen gasket is removed, the gasket should not be cleaned in a degreaser, solvent or any type of detergent solution. To clean the gasket, wipe it off with a lint-free cloth.

FORD C4—C4S

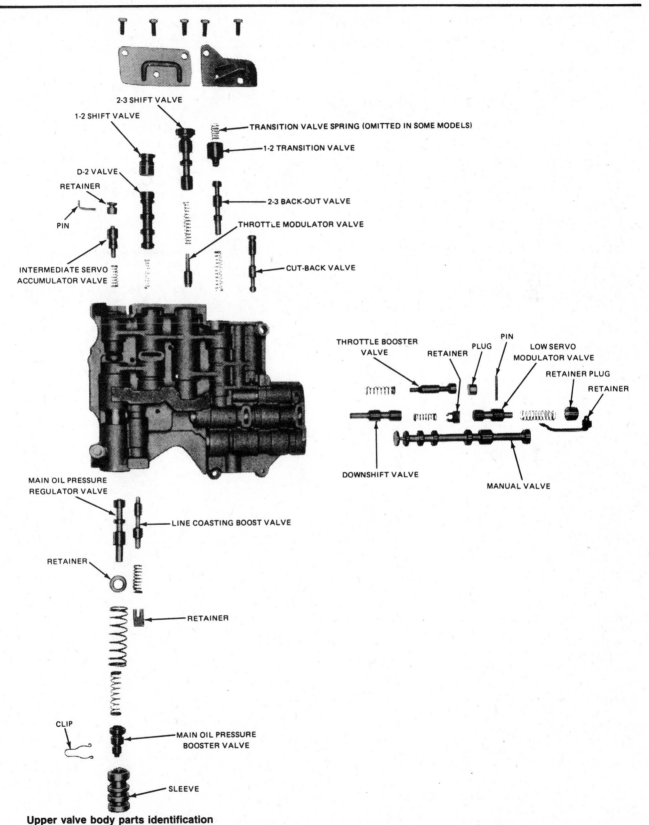

Upper valve body parts identification

1 Remove the eight 10–24 × 1⅜-inch screws that attach the oil screen to the body and remove the screen and gasket. Be careful not to lose the throttle pressure limit valve and spring when separating the oil screen from the valve body.

2 Remove the two ¼–20 × 1½-inch attaching screws from the upper valve body and nine 10 × 24 × ⅞-inch attaching screws from the underside of the lower valve body. Separate the lower valve body, gasket and separator plate from the upper valve body. **Be careful not to lose the upper valve**

Component Removal and Overhaul

FORD C4—C4S

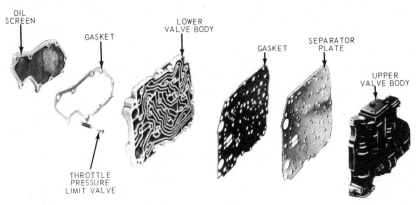

Upper and lower valve bodies

body shuttle valve and check valve when separating the upper and lower valve bodies.

3 Remove the manual valve retaining ring and slide the manual valve out of the body. The retaining ring is used to hold the manual valve in its bore during shipment and should be removed and discarded before sliding the manual valve out of the control valve body.

4 Carefully pry the low servo modulator valve retainer from the body and remove the retainer plug, spring and valve from the body. While working in the low servo modulator valve bore, pry the downshift valve retainer from the body and remove the spring and downshift valve.

5 Using side cutters, carefully pry the throttle booster valve plug retaining pin from the valve body. Remove the plug, valve and spring.

6 Remove the cut-back valve and transition valve cover plate from the valve body.

7 Remove the cut-back valve from the body.

8 Remove the transition valve spring, transition valve, 2–3 back-out valve and spring from the body.

9 Remove the 1–2 shift valve and 2–3 shift valve cover plate from the body.

10 Remove the 2–3 shift valve, spring and throttle modulator valve from the body.

11 Remove the 1–2 shift valve, D-2 valve and spring from the body.

12 Remove the intermediate servo retaining pin and remove the intermediate accumulator retainer, valve and spring from the body.

13 Press the main oil pressure booster valve inward and remove the retaining pin. Remove the main oil pressure booster valve, sleeve, springs, retainer and the main oil pressure regulator valve.

14 Remove the line coasting boost valve retainer from the body and remove the spring and line coasting boost valve.

Assembly

1 Place the shuttle valves in the upper and lower bodies. Position a new gasket and the separator plate on the lower body and install but do not tighten the attaching screw.

2 Place the lower body and plate assembly on the upper valve body and install the 11 attaching screws finger tight.

3 Install the oil screen screws loosely, without the screen, to properly align the upper and lower valve bodies, gasket and separator plate.

4 Torque the four bolts that are covered by the screen to specification.

5 Position the throttle pressure limit valve and spring in the lower valve body. Remove the oil screen attaching screws and place the gasket and oil screen in position on the lower valve body. Reinstall the screen attaching screws. Torque all the valve body and screen attaching screws to specification.

6 **Insert the downshift valve into the body with the small diameter facing inward.** Install the downshift valve spring and retainer. Insert the low servo modulator valve, spring and retainer plug in the body. Depress the plug and install the retainer.

7 Place the throttle booster valve spring, valve (small diameter end into spring) and plug into the body. Depress the plug and press in the retaining pin. **Be sure the three grooves are at the top of the pin as it is installed.**

8 Place the spring, 2–3 back-out valve and the transition valve and spring in the body.

9 Place the cut-back valve in the body. Secure the cut-back and the transition valve cover plate to the body with the two attaching screws. Torque the screws to specification.

10 Place the throttle modulator valve, spring and 2–3 shift valve in the body.

11 Place the spring, D-2 valve and the 1–2 shift valve in the body.

12 Secure the 1–2 shift valve and the 2–3 shift valve cover plate to the body with the three attaching screws. Torque the screws to specification.

13 Place the spring, intermediate servo accumulator valve and retainer in the body. Depress the retainer and install the retaining pin.

14 Insert the line coasting boost valve and spring in the body. Depress the spring and install the retainer.

15 Insert the main oil pressure regulator valve and spring retainer in the body. Install the two springs, sleeve and the main oil pressure booster valve in the body.

16 Hold the main oil pressure booster valve in place and install the retaining pin.

17 Slide the manual valve into the valve body. **Make sure that the end with the two lands closest together is inserted first.**

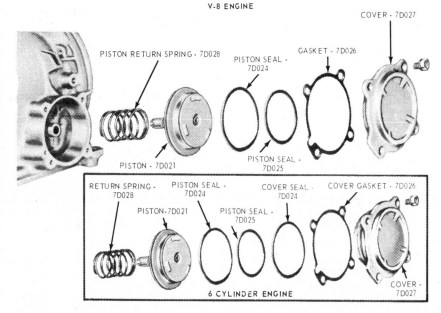

Intermediate servo

144

FORD C4—C4S

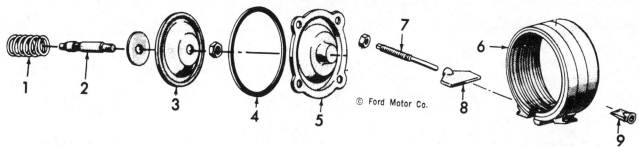

© Ford Motor Co.

1. PISTON SPRING
2. PISTON ROD
3. SERVO PISTON

4. PISTON SEAL
5. COVER
6. REVERSE BAND

7. ADJUSTING SCREW
8. ANCHOR STRUT
9. ACTUATING STRUT

Rear servo parts breakdown

Intermediate Servo

Disassembly

1 Remove the transmission identification tag and the four servo cover-to-case attaching bolts.
2 Remove the servo cover, gasket, servo piston, and piston return spring from the case.
3 Remove the intermediate servo piston from the cover.
4 Remove the seal rings from the servo piston and cover.

Assembly

1 Install a new seal on the cover and servo piston. Lubricate the seals with clean transmission fluid. Install the piston into the cover. Be careful not to damage the piston seal.
2 Install the piston return spring in the servo bore of the case.
3 Place a new gasket on the servo cover. **Position the servo piston and cover assembly into the case with the piston stem slot in a horizontal position to engage the strut.** Use two 5/16–18 bolts, 1¼ inches long, 180 degrees apart, to position the cover against the case. Install two cover attaching bolts. Remove the two 1¼-inch bolts and install the transmission identification tag and the other two cover attaching bolts. Torque the bolts to specification.

Low-Reverse Servo

Disassembly

1 Remove the four servo cover-to-case attaching bolts.

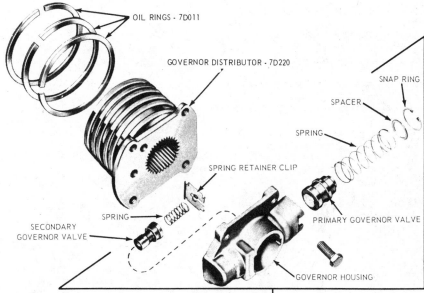

OIL RINGS - 7D011

GOVERNOR DISTRIBUTOR - 7D220

SNAP RING

SPACER

SPRING

SPRING RETAINER CLIP

PRIMARY GOVERNOR VALVE

SPRING

SECONDARY GOVERNOR VALVE

GOVERNOR HOUSING

GOVERNOR ASSEMBLY - 7C063

Governor

2 Remove the servo cover, cover seal, servo piston and piston return spring from the case.
3 The servo piston seal is bonded to the piston. If the seal has to be replaced, replace the piston assembly which includes the seal.
4 On Pinto, Bobcat and Mustang, disassemble the servo piston from the piston stem by inserting a small screwdriver in the hole of the piston stem

and removing the piston attaching nut.

Assembly

1 On Pinto, Bobcat and Mustang, position the accumulator spring and spacer on the piston stem if they were previously removed. **Make sure the crowned side of the spring is facing toward the rod.** Install a new servo piston on the stem and install the at-

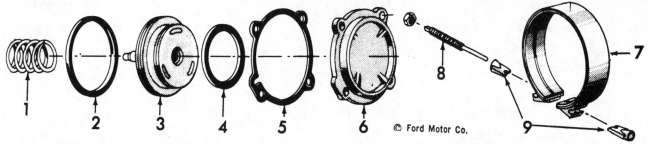

© Ford Motor Co.

1. PISTON SPRING
2. PISTON SEAL
3. SERVO PISTON

4. PISTON SEAL
5. GASKET
6. COVER

7. BAND
8. ADJUSTING SCREW
9. ANCHOR STRUT

Front servo parts breakdown

Component Removal and Overhaul

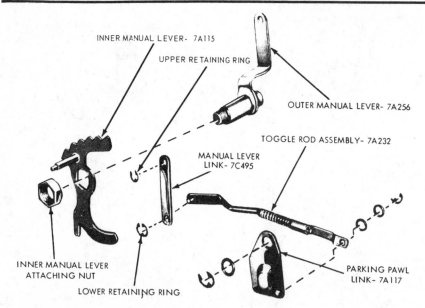

INNER MANUAL LEVER- 7A115

UPPER RETAINING RING

OUTER MANUAL LEVER- 7A256

TOGGLE ROD ASSEMBLY- 7A232

MANUAL LEVER LINK- 7C495

INNER MANUAL LEVER ATTACHING NUT

LOWER RETAINING RING

PARKING PAWL LINK- 7A117

Transmission case internal linkage

taching nut. Torque the nut to specification.
2 Place the piston return spring in the servo bore of the case. Lubricate the piston seal with clean transmission fluid and install the servo piston into the bore of the case.
3 Place a new cover seal on the cover and install the servo cover. Install the four cover attaching bolts. Torque the cover-to-case retaining bolts to specification.

Governor

Disassembly
1 Remove the oil rings from the governor oil distributor.
2 Remove the governor housing-to-distributor attaching bolts. Remove the governor from the oil distributor.
3 Remove the primary governor valve snap ring. Remove the washer, spring, and primary governor valve from the housing.
4 Remove the secondary governor valve spring retaining clip, spring, and governor valve from the housing.

Assembly
1 Install the secondary governor valve in the housing. Install the spring and retaining clip. **Make sure the clip is installed with the small concaved area facing downward, to hold the spring in the correct position.**
2 Install the primary governor valve in the housing. Install the spring washer and snap ring. **Make sure the washer is centered in the housing on top of the spring and the snap ring is fully seated in the ring groove of the housing.**
3 Install the governor assembly on the oil distributor and torque the attaching bolts to specification as listed at the end of this Part.

4 Install the oil rings on the distributor. Check the oil rings for free rotation in the ring grooves on the oil distributor.

Downshift and Manual Linkage
Disassembly
1 Apply penetrating oil to the outer lever attaching nut to prevent breaking the inner lever shaft. Remove the downshift outer lever nut and remove the downshift outer and inner levers.
2 On vehicles equipped with a case mounted neutral start switch, place a screwdriver behind the switch and carefully pry the switch off the lever.
3 From inside the transmission case, remove the upper retaining ring from the manual lever link. Remove the upper end of the lever link from the case retaining pin.
4 From the back of the transmission case, remove the upper retaining ring and flat washer from the parking pawl link. Remove the pawl link from the case retaining pin.
5 From the back of the transmission case, remove the parking pawl link, toggle rod, and manual lever link as an assembly.
6 Remove the rear parking pawl link lower retaining ring, flat washer and link from the toggle rod.
7 Remove the manual lever link, lower retaining ring, flat washer, and link from the toggle rod.
8 Remove the inner manual lever attaching nut and lever. Remove the outer manual lever from the case.
9 To remove the manual lever seal. To install the new seal, use a driver that fits the ID of the seal.

Assembly
1 Install the manual and parking pawl links, flat washers and retaining rings to the toggle rods.

2 Install the outer manual lever in the case. Install the inner manual lever and attaching nut with the chamfer facing toward the lever. Torque the nut to specification.
3 From the back of the transmission case, install the parking toggle rod and link assembly into the case.
4 Install the parking pawl link on the case retaining pin. Install the flat washer and link retaining ring.
5 Position the inner manual lever behind the manual lever link, with the cam on the lever contacting the lower link pin.
6 Install the upper end of the manual lever link on the case retaining pin. Install the retaining ring.
7 Operate the manual lever and check for correct linkage operation.
8 On vehicles equipped with a case mounted neutral start switch, reinstall the switch. Adjust the switch after the control valve body has been installed. Torque the attaching bolts to specification.
9 Install the inner and outer downshift levers. Torque the attaching nut to specification.

Front Pump
Disassembly
1 Remove the four seal rings from the stator support.
2 Remove the five bolts that attach the stator support to the front pump housing. Remove the stator support from the pump housing.
3 Remove the front and rear stator bushings if they are worn or damaged. Use the cape chisel and cut along the bushing seam until the chisel breaks through the bushing wall. Pry the loose ends of the bushing up with an awl and remove the bushing.
4 Remove the drive and driven gears from the front pump housing.
5 Press the bushing from the front pump housing.

Assembly
1 Press a new bushing into the pump housing. **Make sure the bushing is installed with the slot and groove positioned to the rear of the pump body and 60 degrees below the horizontal center line.**
2 Install the drive and driven gears in the pump housing. **Each gear has an identification mark on the side of the gear teeth that are chamfered. The chamfered side with the identification mark has to be positioned downward against the face of the pump housing.**
3 Press new bushings into the stator support. Use the long end of the tool for the front bushing and the short end for the rear bushing. **When installing the rear bushing, be sure the hole in the bushing is lined up with the lube hole in the stator support.**
4 Place the stator support in the pump

FORD C4—C4S

housing and install the five attaching bolts. Torque the bolts to specification.

5 Install the four seal rings on the stator support. **The two large oil rings are assembled first in the oil ring grooves toward the front of the stator support.**

6 Check the pump for free rotation by placing the pump on the converter drive hub in its normal running position and turning the pump housing.

7 If the front pump seal must be replaced, mount the pump in the transmission case and remove the seal.

Reverse-High Clutch

Disassembly

1 Remove the pressure plate snap ring.

2 Remove the pressure plate, and the drive and driven clutch plates. **If the composition clutch plates are to be reused, do not clean them in a vapor degreaser or with a detergent solution. Wipe the plates with a lint-free cloth.**

3 Place the reverse-high clutch drum in an arbor press. Compress the piston return springs and remove the snap ring. When the arbor press ram is released, guide the spring retainer to clear the snap ring groove of the drum.

4 Remove the spring retainer and piston return spring.

5 Remove the piston by inserting air pressure in the piston hole of the clutch drum.

6 Remove the piston outer seal from the piston and the piston inner seal from the clutch drum.

Assembly

1 Install a new inner seal in the clutch drum and a new outer seal on the clutch piston. Lubricate the seals with clean transmission fluid and install the piston into the clutch drum.

2 Place the clutch piston spring into position on the clutch piston. Place the spring retainer on top of the spring, and install assembled parts in reverse-high clutch drum. To install the snap ring, use the tools shown in and place assembled clutch drum in an arbor press.

As the press ram is moved downward, make sure the spring retainer is centered to clear the snap ring groove. Install the snap ring.

3 When new composition clutch plates are used, soak the plates in transmission fluid for fifteen minutes before installing them. Install the clutch plates alternately starting with a steel plate, then a non-metallic plate. The last plate installed is the pressure plate. For the correct number of clutch plates required for each transmission model, refer to specifications.

4 Install the pressure plate snap ring. Make sure the snap ring is fully seated in the snap ring groove of the clutch hub.

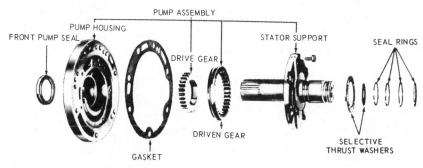

Reverse-high clutch

5 With a feeler gauge, check the clearance between the snap ring and the pressure plate.

6 The pressure plate should be held downward as the clearance is checked. The clearance should be 0.-050 to 0.071 inch. If the clearance is not within limits, selective snap rings are available in the following thicknesses: 0.050 to 0.054, 0.064 to 0.068, 0.078 to 0.082 and 0.092 to 0.096 inch. Install the correct size snap ring and recheck the clearance.

Forward Clutch

Disassembly

1 Remove the clutch pressure plate snap ring.

2 Remove the pressure plate, and the drive and driven clutch plates from the clutch hub.

3 Remove the disc spring snap ring.

4 Apply air pressure at the clutch piston pressure hole to remove the piston from the clutch hub.

5 Remove the clutch piston outer seal and the inner seal from the clutch hub.

Assembly

1 Install new clutch piston seals on the clutch piston and drum. Lubricate the seals with clean transmission fluid.

2 Install the clutch piston into the clutch hub. Install the disc spring and retaining snap ring.

3 **Install the lower pressure plate with the flat side up and the radius side downward.**

4 Install one non-metallic clutch plate and alternately install the drive and driven plates. Before installing new

Front pump and stator support

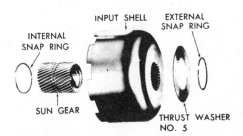

Input shell and sun gear

composition clutch plates, soak them in transmission fluid for fifteen minutes. The last plate installed will be the upper pressure plate. Refer to Specifications for the correct number of clutch plates for the applicable model transmission.

5 Install the pressure plate snap ring. Make sure the snap ring is fully seated in the ring groove of the clutch hub.

6 With a feeler gauge, check the clearance between the snap ring and the pressure plate. Downward pressure on the plate should be used when making this check. The clearance should be 0.025 to 0.050 inch.

7 If the clearance is not within limits, selective snap rings are available in the following thicknesses: 0.050 to 0.054, 0.064 to 0.068, 0.078 to 0.082 and 0.-092 to 0.096 inch. Insert the correct size snap ring and recheck the clearance.

Forward Clutch Hub and Ring Gear

Disassembly

1 Remove the forward clutch hub snap ring.

Component Removal and Overhaul

2 Remove the forward clutch hub from the ring gear.

3 Press the bushing from the clutch hub.

Assembly

1 Install a new bushing into the clutch hub.

2 Install the forward clutch hub in the ring gear. Make sure the hub is bottomed in the groove of the ring gear.

3 Install the front clutch hub snap ring. Make sure the snap ring is fully seated in the snap ring groove of the ring gear.

Input Shell and Sun Gear

Disassembly

1 Remove the external snap ring from the sun gear.

2 Remove thrust washer No. 5 from the input shell and sun gear.

3 From inside the input shell, remove the sun gear. Remove the internal snap ring from the sun gear.

4 If the sun gear bushings are to be replaced, press both bushings through the gear.

Assembly

1 Press a new bushing into each end of the sun gear.

2 Install the internal snap ring in the sun gear. Install the sun gear in the input shell.

3 Install thrust washer No. 5 on the sun gear and input shell.

4 Install the external snap ring on the sun gear.

Reverse Ring Gear and Hub

Disassembly

1 Remove the hub snap ring from the reverse ring gear.

2 Remove the hub from the reverse ring gear.

Assembly

1 Install the hub in the reverse ring gear. Make sure the hub is fully seated in the groove of the ring gear.

2 Install the snap ring in the reverse ring gear. Make sure the snap ring is fully seated in the snap ring groove of the ring gear.

Low-Reverse Brake Drum Bushing Removal and Installation

1 Replace the low-reverse brake drum bushing if it is worn or damaged. To remove the bushing, use the cape chisel and cut along the bushing seam until the chisel breaks through the bushing wall. Pry the loose ends of the bushing up with an awl and remove the bushing.

2 Install a new bushing.

Ford C4 & C4S Torque Specifications

Item	FT./LBS.	IN./LBS.
Converter to flywheel	23-28	
Converter hsg. to trans.	28-40	
Front pump to trans. case	28-38	
Outer race to case	13-20	
Oil pan to case	12-16	
Rear servo cover to case	12-20	
Stator support to pump	12-20	
Converter cover to converter hsg.	12-16	
Intermediate servo cover to case	16-22	
Extension hsg. to case	28-40	
Converter drain plug	20-30	
Pressure gauge tap	9-15	
Manual control lever nut	8-12	
Downshift lever to shaft	12-16	
Filler tube to engine	20-25	
Filler tube to pan	32-42	
Diaphragm assy to case	15-23	
Distributor sleeve to case	12-20	
Reverse servo piston to rod	12-20*	
Transmission to engine	40-50	
Transmission to engine: Falcon, Mustang, Pinto	28-38	
Cooler bracket & oil pan to case	12-16	
Band adjust screws to case	35-45	
Yoke to output shaft	60-120	
T.R.S. switch to case	4-8	
End plate to valve body screw		25-40
Lower to upper valve body bolts		40-60
Screen to valve body screws		40-60
Neutral switch to case screws		55-75
Screen & valve body to case bolts		80-120
Screen to valve body bolts		40-55
Governor body to distributor (collector) body bolts		80-120
Cooler line fittings		80-120
Reinforcement plate to body		40-55

*On 1974 Pintos, tighten to 10 ft./lbs. and back off 5/8 turn.

Intermediate Servo Piston—Cover and Spring Identification

Vehicle	Engine	Cover Piston Ident.	Spring	
			Free Length	Paint Stripe
Versailles	302-2V	A	2.48	White
Versailles	351-2V	R		

148

FORD C6

TROUBLESHOOTING
FORD C6

	ITEMS TO CHECK	
THE PROBLEM	**IN THE CAR**	**OUT OF CAR**
NO DRIVE IN D, 2 & 1	Manual linkage. Control pressure. Valve body. Air pressure.	Forward clutch. Leakage in hydraulic system.
ROUGH INITIAL ENGAGEMENT IN D or 2	Engine idle speed. Vacuum modulator or lines restricted—leaking—adjustment. Control pressure. Pressure regulator. Valve body.	Forward clutch.
1-2 or 2-3 SHIFT POINTS INCORRECT or ERRATIC	Fluid level. Vacuum modulator or lines restricted—leaking—adjustment. Downshift linkage—including inner lever position. Manual linkage. Governor. Control pressure. Valve body. Air pressure.	
ROUGH 1-2 UPSHIFTS	Vacuum modulator or lines restricted—leaking—adjustment. Intermediate servo. Intermediate band. Control pressure. Valve body. Pressure regulator.	
ROUGH 2-3 SHIFTS	Vacuum modulator or lines restricted—leaking—adjustment. Intermediate servo. Control pressure. Pressure regulator. Intermediate band. Valve body. Air pressure.	Reverse—high clutch. Reverse—high clutch piston air bleed valve
DRAGGED OUT 1-2 SHIFT	Fluid level. Vacuum modulator or lines restricted—leaking—adjustment. Intermediate servo. Control pressure. Intermediate band. Valve body. Pressure regulator. Air pressure.	Leakage in hydraulic system.
ENGINE OVERSPEEDS ON 2-3 SHIFT	Manual linkage. Fluid level. Vacuum modulator or lines restricted—leaking—adjustment. Intermediate servo. Control pressure. Valve body. Pressure regulator intermediate band.	Reverse—high clutch. Reverse—high clutch piston air bleed valve.
NO 1-2 or 2-3 SHIFT	Manual linkage. Downshift linkage—including inner lever position. Vacuum modulator or lines restricted—leaking—adjustment. Governor. Control pressure. Valve body. Intermediate band. Intermediate servo.	Leakage in hydraulic system.
NO 3-1 SHIFT IN D	Governor. Valve body.	
NO FORCED DOWNSHIFTS	Downshift linkage—including inner lever position. Valve body. Vacuum modulator or lines restricted—leaking—adjustment.	
RUNAWAY ENGINE ON FORCED 3-2 DOWNSHIFT	Control pressure. Intermediate servo. Intermediate band. Pressure regulator. Valve body. Vacuum modulator or lines restricted—leaking—adjustment.	Leakage in hydraulic system.
ROUGH 3-2 or 3-1 SHIFT AT CLOSED THROTTLE	Engine idle speed. Vacuum modulator or lines restricted—leaking—adjustment. Intermediate servo. Valve body. Pressure regulator.	
SHIFTS 1-3 IN D	Intermediate band. Intermediate servo. Vacuum modulator or lines restricted—leaking—adjustment. Valve body. Governor. Air pressure.	
NO ENGINE BRAKING IN 1st GEAR 1 RANGE	Manual linkage. Low-reverse clutch. Valve body. Governor. Air pressure.	
CREEPS EXCESSIVELY	Engine idle speed.	
SLIPS or CHATTERS IN 1st GEAR D	Fluid level. Vacuum modulator or lines restricted—leaking—adjust. Control pressure. Pressure regulator. Valve body.	Forward clutch. Leakage in hydraulic system. Planetary one-way clutch.

FORD C6

Ford C6
Identification

The identification tag is located under the left side extension housing bolt. The C6 transmission is coded U on the vehicle certification label. The illustrations below show the transmission identification tag and the code interpretation.

Ford C6 transmission identification tag 1974 and earlier units

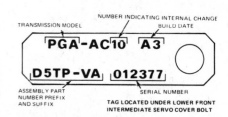

Ford C6 transmission identification tag 1975 and later units

FORD C6

TROUBLESHOOTING
FORD C6

THE PROBLEM	ITEMS TO CHECK	
	IN THE CAR	OUT OF CAR
SLIPS or CHATTERS IN SECOND GEAR	Fluid level. Vacuum modulator or lines restricted—leaking—adjustment. Intermediate servo. Intermediate band. Control pressure. Pressure regulator. Valve body. Air pressure.	Forward clutch. Leakage in hydraulic system.
SLIPS or CHATTERS IN R	Fluid level. Vacuum modulator or lines restricted—leaking—adjustment. Low-reverse clutch. Control pressure. Pressure regulator. Valve body. Air pressure.	Front pump. Leakage in hydraulic system. Reverse—high clutch piston air bleed valve.
NO DRIVE IN D ONLY	Manual linkage. Control pressure. Valve body.	Planetary one-way clutch.
NO DRIVE IN 2 ONLY	Fluid level. Manual linkage. Control pressure. Intermediate servo. Valve body. Air pressure.	Leakage in hydraulic system.
NO DRIVE IN 1 ONLY	Fluid level. Manual linkage. Control pressure. Valve body. Air pressure.	Leakage in hydraulic system.
NO DRIVE IN R ONLY	Fluid level. Manual linkage. Low-reverse clutch. Control pressure. Valve body. Air pressure.	Reverse—high clutch. Leakage in hydraulic system. Reverse—high clutch piston air bleed valve.
NO DRIVE IN ANY SELECTOR LEVEL POSITION	Fluid level. Manual linkage. Control pressure.	Leakage in hydraulic system. Front pump.
LOCKUP IN D ONLY		Parking linkage. Leakage in hydraulic system.
LOCKUP IN 2 ONLY	Low-reverse clutch.	Reverse—high clutch. Parking linkage. Leakage in hydraulic system. Planetary one-way clutch.
LOCKUP IN 1 ONLY		Parking linkage. Leakage in hydraulic system.
LOCKUP IN R ONLY		Forward clutch. Parking linkage. Leakage in hydraulic system.
PARKING LOCK BINDS or DOES NOT HOLD	Manual linkage.	Parking linkage.
TRANSMISSION OVERHEATS	Oil cooler and connections. Pressure regulator. Vacuum modulator or lines restricted—leaking—adjustment. Control pressure.	Converter one-way clutch. Converter pressure check valves.
MAXIMUM SPEED TOO LOW, POOR ACCELERATION	Engine performance. Vehicle brakes.	Converter one-way clutch.
TRANSMISSION NOISY IN N & P	Fluid level. Pressure regulator.	Front pump.
TRANSMISSION NOISY IN 1st, 2nd, 3rd, or REVERSE GEAR	Fluid level. Pressure regulator.	Planetary assembly. Forward clutch. Front pump. Planetary one-way clutch.
FLUID LEAKS	Fluid level. Converter drain plugs. Oil pan gasket, filler tube or seal. Oil cooler and connections. Manual or downshift lever shaft seal ⅛ inch pipe plugs in case. Extension housing-to-case gasket. Extension housing rear oil seal. Speedometer driven gear adapter seal. Vacuum modulator or lines restricted—leaking—adjustment. Intermediate servo.	Engine rear oil seal. Front pump oil seal. Front pump to case gasket or seal.
CAR MOVES FORWARD IN N	Manual linkage.	Forward clutch.

Vacuum Diaphragm Unit Check Altitude-Compensating Type

The vacuum diaphragm unit may be checked for damaged or ruptured bellows as follows:

1 Remove the diaphragm and the throttle valve rod from the transmission.
2 Insert a rod into the diaphragm unit until it is seated in the hole. Make a reference mark on the rod where it enters the diaphragm.
3 Place the diaphragm unit on a scale with the end of the rod resting on the weighing pan and gradually press down on the diaphragm unit.
4 Note the force (in pounds) at which the reference mark on the rod moves into the diaphragm. If the reference mark is still visible at 12 pounds of pressure on the scale, the diaphragm bellows is good. If the reference mark on the rod disappears before four pounds of force, the diaphragm bellows is damaged and the diaphragm unit must be replaced.

Vacuum Diaphragm

A screw is provided in the inlet tube of the vacuum diaphragm assembly, to permit small adjustments in control pressure. If control pressure is uniformly high or low, in all ranges, it may be brought within specifications by turning this screw. Control pressure may also be varied to alter shift feel, but in no case should

FORD C6

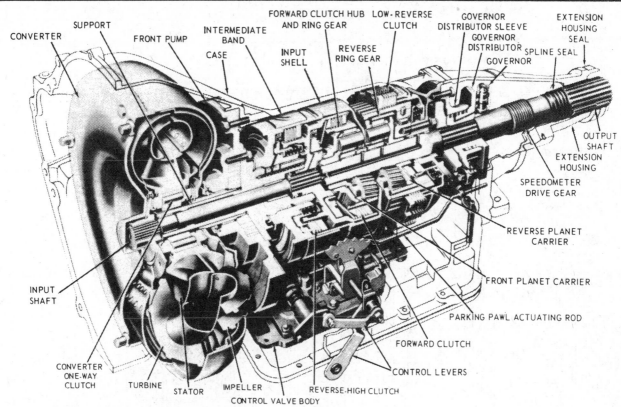

CONVERTER · SUPPORT · FRONT PUMP · INTERMEDIATE BAND · CASE · INPUT SHELL · FORWARD CLUTCH HUB AND RING GEAR · REVERSE RING GEAR · LOW-REVERSE CLUTCH · GOVERNOR DISTRIBUTOR SLEEVE · GOVERNOR DISTRIBUTOR · GOVERNOR · SPLINE SEAL · EXTENSION HOUSING SEAL · OUTPUT SHAFT · EXTENSION HOUSING · SPEEDOMETER DRIVE GEAR · REVERSE PLANET CARRIER · FRONT PLANET CARRIER · PARKING PAWL ACTUATING ROD · CONTROL LEVERS · FORWARD CLUTCH · REVERSE-HIGH CLUTCH · CONTROL VALVE BODY · IMPELLER · STATOR · TURBINE · CONVERTER ONE-WAY CLUTCH · INPUT SHAFT

Cutaway view Ford C-6 automatic transmission

it go beyond the specified minimum or maximum.

Control pressure is increased by turning the adjusting screw clockwise, and reduced by turning counter-clockwise. One full turn will change control pressure approximately 2–3 psi.

CHILTON CAUTION: *Always check neutral safety switch after linkage adjustments.*

Intermediate Band

1 Raise the car on a hoist or place it on jack stands.
2 Clean threads of the intermediate band adjusting screw.
3 Loosen adjustment screw locknut.
4 Tighten the adjusting screw to 10 ft./lbs., and back the screw off exactly 1½ turns. Tighten the adjusting screw locknut.

Throttle Linkage Adjustment

ALL FORD & MERCURY CARS (EXCEPT CONTINENTAL, MARK III, MARK IV)

1 Disconnect throttle and downshift return spring.
2 Hold the carburetor throttle lever in wide open position, (against stop), and hold the transmission linkage in full downshift position against its internal stop.
3 Turn the carburetor downshift lever adjustment screw to within .010–.080 in. of the contacting pickup surface of the throttle lever.

TOOL—T71P-77370-B, -C, OR -D

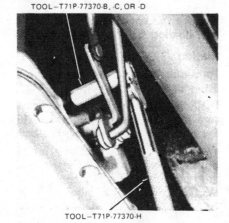

TOOL—T71P-77370-H

Adjusting intermediate band

4 Connect return spring.

Downshift Rod Adjustment

CONTINENTAL, MARK III, MARK IV, MARK V

1 Loosen the locknut on the rod, and disconnect it from the ballstud on the bellcrank by sliding the spring clip off the end of the rod.
2 Pull the rod upward, and hold it tightly against the downshift rod against the internal stop. Adjust the length of the rod so the hole in the link is aligned with the stud on the bellcrank assembly.
3 Lengthen the rod one turn, and position it on the bellcrank. Slide the

spring clip over the end, and tighten the locknut.
4 Be sure the outer bracket of the bellcrank is against the stop pin. If not, lengthen the rod one more turn.

Transmission Shift Points

Each individual model of transmission has specific limits on transmission shift points for all its possible applications. Since shift point limits are highly specific, the following are general guidelines that apply to transmission shift points:
1 The transmission should shift smoothly and crisply; it should never hunt or shift indecisively.
2 The transmission should be responsive. Full throttle should give a 3–2 or 2–1 downshift, depending on road speed; greater amounts of acceleration throttle should produce higher upshift speeds.
3 You should be able to feel a positive detent at the bottom of the accelerator pedal travel. Going through detent should greatly increase upshift points; this may also produce downshifts under some conditions.
4 Most transmissions should shift from 3 to 2 range at less than full throttle under 30 mph.
5 Most shift point problems fall into one of the following categories:
   a. an improper adjustment or other

FORD C6

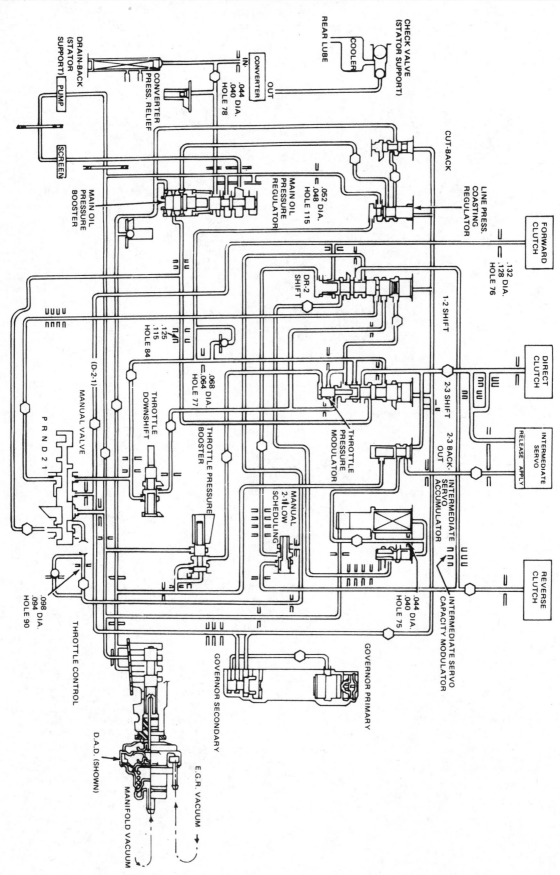

C-6 transmission hydraulic schematic

FORD C6

C6 with Altitude-type Vacuum Diaphragm
Control Pressure at Zero Governor RPM

Engine speed			Idle			As required		As required		
Throttle			Closed			As required		As required		
Manifold vacuum (inches Hg)			Above 18 ①			10		Stall (below 1.0)		
Range			Control Pressure (psi)		TV Pressure (psi)	Control Pressure (psi)	TV Pressure (psi)	Control Pressure (psi)		TV Pressure (psi)
			P, N, D, 2, 1	R		D, 2, 1		D, 2, 1	R	
	Barometric Pressure in Inches HG	Nominal Altitude (Feet)								
psi @ Barometric Pressure ②	29.5	Sea Level	56–62	71–86	7–10 ⑤	100–115	40–44	160–190	240–300	77–84
psi @ barometric pressure ③	28.5	1000	49–59	65–80	4–7	99–114	37–41	158–176	233–290	74–80
	27.5	2000	49–56	60–75	2–5	96–111	35–39	156–174	228–284	72–78
	26.5	3000	49–56	56–71	0–3	91–106	32–36	151–169	222–277	69–75
	25.5	4000	49–56	56–65	0	88–103	30–34	146–164	215–269	66–72
	24.5	5000	49–56	56–65	0	84–98	27–31	143–161	211–264	64–70
	23.5	6000	49–56	56–65	0	80–95	25–29	138–156	204–256	61–67

① It may not be possible to obtain 18 inches of engine vacuum at idle. For idle vacuums of less than 18 inches the following table provides idle speed pressure specifications in D range:

Manifold Vacuum	Barometric Pressure at 29.5 in. Hg ②		Barometric Pressure at 24.5 in. Hg. ④	
	T.V.	Cont.	T.V.	Cont.
17	11–14	56–69	0–1	49–56
16	15–18	56–75	2–5	49–56
15	20–22	56–84	7–9	49–61
14	23–26	56–92	10–13	56–67
13	28–31	56–98	15–18	56–75
12	32–35	56–105	19–22	56–84
11	36–40	56–111	23–27	56–92

② These specifications (with altitude compensating diaphragm) apply at observed barometric pressure of 29.5 inches (nominal sea level)

③ Specifications for barometric pressures of less than 29.5 inches.

④ At barometric pressures between 29.5 inches and 24.5 inches idle, pressures should fall between the values shown.

Ford C6
Control Pressure at Zero Governor RPM

Engine Speed	Throttle	Manifold Vac. In. Hg.	Range	Psi
As required	As required	12 & above	P, N, D	55–86
			2, 1	55–122
			R	55–197
As required	As required	10	D	98–110
Stall	Thru detent	Below 1.0	D, 2, 1	143–164
			R	239–272

C6
Selective Thrust Washers

Identification No.	Thrust Washer Thickness—Inch	Identification No.	Thrust Washer Thickness—Inch
1	0.056-0.058	4	0.103-0.105
2	0.073-0.075	5	0.118-0.120
3	0.088-0.090		

CHECKS AND ADJUSTMENTS
C6 TRANSMISSION

Operation	Specification
Transmission end play	0.008-0.044 (selective thrust washers available)
Turbine and stator end play	New or rebuilt 0.021 in. max. Used 0.030 in. max.①
Intermediate band adjustment	Remove and discard locknut. Adjust screw to 10 ft./lbs. torque, then back off 1 turn, install new lock nut and tighten locknut to specification. ②

① To check end play, exert force on checking tool to compress turbine to cover thrust washer wear plate. Set indicator at zero.

② 1974–77 models: back off adjusting screw 1½ turns.

FORD C6

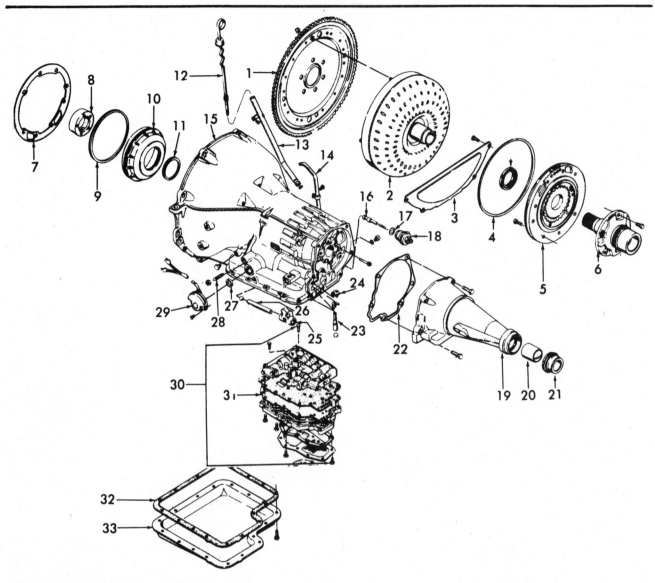

C-6 Automatic transmission and parts breakdown

1. FLYWHEEL AND RING GEAR
2. TORQUE CONVERTER
   ASSEMBLY
3. COVER
4. OIL SEAL
5.-6. FRONT OIL PUMP
   ASSEMBLY
7. GASKET
8. RACE
9. OIL SEAL
10. CLUTCH PISTON
11. SEAL

12. INDICATOR—OIL LEVEL
13. OIL FILTER TUBE
14. BREATHER TUBE
15. TRANSMISSION CASE
16. VALVE ROD
17. GASKET
18. VACUUM DIAPHRAGM
19. REAR EXTENSION HOUSING
20. REAR BUSHING
21. REAR OIL SEAL
22. GASKET

23. SHAFT
24. PLATE
25. BAND LEVER
26. LEVER SHAFT
27. OIL SEAL
28. ADJUSTING SCREW
29. NEUTRAL SAFETY SWITCH
30. MAIN CONTROL VALVE
31. GASKET
32. OIL PAN GASKET
33. OIL PAN

malfunction of an external mechanical or electrical linkage.
   b. an internal transmission problem marked by improper shift feel along with improper shift points.

Overhaul

Disassembly

1 Mount the transmission in a holding stand.

2 Remove the 17 fluid pan attaching bolts. Remove the pan and gasket.

3 Remove the 8 valve body attaching bolts. Lift the valve body from the transmission case.

4 Attach a dial indicator to the front pump. Install tool T61L-7657-B in the extension housing to center the shaft.

5 Pry the gear train to the rear of the case and at the same time, press the input shaft inward until it bottoms. Set the dial indicator to read zero.

6 Pry the gear train forward and note the amount of gear train end play on the dial indicator. Record the end play to facilitate assembling the transmission. Remove the dial indicator from the pump and the tool from the extension housing.

7 Remove the vacuum diaphragm, rod

FORD C6

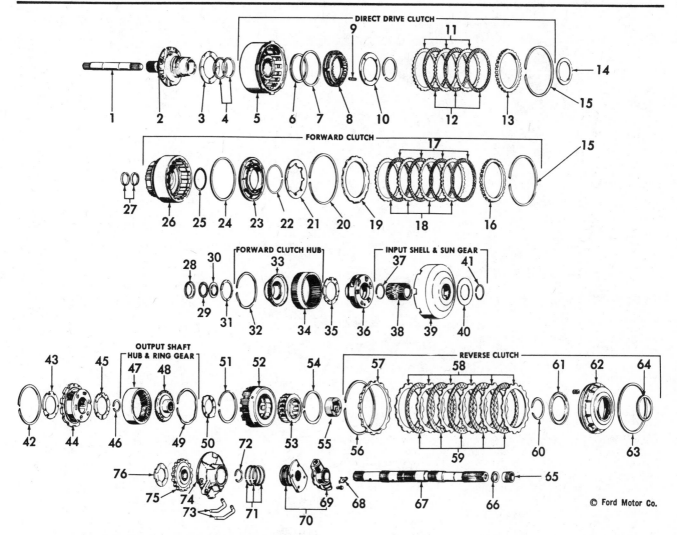

Gear train parts identification C-6 automatic transmission

1. INPUT SHAFT
2. STATOR SUPPORT
3. THRUST WASHER
4. OIL SEAL
5. INTERMEDIATE DRUM
6. OIL SEAL
7. OIL SEAL
8. CLUTCH PISTON
9. RETURN SPRING
10. SPRING RETAINER
11. CLUTCH PLATE—DRIVEN
12. CLUTCH PLATE—DRIVE
13. PRESSURE PLATE
14. THRUST WASHER
15. SNAP RING
16. PRESSURE PLATE
17. CLUTCH PLATE—DRIVE
18. CLUTCH PLATE—DRIVEN
19. PRESSURE PLATE
20. SNAP RING
21. DISC SPRING
22. RING
23. CLUTCH PISTON
24. OIL SEAL
25. OIL SEAL
26. CLUTCH CYLINDER

27. OIL SEAL
28. FRONT RACE
29. THRUST BEARING
30. REAR RACE
31. THRUST WASHER
32. SNAP RING
33. CLUTCH HUB
34. FORWARD RING GEAR
35. THRUST WASHER
36. FORWARD DRUM AND
    PLANETARY ASSEMBLY
37. SNAP RING
38. SUN GEAR
39. INPUT SHELL
40. THRUST WASHER
41. SNAP RING
42. SNAP RING
43. THRUST WASHER
44. REVERSE PLANET ASSEMBLY
45. THRUST WASHER
46. SNAP RING
47. RING GEAR
48. OUTPUT SHAFT HUB
49. SNAP RING
50. THRUST WASHER
51. SNAP RING

52. REVERSE CLUTCH HUB
53. OVERRUNNING CLUTCH
54. SNAP RING
55. INNER RACE
56. SNAP RING
57. PRESSURE PLATE
58. CLUTCH PLATE—DRIVEN
59. CLUTCH PLATE—DRIVE
60. SNAP RING
61. SPRING RETAINER
62. REVERSE PISTON
63. OUTER SEAL
64. INNER SEAL
65. OIL SEAL
66. SNAP RING
67. OUTPUT SHAFT
68. SPRING RETAINER
69. GOVERNOR BODY
70. GOVERNOR ASSEMBLY
71. SEAL RING
72. SNAP RING
73. OIL TUBE
74. OIL DISTRIBUTOR SLEEVE
75. PARKING GEAR
76. THRUST WASHER

FORD C6

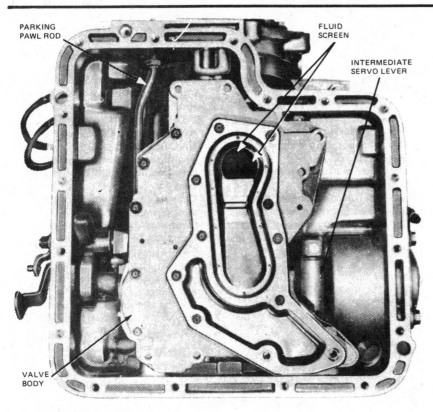

PARKING PAWL ROD
FLUID SCREEN
INTERMEDIATE SERVO LEVER
VALVE BODY

Transmission with pan removed

and the primary throttle valve from the case. Slip the input shaft out of the front pump.

8 Remove the front pump attaching bolts. Pry the gear train forward to remove the pump.

9 Loosen the band adjustment screw and remove the two struts.

10 Rotate the band 90 degrees counter-clockwise to align the ends with the slot in the case. Slide the band off the reverse-high clutch drum.

11 Remove the forward part of the gear train as an assembly.

12 Remove the bolts that attach the servo cover to the transmission case.

13 Remove the cover, piston, spring and gasket from the case.

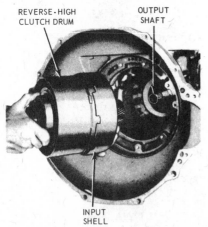

REVERSE-HIGH CLUTCH DRUM
OUTPUT SHAFT
INPUT SHELL

Removing forward part of gear train

14 Remove the large snap ring that secures the reverse planet carrier in the low-reverse clutch hub. Lift the planet carrier from the drum.

15 Remove the snap ring that secures the reverse ring gear and hub on the output shaft. Slide the ring gear and hub off the shaft.

16 Rotate the low-reverse clutch hub in a clockwise direction and at the same time, withdraw it from the case.

17 Remove the reverse clutch snap ring from the case, then remove the clutch discs, plates and pressure plate from the case.

18 Remove the extension housing attaching bolts and vent tube from the case. Remove the extension housing and gasket.

19 Slide the output shaft assembly from the transmission case.

20 Remove the distributor sleeve attaching bolts and remove the sleeve, parking pawl gear and the thrust washer. If the thrust washer is staked in place, use a sharp chisel and cut off the metal from behind the thrust washer. Be sure to clean the rear of the case with air pressure or a suitable solvent to remove any metal particles.

21 Compress the reverse clutch piston release spring. Remove the snap ring. Remove the tool and the spring retainer.

22 Remove the one-way clutch inner race attaching bolts from the rear of the case. Remove the inner race from inside of the case.

23 Remove the reverse clutch piston from the case.

Assembly

1 Place the transmission case in a holding fixture.

2 Position the low-reverse clutch piston so that the check ball is in the 6 o'-clock position (toward the bottom of case next to oil pan) and tap the piston into place in the case with a clean rubber hammer.

3 Hold the one-way clutch inner race in position and install the attaching bolts. Torque bolts to limits as given in the Specifications.

4 Install a low-reverse clutch return spring in each pocket in the clutch piston. Press the springs firmly into the piston to prevent them from falling out.

5 Position the spring retainer over the springs and position the retainer snap ring in place on the one-way clutch inner race.

6 Install the compressing tool and compress the springs just enough to install the low-reverse clutch piston retainer snap ring.

7 Install the snap ring, then remove the compressing tool.

8 Place the transmission case on the bench with the front end facing downward.

9 Position the parking gear thrust washer and the gear on the case. **Do not restake the thrust washer.**

10 Position the oil distributor and tubes in place on the rear of the case. Install attaching bolts and torque to specification.

11 Install the output shaft and governor as an assembly.

12 Place a new gasket on the rear of the transmission case. Position the extension housing on the case and install the attaching bolts. Torque the attaching bolts to specification.

13 Place the transmission case in the holding fixture.

14 Coat two new gaskets with petroleum jelly and position them on the servo cover.

15 Position the servo spring on the piston rod.

16 Insert the servo piston rod in the case. Install the attaching bolts, making sure that the identification tag is in place. Torque the attaching bolts to specifications.

17 Align the low-reverse clutch hub and one-way clutch with the inner race at the rear of the case. Rotate the low-reverse clutch hub clockwise while applying pressure to seat it on the inner race.

18 Install the low-reverse clutch plates, starting with a steel plate and following with friction and steel plates alternately. Retain them with vaseline. Refer to the Specifications for the number of plates required. If new

FORD C6

composition plates are being used, soak them in clean transmission fluid for fifteen minutes before installation. Install the pressure plate and the snap ring. Test the operation of the low-reverse clutch by applying air pressure at the clutch pressure apply hole in the case.

19 Install the reverse planet ring gear thrust washer and the ring gear and hub assembly. Insert the snap ring in the groove in the output shaft.

20 Assemble the front and rear thrust washers onto the reverse planet assembly; retain with vaseline. Insert the assembly into the ring gear and install the snap ring.

21 Set the reverse-high clutch on the bench, with the front end facing down. Install the thrust washer on the rear end of the reverse-high clutch assembly. Retain the thrust washer with vaseline and insert the splined end of forward clutch into the open end of the reverse-high clutch with splines engaging the direct clutch friction plates.

22 Install the thrust washer and retain it with vaseline, on the front end of the forward planet ring gear and hub. Insert the ring gear into the forward clutch.

23 Install the thrust washer on the front end of the forward planet assembly. Retain the washer with vaseline and insert the assembly into the ring gear. Install the input shell and sun gear assembly.

24 Install the reverse-high clutch, forward clutch, forward planet assembly and drive input shell, and sun gear as an assembly into the transmission case.

25 Insert the intermediate band into the case around the direct clutch cylinder. Install the struts and tighten the band adjusting screw sufficiently to retain the band.

26 Place a selective thickness bronze thrust washer on the rear shoulder of the stator support and retain it with vaseline. If the end play was not within specification when checked prior to disassembly, replace the washer with one of proper thickness. Refer to Specifications Part for selective thrust washer thicknesses. Using two 5/16-inch bolts three inches long, make two alignment studs. Cut the heads from the bolts and grind a taper on the cut end. Temporarily install the two studs opposite each other in the mounting holes of the case. Slide a new gasket onto the studs. Position pump on case, being careful not to damage the large seal on the OD of the pump housing (Removing the aligning studs). Install six of the seven mounting bolts and torque to specification.

27 Adjust the intermediate band as detailed under Adjustments, and install the input shaft with the long splined

end inserted into the forward clutch assembly.

28 Install tool 4201-C at the seventh pump mounting bolt and check the transmission end play. Remove the tool. Install the seventh mounting bolt and torque to specifications.

29 Install the control valve in the case, making sure that the levers engage the valves properly. Install the primary throttle valve, rod, and the vacuum diaphragm in the case.

30 Install a new pan gasket and the pan. Torque the bolts to specifications.

31 Install the converter assembly.

32 Install the transmission in the vehicle as detailed under Removal and Installation.

Control Valve Body

Disassembly

When the main control is disassembled and the valve body-to-screen gasket is removed, the gasket should not be cleaned in a degreaser solvent or any type of detergent solution. To clean the gasket, wipe it off with a lint-free cloth.

1 Remove the nine screws that attach the screen to the lower valve body and remove screen and gasket.

2 Remove the five upper-to-lower valve body and hold-down plate attaching

screws. Remove the seven attaching screws from the underside of the lower valve body.

3 Separate the bodies and remove the separator plate and gasket. Be careful not to lose the check valves and springs. Remove and clean the separator plate screen if necessary.

4 Remove the manual valve plunger retaining pin from the upper valve body and remove the plunger.

5 Slide the manual valve out of the valve body.

6 Cover the downshift valve port with a finger; then, working from the underside of the body, remove the downshift valve retainer. Remove the spring and downshift valve.

7 Apply pressure on the pressure boost valve retaining plate and remove the two attaching screws. Slowly release the pressure and remove the plate, sleeve and the pressure boost valve. Remove the two springs and the main regulator valve from the same bore.

8 Apply pressure on the throttle boost valve retaining plate and remove the two attaching screws. Slowly release the pressure and remove plate, throttle pressure boost valve and spring, and the manual low 2–1 scheduling valve and spring from the body.

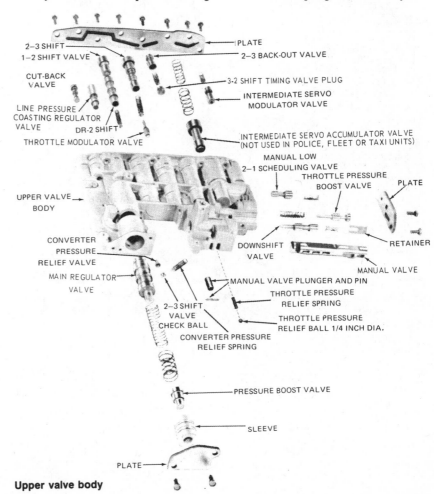

Upper valve body

FORD C6

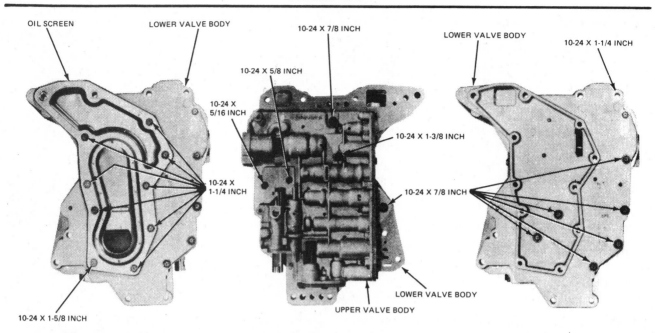

OIL SCREEN   LOWER VALVE BODY   10-24 X 7/8 INCH   LOWER VALVE BODY   10-24 X 1-1/4 INCH

10-24 X 5/8 INCH

10-24 X 5/16 INCH

10-24 X 1-3/8 INCH

10-24 X 1-1/4 INCH

10-24 X 7/8 INCH

10-24 X 1-5/8 INCH

LOWER VALVE BODY

UPPER VALVE BODY

OIL SCREEN-TO-LOWER VALVE BODY ATTACHING SCREWS   **UPPER-TO-LOWER VALVE BODY AND HOLD-DOWN PLATE ATTACHING SCREWS**   **LOWER VALVE BODY-TO-UPPER VALVE BODY ATTACHING SCREWS**

Control valve body and screen attaching screws

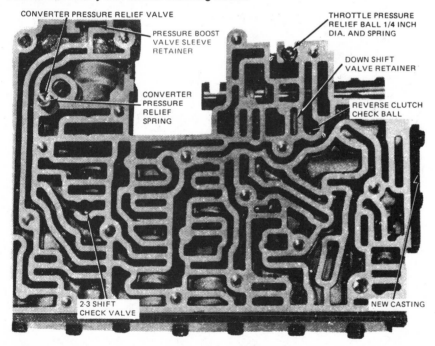

CONVERTER PRESSURE RELIEF VALVE

PRESSURE BOOST VALVE SLEEVE RETAINER

CONVERTER PRESSURE RELIEF SPRING

THROTTLE PRESSURE RELIEF BALL 1/4 INCH DIA. AND SPRING

DOWN SHIFT VALVE RETAINER

REVERSE CLUTCH CHECK BALL

2-3 SHIFT CHECK VALVE

NEW CASTING

Converter pressure relief valve, throttle pressure relief valve, and 2–3 shift check valve locations

9 Apply pressure on the remaining valve retaining plate and remove the eight attaching screws.

10 Hold the valve body so that the plate is facing upward. Slowly release the pressure and remove the plate.

11 When removing the various valves from the control valve body, keep all ports covered with your fingers except the bore the valve is being removed

from. Remove the spring and the intermediate servo modulator valve from the valve body.

12 Remove the intermediate servo accumulator valve and springs.

13 Remove the 2–3 back-out valve, spring, and 3–2 shift timing valve plug. Do not remove the 3–2 shift timing valve plug if it is found to be peened, preventing its removal. This

condition will not affect transmission operation and is not cause for replacement of the main control.

14 Remove the 2–3 shift valve, spring and the throttle modulator valve.

15 Remove the 1–2 shift valve, DR-2 shift valve and the spring from the valve body.

16 Remove the coasting regulator valve from the body.

17 Remove the cutback control valve to complete the disassembly of the control valve.

Assembly

1 Place the 3–2 shift timing valve plug in the valve body if it was previously removed.

2 Place the downshift valve and spring in the valve body. Compress the spring and install the retainer from the underside of the body.

3 Place the valve body on a clean surface with the passage side facing up. Place the converter relief valve spring in its bore. Coat the converter relief valve check valve with vaseline and place it on top of the spring. Place the 2–3 shift check valve ball in its cavity. Place the throttle pressure relief valve spring in its bore. Coat the throttle pressure relief valve check ball with vaseline and place it on top of the spring.

4 Install the separator screen in the separator plate if it was previously removed. **Be sure the screen tabs are flush with the separator plate surface.** Carefully position the separator plate and new gasket on the lower valve body. Place the two hold-down plates on the separator plate and install the attaching screws finger tight.

FORD C6

5 Place the lower body and plate assembly on the upper valve body and install the attaching screws finger tight.

6 Install the oil screen screws loosely, without the screen, to properly align the upper and lower valve bodies, gasket and separator plate.

7 Torque the two bolts that are covered by the screen to specification.

8 Remove the oil screen attaching screws and place the gasket and oil screen in position on the lower valve body. Reinstall the screen attaching screws.

9 Torque all the valve body and screen attaching screws to specification.

10 Place the cutback valve and coasting regulator valve in the valve body.

11 Place the spring DR-2 shift valve and the 1–2 shift valve in the body.

12 Place the throttle modulator valve and spring and the 2–3 shift valve in the valve body.

13 Place the spring and the 2–3 backout valve in the valve body.

14 Place the two springs and intermediate servo accumulator valve in the valve body.

15 Place the intermediate servo modulator valve and spring in the body.

16 Carefully place the valve retaining plate on the body and secure it with the eight attaching screws. Tighten the screws to specification.

17 Place the throttle pressure boost valve and spring in the valve body. Place the manual low 2–1 scheduling valve and spring in the valve body and install the retaining plate. Torque the attaching screws to specification.

18 Place the main regulator, two springs, pressure boost valve and the sleeve in the valve body.

19 Install the pressure booster plate and torque the two attaching screws to specification.

20 Place the manual valve in the valve body and install the plunger and the retaining pin in the body.

Intermediate Servo

Disassembly

1 Apply air pressure to the port in the servo cover to remove the piston and rod.

2 Replace the complete piston and rod assembly if the piston or piston sealing lips are unserviceable or damaged.

3 Remove the seal from the cover.

Assembly

1 Dip the new seals in transmission fluid.

2 Install new seals on the cover.

3 Dip the piston in transmission fluid and install it in the cover.

Governor

Disassembly

1 Remove the governor attaching bolts and remove the governor.

1. BAND LEVER
2. ADJUSTING SCREW
3. ANCHOR STRUT
4. INTERMEDIATE BAND
5. LEVER SHAFT
6. PISTON SPRING
7. OIL SEAL
8. SERVO PISTON
9. OIL SEAL
10. COVER SEAL
11. GASKET
12. SERVO COVER

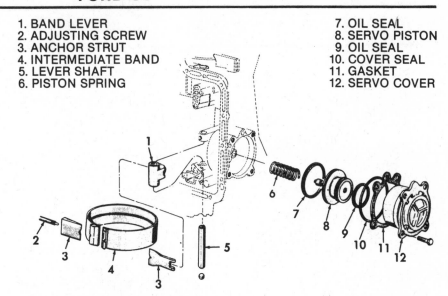

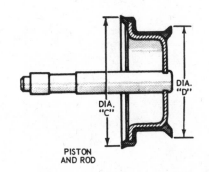

Servo and band parts identification

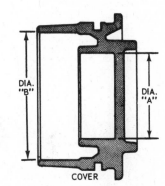

INTERMEDIATE SERVO PISTON AND COVER IDENTIFICATION

Transmission	Piston		Cover		
	Dia. C	Dia. D	Dia. A	Dia. B	Letter Ident.
PJA-C6	3.022	2.117	2.077	2.982	L
PJA-G4-H5	3.004	2.009	2.074	2.978	
PJD-S2-U1	2.907	2.140	2.100	2.866	P
	2.889	2.122	2.097	2.863	
PJD-R1-T1					
PJD-G4-P1					
PJD-H7-K					
PJD-F5-N1					
PJD-E5-M1					
PJA-V-W	2.958	1.987	1.947	2.917	N
	2.939	1.969	1.944	2.914	
① Letter cast in servo cover					

159

2 Remove the snap ring that secures the governor distributor on the output shaft and slide it off the front of the shaft.

3 Remove the seal rings from the distributor.

Assembly

1 Carefully install new seal rings on the distributor.

2 Working from the front end of the output shaft, slide the governor distributor into place on the shaft. Install

the snap ring to secure it. Make sure that the snap ring is seated in the groove.

3 Position the governor on the distributor and secure with the attaching screws.

Downshift and Manual Linkage

Disassembly

1 Remove the nut and lock washer that secures the outer downshift lever to the transmission and remove the lever.

2 Slide the downshift lever out from the inside of the case. Remove the seal from the recess in the manual lever shaft.

3 Remove the C-ring that secures the parking pawl actuating rod to the manual lever. Remove the rod from the case.

4 Remove the nut that secures the inner manual lever to the shaft. Remove the inner lever from the shaft. Slide the outer lever and shaft from the case.

5 Remove the seal from the case with Tools T59L-100-B and T58L-101-A or 7600-E.

6 Dip the new seal in transmission fluid and install it in the case.

Assembly

1 Slide the outer manual lever and shaft in the transmission case.

2 Position the inner lever on the shaft, making sure the leaf spring roller is positioned in the inner manual lever detent. Install the attaching nut. Tighten the nut to specification listed at the end of this Part. Install the parking pawl actuating rod and secure it to the inner manual lever with a C-washer.

3 Install a new downshift lever seal in the recess of the outer lever shaft. Slide the downshift lever and shaft into position.

4 Place the outer downshift lever on the shaft and secure it with a lockwasher and nut.

Parking Pawl Linkage

Disassembly

1 Remove the bolts that secure the parking pawl guide plate to the case. Remove the plate.

2 Remove the spring, parking pawl and shaft from the case.

3 Working from the pan mounting surface, drill a ⅛-inch diameter hole through the center of the cupped plug. Pull the plug from the case with a wire hook.

4 Unhook the end of the spring from the park plate slot to relieve the tension.

5 Thread a ¼-20 inch or 8–32 × 1-¼ inch screw into the park plate shaft. Pull the shaft from the case with the screw.

Assembly

1 Position the spring and park plate in

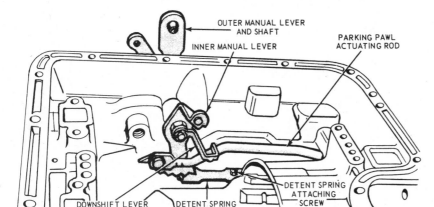

Downshift and manual linkage

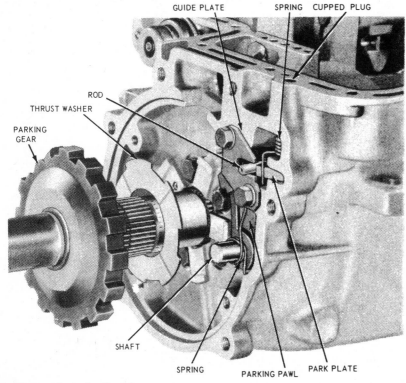

Parking pawl mechanisms

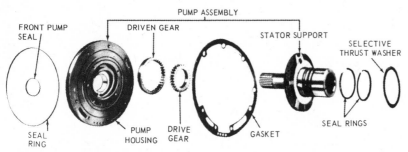

Front pump disassembled

FORD C6

the case and install the shaft. Place the end of the spring into the slot of the park plate.

2 Install a new cupped plug to retain the shaft.

3 Install the parking pawl shaft in the case. Slip the parking pawl and spring into place on the shaft.

4 Position the guide plate on the case, making sure that the actuating rod is seated in the slot of the plate. Secure the plate with two bolts and lockwashers.

Servo Apply Lever

Disassembly

1 Working from inside of the transmission case, carefully drive on the servo apply lever shaft to remove the cup plug. The shaft can be withdrawn from the case by hand.

Assembly

1 Hold the servo apply lever in position and install the new shaft.

2 Using a fabricated drive the cup plug into position in the case. Be sure the plug is flush with the shoulder of the counterbore. The cup plug may be coated with Loctite, Part NO. C3AZ-19554-A, before installation.

Front Pump

The front seal can be replaced after the pump has been installed on the transmission.

Disassembly

1 Remove the two seal rings and the selective thrust washer.

2 Remove the large square-cut seal from the O.D. of the pump housing.

3 Remove the 5 bolts that secure the stator support to the pump housing. Lift the support from the housing.

4 Remove the drive and the driven gear from the housing.

5 If the pump housing bushing is worn or damaged, replace it.

   Place the new bushing in position, making sure the half moon slot in the bushing is on top and in line with the oil lube hole near the seal bore. Press the bushing in 0.060–0.080 inches below the front face of the bushing bore. Use Tool T66L-7003-B9 and handle to seat the bushing properly. After assembly, the half moon slot must be in past the lube hole to provide proper lubrication.

Assembly

1 Install the drive and driven gears in the pump housing. Each gear has either an identification mark or chamfered teeth on one face. **The identification mark or the chamfered surface on each gear must be installed toward the front of the pump housing.**

2 Position the stator support in the pump housing and install the five at-

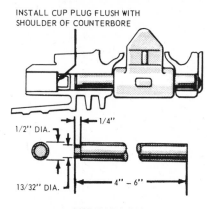

INSTALL CUP PLUG FLUSH WITH SHOULDER OF COUNTERBORE

1/2" DIA.
13/32" DIA.
1/4"
4" – 6"

INSTALLING TOOL
MAKE FROM 1/2" DIA. DRILL ROD

Servo apply lever installation

taching bolts. Torque the bolts to specification.

3 Carefully install two new seal rings on the pump support. Make sure that the ends of the rings are engaged to lock them in place. Install a new square-cut seal on the O.D. of the pump housing.

4 Install the selective thrust washer. Make sure that the correct thickness selective washer is being used to obtain the specified end play. Refer to Specifications.

5 Place the pump on the converter, making sure that the drive gear engages the converter hub. Rotate the pump to make sure that the gears rotate freely.

Reverse-High Clutch

Disassembly

1 Separate the drive train as shown in Fig. 36. Remove the pressure plate retaining snap ring.

2 Remove the pressure plate and the drive and driven clutch plates.

3 Install Tool T65L-77515-A on the reverse-high clutch drum. Make sure that the legs clear the snap ring enough to permit expanding it for removal. Remove the snap ring and remove the tool.

4 Remove the spring retainer and the piston return springs.

5 Apply air pressure to the piston apply hole in the clutch hub and remove the piston.

6 Remove the piston outer seal from the piston and the inner seal from the clutch drum.

7 Remove the front and rear bushings from the clutch drum if they are worn or damaged. To remove the front bushing, use a cape chisel and cut along the bushing seam until the chisel breaks through the bushing wall. Pry the loose ends of the bushing up with an awl and remove the bushing.

   Remove the rear bushing, and press the bushing from the drum.

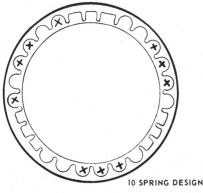

SPRINGS MUST BE INSTALLED IN POCKETS MARKED **X** ONLY

10 SPRING DESIGN

Reverse-high clutch piston return spring locations

Assembly

1 If the clutch drum bushings were removed, position the drum in a press and press new bushings into the drum.

2 Dip the new seals in transmission fluid and install one on the drum and one on the piston.

3 Install the piston in the clutch drum.

4 Position the piston return springs in the piston sockets. Place the spring retainer on the springs.

5 Install Tool T65L-77515-A and compress the springs. Make certain that the spring retainer is centered while compressing the springs. Install the snap ring. Before releasing the pressure on the tool, make certain that the snap ring is positioned inside of the four snap ring guides on the spring retainer.

6 Clutch plate usage varies with each model, refer to the Specifications for the number of plates required. Dip the clutch plates in clean transmission fluid. Install the clutch plates alternately starting with a steel drive plate. **When new composition clutch plates are used, soak the plates in automatic transmission fluid for 15 minutes before they are assembled.**

7 After all clutch plates have been installed, position the pressure plate in the clutch drum. Install the pressure plate snap ring.

8 With a feeler gauge, check the clearance between the pressure plate and snap ring.

9 The pressure plate should be held downward as the clearance is checked. Refer to Specifications at end of this Part for the proper clearance. If the clearance is not within specifications, selective thickness snap rings are available in the following thicknesses: 0.056–0.060, 0.065–0.069, 0.-074–0.078, 0.083–0.087 and 0.092–0.-096 inch. Install the correct size snap ring and recheck the clearance.

Forward Clutch

Disassembly

1 Remove the clutch pressure plate snap ring.

FORD C6

C6 Torque Specifications

	Ft./Lbs.
Converter to flywheel	20–30
Front pump to trans. case	16–30
Overrunning clutch race to case	18–25
Oil pan to case	12–16
Stator support to pump	12–16
Converter cover to converter hsg.	12–16
Guide plate to case	15–22
Intermediate servo cover to case	15–22
Diaphragm assy. to case	15–23
Distributor sleeve to case	12–16
Extension assy. to trans. case	25–35
Pressure gauge tap	9–15
Band adj. screw locknut to case	35–45
Cooler tube connector lock	20–35
Converter drain plug	14–28
Manual valve inner lever to shaft	30–40
Downshift lever to shaft	12–16
Filler tube to engine	20–25
Transmission to engine	40–50
Steering col. lock rod adj. nut	10–20
Neutral start switch actuator lever bolt	6–10
T.R.S. switch-to-case	4–8

	In./Lbs.
End plates to body	20–45
Inner downshift lever stop	20–45
Reinforcement plate to body	20–45
Screen and lower to upper valve body	40–55
Neutral switch to case	55–75
Neutral switch-to column	20
Control assy. to case	90–125
Gov. body to collector body	60–120
Oil tube connector	80–145

2 Remove the rear pressure plate, the drive and driven plates and the forward pressure plate from the clutch hub.

3 Remove the snap ring that secures the disc spring in the clutch cylinder. Remove the disc spring.

4 Apply air pressure to the clutch apply passage in the cylinder to remove the piston.

5 Remove the seal from the piston and the seal from the clutch hub.

Assembly

1 Dip two new seals in transmission fluid. Install the smaller seal on the clutch hub and the other seal on the clutch piston.

2 Install the clutch piston in the cylinder.

3 Make sure that the steel pressure ring is in the groove on the piston. **Position the disc spring in the cylinder with the dished face downward.** Install the spring. Secure the disc with the retaining snap ring.

4 **Install the forward pressure plate with the flat side up and the beveled side downward.** Dip the clutch plates in clean transmission fluid. Install first a composition driven plate and a steel drive plate. Install the remaining plates in this sequence. Refer to the Specification for the number of plates required. The last plate installed will be the rear pressure plate. Install the snap ring and make certain that it seats fully in the groove. When new composition clutch plates are used, soak the plates in automatic transmission fluid for 15 minutes before they are assembled.

5 With a feeler gauge, check the clearance between the snap ring and the pressure plate. Downward pressure on the plate should be maintained when making this check. Refer to Specifications for the proper clearance.

6 If the clearance is not within specifications, selective snap rings are available in the following thicknesses: 0.056 to 0.060, 0.065 to 0.069, 0.074 to 0.078, 0.083 to 0.087 and 0.092 to 0.096 inch. Insert the correct size snap ring and recheck the clearance.

Input Shell and Sun Gear

Disassembly

1 Remove the external snap ring from the sun gear.

2 Remove the thrust washer from the input shell and sun gear.

3 Working from inside the input shell remove the sun gear. Remove the internal snap ring from the gear.

Assembly

1 Install the forward snap ring on the forward end (short end) of the sun gear. Working from inside the input shell, slide the sun gear and snap ring into place making sure that the longer end is at the rear.

2 Place the No. 6 thrust washer on the sun gear and install the rear snap ring.

Output Shaft Hub and Ring Gear

Disassembly

1 Remove the hub snap ring from the ring gear.

2 Lift the hub from the ring gear.

Assembly

1 Position the hub in the ring gear.

2 Secure the hub with the snap ring. Make certain that the snap ring is fully engaged with the groove.

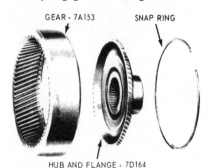

GEAR - 7A153   SNAP RING

HUB AND FLANGE - 7D164

Output shaft hub and ring gear

One-Way Clutch

Disassembly

1 Remove the snap ring and bushing from the rear of the low-reverse clutch hub.

2 Remove the rollers from the spring assembly and lift the spring assembly from the hub.

3 Remove the remaining snap ring from the hub.

Assembly

1 Install a snap ring in the forward snap ring groove of the low-reverse clutch hub.

2 Place the low-reverse clutch hub on the bench with the forward end down.

3 Install the one-way clutch spring assembly on top of the snap ring.

4 Install a roller into each of the spring assembly compartments.

5 Install the bushing on top of the spring assembly.

6 Install the remaining snap ring at the rear of the low-reverse clutch hub to secure the assembly.

FORD C6

Low-Reverse Clutch Piston

Disassembly
1 Remove the inner and the outer seal from the reverse clutch piston.

Assembly
1 Dip the two new seals in clean transmission fluid.
2 Install the seals on the piston.

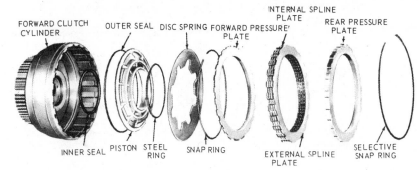

Forward clutch disassembled

APPROXIMATE REFILL CAPACITIES—QTS.

1974   Torino,   Montego,   Cougar
        PGA-AF2, AG2—10½
        All others—12½
1975–80   All—12½

FORD C6 APPROXIMATE REFILL CAPACITIES

'74–76 Lincoln Continental, Mark III, Mark IV	13 qt.
'74–76 Ford Mustang; Torino, Mercury Cougar, Montego (Models PGA-AF, AG, Av only)	10½ qt.
'74–80 All Other Models	12½ qt.

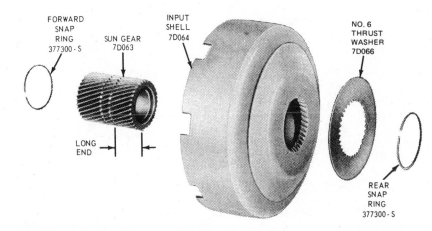

Input shell and sun gear disassembled

Output shaft disassembled

CHECKS AND ADJUSTMENTS
C6 TRANSMISSION

Operation	Specification	
Transmission End Play	0.008-.042 (Selective Thrust Washers Available)	
Turbine and Stator End Play	New or rebuilt 0.021 max. Used 0.040 max. ①	
Intermediate Band Adjustment	Remove and discard lock nut. Adjust screw to 10 ft-lbs torque, then back off 1-1/2 turns, install new lock nut and tighten lock nut to 40 ft-lb.	
Forward Clutch Pressure Plate to Snap Ring Clearance	0.021-0.046	
Selective Snap Ring Thicknesses	0.056-0.060, 0.065-0.069, 0.074-0.078, 0.083-0.087, 0.092-0.096, 0.110-0.114, 0.128-0.132	
Reverse High Clutch Pressure		
Plate to Snap Ring Clearance	0.022-0.036	0.027-0.043
Selective Snap Ring Thicknesses	0.056-0.060, 0.065-0.069, 0.074-0.078, 0.083-0.087, 0.092-0.096	
① To check end play, exert force on checking tool to compress turbine to cover thrust washer wear plate. Set indicator at zero.		

FORD CW & FMX

TROUBLESHOOTING
FORD CW & FMX

	CHECK THE FOLLOWING ITEMS	
THE PROBLEM	IN THE CAR	OUT OF CAR
ROUGH INITIAL ENGAGEMENT IN D1 OR D2	Engine idle speed. Vacuum modulator or lines. Control pressure. Pressure regulator. Valve body. Front band.	
1-2 OR 2-3 SHIFT POINTS INCORRECT OR ERRATIC	Fluid level. Vacuum modulator or lines. Manual linkage. Governor. Control pressure. Valve body. Downshift linkage.	
ROUGH 2-3 SHIFTS	Vacuum modulator or lines. Front band. Pressure regulator. Front servo.	
ENGINE OVERSPEEDS ON 2-3 SHIFT	Vacuum modulator or lines. Front band. Valve body. Pressure regulator	Rear clutch piston air bleed valve.
NO 1-2 OR 2-3 UPSHIFT	Governor. Vale body. Manual linkage. Front servo. Front band.	Rear clutch. Hydraulic system leakage. Fluid distributor sleeve—output shaft.
NO 3-1 DOWNSHIFT	Engine idle speed. Vacuum diaphragm unit or tubes. Valve body.	
NO FORCED DOWNSHIFT	Downshift linkage. Control pressure. Valve body.	
RUNAWAY ENGINE ON FORCED DOWNSHIFT	Front band. Pressure regulator. Valve body. Front servo. Vacuum diaphragm unit or tubes.	Hydraulic system leakage.
ROUGH 3-2 OR 3-1 SHIFT AT CLOSED THROTTLE	Engine idle speed. Vacuum diaphragm unit or tubes. Valve body.	
CREEPS EXCESSIVELY	Engine idle speed. Vehicle brakes.	
SLIPS OR CHATTERS IN FIRST GEAR, D1	Fluid level. Vacuum modulator or lines. Control pressure. Valve body.	Fluid distributor sleeve—output shaft. Planetary one-way clutch. Front clutch. Hydraulic system leakage.
SLIPS OR CHATTERS IN SECOND GEAR	Fluid level. Vacuum modulator or lines. Front band. Control pressure. Pressure regulator. Valve body. Front servo.	Front clutch. Hydraulic system leakage.
SLIPS OR CHATTERS IN R	Fluid level. Rear band. Control pressure. Pressure regulator. Valve body. Rear servo. Vacuum modulator or lines.	Rear clutch. Hydraulic system leakage. Fluid distributor sleeve—output shaft.
NO DRIVE IN D1	Manual linkage. Valve body.	Planetary one-way clutch.
NO DRIVE IN D2	Valve body. Perform air pressure checks. Manual linkage.	Front clutch. Hydraulic system leakage. Fluid distributor sleeve—output shaft.

FORD CW-FMX

Ford CW and FMX

Identification

The identification tag is located under the left side extension housing bolt. The CW transmission is coded Y on the vehicle certification label; the FMX transmission is coded X. The illustrations below show the transmission identification tags and the code interpretation.

In-Car Testing Procedures

NOTE: The Ford Rotunda ARE-2905 Automatic Transmission Tester is shown here because it is a convenient tool for quick testing. A vacuum gauge, tachometer, and 400 psi pressure gauge may be substituted.

Control Pressure Check

When the vacuum diaphragm unit operates properly and the downshift linkage is adjusted correctly, all transmission

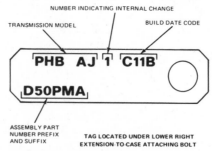

Ford FMX Identification Tag-1975 and later models

shifts (automatic and kickdown) should occur within the specified road speed limits. If these shifts do not occur within the limits or if the transmission slips during a shift point, perform the following procedure to locate the problem:

1 Connect a Ford Rotunda ARE-2905

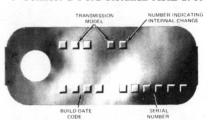

Ford FMX Identification Tag-1974 and earlier models

TROUBLESHOOTING
FORD CW & FMX

THE PROBLEM	CHECK THE FOLLOWING ITEMS	
	IN THE CAR	OUT OF CAR
NO DRIVE IN L	Manual linkage. Valve body. Perform air pressure checks.	Front clutch. Hydraulic system leakage. Fluid distributor sleeve—output shaft.
NO DRIVE IN R	Rear band. Rear servo. Valve body. Perform air pressure checks. Manual linkage.	Rear clutch. Rear pump, no longer used. Fluid distributor sleeve—output shaft.
NO DRIVE IN ANY SELECTOR POSITION	Fluid level. Manual linkage. Control pressure. Pressure regulator. Valve body. Perform air pressure checks.	Hydraulic system leakage. Front pump.
LOCKUP IN D1	Manual linkage. Rear servo. Front servo.	Rear clutch. Parking linkage. Hydraulic system leakage.
LOCKUP IN D2	Manual linkage. Rear band. Rear servo.	Rear clutch. Parking linkage. Hydraulic system leakage. Planetary one-way clutch.
LOCKUP IN L	Front band. Front servo. Valve body.	Rear clutch. Engine rear oil seal. Hydraulic system leakage.
LOCKUP IN R	Front band. Front servo.	Fluid level. Parking linkage. Hydraulic system leakage.
PARKING LOCK BINDS OR DOES NOT HOLD	Manual linkage.	Parking linkage.
TRANSMISSION OVERHEATS	Fluid level. Oil cooler and/or connections. Pressure regulator. Front band.	Front pump-to-case seal or gasket.
ENGINE WILL NOT PUSH-START	Fluid level. Manual linkage. Pressure regulator. Valve body.	Rear pump not used. Hydraulic system leakage.
MAXIMUM SPEED TOO LOW, POOR ACCELERATION	Engine performance.	Front pump-to-case seal or gasket
TRANSMISSION NOISY IN N AND P	Pressure regulator.	Front clutch. Front pump.
NOISY TRANSMISSION DURING COAST 30-20 MPH WITH ENGINE STOPPED		Rear pump not used.
TRANSMISSION NOISY IN ANY DRIVE POSITION	Pressure regulator.	Planetary assembly. Rear clutch. Front clutch. Front pump.
FLUID LEAKS	Converter drain plug. Oil pan, filler tube and/or seals. Oil cooler and/or connections. Manual or throttle shaft seals. Pipe plug, side of case. Extension housing-to-case gasket or washer. Center support bolt lock. Extension housing rear oil seal. Speedometer drive gear adaptor seal.	Engine rear oil seal. Front oil pump seal. Front pump-to-case seal or gasket.

Automatic Transmission Tester as follows:
a. Tachometer cable to engine.
b. Vacuum gauge hose with a T-fitting between the vacuum hose and vacuum diaphragm.

NOTE: On vehicles equipped with a dual area diaphragm (DAD), check the control pressure at 10 in. of vacuum by removing the exhaust gas recirculation (EGR) control hose from the diaphragm and plugging the hose. Do not plug the EGR port in the diaphragm; this port must be left open to atmospheric pressure. When checking the control pressure at stall and idle, keep the hose connected.

  c. Pressure gauge to the control pressure outlet on the transmission.
2 Apply the parking brake and start the engine. On a car equipped with a vacuum brake release, use the service brakes since the parking brake will release automatically when the transmission is put in any Drive position.

3 Check the transmission diaphragm unit for leaks (see below).
4 Check control pressure in all selector lever positions at specified manifold vacuum (see specifications). Record readings and compare to specifications.

Vacuum Diaphragm Unit Check

1 Remove the vacuum diaphragm unit from the transmission using crowfoot wrench, after disconnecting the vacuum hose (see illustration).
2 Adjust a vacuum pump until the vac-

Ford FMX
Control Pressure at Zero Governor RPM

Transmission Model	Range	Control Pressure (psi) Manifold Vacuum (in. Hg)			
		12 and Above	16-13	10	Below 1.0 (Stall)
PHA-F, PHB	P,N,D,2,1	61–107	—	75–120	—
	R	90–156	—	—	185–225
	D,2,1	—	—	—	154–188

Component Removal and Overhaul

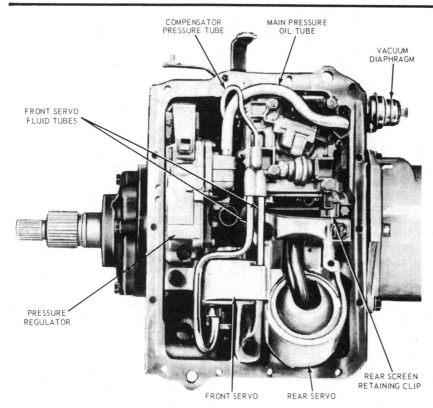

COMPENSATOR PRESSURE TUBE

MAIN PRESSURE OIL TUBE

VACUUM DIAPHRAGM

FRONT SERVO FLUID TUBES

PRESSURE REGULATOR

FRONT SERVO

REAR SERVO

REAR SCREEN RETAINING CLIP

Hydraulic control system components

uum gauge shows 18 in./Hg. with the vacuum hose blocked.

3 Connect the vacuum hose to the vacuum diaphragm unit and note the reading on the vacuum gauge. If the reading is 18 inches of vacuum, the vacuum diaphragm unit is good. While removing the vacuum hose from the vacuum diaphragm unit, hold a finger over the end of the control rod. As the vacuum is released, the internal spring of the vacuum diaphragm unit will push the control rod out.

Shift Point Checks for Automatic Transmissions

To determine if the transmission is shifting at the proper road speeds, use the following procedure:

1 Check the minimum throttle upshifts by placing the transmission selector lever in the Drive position and noting the road speeds at which the transmission shifts from first gear to second gear to third gear. All shifts should occur within the specified limits.

2 While driving in third gear, depress the accelerator pedal past the detent (to the floor). Depending on vehicle speed, the transmission should downshift from third gear to second gear or from second gear to first gear.

3 Check the closed-throttle downshift from third gear to first gear by coasting down from about 30 mph in third gear. This downshift should occur at the specified road speed.

4 With the transmission in third gear and the car moving at a road speed of 35 mph, the transmission should downshift to second gear when the selector lever is moved from D to 2 to 1. This check will determine if the governor pressure and shift control valves are operating properly. If the transmission does not shift within the specified limits or certain gears cannot be obtained, refer to the Trouble Diagnosis chart at the beginning of this section.

Air Pressure Checks

If the car will not move in one or more ranges, or, if it shifts erratically, the items at fault can be determined by using air pressure at the indicated passages.

Drain the transmission and remove the oil pan and the control valve assembly.

Front Clutch

Apply sufficient air pressure to the front clutch input passage. A dull thud can be heard when the clutch piston moves.

Governor

Remove the governor inspection cover from the extension housing. Apply air to the front clutch input passage. Listen for a sharp click and watch to see if the governor valve snaps inward as it should.

Rear Clutch

Apply air to the rear clutch passage and listen for the dull thud that will indicate

that the rear clutch piston has moved. Listen for leaks.

Front Servo

Apply air pressure to the front servo apply tube and note if front band tightens. Shift the air to the front servo release tube, which is next to the apply tube, and watch band release.

Rear Servo

Apply air pressure to the rear servo apply passage. The rear band should tighten around the drum.

Conclusions

If the operation of the servos and clutch is normal with air pressure, the no-drive condition is due to the control valve and pressure regulator valve assemblies, which should be disassembled, cleaned and inspected.

If operation of the clutches is not normal; (e.g., both clutches apply from one passage or one fails to move), the aluminum sleeve (bushing) in the output shaft is out of position or badly worn.

Use air pressure to check the passages in the sleeve, shaft, and primary sun gear shaft.

If the passages in the two shafts and the sleeve are clean, remove the clutch assemblies, clean and inspect the parts.

Erratic operation can also be caused by loose valve body screws. When reinstalling the valve body be careful to tighten all screws and bolts to specifications.

In-Car Service Procedures

Shift Linkage Adjustment

FORD FULL-SIZE CARS
FORD THUNDERBIRD
MERCURY FULL-SIZE CARS

1 With engine off, loosen clamp at shift lever so shift rod is free to slide. On models with a shift cable, remove the nut from the transmission manual lever and disconnect the cable from the transmission.

2 Position lever in D position tightly against the D stop.

3 Shift lever at transmission into D position.

NOTE: D position is third from rear on all column shift Select Shift transmissions, fourth from rear on all console shift Select Shift transmissions.

4 Tighten clamp and torque nut to 10–20 ft./lbs.

Shift Linkage Adjustment

ALL FORD MID-SIZE CARS WITH COLUMN SHIFT

1 Turn engine off. Loosen shift rod clamp on transmission. On cable type units, remove nuts at transmission shift rod and at manual lever stud.

2 Shift transmission into D.

FORD CW & FMX

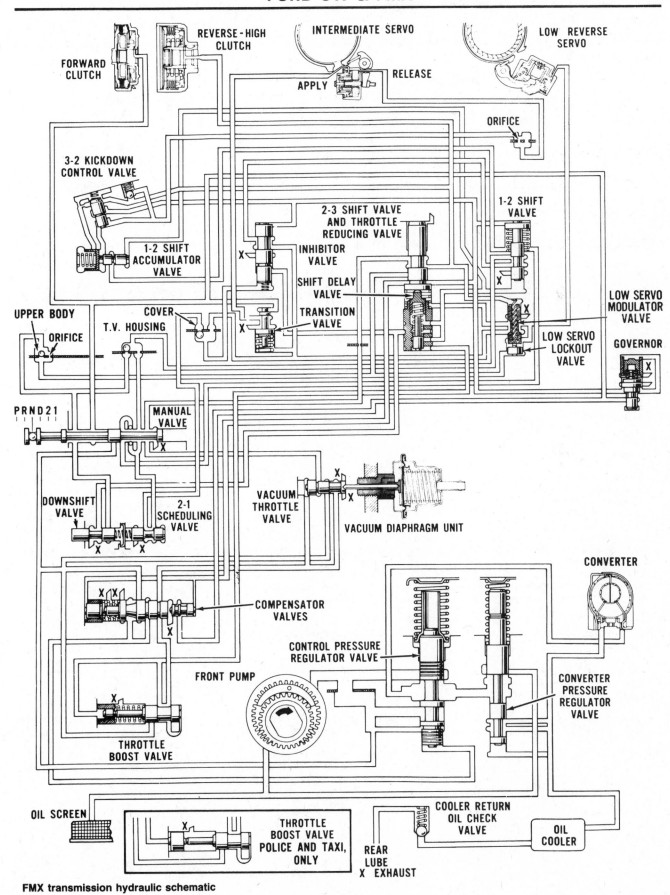

FMX transmission hydraulic schematic

Component Removal and Overhaul

CHILTON CAUTION: *Be sure to remove the tube in this manner. Otherwise, the tube could be kinked or bent causing improper fluid pressures and possible damage to the transmission.*

7 Loosen the front servo attaching bolts about three turns.

8 Remove the three control valve body attaching bolts and carefully lower the valve body, sliding it off the front servo tubes. *Do not damage the valve body or the tubes.*

9 Disassemble the control valve body into the various parts as shown in the illustration in the overhaul section.

10 When installing the control valve body, align the front servo tubes with the holes in the valve body. Shift the manual lever to the 1 detent and place the inner downshift lever between the downshift lever stop and the downshift valve. Be sure the manual lever engages the actuating pin in the manual detent lever.

11 Loosely install the control valve body attaching bolts and move the control valve body toward the center of the transmission case until there is a clearance of 0.050 in. between the manual valve and the actuating pin on the manual detent lever.

12 Tighten the attaching bolts to 8–10 ft./lbs. torque. Ensure that the rear fluid filter retaining clip is installed under the valve body.

13 Install the main pressure oil tube, connecting the end to the pressure regulator unit first and then connecting the other end to the main control valve assembly by gently tapping it with a soft-faced hammer.

14 Install the compensator pressure tube on the pressure regulator and control valve body.

15 Check the manual lever for free motion in each detent position by rotating it one full turn. If the manual lever binds in any detent position, loosen the valve body attaching bolts and move the valve body away from the center of the transmission case until the binding is relieved. Retighten the attaching bolts according to Step 12.

16 Place the pushrod in the bore of the vacuum diaphragm unit and install the vacuum diaphragm unit.

17 Tighten the front servo attaching bolts.

18 Adjust the front band.

19 Install the fluid filter and its retaining clip.

20 Adjust the rear band.

21 Install the oil pan with a new pan gasket.

22 Fill transmission with fluid. Start and run engine for a few minutes and check the fluid level after shifting the transmission through all positions. *Do not overfill the transmission.*

23 Check the adjustment of the transmission control linkage.

Front Servo R&R

Removal

1 Drain fluid and remove pan and screen.

2 Remove vacuum diaphragm.

3 Loosen the three valve body attaching bolts.

4 Remove the front servo attaching bolts, hold the strut and remove the screw.

Installation

1 Position the front band forward with the end of the band facing down.

2 Make sure the front servo anchor pin is in position in the case web. Align the large end of the servo strut with the servo actuating lever. Align the small end with the band lever.

3 Rotate the band, strut and servo to align the anchor end of the band with the anchor in the case. Push the servo body onto the control valve body tubes.

4 Install the bolts and torque to 30–35 ft./lbs.

5 Torque the control valve body attaching bolts to 8–10 ft./lbs.

6 Check the manual valve to manual lever actuating pin clearance: .050″.

7 Adjust the front band.

8 Install the vacuum diaphragm unit and rod.

9 Install the fluid screen and pan. Fill the transmission.

10 Adjust the downshift and manual linkage.

Rear Servo R&R

Removal

1 Drain the transmission. Remove the pan and screen.

2 Remove the bacuum diaphragm unit.

3 Remove the control valve body and two front servo tubes.

4 Remove the rear servo bolts, hold the actuating and anchor struts and remove the servo.

Installation

1 Position the servo anchor strut on the servo band and rotate the band to engage the strut.

2 Hold the servo anchor strut in position with the fingers, position the actuating lever strut and install the servo.

3 Install but do not tighten the servo attaching bolts. The longer bolt must be installed in the inner bolt hole.

4 Move the rear servo toward the center of the case, against the servo attaching bolts. Torque the bolts to 40–45 ft./lbs.

5 Install the front servo tubes and control valve body.

6 Check the clearance between the manual valve and the manual lever actuating pin: .050″.

7 Adjust the rear band.

8 Install the screen and pan and fill the transmission.

Pressure Regulator R&R

1 Drain the transmission of fluid and remove the pan, fluid filter screen, and its retaining clip. Discard the used pan gasket.

2 Remove the compensator pressure tube from between the control valve body and the pressure regulator. See illustration for location of components.

3 Remove the main pressure oil tube by gently prying off the end connected to the control valve first and then disconnect the other end from the pressure regulator. *Be sure to remove the tube in this order to prevent kinking or bending it.*

4 Loosen the spring retainer clip and carefully release the spring tension on the pressure springs. Remove the valve springs, retainer and valve stop, and the valves from the pressure regulator body.

5 Remove the pressure regulator attaching bolts and washers and take the regulator body out of the transmission case.

6 After cleaning, inspection, and reassembly, install the pressure regulator unit in the transmission by reversing the procedures in steps 1 through 5.

Overhaul

Disassembly

1 After the transmission has been removed from the vehicle, place the assembly in the transmission holder.

2 Remove the transmission pan, gasket, and screen retainer clip.

3 Lift the screen from the case.

4 Remove the spring seat from the pressure regulator. **Maintain constant pressure on the seal to prevent distortion of the spring seat and the sudden release of the springs.**

5 Remove the pressure regulator springs and pilots, but do not remove the valves.

6 Remove the small compensator pressure tube from the pressure regulator and the control valve body.

7 Remove the main pressure oil tube first, by gently prying up the end that connects to the main control valve assembly, then, remove the other end of the tube from the pressure regulator. **Be sure to remove the tube in this manner. Failure to do so, could kink or bend the tube causing excessive transmission internal leakage.**

8 Loosen the front and rear servo band adjusting screws five turns. Loosen the front servo attaching bolts three turns.

9 Remove the vacuum diaphragm unit and push rod.

10 Remove the control valve body attaching bolts. Align the levers to permit removal of the valve body. Then lift the valve body clear of the transmission case. Pull the body off the servo tubes and remove from the case.

FORD CW & FMX

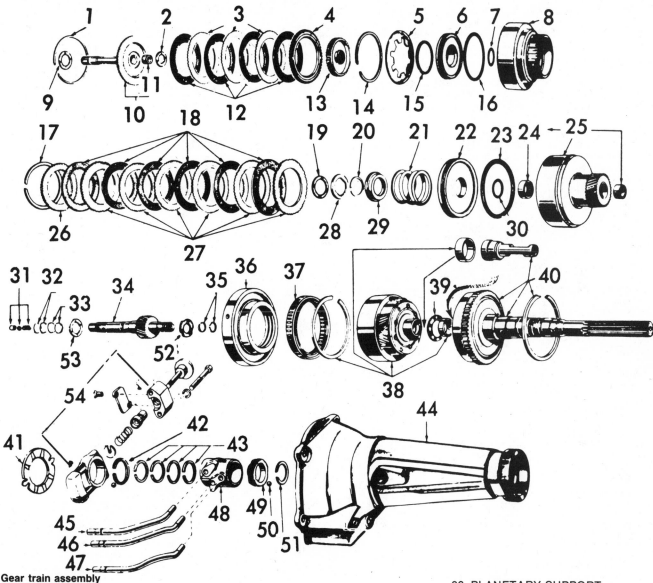

Gear train assembly

1. SNAP RING
2. THRUST WASHER
3. CLUTCH PLATE
4. CLUTCH PRESSURE PLATE
5. DISC SPRING
6. CLUTCH PISTON
7. OIL SEAL
8. FRONT CLUTCH CYLINDER
9. THRUST WASHER
10. INPUT SHAFT
11. BUSHING
12. CLUTCH PLATE
13. FRONT CLUTCH HUB
14. SNAP RING
15. RING
16. OIL SEAL
17. SNAP RING
18. CLUTCH PLATE

19. THRUST WASHER
20. SNAP RING
21. PISTON SPRING
22. CLUTCH PISTON
23. PISTON OIL SEAL
24. BEARING
25. FRONT BRAKE DRUM AND
    SUN GEAR
26. CLUTCH PRESSURE PLATE
27. CLUTCH PLATE
28. THRUST WASHER
29. PISTON SPRING
30. OIL SEAL
31. OIL SEAL
32. OIL SEAL
33. OIL SEAL
34. PRIMARY SUN GEAR
35. OIL SEAL

36. PLANETARY SUPPORT
37. SPRAG CLUTCH
38. PLANETARY ASSEMBLY
39. THRUST WASHER
40. OUTPUT SHAFT
41. THRUST WASHER
42. SNAP RING
43. OIL SEAL
44. EXTENSION HOUSING
45. FRONT OIL TUBE
46. IMMEDIATE OIL TUBE
47. REAR OIL TUBE
48. OIL DISTRIBUTOR SLEEVE
49. DRIVE GEAR
50. CHECK BALL
51. SNAP RING
52. THRUST WASHER
53. THRUST WASHER
54. GOVERNOR ASSEMBLY

11 Remove the regulator from the case. **Keep the control pressure valve and the converter pressure regulator valve in the pressure regulator to avoid damage to the valves.**

12 Remove the front servo apply and release tubes by twisting and pulling at the same time. Remove the front servo attaching bolts. Hold the front servo strut with the fingers, and lift the servo assembly from the case.

13 Remove the rear servo attaching bolts. Hold the actuating and anchor struts with the fingers, and lift the servo from the case.

FORD CW & FMX

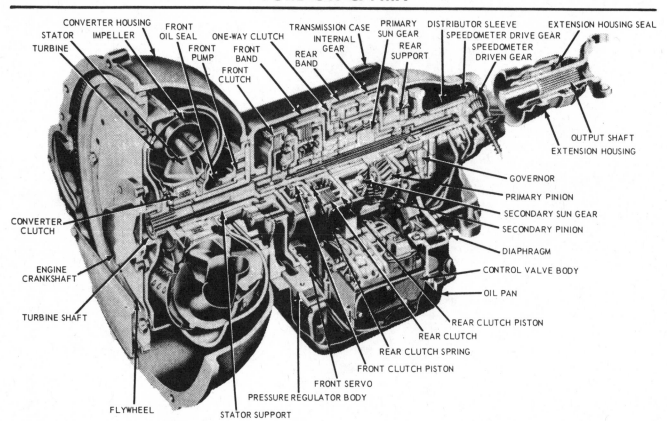

Cutaway view FMX transmission

3 Shift the transmission lever into the second detent from the rear.
4 Tighten the shift rod clamp to 10–20 ft./lbs. Position the cable onto the transmission manual lever stud, aligning the flats. Replace nuts at either end.
5 Check for proper pointer alignment and operation in all positions.

Shift Linkage Adjustment

ALL FORD MID-SIZE CARS WITH FLOOR/CONSOLE SHIFT
1 Place shift lever in D.
2 Loosen the transmission manual lever retaining nut.
3 Shift transmission manual lever into second detent·from the rear.
4 Tighten manual transmission lever nut to 10–20 ft./lbs.
5 Check operation in all positions.

Neutral Start Switch Adjustment

ALL FORD MOTOR CO. CARS

COLUMN SHIFT
Models which are equipped with a column mounted shift lever are not equipped with neutral start switch. Instead, an ignition lock cylinder-to-shift lever interlock prevents these models from being started in any gear other than Park or Neutral.

Throttle Linkage Adjustment

ALL FORD MOTOR CO. CARS
1 Disconnect downshift return spring.
2 Hold the carburetor throttle lever in wide open position (against stop); and hold the transmission linkage in full downshift position against its internal stop.
3 Turn the carburetor downshift lever adjustment screw to within .010–.080 in. of the contacting pickup surface of the throttle lever.
4 Connect return spring.

Band Adjustments

FRONT BAND ADJUSTMENT
When it is necessary to adjust the front band of the transmission, perform the following procedure:
1 Drain transmission fluid and remove oil pan, fluid filter screen, and clip.
2 Clean the pan and filter screen; remove the old gasket.
3 Loosen the front servo adjusting screw locknut. See illustration.

NOTE: Special band adjusting wrenches are recommended to do this operation correctly and quickly.

4 Pull back the actuating rod and insert a ¼ in. spacer bar between the adjusting screw and the servo piston stem. Tighten the adjusting screw to 10 in./lbs. torque. Remove the spacer bar

and tighten the adjusting screw an additional ¾ turn, except on CW. Hold the adjusting screw and tighten the locknut securely (20–25 ft./lbs.).
5 Install the transmission fluid filter screen and clip. Install the pan with a new pan gasket.
6 Refill the transmission to the FULL mark on the Dipstick. Start the engine, run for a few minutes, shift the selector lever through all positions, and place it in Park. Recheck the fluid level again and add fluid to proper level if necessary.

REAR BAND ADJUSTMENTS
The rear band of the FMX transmission may be adjusted by any of the methods given below. On most cars the basic external band adjustment is satisfactory. The internal adjustment may be performed in cases where the adjustment required is outside the range of the external adjustment. On certain cars with a console floor shift, the entire console, shift lever and linkage will have to be removed to gain access to the rear band external adjusting screw.

REAR BAND EXTERNAL ADJUSTMENT
The procedure for adjusting the rear band externally is as follows:
1 Locate the external rear band adjusting screw on the transmission case, clean all dirt from the threads, and coat the threads with light oil.

FORD CW & FMX

NOTE: The adjusting screw is located on the upper right side of the transmission case. Access is often through a hole in the front floor to the right of center under the carpet.

2 Loosen the locknut on the rear band external adjusting screw.
3 Tighten the adjusting screw to 10 ft./lbs. torque. If the adjusting screw is tighter than 10 ft./lbs. torque, loosen the adjusting screw and re-tighten to the proper torque.
4 Back off adjusting screw 1½ (1¼ on CW) turns. Hold the adjusting screw steady while tightening the locknut to the proper torque (35–40 ft./lbs.). *Severe damage may result if adjusting screw is not backed off exactly as specified.*

REAR BAND INTERNAL ADJUSTMENT

The rear band is adjusted internally as follows:
1 Drain the transmission fluid. If it is to be reused, filter it as it drains from the transmission. Reuse the transmission fluid only if it is in good condition.
2 Remove and clean the pan, fluid filter, and clip.
3 Loosen the rear servo adjusting locknut.
4 Pull the adjusting screw end of the actuating lever away from the servo body and insert the spacer tool between the servo accumulator piston and the adjusting screw. *Be sure the flat surfaces of the tool are placed squarely between the adjusting screw and the accumulator piston. Tool must not touch servo piston and the handle must not touch the servo piston spring retainer.*
5 Using a torque wrench with an allen head socket adapter tighten the adjusting screw to 24 in./lbs.
6 Back off the adjusting screw exactly 1½ turns. Hold adjusting screw steady and tighten the locknut securely. Remove the spacer tool.
7 Install the fluid filter, clip, and pan with a new gasket.
8 Fill the transmission with the correct amount of fluid.

Major Component R&R Procedures

Governor Assembly R&R

The extension housing must be removed to remove the governor assembly from the output shaft. It may be necessary to remove the entire transmission from the car to do this.

Extension Housing R&R

1 Raise the car high enough for easy access to the extension housing.
2 Drain the transmission.
3 Disconnect the driveshaft from the rear axle and slide the front yoke from the extension housing.

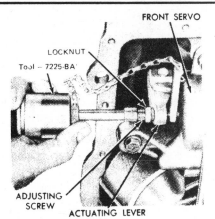
Front band adjustment

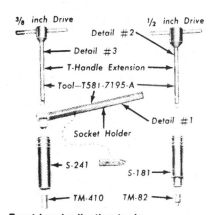
Front band adjusting tools

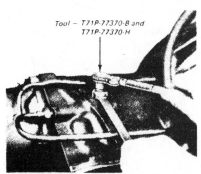
Adjusting rear band-except LTD 11/Thunderbird/Cougar

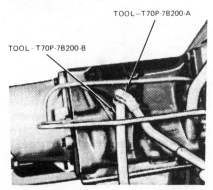
Adjusting rear band-LTD 11/Thunderbird/Cougar

4 Disconnect the speedometer cable from the extension housing.
5 Remove the rear engine support nuts. Place a transmission jack under the transmission and raise it enough to clear the crossmember.
6 Remove the bolts and nuts securing the crossmember to the side rails of the frame. Move the crossmember out of the way.
7 Remove the attaching bolts holding the engine rear support to the extension housing and remove the rear support from the extension housing.
8 Remove all the extension housing attaching bolts, slide the housing off the output shaft, and discard the gasket.
9 Installation is the reverse of the removal procedures. Tighten the housing attaching bolts and check that the output shaft rotates freely by hand. If it binds or feels tight, check the needle bearing and race for correct position.

Extension Housing Bushing & Rear Seal R&R

1 Disconnect the driveshaft from the extension housing.
2 Carefully remove the rear seal from the housing.
3 Remove the extension housing bushing using a bushing remover tool. *Be careful not to damage the spline seal.*
4 Install new bushing in the extension housing.
5 Inspect the universal joint yoke sealing surface for scoring or gouges. Replace the yoke if damaged.
6 Inspect the housing counterbore for burrs and, if necessary, smooth with crocus cloth.
7 Install the new rear seal into the end of the extension housing. Check that the seal is firmly seated.

Control Valve Body & Oil Pan R&R

1 Raise the car on a hoist or jackstands and place a drain pan under the transmission.
2 Drain the transmission.
3 Remove the oil pan, fluid filter screen, and clip and clean them thoroughly. Discard the old pan gasket.
4 Remove the vacuum diaphragm assembly using a crowfoot wrench (Snap-On tool S8696-A or FCO-24). *Do not use pliers, pipe wrenches, etc. to remove the vacuum diaphragm unit. Do not let any solvents enter the vacuum diaphragm unit.* Remove the push rod, the fluid screen and its retaining clip.
5 Remove the small compensator pressure tube. See illustration.
6 Disconnect the main pressure oil tube by carefully loosening the end connected to the control valve body first and then removing the tube from the pressure regulator unit.

Component Removal and Overhaul

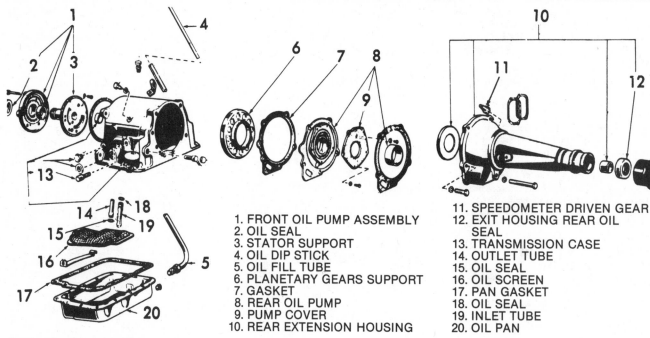

Case and parts assemblies

1. FRONT OIL PUMP ASSEMBLY
2. OIL SEAL
3. STATOR SUPPORT
4. OIL DIP STICK
5. OIL FILL TUBE
6. PLANETARY GEARS SUPPORT
7. GASKET
8. REAR OIL PUMP
9. PUMP COVER
10. REAR EXTENSION HOUSING
11. SPEEDOMETER DRIVEN GEAR
12. EXIT HOUSING REAR OIL SEAL
13. TRANSMISSION CASE
14. OUTLET TUBE
15. OIL SEAL
16. OIL SCREEN
17. PAN GASKET
18. OIL SEAL
19. INLET TUBE
20. OIL PAN

Transmission End Play Check

1 Remove one of the front pump attaching bolts. Mount the dial indicator support Tool 77067 in the front pump bolt hole. Mount a dial indicator on the support so that the contact rests on the end of the turbine shaft.
2 Install the extension housing seal replacer on the output shaft to provide support for the shaft.
3 Pry the front clutch cylinder to the rear of the transmission with a large screwdriver. Set the dial indicator at zero while maintaining a slight pressure on the screwdriver.
4 Remove the screwdriver and pry the units toward the front of the transmission by inserting the screwdriver between the large internal gear and the transmission case.
5 Record the indicator reading for use during transmission assembly. End play should be 0.010 to 0.029 inch (minimum end play is preferred). If end play is not within specifications, a new selective thrust washer must be used when the transmission is assembled.
6 Remove the indicator support, and then remove the seal replacer from the output shaft.

Removal of Case and Extension Housing Parts

1 Remove the remaining front pump attaching bolts. Then remove the front pump and gasket. **If necessary, tap the screw bosses with a soft-faced hammer to loosen the pump from the case.**
2 Remove the lubrication tube from the case. Remove the five transmission to extension housing bolts. These bolts also attach the rear support to the case. Remove the extension housing.
3 Remove the output shaft assembly. To facilitate output shaft removal, insert a screwdriver between the output shaft ring gear and pinion carrier and pry the output shaft rearward. **Be careful not to bend the pressure tubes between the rear support or distributor sleeve and case as the tubes are removed from the case.**
4 Remove the four seal rings from the output shaft with the fingers, to prevent breaking the rings.
5 Remove the governor snap ring from the output shaft. Using a soft-faced hammer, tap the governor assembly off the output shaft. Remove the governor drive ball.
6 Remove the rear support and gasket from the output shaft. Remove the needle bearings and race from the rear support.
7 Remove the selective thrust washer from the rear of the pinion carrier and remove the pinion carrier.
8 Remove the primary sun gear rear thrust bearing and races from the pinion carrier.
9 **Note the rear band position for reference in assembly.** The end of the band next to the adjusting screw has a depression (dimple) in the center of the boss. Squeeze the ends of the rear band together, tilt the band to the rear, and remove the rear band from the case.
10 Remove the two center support outer bolts (one each side) from the transmission case.
11 Exert enough pressure on the end of the input shaft to hold the clutch units together. Then remove the center sup-

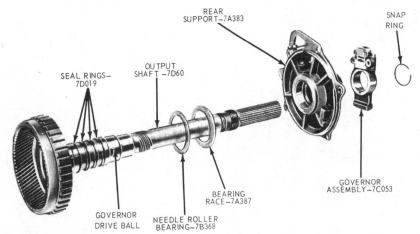

Output shaft disassembled

FORD CW & FMX

port and the front and rear clutch assemblies as a unit.

12 Install the clutch assemblies in the bench fixture.

13 Remove the thrust washer from the front of the input shaft.

14 To remove the front band, position the band ends between the case webbing and tilt the bottom of the band rearward. Then, squeeze the ends of the band together and remove from the rear of the case.

15 Lift the front clutch assembly from the primary sun gear shaft.

16 Remove the bronze and the steel thrust washers from the rear clutch assembly. Wire the thrust washers together to assure correct installation.

17 Remove the front clutch seal rings from the primary sun gear shaft. Lift the rear clutch assembly from the primary sun gear shaft. Remove the rear clutch seal rings from the primary sun gear shaft. **Do not break the seal rings.** Remove the primary sun gear front thrust washer.

18 If the transmission case bushing is to be replaced, press the bushing out of the case.

19 Install a new transmission case bushing.

20 If the rear brake drum support bushing is to be replaced, press the bushing from the support. Press a new bushing into the brake drum support.

21 Remove the output shaft bushing if it is worn or damaged. Use the cape chisel and cut along the bushing seam

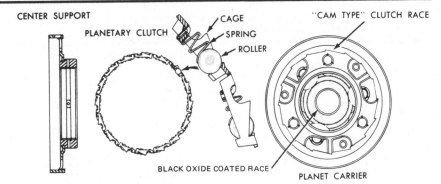

Planetary clutch, carrier and center support

until the chisel breaks through the bushing wall. Pry the loose ends of the bushing up with an awl and remove the bushing.

22 Insert a new bushing into the installation tool and position the tool and bushing over the output shaft hub. Then, press the bushing on the output shaft hub.

Assembly

Do not use force to assemble mating parts. If the parts do not assemble freely, examine them for the cause of the difficulty. Always use new gaskets and seals during the assembly operations.

1 Install the front band in the transmission case with the anchor end aligned with the anchor in the case.

2 Make sure the thrust washer is in place on the input shaft. Lift the

clutch assemblies out of the holding block. **Do not allow the clutches to separate.**

3 Install the clutch subassemblies in the transmission case while positioning the servo band on the drum. Hold the units together while installing them.

Installation—Center Support

a. Install the center support and the rear band in the case.

b. Install the one-piece inner bearing and race assembly in the planet carrier, being sure that the black oxide-coated race is up (or toward the front of the transmission).

c. Lubricate the bearing surface on the center support, the rollers of the planetary clutch, and the cam race in the carrier with petroleum jelly.

d. Install the planetary clutch in the carrier.

e. Carefully position the planet carrier on the center support. Move the carrier forward until the clutch rollers are felt to contact the bearing surface of the center support.

f. While applying forward pressure on the planet carrier, rotate it counterclockwise, as viewed from the rear. The clutch rollers will roll toward the large opening end of the cams in the race, compressing the

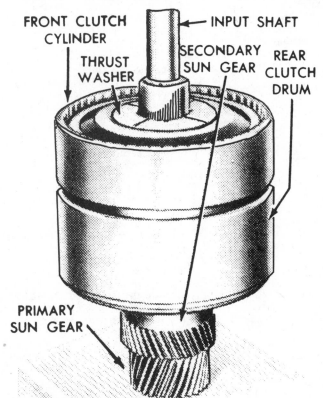

Input shaft and clutch

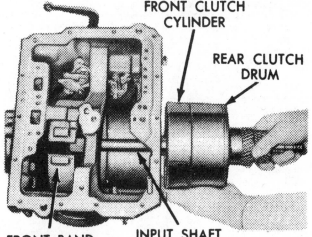

FRONT BAND
INPUT SHAFT
Input shaft placement

Component Removal and Overhaul

FORD CW & FMX

spring slightly, and the rollers will ride up the chamfer on the planetary support and onto the inner race.

g. Push the planet carrier all the way forward.

h. Check the operation of the planetary clutch by rotating the carrier counterclockwise. It should rotate counterclockwise (viewed from the rear) with a slight drag, and should lock up when attempting to rotate in a clockwise direction.

i. Install the selective thrust washer on the pinion carrier rear pilot. **If the end play was not within specifications when checked prior to disassembly, replace the washer with one of proper thickness.** Refer to the Specifications at end of this Part, for selective thrust washer thicknesses.

j. Install the output shaft, carefully meshing the internal gear with the pinions.

4 With the center support properly assembled, position the needle bearing and retainer on the rear support.

5 Position a new rear support to case gasket on the rear support. Retain the gasket with transmission fluid.

6 Install the rear support. As the support is installed, insert the tubes into the case.

7 Position the governor drive ball in the pocket in the output shaft. Retain the ball with transmission fluid.

8 Install the governor assembly, aligning the groove with the ball in the output shaft.

9 Install the governor with the governor body plate toward the front of the vehicle. Install the governor snap ring.

10 Insert the extension housing oil seal replacer and pilot in the housing. Position a new gasket to the extension housing and install the extension housing on the transmission case. Coat the bolt threads with sealer and install the extension housing attaching bolts, vacuum tube clip, and external tooth lockwasher. **The lockwashers must be installed with the rolled edge toward the transmission case to insure a tight seal.**

11 Torque the extension housing attaching bolts to specification as listed at the end of this Part. Install the lubrication tube.

12 Position a new front pump gasket in the counterbore of the transmission case.

13 Install the front pump, aligning the pump bolt holes with the holes in the case. Install three of the pump attaching bolts and torque to specification.

14 Mount the dial indicator support in a front pump bolt hole. Mount a dial indicator on the support so that the contact rests on the end of the turbine shaft.

15 Use a large screwdriver to pry the front of the clutch drum toward the rear of the transmission. Set the dial indicator at zero.

16 Remove the screwdriver and pry the units toward the front of the transmission by inserting a screwdriver between the large internal gear and the transmission case. Note the indicator reading. End play should be 0.010–0.029 inch (minimum end play is preferred). If the end play is not within specifications, a new selective thrust washer must be used.

17 Remove the indicator and the tool from the extension housing.

18 Install the one remaining front pump attaching bolt. Torque bolt to specification.

19 Position the front band forward in the case with the band ends up.

20 Position the front servo strut with the slotted end aligned with the servo actuating lever, and with the small end aligned with the band end. Rotate the band, strut, and servo into position engaging the anchor end of the band with the anchor pin in the case.

21 Locate the servo on the case, and install the attaching bolts. **Tighten the attaching bolts only two or three threads.**

22 Install the servo release tube.

23 Position the rear servo anchor strut, and rotate the rear band to engage the strut.

24 Position the rear servo actuating lever strut with a finger, and then install the servo and attaching bolts. Move the rear servo (with reasonable force) toward the centerline of the transmission case, against the servo attaching bolts. While holding the servo in this position, torque the attaching bolts to specification.

25 Install the front servo apply tube.

26 Install the pressure regulator body and attaching bolts. Torque the bolts to specification.

27 Install the control and converter valve guides and springs. Install the spring retainer.

28 Install the control valve assembly, carefully aligning the servo tubes with the control valve. Align the inner downshift lever between the stop and the downshift valve. Shift the manual lever to the 1 position. **Align the manual valve with the actuating pin in the manual detent lever. Do not tighten the attaching bolts.**

29 Move the control valve body toward the center of the case, until the clearance is less than 0.050 inch between the manual valve and the actuating pin on the manual detent lever.

30 Torque the attaching bolts to specification as listed at the end of this Part. Be sure that the rear fluid screen retaining clip is installed under the valve body bolt.

31 Install the main pressure oil tube. **Be sure to install the end of the tube that connects to the pressure regulator assembly first. Then, install the other end of the tube into the main control assembly by tapping it gently with a soft hammer.**

32 Install the small control pressure compensator tube in the valve body and regulator.

33 Turn the manual valve one full turn in each manual lever detent position. If the manual valve binds against the actuating pin in any detent position, loosen the valve body attaching bolts and move the body away from the center of the case. Move the body only enough to relieve the binding. Torque the attaching bolts to specification. Check the manual valve for binding.

34 Torque the front servo attaching bolts to specification.

35 Adjust the front and rear bands.

36 Position the control rod in the bore of the vacuum diaphragm unit and install the diaphragm unit. **Make sure the control rod enters the throttle valve as the vacuum unit is installed.**

37 Torque the diaphragm unit to specification.

38 Position the fluid screen on the rear clip so that the tang enters the hole in the screen flange. Then, rotate the screen (clockwise) until the grommet is over the pump inlet port of the valve body regulator and press the screen down firmly. Install the screen retaining clip.

39 Place a new gasket on the transmission case and install the pan. Install the attaching bolts and lockwashers. Torque the bolts to specification.

If the converter and converter housing were removed from the transmission, install these components.

Control Valve Body

During the disassembly of the control valve assembly, avoid damage to valve parts and keep the valve parts clean. Place the valve assembly on a clean shop towel while performing the disassembly operation. **Do not separate the upper and lower valve bodies and cover until after the valves have been removed.**

Disassembly

1 Remove the manual valve.

2 Remove the throttle valve body and the separator plate. **Be careful not to lose the check valve when removing the separator plate from the valve body.** Remove the throttle valve and plug.

3 Remove one screw attaching the separator plate to the lower valve body. Remove the upper body front plate. **The plate is spring-loaded. Apply pressure to the plate while removing the attaching screws.**

4 Remove the compensator sleeve and plug, and remove the compensator valve springs. Remove the compensator valve.

5 Remove the throttle boost short valve and sleeve. Remove the throttle boost valve spring and valve.

174

FORD CW & FMX

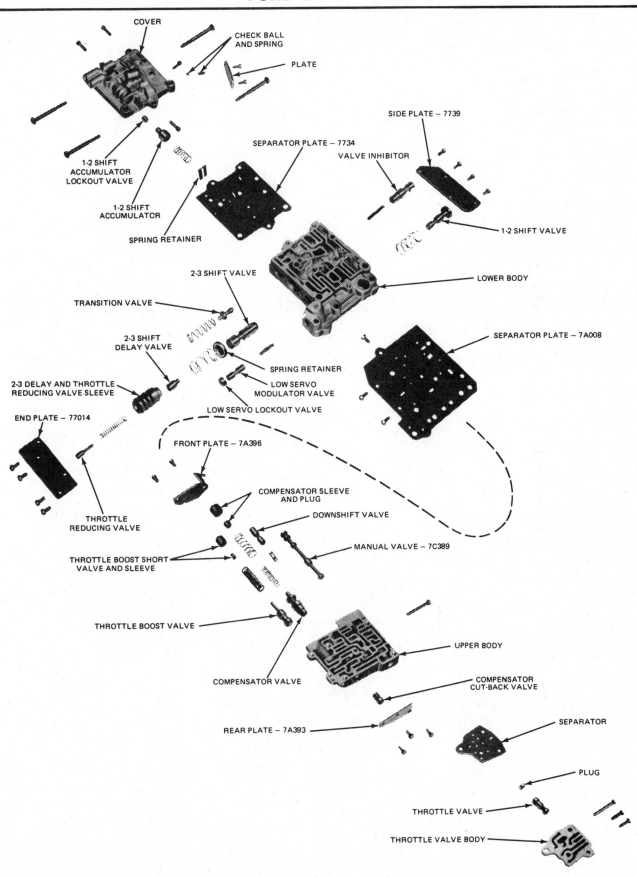

COVER

CHECK BALL AND SPRING

PLATE

SIDE PLATE – 7739

SEPARATOR PLATE – 7734

VALVE INHIBITOR

1-2 SHIFT ACCUMULATOR LOCKOUT VALVE

1-2 SHIFT ACCUMULATOR

SPRING RETAINER

1-2 SHIFT VALVE

LOWER BODY

2-3 SHIFT VALVE

TRANSITION VALVE

SEPARATOR PLATE – 7A008

2-3 SHIFT DELAY VALVE

SPRING RETAINER

LOW SERVO MODULATOR VALVE

2-3 DELAY AND THROTTLE REDUCING VALVE SLEEVE

LOW SERVO LOCKOUT VALVE

END PLATE – 77014

FRONT PLATE – 7A396

COMPENSATOR SLEEVE AND PLUG

DOWNSHIFT VALVE

THROTTLE REDUCING VALVE

MANUAL VALVE – 7C389

THROTTLE BOOST SHORT VALVE AND SLEEVE

THROTTLE BOOST VALVE

UPPER BODY

COMPENSATOR VALVE

COMPENSATOR CUT-BACK VALVE

REAR PLATE – 7A393

SEPARATOR

PLUG

THROTTLE VALVE

THROTTLE VALVE BODY

Control valve body parts assembly

Component Removal and Overhaul

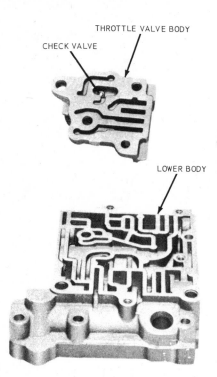

THROTTLE VALVE BODY

CHECK VALVE

LOWER BODY

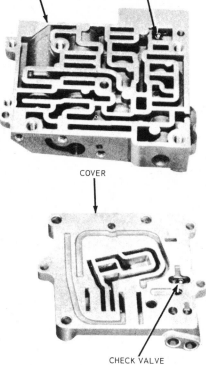

UPPER BODY

CHECK VALVE

COVER

CHECK VALVE

Control valve check valve locations

6 Remove the downshift valve and spring.
7 Remove the upper valve body rear plate.
8 Remove the compensator cut back valve.
9 Remove the lower body side plate. **The plate is spring-loaded. Apply pressure to the plate while removing the attaching screws.**
10 Remove the 1–2 shift valve and spring. Remove the inhibitor valve and spring.
11 Remove the two screws attaching the separator plate to the cover. Remove the lower body end plate. **The end plate is spring-loaded. Apply pressure to the plate while removing the attaching screws.**
12 Remove the low servo lockout valve, low servo modulator valve, and spring.
13 Remove the 2–3 delay and throttle reducing valve sleeve, the throttle reducing valve, spring, and the 2–3 shift delay valve. The reducing valve sleeve is lightly staked in the valve body bore. To remove the sleeve, use a blunt instrument against the end of the 2–3 shift valve and push the sleeve from its bore. Remove the 2–3 shift valve spring, spring retainer, and valve.
14 Remove the transition valve spring and valve.
15 Remove the plate from the valve body cover.
16 Remove the check ball spring and check ball. Remove the 3–2 kickdown control valve spring and valve.

17 Remove the 1–2 shift accumulator valve spring retainer from the cover. Remove the spring, 1–2 shift accumulator valve and 1–2 shift accumulator lockout valve.
18 Remove the through bolts and screws. Then, separate the bodies. Remove the separator plates from the valve bodies and cover. Be careful not to lose the check valves.

Assembly
1 Arrange all parts in their correct positions. Rotate the valves and plugs when inserting them in their bores, to avoid shearing of soft body castings.
2 Place the check valve in the upper body. Then, position the separator plate on the body.
3 Position the lower body on the upper body, and start **but do not tighten the attaching screws.**
4 Position the cover and separator plate on the lower body. Start the four through bolts.
5 Align the separator with the upper and lower valve body attaching bolt holes. Install and torque the four valve body bolts to specification. **Excessive tightening of these bolts may distort the valve bodies, causing valves or plugs to stick.**
6 Install the check ball and spring in the cover. Install the plate.
7 Insert the 1–2 shift accumulator lockout valve, 1–2 shift accumulator valve, and spring in the cover. Install the valve spring retainer.

8 Install the transition valve and spring in the lower body.
9 Install the 2–3 shift valve spring retainer and spring. Install the 2–3 shift delay valve, spring, and throttle reducing valve in the sleeve. Slide the assembly into position in the lower body. Do not restake the sleeve.
10 Install the low servo lockout valve spring. Install the low servo modulator and low servo lockout valves. Install the lower body end plate.
11 Install the inhibitor valve spring and valve in the lower body.
12 Install the 1–2 shift valve spring and valve. Install the lower body side plate.
13 Install the compensator cutback valve in the upper body. Install the upper body rear plate.
14 Install the downshift valve and spring in the body.
15 Install the throttle boost valve and spring. Install the throttle boost short valve and sleeve.
16 Install the compensator valve, inner and outer compensator springs, and the compensator sleeve and plug.
17 Position the front plate. Apply pressure to the plate while installing the two attaching screws.
18 Install the throttle valve, plug, and check valve in the throttle valve body. Position the separator on the upper body and install the throttle valve body. Install the three attaching screws.
19 Install four screws attaching the cover to the lower body, two screws attaching the separator plate to the upper body, and one screw attaching the separator plate to lower body. Torque the cover and body screws to specification.
20 Install the manual valve.

Front Servo

Disassembly
1 Remove the servo piston retainer snap ring. The servo piston is spring-loaded. **Apply pressure to the piston when removing the snap ring.**
2 Remove the servo piston retainer, servo piston, and the return piston from the servo body. It may be necessary to tap the piston stem lightly with a soft-faced hammer to separate the piston retainer from the servo body.
3 Remove all the seal rings, and remove the spring from the servo body.
4 Inspect the servo body for cracks, and the piston bore and the servo pistons stem for scores. Check fluid passages for obstructions.
5 Check the actuating lever for free movement, and inspect it for wear. To replace the actuating lever shaft, it will be necessary to press the shaft out of the bracket. The shaft is retained in the body by serrations on one end of the shaft. These serrations cause a press fit at that end. To remove the

shaft, press on the end opposite the serrations. Inspect the adjusting screw threads and the threads in the lever.

6 Check the servo spring and servo band strut for distortion.

7 Inspect the servo band lining for excessive wear and bonding to the metal. **The band should be replaced if worn to a point where the grooves are not clearly evident.**

8 Inspect the band ends for cracks and check the band for distortion.

Assembly

1 Lubricate all parts of the front servo with transmission fluid to facilitate assembly.

2 Install the inner and outer O-rings on the piston retainer. Install a new O-ring on the return piston and on the servo piston.

3 Position the servo piston release spring in the servo body. Install the servo piston, retainer, and return piston in the servo body as an assembly. Compress the assembly into the body, and secure it with the snap ring. **Make sure the snap ring is fully seated in the groove.**

4 Install the adjusting screw and locknut in the actuating lever if they were previously removed.

Rear Servo

Disassembly

1 Remove the servo actuating lever shaft retaining pin with a ⅛ inch punch. Remove the shaft and actuating lever.

2 Press down on the servo spring retainer, and remove the snap ring. **Release the pressure on the retainer slowly to prevent the spring from flying out.**

3 Remove the retainer and servo spring.

4 Force the piston out of the servo body with air pressure. Hold one hand over the piston to prevent damage.

5 Remove the piston seal ring.

Assembly

1 Install a new seal ring on the servo piston.

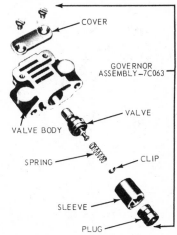

Governor

FRONT SERVO COMPONENTS IDENTIFICATION – FMX AUTOMATIC TRANSMISSION

Model Usage	Retainer	Piston	Dim. A	Dim. B	Dim. C	Dim. D
PHB- (Except PHB-AR) PHC-A2, B2	C9AP-7A269-A	C9ZP-7A260-A	1.314 1.312	2.372 2.369	1.300 1.295	2.361 2.356
PHB-AR	1P-77364-A	1P-77359-B	1.439 1.437	2.372 2.369	1.425 1.420	2.361 2.356

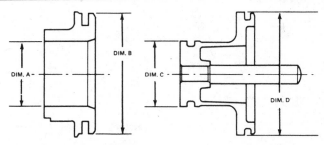

Front servo component identification

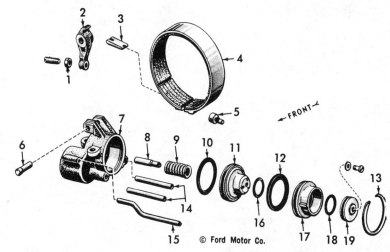

1. NUT
2. LEVER
3. STRUT
4. FRONT BAND
5. PIN
6. SHAFT
7. SERVO BODY
8. PIN
9. RETURN SPRING
10. OIL SEAL
11. SERVO PISTON
12. OIL SEAL
13. SNAP RING
14. OIL TUBE
15. ACCUMULATOR OIL TUBE
16. OIL SEAL
17. PISTON GUIDE
18. OIL SEAL
19. RELEASE PISTON

Front servo parts breakdown

2 Install the piston in the servo body. **Lubricate the parts to facilitate assembly.** Install the servo spring with the small coiled end against the servo piston.

3 Install the spring retainer. Compress the spring with a C-clamp. Then install the snap ring. **The snap ring must be fully seated in the groove.**

4 Install the actuating lever, with the socket in the lever bearing on the piston steam. Install the actuating lever shaft, aligning the retaining pin holes, and install the pin.

5 Check the actuating lever for free movement.

Governor

Disassembly

1 Remove the governor valve body cover.

2 Remove the valve body from the counterweight.

3 Remove the plug, sleeve, and the valve and spring from the body.

4 Remove the screen from its bore in the valve body.

Assembly

1 Install the governor valve and spring assembly in the bore of the valve body. Install the sleeve and plug.

FORD CW & FMX

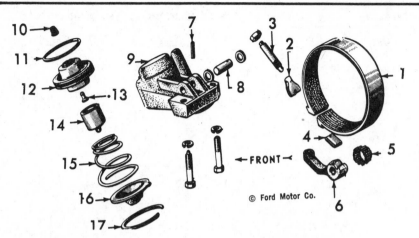

1. REAR SERVO BAND
2. ANCHOR STRUT
3. BAND ADJUSTING SCREW
4. STRUT
5. NEEDLE BEARING
6. LEVER
7. PIN
8. SHAFT
9. SERVO BODY
10. SPRING
11. OIL SEAL
12. SERVO PISTON
13. CHECK VALVE
14. ACCUMULATOR
15. PISTON SPRING
16. SPRING RETAINER
17. SNAP RING

Rear servo parts breakdown

2 Install the screen.
3 Install the body on the counterweight. **Make sure the fluid passages in the body and the counterweight are aligned.**
4 Position the valve body cover on the body, and install the screws.

Pressure Regulator

Disassembly
1 Remove the valves from the regulator body.
2 Remove the regulator body cover attaching screws, and remove the cover.
3 Remove the separator plate.
4 Wash all parts thoroughly in clean solvent and blow dry with moisture-free compressed air.
5 Inspect the regulator body and cover mating surfaces for burrs.
6 Check all fluid passages for obstructions.
7 Inspect the control pressure and converter pressure valves and bores for burrs and scores. Remove all burrs carefully with crocus cloth.
8 Check the free movement of the valve in their bores. Each valve should fall freely into its bore when both the valve and bore are dry.
9 Inspect the valve springs for distortion.

Assembly
1 Position the separator plate on the regulator cover.
2 Position the regulator cover and separator plate on the regulator body, and install the attaching screws. Torque the screws to specification.
3 Insert the valves in the pressure regulator body.

Downshift and Manual Linkage

Disassembly
1 Remove the inner downshift lever shaft nut. Then remove the inner downshift lever.
2 Remove the outer downshift lever and shaft. Remove the downshift shaft seal from the counterbore in the manual lever shaft.
3 Remove the cotter pin from the parking pawl toggle operating rod and remove the clip from the parking pawl operating lever. Remove the parking pawl operating rod.
4 Rotate the manual shift until the detent lever clears the detent plunger. Then remove the detent plunger and spring. **Do not allow the detent plunger to fly out of the case.**
5 Remove the manual lever shaft nut, and remove the detent lever. Remove the outer manual lever and shaft from the transmission case.
6 Tap the toggle lever sharply toward the rear of the case to remove the plug and pin.
7 Remove the pawl pin by working the pawl back and forth. Remove the pawl and toggle lever assembly, and then disassemble.
8 Remove the manual shaft seal and case vent tube. Remove the oil cooler return check valve from the back of the case.

Assembly
1 Coat the outer diameter of a new manual shaft seal with sealer, then install the seal in the case with a driver.
2 Install the vent tube in the transmission case.
3 Assemble the link to the pawl with the pawl link pin, washer, and pawl return spring. Assemble the toggle lever

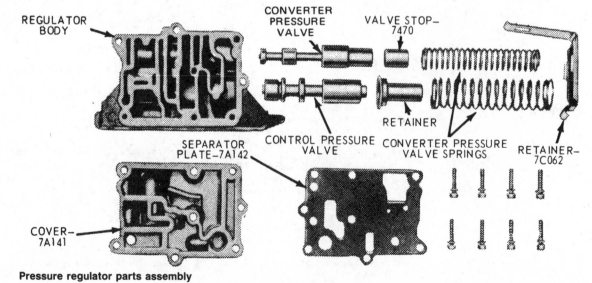

Pressure regulator parts assembly

FORD CW & FMX

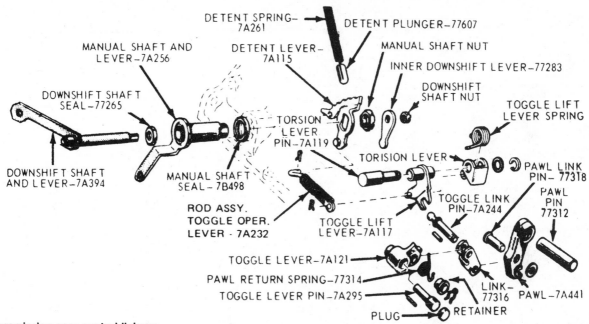

Transmission case control linkage

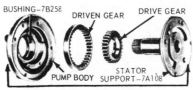

PUMP ASSEMBLY – 7A103

Front pump

to the link with the toggle link pin. Position the pawl return spring over the toggle link pin, and secure in place with the washer and the small retainer clip. Install the assembly in the transmission case by installing the pawl pin and the toggle lever pin. Press in the retaining plug tightly against the toggle lever pin. Install the torsion lever assembly. Position the spring on the torsion lever with a screwdriver. Make certain that the short side of toggle lift lever does not extend beyond the largest diameter of the ball on the toggle link pin. Tap the toggle lift lever in or out as necessary to center the toggle lift lever on the ball.

4 Install the manual lever and shaft in the transmission case. Position the detent lever on the shaft, and secure with a nut. Tighten the nut to 20–30 ft./lbs. torque. Rotate the manual lever to the rear of the case. Position the detent spring in the case. Hold the detent plug on the spring with a 3/16 inch socket wrench, then depress the spring until the plug is flush with the case. Carefully rotate the manual lever to the front of the case to secure the plug. **A piece of thin walled tubing may be used to depress the plug if a small socket wrench is not available.**

5 Position the ends of the parking pawl operating rod in the detent lever and toggle lift lever, and secure with the two small retaining pins.

6 Install a new seal on the downshift lever shaft, then install the lever and shaft in the case. Position the inner downshift lever on the inner end of the shaft with the mark O facing toward the center of the case. Install the lockwasher and nut, then tighten the nut to 17–20 ft./lbs. torque.

7 Check the operation of the linkage. The linkage should operate freely without binding.

Front Pump
Disassembly
1 Remove the stator support attaching screws and remove the stator support. **Mark the top surface of the pump driven gear with Prussian blue to assure correct assembly. Do not scratch the pump gears.**

2 Remove the drive and driven gears from the pump body.

3 If the pump housing bushing is to be replaced, press the bushing from the front housing.

4 Press a new bushing into the pump housing.

5 If any parts other than the stator support, bushings or oil seal are found defective, replace the pump as a unit. Minor burrs and scores may be removed with crocus cloth. The stator support is serviced separately.

6 If the oil seal requires replacement, bolt the front pump to the transmission case with capscrews, and install an oil seal remover. Then pull the front seal from the pump body.

7 Clean the pump body counterbore. Then inspect the bore for rough spots. Smooth up the counterbore with crocus cloth.

8 Remove the pump body from the transmission case.

Assembly
1 If the oil seal was removed, coat the outer diameter of a new seal with Motorcraft Sealing Compound, or its equivalent. Then position the seal in the pump body. Drive the seal into the pump body until the seal is firmly seated in the body.

2 **Place the pump driven gear in the pump body with the mark on the gear or tooth gear chamfer facing down. Install the drive gear in the pump body with the chamfered side of the flats facing down.**

3 Install the stator support and attaching screws. Check the pump gears for free rotation.

Rear Support Bushing Removal and Installation
1 Remove the three pressure tubes from the support housing.

2 Remove the rear support bushing if it is worn or damaged. Use a cape chisel and cut along the bushing seam until the chisel breaks through the bushing wall. Pry the loose ends of the bushing

Rear Clutch
Disassembly
1 Remove the clutch pressure plate snap ring, and remove the pressure plate from the drum. Remove the waved cushion spring. Remove the composition and steel plates.

2 Compress the spring, and remove the snap ring.

3 Guide the spring retainer while releasing the pressure, to prevent the re-

FORD CW & FMX

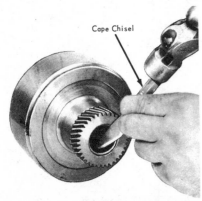

Removing rear clutch sun gear bushing

tainer from locking in the snap ring grooves.
4 Position the primary sun gear shaft in the rear clutch. Place an air hose nozzle in one of the holes in the shaft, and place one finger over the other hole. Then force the clutch piston out of the clutch drum with air pressure. **Hold one hand over the piston to prevent damage to the piston.**
5 Remove the inner and outer seal rings from the clutch piston.

6 Remove the rear clutch sun gear bushing if it is worn or damaged.
    Use the cape chisel and cut along the bushing seam until the chisel breaks through the bushing wall. Pry the loose ends of the bushing up with an awl and remove the bushing.

Assembly
1 If the rear clutch sun gear bushing was removed, press a new bushing into the rear clutch sun gear.
2 Install new inner and outer seal rings on the piston.
3 To install the piston in the clutch drum, lubricate the piston seals and tools with clean transmission fluid. Push the small fixture down over the cylinder hub. Insert the piston into the large fixture, with the seal toward the thin-walled end. Hold the piston and large fixture, and insert as a unit into the cylinder. Push down over the small fixture until the large tool stops against the shoulder in the cylinder; then push the piston down, out of the tool, until it bottoms in the cylinder. Remove the tools.
4 Install the clutch release spring, and position the retainer on the spring.

5 Install the tool on the spring retainer. Compress the clutch spring, and install the snap ring. **While compressing the spring, guide the retainer to avoid interference of the retainer with the snap ring groove. Make sure the snap ring is fully seated in the groove.**
    When new composition clutch plates are used, soak the plates in automatic transmission fluid for 15 minutes before they are assembled.
6 Install the external tabbed waved cushion spring. Then, install steel and friction plates alternately starting with a steel plate.
7 Install the clutch pressure plate with the bearing surface down. Then install the clutch pressure plate snap ring. Make sure the snap ring is fully seated in the groove.
8 Check the free pack clearance between the pressure plate and the first internal plate with a feeler gauge. The clearance should be 0.030 to 0.055 inch. If the clearance is not within specifications, selective snap rings are available in the following thicknesses: 0.060 to 0.064, 0.074 to 0.078, 0.088 to 0.092 and 0.102 to 0.106 inch. Insert the correct size snap ring and recheck the clearance.
9 Install the thrust washer on the primary sun gear shaft. **Be sure the thrust washer is installed with the tabs of the washer away from the sun gear thrust face.** Lubricate all parts with automatic transmission fluid or petroleum jelly. Install the two center seal rings.
10 Install the rear clutch on the primary sun gear shaft. **Be sure all of the needles are in the hub if the unit is equipped with loose needle bearings.** Assemble two seal rings in the front grooves.
11 Install the steel and the bronze thrust washers on the front of the secondary sun gear assembly. **If the steel washer is chamfered, place the chamfered side down.**

Front Clutch
Disassembly
1 Remove the clutch cover snap ring with a screwdriver, and remove the input shaft from the clutch drum.
2 Remove the thrust washer from the thrust surface of the clutch hub. Insert one finger in the clutch hub, and lift the hub straight up to remove the hub from the clutch drum.
3 Remove the composition and the steel clutch plates, and then remove the pressure plate from the clutch drum.
4 Place the front clutch spring compressor on the release spring, position the clutch drum on the bed of an arbor press, and compress the release spring with the arbor press until the release spring snap ring can be removed.
5 Remove the clutch release spring from the clutch drum.
6 Install a special nozzle on an air hose.

FORD FMX TORQUE SPECIFICATIONS

	Ft./Lbs.
Converter to flywheel	23-28
Converter hsg. to trans. case	40-50
Front pump to trans. case	17-22
Front servo to trans. case	30-35
Rear servo to trans. case	40-45
Upper valve body to lower valve body	4-6
Oil pan to case	10-13
Converter cover to converter hsg.	12-16
Regulator to case	17-22
Planetary support to trans. case	20-25
Control valve body to trans. case	8-10
Diaphragm assy. to case	20-30
Cooler return check valve	9-12
Extension assy. to trans. case	30-40
Pressure gauge tap	7-15
Converter drain plug	15-28
Rear band adjusting screw to case	35-40
Front band adjusting screw locknut	20-25
Manual valve inner lever to shaft	20-30
Downshift lever to shaft	17-20
Filler tube to engine	20-25
Transmission to engine	40-50
Neutral start switch actuator lever bolt	6-10
Steering col. lock rod adj. nut	10-20
T.R.S. switch to case	4-8
	In./Lbs.
Pressure regulator cover screws	20-30
Control valve body screws (10-24)	20-30
Front servo release piston	20-30
End plates to body	20-30
Stator support to pump	25-35
Lower body and cover plate to valve body	20-30
T.V. body to valve body	20-30
Lower valve body cover and plate to valve body	48-72

FORD CW & FMX

Place the nozzle against the clutch apply hole in the front clutch housing, and force the piston out of the housing.

7 Remove the piston inner seal from the clutch housing. Remove the piston outer seal from the groove in the piston.

8 Remove the input shaft bushing if it is worn or damaged. Use the cape chisel and cut along the bushing seam until the chisel breaks through the bushing wall. Pry the loose ends of the bushing up with an awl and remove the bushing.

Assembly

1 If the input shaft bushing was removed, slip a new bushing over the end of the installation tool and position the tool and bushing to the bushing hole. Then, press the bushing into the input shaft.

2 Lubricate all parts with transmission fluid. Install a new piston inner seal ring in the clutch cylinder. Install a new piston outer seal in the groove in the piston.

3 Install the piston in the clutch housing. **Make sure the steel bearing ring is in place on the piston.**

4 Position the release spring in the clutch cylinder with the concave side up. Place the release spring compressor on the spring, and compress the spring with an arbor press. Then install the snap ring. **Make sure the snap ring is fully seated in the groove.**

5 Install the front clutch housing on the primary sun gear shaft by rotating the clutch units to mesh the rear clutch plates with the serrations on the clutch hub. **Do not break the seal rings.**

6 Install the clutch hub in the clutch cylinder with the deep counterbore down. Install the thrust washer on the clutch hub.

7 Install the pressure plate in the clutch cylinder with the bearing surface up. Install the composition and the steel

FORD CW TORQUE SPECIFICATIONS

	Ft./Lbs.
Line pressure plug	10-20
Front pump to case	16-24
Manual control lever to shaft	35-50
Throttle valve lever to shaft	15-25
Center support to case	20-30
Extension to case	25-40
Rear servo to case	37-52
Rear servo adjusting screw locknut	35-50
Front servo to case	25-40
Oil pan to case	10-20
Vacuum diaphragm to case	10-20

	In./Lbs.
Stator support to pump body	20-35
Auto. Adj. cam plate to servo body	15-35
Upper valve body to lower valve body	50-100
Valve body assembly to case	50-100
Oil screen to valve body	20-35

clutch plates alternately, starting with a composition plate. When new composition clutch plates are used, soak the plates in automatic transmission fluid for 15 minutes before they are assembled.

8 The final friction plate to be installed is selective. Install the thickest plate that will be a minimum of 0.010 inch below input shaft shoulder in cylinder. For all other plates, use the thinnest available.

9 Install the turbine shaft in the clutch cylinder, and then install the snap ring. **Make sure the snap ring is fully seated in the groove.**

10 Install the thrust washer on the turbine shaft.

Primary Sun Gear Shaft

Disassembly

1 Position the primary sun gear shaft in the clutch bench fixture.

2 Check the fit of the seal rings in their respective bores. If equipped with cast iron seal rings, a clearance of 0.002 to 0.009 inch should exist between the ends of the rings.

If equipped with teflon seals that are worn or damaged, cut the seals from the shaft with a sharp knife. Be careful not to score the ring grooves.

Assembly

Replace the teflon seals with cast iron seal rings, and check for free movement in the groove.

CHECKS AND ADJUSTMENTS FMX TRANSMISSION

Operation	Specification
Transmission End Play Check	0.010-0.029 (Selective Thrust Washers Available)
Turbine and Stator End Play Check	New or rebuilt 0.023 max., Used 0.050 max.
Front Band Adjustment (Use 1/4 inch spacer between adjustment screw and servo piston stem)	Adjust screw to 10 in-lbs torque. Remove spacer, hold screw and then tighten screw an additional 3/4 turn. Torque locknut to 20-25 ft-lb.
Rear Band Adjustment	Loosen locknut. Adjust screw to 10 ft-lb torque, then back off exactly 1-1/2 turns. Hold screw and tighten locknut to 35-40 ft-lb.
Primary Sun Gear Shaft Ring End Gap Check	0.002-0.009
Rear Clutch Selective Snap Ring Thickness	0.060-0.064, 0.074-0.078, 0.088-0.092, 0.102-0.106

FORD JATCO

Ford Jatco Automatic Transmission

Identification

The service identification tag is attached to the right side of the transmission case to the rear of the filler tube. JATCO transmission is identified by the letter "S" on the Vehicle Certification Label.

Application

Granada and Monarch

Adjustments

INTERMEDIATE BAND

1 Raise the vehicle, and remove the servo cover.
2 Loosen the intermediate band adjusting screw lock nut and tighten the adjusting screw to 10 ft./lbs.
3 Back off the adjusting screw two turns. While holding the adjusting screw stationary, tighten the adjusting screw lock nut to 22–29 ft./lbs. (30–39 N.m)
4 Install the servo cover using a new gasket, lower the vehicle and check transmission fluid level.

Component Removal and Installation

NOTE: These operations can be accomplished without removing the transmission from the vehicle.

CONTROL VALVE BODY

REMOVAL

1 Raise the vehicle and drain the transmission fluid by removing the transmission oil pan.
2 Remove the downshift solenoid, vacuum diaphragm, vacuum diaphragm rod and O-rings.
3 Remove the valve body-to-case attaching bolts. Hold the manual valve to keep it from sliding out of the valve body and remove the valve body from the case. Failure to hold the manual valve while removing the control assembly, could cause the manual valve to become bent or damaged.

INSTALLATION

1 Position the valve body to the case and install the attaching bolts.
2 Using a *new* gasket, secure the transmission oil pan to the case and tighten the bolts to correct specifications.
3 Lower the vehicle, and fill the transmission to the correct level with the specified automatic transmission fluid.

SERVO

REMOVAL

1 Raise the vehicle and drain the transmission fluid by removing the transmission oil pan.

2 Remove the control valve body from the transmission assembly.
3 Remove the servo cover.
4 Remove the three bolts securing the servo retainer to the transmission case and remove the retainer and servo piston as an assembly. Remove the return spring from the case.
5 Remove the intermediate band apply strut through the bottom of the case.

INSTALLATION

1 Install the piston return spring in the case.
2 Install the intermediate band apply strut in band end.
3 Install the piston and servo retainer into the transmission case and align piston rod with the band strut. While holding the retainer in place, install the three attaching bolts and tighten the bolts to specifications.
4 Correctly adjust the intermediate band, and install the control valve body.
5 Using a *new* gasket, install the transmission oil pan, and tighten the mounting bolts to the correct specifications.
6 Using a *new* gasket, install the servo cover and tighten attaching bolts to the correct specifications.
7 Lower the vehicle and refill the transmission to the correct level.

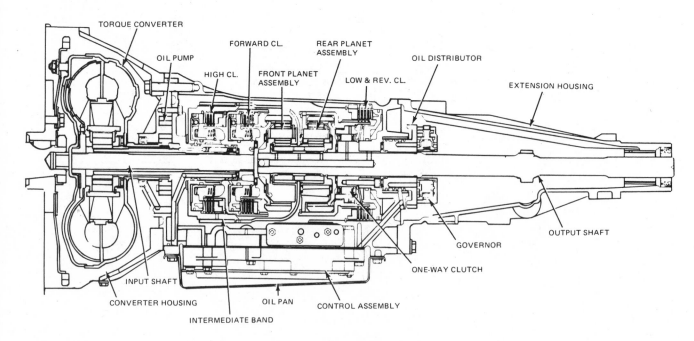

Sectional view

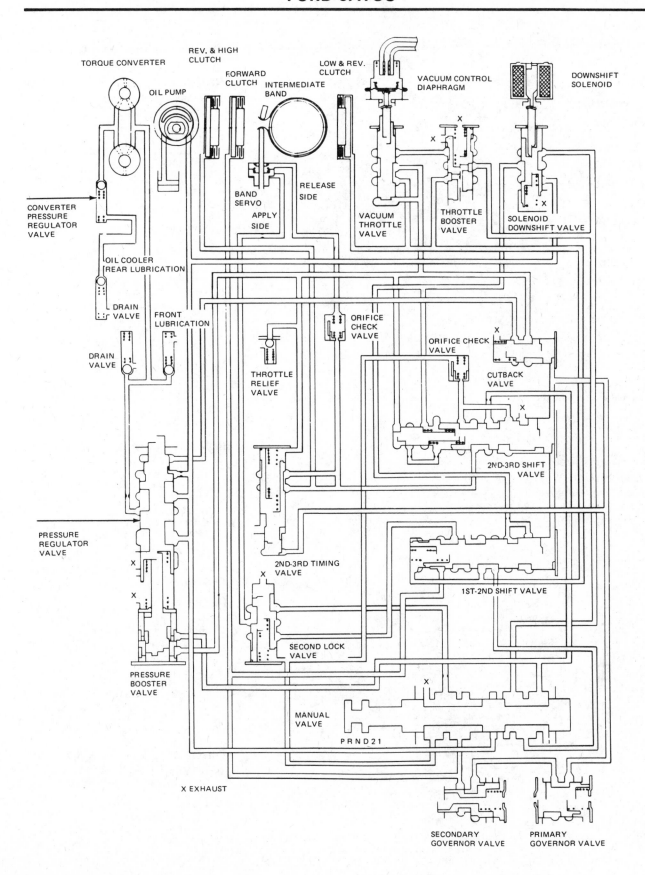

Hydraulic schematic

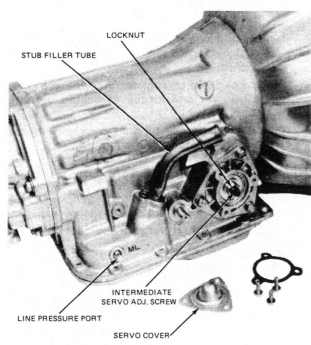

Intermediate band adjustment

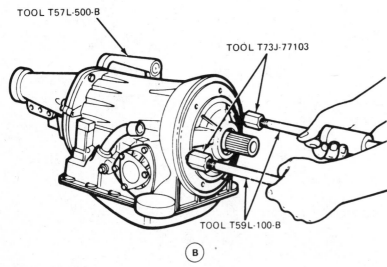

Oil pump removal

EXTENSION HOUSING REAR SEAL

REMOVAL

1 Remove the drive shaft and remove the old seal using a sharp chisel.

INSTALLATION

1 Install a new seal using Tool T61L-7657-A.
2 Install the drive shaft and check transmission fluid level.

EXTENSION HOUSING

REMOVAL

1 Remove the drive shaft.
2 Disconnect the speedometer cable from the extension housing.
3 Remove the transmission rear support-to-crossmember bolts and nuts.
4 Raise the transmission sufficiently to remove the rear support.
5 Loosen the extension housing attaching bolts to drain the transmission fluid.
6 Remove the eight extension housing-to-case attaching bolts and remove the extension housing.

NOTE: Do not lose washer from parking panel shaft.

INSTALLATION

1 Using a *new* gasket, install the extension housing and tighten the attaching bolts to correct specifications.
2 Install the rear support and lower the transmission.
3 Connect the speedometer cable.
4 Install the drive shaft, and fill the transmission to the correct level with the specified transmission fluid.

GOVERNOR

REMOVAL

1 Remove the extension housing.
2 Remove the four governor housing-to-oil distributor attaching bolts, and remove the governor housing.

INSTALLATION

1 Install the governor housing on the oil distributor and tighten the bolts to correct specifications.
2 Install the extension housing.

Transmission Disassembly

1 Remove the torque converter from the transmission front pump and converter housing.
2 Remove the converter housing after removing its attaching bolts.
3 Remove the transmission oil pan.
4 Unscrew the downshift solenoid and vacuum diaphragm unit and remove the vacuum diaphragm rod from the case.
5 Remove the control valve body.

FORD JATCO

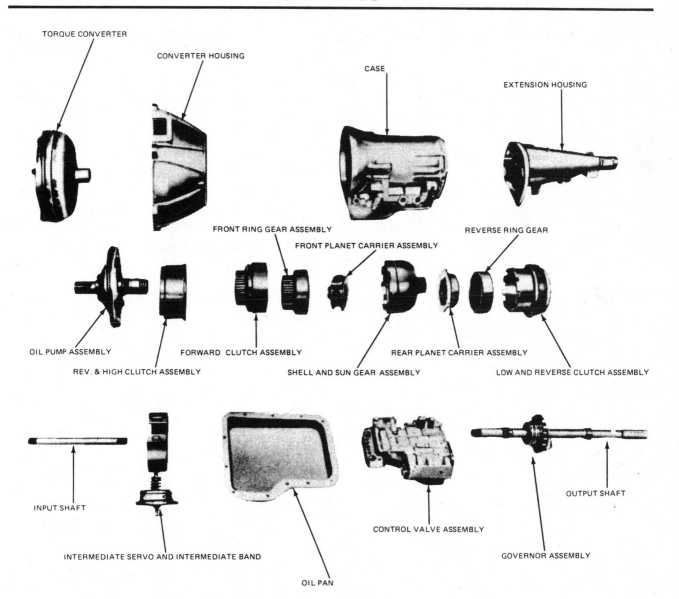

TORQUE CONVERTER

CONVERTER HOUSING

CASE

EXTENSION HOUSING

FRONT RING GEAR ASSEMBLY

FRONT PLANET CARRIER ASSEMBLY

REVERSE RING GEAR

OIL PUMP ASSEMBLY

FORWARD CLUTCH ASSEMBLY

REAR PLANET CARRIER ASSEMBLY

REV. & HIGH CLUTCH ASSEMBLY

SHELL AND SUN GEAR ASSEMBLY

LOW AND REVERSE CLUTCH ASSEMBLY

INPUT SHAFT

INTERMEDIATE SERVO AND INTERMEDIATE BAND

OIL PAN

CONTROL VALVE ASSEMBLY

OUTPUT SHAFT

GOVERNOR ASSEMBLY

Transmission major components

6 Loosen the intermediate band adjusting screw lock nut and tighten the adjusting screw to prevent the reverse and high clutch drum from falling when the oil pump is removed.

7 Remove the input shaft from the front pump.

8 Remove the oil pump using slide hammers (Tools T73J-77103 and T59L-100-B).

9 Loosen the intermediate band adjusting screw and remove the band strut.

10 Remove the band and forward part of the gear train as an assembly.

11 Remove the large snap ring that secures the rear planet carrier to the connecting drum, and remove the planet carrier from the drum.

12 Remove the output shaft snap ring, and remove the internal drive flange.

13 Turn the connecting drum counterclockwise as far as it will go. Next turn the connecting drum clockwise and remove the drum and one-way clutch from the case as an assembly.

14 Remove the extension housing.

NOTE: Do not lose the washer from the parking panel shaft.

15 Slide the output shaft assembly from the transmission case.

16 Remove the governor and oil distributor from the case.

17 Remove the oil distributor needle bearing and race from inside the transmission case.

18 Remove the retaining ring from the park rod actuating lever, and remove the lever and park pawl actuating rod.

19 Remove the manual shaft attaching nut and remove the manual plate and spacer. Remove the manual shaft from the case.

20 Remove the snap ring securing the low and reverse clutch assembly to the case. Remove the retaining plate, steel and friction plates and dished plate from the case.

21 Remove the eight one-way clutch inner race attaching bolts and washers from the rear of the case. Remove the inner race, snap ring, piston return spring and support ring from inside the case.

22 Remove the low and reverse clutch piston from the case.

FORD JATCO

Component Disassembly and Reassembly

CONTROL VALVE BODY

DISASSEMBLY

1 Remove the oil screen-to-control valve mounting bolts, and remove the screen.
2 Remove the attaching bolts, and separate the upper and lower valve bodies. Be careful not to lose the lower valve body check valves and springs.
3 Slide the manual valve out of the valve body.
4 Remove the 1–2 shift valve and 2–3 shift valve cover plate from the body.
5 Remove the 1–2 shift valve and spring from the body.
6 Remove the 2–3 shift valve, spring

7 Remove the throttle modulator valve and spring from the body.
8 Remove the downshift valve cover plate from the body.
9 Remove the downshift valve and spring from the body.
10 Remove the throttle back-up valve spring and valve from the body.
11 Remove the throttle valve from the body.
12 Remove the 2–3 timing valve spring and valve from the body.
13 Remove the pressure regulator valve cover plate from the body.
14 Remove the pressure regulator valve sleeve, plug, spring seat, spring and pressure regulator valve.
15 Remove the second lock valve and spring from the body.
and plug from the body.

ASSEMBLY

1 Place the second lock valve and spring in the body.
2 Place the pressure regulator valve, spring, spring seat, plug and sleeve in the body.
3 Secure the pressure regulator valve cover plate to the body with the three attaching bolts.
4 Place the 2–3 timing valve and spring in the body.
5 Place the throttle valve in the body.
6 Place the throttle back-up valve and spring in the body.
7 Place the downshift valve spring and valve in the body.
8 Secure the downshift valve cover plate to the body with the three attaching bolts. Tighten the bolts to 2–3 ft./lbs. (2.71–4.06 N·m).

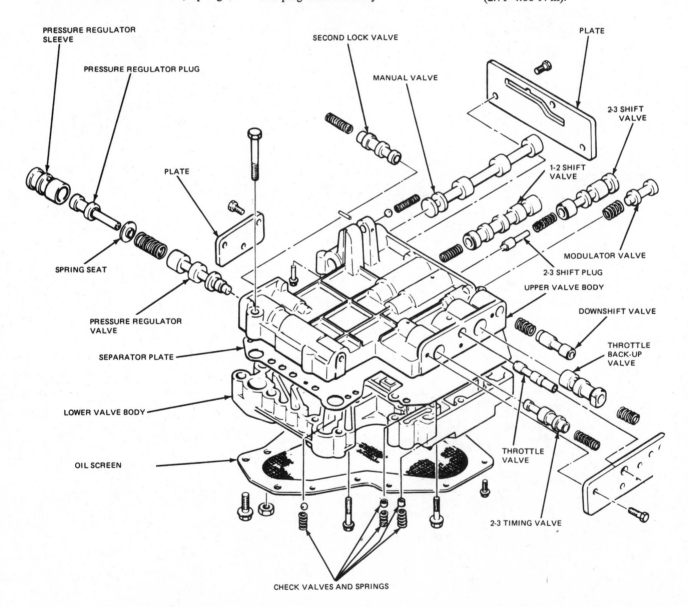

Control valve assembly components

FORD JATCO

9 Place the throttle modulator valve spring and valve in the valve body.
10 Place the 2–3 shift plug, spring and 2–3 shift valve in the body.
11 Place the 1–2 shift valve spring and valve in the valve body.
12 Secure the 1–2 shift valve and 2–3 shift valve cover plate with the three attaching bolts, tighten the bolts to 2–3 ft./lbs. (2.71–4.06 N·m).
13 Place the check valve and springs in the upper body.
14 Position the separator plate on the lower body and place the lower body and plate assembly on the upper valve body. Tighten the attaching bolts to 2–3 ft./lbs. (2.71–4.06 N·m).
15 Position the oil screen to the valve body and install the attaching bolts. Tighten to 2–3 ft./lbs. (2.71–4.06 N·m).
16 Slide the manual valve into the valve body.

SERVO

DISASSEMBLY

1 Apply air pressure to the port in the servo retainer to remove the piston.
2 Remove the seals from the retainer.
3 Remove the seal from the piston.

ASSEMBLY

1 Dip the new seals in transmission fluid.
2 Install new seals on the servo retainer and piston.
3 Dip the piston in transmission fluid and install it in the retainer.

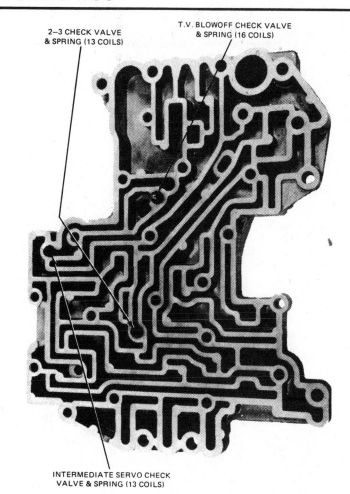

Check valve and spring locations in control valve body

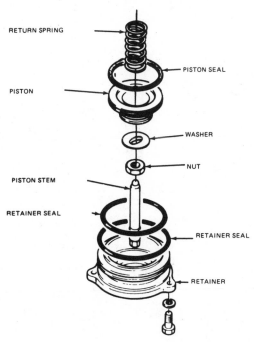

Servo components

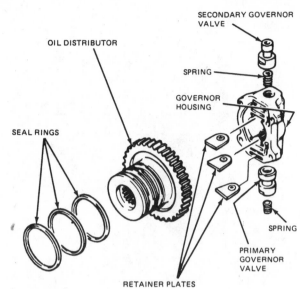

Governor components

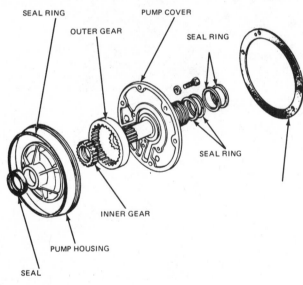

Oil pump components

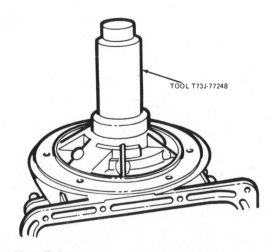

Front pump seal installation

GOVERNOR
DISASSEMBLY

1 Remove the governor housing-to-distributor attaching bolts. Remove the governor from the oil distributor.
2 Remove the secondary governor valve retainer plate and remove the secondary valve and spring from the housing.
3 Remove the primary governor valve retainer plate and remove the primary spring and valve from the housing.

ASSEMBLY

1 Install the primary governor valve and spring in the housing and install the retainer plate.
2 Install the secondary governor valve spring and valve in the housing and install the retainer plate.
3 Install the governor assembly on the oil distributor and tighten the mounting bolts to correct specifications.

OIL PUMP
DISASSEMBLY

1 Remove the four seal rings from the stator support.
2 Remove the large seal ring from the pump housing outside diameter.
3 Remove the five bolts that attach the pump cover to the pump housing. Remove the pump cover from the housing.
4 Mark the front face of the pump inner and outer gears to assure correct assembly. Remove the inner and outer gears from the pump housing.

NOTE: Do not scratch the gears.

ASSEMBLY

1 Install the inner and outer gears in the pump housing. Align the marks made during disassembly.
2 Lubricate the gears with clean automatic transmission fluid. Install the pump housing on the oil pump aligning tool (T73J-77108). Position the pump cover and gasket to the pump housing and install the five attaching bolts. Tighten the bolts to correct specifications, and remove the oil pump assembly from the aligning tool.
3 Carefully install four new seal rings on the stator support. Install a new seal ring on the pump housing outside diameter.
4 If the front pump seal must be replaced, remove it from the housing with a suitable tool, and install a new seal using Tool T73J-77248.

FORD JATCO

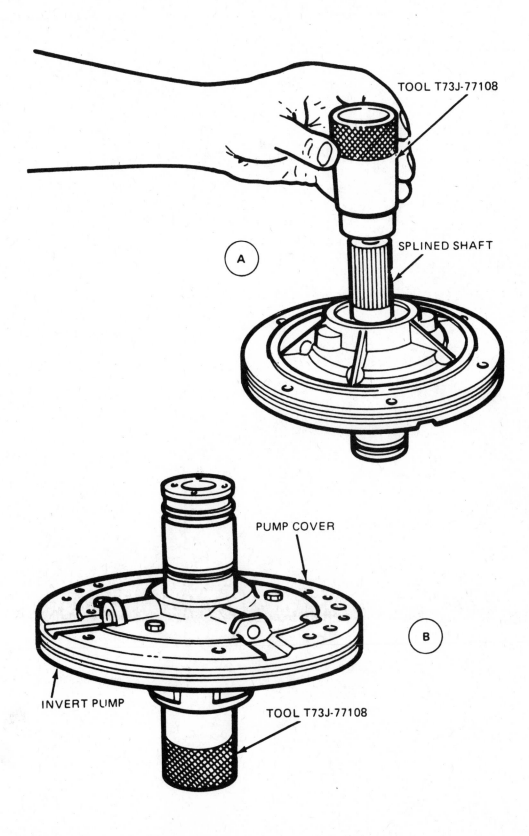

TOOL T73J-77108

SPLINED SHAFT

A

PUMP COVER

B

INVERT PUMP

TOOL T73J-77108

Oil pump alignment

FORD JATCO

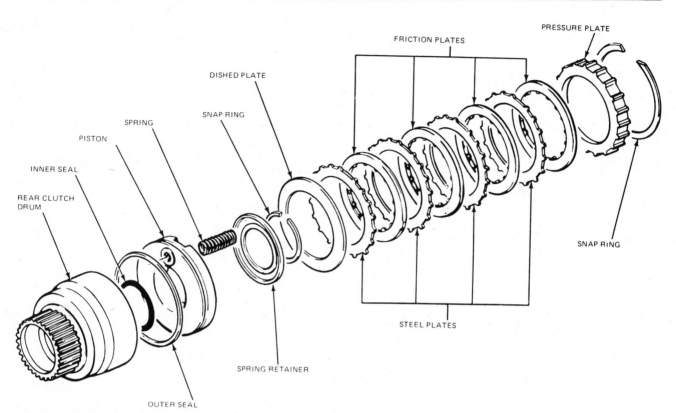

Forward clutch components

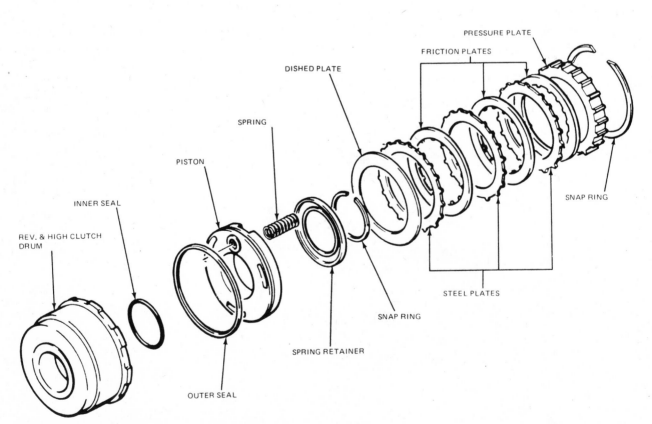

Components of the combined reverse and high gear clutch

FORD JATCO

REVERSE AND HIGH CLUTCH

DISASSEMBLY

1 Remove the clutch pressure plate snap ring.
2 Remove the pressure plate, friction plate and steel plates and dished plate from the clutch hub.
3 Install a spring compressor Tool T65L-77515-A, and remove the coil spring retainer snap ring. Remove the tool, the spring retainer and the ten coil springs.
4 Apply air pressure to the clutch piston pressure hole in the clutch hub and remove the piston.
5 Remove the piston outer seal from the piston and the inner seal from the clutch drum.

ASSEMBLY

1 Dip new seals in transmission fluid, and install one on the drum and one on the piston.
2 Lubricate the piston and install the clutch piston into the clutch hub. Position the ten coil springs in the piston and spring retainer on the springs. Install the clutch spring compressor tool and compress the springs to install the retaining snap ring.
3 Dip the clutch plates in clean transmission fluid. Install the dished plate. Install the clutch plates alternately, starting with a steel plate.
4 After all clutch plates have been installed, position the pressure plate in the clutch drum. Install the pressure plate snap ring.
5 With a thickness gauge, check the clearance between the pressure plate and snap ring. The clearance should be 0.062–0.071 inch. If the clearance is not within specifications, selective pressure plates are available in the following thicknesses: 0.283, 0.291, 0.-299, 0.307, 0.315 and 0.323 inch. Install the correct size pressure plate and recheck the clearance.

FORWARD CLUTCH

DISASSEMBLY

1 Remove the clutch pressure plate snap ring.
2 Remove the pressure plate, friction plates, steel plates and dished plate from the clutch hub.
3 Install spring compressor tool (T65L-77515A) and remove the coil spring retainer snap ring. Remove the tool, the coil spring retainer and ten coil springs.
4 Apply air pressure to the clutch piston pressure hole in the clutch hub and remove the piston.
5 Remove the piston outer seal from the piston and the inner seal from the clutch drum.

ASSEMBLY

1 Dip the new seals in transmission fluid and install one on the drum and one on the piston.
2 Lubricate the piston and install the clutch piston into the clutch hub. Position the ten coil springs in the piston and the spring retainer on the springs. Install the clutch spring compressor tool and compress the springs. Install the snap ring and remove the tool.
3 After all clutch plates have been installed, position the pressure plate in the clutch drum. Install the pressure plate snap ring.

INPUT SHELL AND SUN GEAR

DISASSEMBLY

1 Remove the external snap ring from the sun gear.
2 Remove the thrust washer from the input shell and sun gear.
3 From inside the input shell, remove the sun gear. Remove the internal snap ring from the sun gear.

ASSEMBLY

1 Install the internal snap ring on the sun gear, and install the sun gear in the input shell.
2 Install the thrust washer on the sun gear and input shell.
3 Install the external snap ring on the sun gear.

REVERSE RING GEAR AND HUB

DISASSEMBLY

1 Remove the hub retaining snap ring from the ring gear.
2 Lift the hub from the ring gear.

ASSEMBLY

1 Position the hub in the ring gear.
2 Secure the hub with the snap ring. Make sure that the snap ring is fully engaged in the groove.

ONE-WAY CLUTCH

DISASSEMBLY

1 Remove the snap ring from each end of the one-way clutch and remove the clutch.
2 Remove the snap ring securing the outer race to the connecting drum and remove the race.

ASSEMBLY

1 Position the outer race in the connecting drum.
2 Secure the outer race with the snap ring. Make sure the snap ring is fully seated in the groove.
3 Install a snap ring in the forward snap ring groove of the connecting drum.
4 Install the one-way clutch on top of the snap ring. Be sure the arrow stamped on the one-way clutch is pointed toward the front of the trans-mission.
5 Install the remaining snap ring at the rear of the one-way clutch to secure the assembly.

Transmission Reassembly

When assembling the transmission sub-assemblies, make sure that the correct thrust washer or needle bearing is used between certain sub-assemblies. Lubricate surfaces with transmission fluid. Petroleum jelly should be used to hold the thrust washers and needle bearings in their proper locations. If the end play is not within specifications after the transmission is assembled, either the wrong selective thrust washers and races were used, or a thrust washer or needle bearing came out of position during assembly.

1 Install reverse clutch inner and outer piston seals on piston. Lubricate the low and reverse clutch piston and seals with transmission fluid, and install into the tranmission case.
2 Install the low and reverse clutch support ring, piston return spring, snap ring and one-way clutch inner race in the case.
3 Hold the one-way clutch inner race in position and install and tighten the mount bolts to correct specifications.
4 Position the low and reverse clutch dished plate in the case. Dip the steel and friction plates in clean transmission fluid, and install the plates alternately starting with a steel plate.
5 After all the plates have been installed, position the retaining plate in the case. Install the pressure plate snap ring.
6 With a thickness gauge, check the clearance between the pressure plate and snap ring. The clearance should be 0.032–0.041 inch. If the clearance is not within this specified range, selective pressure plates are available in the following thicknesses: 0.307, 0.-315, 0.323, 0.331, 0.339 and 0.346 inch. Insert the correct pressure plate, and recheck the clearance.
7 While applying pressure, rotate the connecting drum clockwise to mesh the low and reverse clutch plates with the splines of the drum.
8 Install the oil distributor needle bearing and race in the case. Install the governor and oil distributor in the case taking care not to damage the seal rings.
9 Install the output shaft.
10 Position the needle bearings on the front and rear sides of the interal drive flange. Retain the bearings with petroleum jelly. Install the flange on the output shaft. Install the snap ring.
11 Position the needle bearing race on the rear side of the rear planet carrier and the needle bearing on the front side of the carrier. Retain the needle bearing and race with petroleum jelly. Insert the assembly into the internal drive flange, and install the snap ring.
12 Install the manual shaft in the case.

FORD JATCO

Install the manual plate, spacer and attaching nut, and tighten to correct specifications.

13 Install the park actuating lever and rod into the case and secure with the retaining ring.

14 Install the servo return spring, piston and servo retainer into the case. While holding the retainer in place, install the three attaching bolts, and tighten to correct specifications.

15 Place a new gasket on the rear of the transmission case. Install park pawl, spring and spacer washer on shaft. Fit end of parking rod between the two steel balls in the supporter. Position the extension housing on the case and install the mounting bolts; tightening to correct specifications.

16 Position the needle bearing on the rear end of the forward clutch hub and the bearing race on the front end of the front planet carrier. Assemble the forward clutch hub and planet carrier and install the assembly into the input shell and sun gear drum.

17 Set the input shell and sun gear assembly on a bench with the forward clutch hub facing upward. Position the needle bearing on the front end of the forward clutch hub. Install the needle bearing race. Install the forward clutch assembly on the forward clutch hub by rotating the units to mesh the forward clutch plates with the splines of the forward clutch hub.

18 Install the reverse and high clutch assembly on the forward clutch assembly by rotating the units to mesh the reverse and high clutch plates with the splines of the forward clutch.

19 Install the reverse and high clutch, forward clutch front planet carrier and input shell and sun gear as an assembly into the transmission case.

20 Insert the intermediate band into the case around the reverse and high clutch drum. Install the strut and tighten the band adjusting screw sufficiently to retain the band.

21 Place a selective thickness thrust washer and bearing race on the rear end of the oil pump cover and retain them with petroleum jelly.

22 Position the oil pump in the transmission case.

23 Place the converter housing on the case and install the six converter housing-to-case attaching bolts using a sealing compound. Tighten the bolts to correct specifications.

24 Install the input shaft.

25 To check the oil pump body-to-intermediate drum clearance, move the intermediate drum forward. With a thickness gauge, check the clearance between the engaging lugs on the intermediate drum and the input shell. The clearance should be .020–.032 inch. If this clearance is not within specifications, selective thrust washers are available in the follwoing thicknesses: .059, .067, .075, .083, .091,

.098 and .106 inch. Install the correct thrust washer against the pump body and recheck the clearance.

26 Check the transmission total end play.

27 With the dial indicator contacting the end of the input shaft, set the indicator at zero.

28 Insert a screwdriver behind the input shell. Move the input shell and the front part of the gear train forward.

29 Record the dial indicator reading. The end play should be 0.010–0.020 inch. If the end play is not within the specified range, selective races are available in the following thicnesses: 0.047, 0.055, 0.063, 0.071, 0.079 and 0.087 inch. Install the correct race and recheck the end play.

30 Adjust the band, and install the servo cover.

31 Install the control vavle assembly.

32 Install the vacuum diaphragm rod into the transmission case. If the valve body, transmission case or vacuum diaphragm unit was replaced, a new length vacuum throttle valve rod may

be required. With the throttle valve fully depressed, measure the depth of the throttle valve bore. This dimension will determine which selective throttle valve rod to use.

The rod lengths for the various depths are:

Depth of 1.071 inch, use the 1.140 inch rod.

Depth of 1.075–1.091 inch, use the 1.160 inch rod.

Depth of 1.095–1.110 inch, use the 1.180 inch rod.

Depth of 1.114–1.130 inch, use the 1.200 inch rod.

Depth of 1.134 inch, use the 1.220 inch rod.

33 Install the new O-ring on the vacuum unit and install the vacuum unit in the case hand tight.

34 Install the downshift solenoid in the case hand tight.

35 Using a new gasket, secure the oil pan to the transmission case, and torque the mounting bolts to the correct specifications.

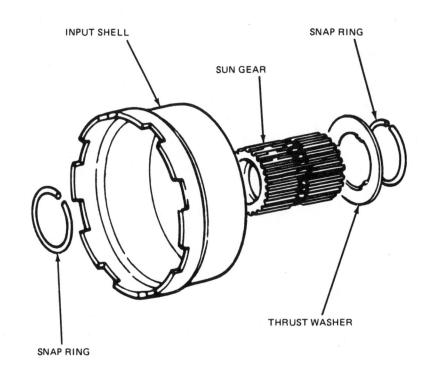

INPUT SHELL SUN GEAR SNAP RING

SNAP RING THRUST WASHER

Input shell and sun gear components

FORD JATCO

REVERSE RING GEAR & HUB
ASSEMBLY

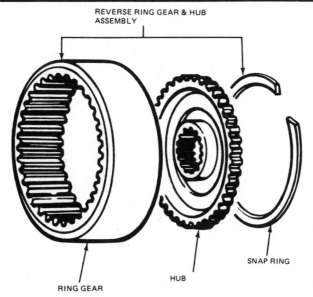

SNAP RING

HUB

RING GEAR

Reverse ring gear and hub

CONNECTING DRUM

ONE-WAY CLUTCH

OUTER RACE

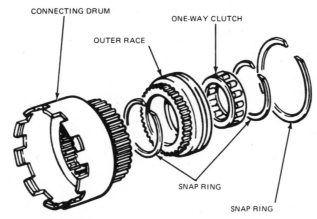

SNAP RING

SNAP RING

One-way clutch components

THROTTLE VALVE FULLY
DEPRESSED

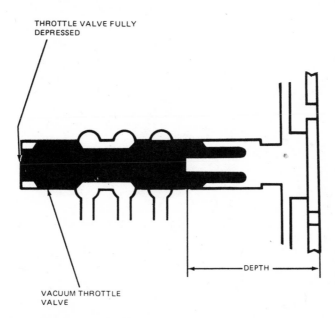

DEPTH

VACUUM THROTTLE
VALVE

Measuring throttle valve rod

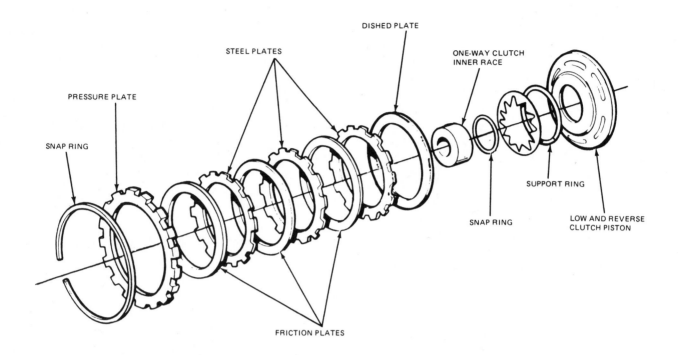

Low and reverse clutch and one-way clutch components

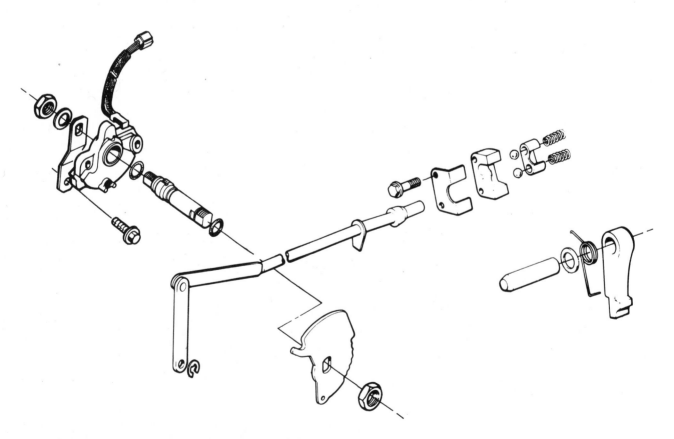

Parking pawl and manual control linkage

FORD JATCO

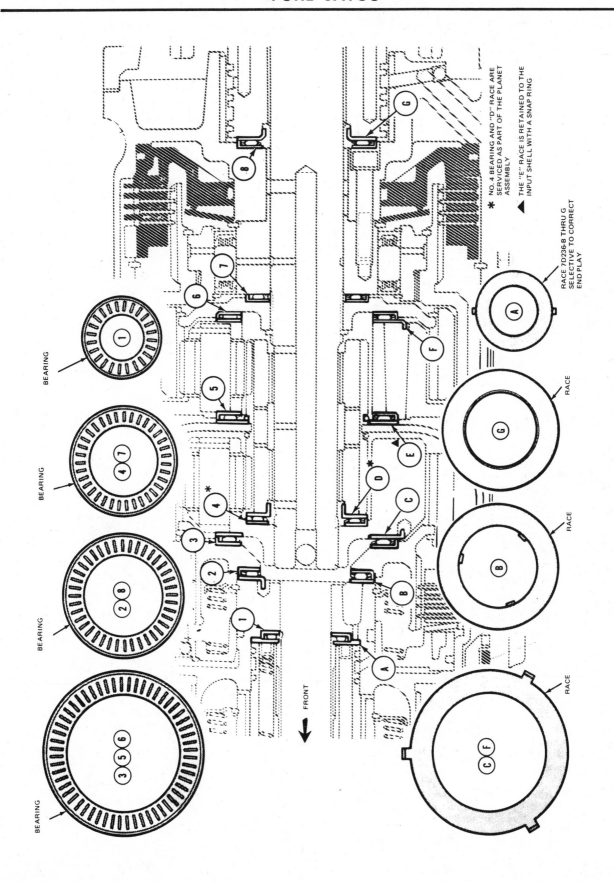

* NO. 4 BEARING AND "D" RACE ARE SERVICED AS PART OF THE PLANET ASSEMBLY

▲ THE "E" RACE IS RETAINED TO THE INPUT SHELL WITH A SNAP RING

RACE 7D236-B THRU G SELECTIVE TO CORRECT END PLAY

Location and identification of thrust bearings and races

Component Removal and Overhaul

FORD JATCO

REVERSE CLUTCH

Steel	Friction	Clearance
7	7	0.032–0.041

CLUTCH PLATE USAGE

Model	FORWARD CLUTCH			HIGH CLUTCH		
	Steel	Friction	Clearance	Steel	Friction	Clearance
PLA-A2	6	6	No Specification	4	4	0.062–0.071

INTERMEDIATE SERVO COMPONENT IDENTIFICATION

Vehicle	Engine	Trans. Model	Servo Piston I.D.	Retainer I.D.	Spring I.D.
Granada/Monarch	250–1V (4.1L)	PLA-A2	F	64–36	None

DIAPHRAGM ASSEMBLY

The Jatco automatic transmission uses a
SINGLE AREA vacuum diaphragm.

MAIN CONTROL AND GOVERNOR IDENTIFICATION

Vehicle	Engine	Transmission Model	Main Control I.D.	Governor I.D.
Granada/Monarch	250–1V (4.1L)	PLA-A2	FDI	FD

GENERAL SPECIFICATIONS

Operation	Specifications (inches)
Total Gear Train End Play	0.010–0.020
Selective Bearing Race Thickness	0.047, 0.055, 0.063, 0.071, 0.079, 0.087
Reverse and High Clutch Pressure Plate-To-Snap Ring Clearance	0.063–0.079
Low and Reverse Clutch Pressure Plate To Snap Ring Clearance	0.032–0.049

FORD JATCO

GEAR RATIOS

First	2.458:1
Second	1.458:1
Third	1.000:1
Reverse	2.181:1
Stall Speed	1750–2000

END PLAY SPECIFICATION

Oil Pump Cover to Front Clutch Drum	Total End Play
0.020–0.032	0.010–0.020

TORQUE SPECIFICATIONS

Item	Ft-lbs	(N·m)
Flywheel to crankshaft	100–115	(135–155)
Flywheel to torque converter	29–36	(40–48)
Converter housing to engine	29–36	(40–48)
Converter housing to transmission case	23–24①	(32–32)
Extension housing to transmission case	15–18	(21–24)
Oil pan to transmission case	4–5	(5.42–6.77)
Servo piston rod (when adjusting intermediate band)	9–10	(12–13)
Servo piston rod lock nut	22–29	(30–39)
Servo piston retainer to transmission case	4–5	(5.42–6.77)
Servo cover to retainer	4–5	(5.42–6.77)
One-way clutch inner race to transmission case	11–13	(15–17)
Control valve body to transmission case	4–5	(5.42–6.77)
Lower valve body to upper valve body	2–3	(2.71–4.06)
Side plate to control valve body	2–3	2.71–4.06)
Stud and nut upper to lower control valve body attaching	4–5	(5.42–6.77)
Oil screen to lower valve body	2–3	(2.71–4.06)
Governor valve body to oil distributor	4–6	(5.42–8.13)
Oil pump cover to oil pump housing	4–6	(5.42–8.13)
Manual shaft lock nut	22–29	(30–39)
Oil pressure test plug	7–11	(10–14)
Actuator for parking rod to extension housing	6–8	(9–10)
Down shift solenoid	Hand Tight	
Vacuum diaphragm	Hand Tight	
Outer manual lever to shaft nut	14–25	(19–33)
Converter housing cover to converter housing	12–16	(16–22)
Filler tube to case	4–5	(5.5–7)

① Use thread sealant when installing

FORD OVERDRIVE

Ford Automatic Overdrive Transmission

Description

This automatic transmission differs from conventional automatic transmissions since it uses a sophisticated method for providing mechanical "direct drive" rather than hydraulic in its higher gear ranges. This is accomplished through a direct drive input shaft which couples the engine to the direct clutch and bypasses the torque converter. This shaft is driven by the torque converter cover through the damper assembly.

The transmission consists basically of a torque converter, compound planetary gear train and a hydraulic control system. The torque converter incorporates a damper assembly and the input shaft for high and overdrive gears. A FMX type gear train is used. First, second and high gears have the same power flow as the FMX except that in third gear the input torque is split between two shafts.

Overdrive is accomplished by the addition of a band to lock the reverse sun gear while driving the planet carrier. Engine torque flows through the damper assembly to bypass the torque converter. For gear control, the automatic overdrive has four friction clutches, two one-way roller clutches and two bands. The overdrive servo and direct clutch are unique to this unit, and an intermediate clutch replaces the front band.

The hydraulic system oil pump and stator support are similar to the Ford C6 automatic transmission. The band servo pockets and the valve body upper passages are cast into the case.

Throttle Valve Adjustment

NOTE: T.V. adjustment is critical, and most shift problems will be related to this adjustment. Anytime the curb idle speed is changed by more than 50 rpm, check for correct T.V. linkage adjustment.

Linkage

1 Set the screw at the T.V. linkage lever (at engine) at its midpoint and be sure the throttle is against the idle stop.
2 Place shift lever in NEUTRAL and firmly set parking brake.
3 Loosen the bolt on the T.V. control rod sliding trunnion block.
4 Push up on the lower end of the T.V. rod to assure that the carburetor linkage lever is held firmly against the throttle lever.
5 Release the rod and make sure it stays in position. If it does not, the return spring at the carburetor is disconnected or defective.
6 Push the transmission T.V. lever up against its internal stop, and tighten bolt on trunnion block.
7 Recheck that the throttle lever is at the idle stop.

Pressure

1 Connect a 100 psi pressure gauge to the T.V. limit pressure tap on the right side of the case.
2 Verify that the throttle lever is at the idle stop. Idle the engine in neutral and set parking brake.
3 Place a 1/16 inch drill or gauge between the linkage lever adjustment screw and throttle lever. Check that T.V. pressure is 5 psi or lower.
4 Replace the 1/16 inch gauge with a 5/16 inch gauge. With the engine idling in neutral, T.V. pressure must be at least 22 psi. Less than 22 psi indicates a too short control linkage. Turn the screw in to increase pressure.

Air Pressure Tests

The clutch and servo circuits through the case, pump assembly and output shaft can be pneumatically tested through the passages identified in the illustration. Compressed air, at 25 psi, should be filtered and dry to avoid fluid contamination.

Clutches

The clutches should apply with a dull thud when compressed air is blown into their pressure ports. A hissing sound and the absence of application noise indicates a leak.

Servos

The low/reverse servo application should be heard as the piston bottoms

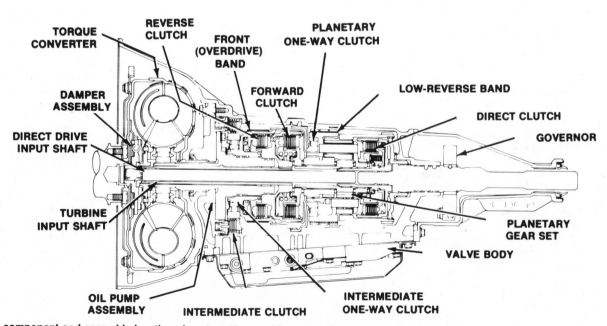

Main component and assembly locations in automatic overdrive transmission.

FORD OVERDRIVE

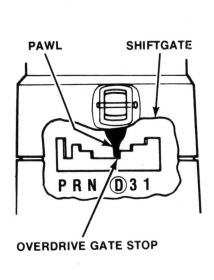

PAWL   SHIFTGATE

P R N D 3 1

OVERDRIVE GATE STOP

GATE STOPS

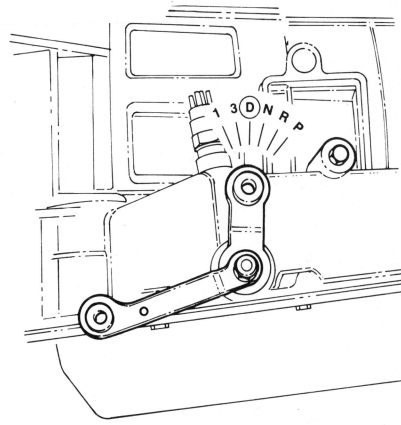

TRANSMISSION LEVER POSITIONS

Manual linkage gate stops and lever positions

with air applied. When air pressure is removed, the piston should be felt to move back against the cover.

For the overdrive servo test, first pressurize the apply side. Hold the pressure and pressurize the release side. The piston should be heard stroking to apply the band, and also to bottom in the cover when release pressure is applied.

Governor
Air blown into the "line to governor" passage should cause the governor to whistle. The governor may be stuck if there is no noise.

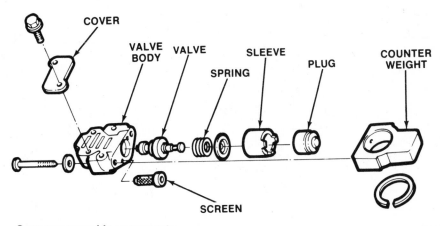

COVER · VALVE BODY · VALVE · SPRING · SLEEVE · PLUG · COUNTER WEIGHT · SCREEN

Governor assembly components

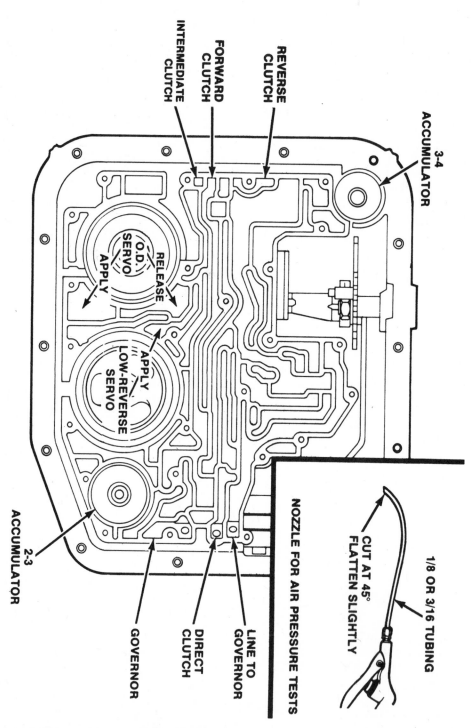

Transmission case air test pressure passage identification

INTERMEDIATE CLUTCH

FORWARD CLUTCH

REVERSE CLUTCH

3-4 ACCUMULATOR

O.D. SERVO

APPLY

RELEASE

APPLY

LOW-REVERSE SERVO

2-3 ACCUMULATOR

GOVERNOR

DIRECT CLUTCH

LINE TO GOVERNOR

NOZZLE FOR AIR PRESSURE TESTS

CUT AT 45° FLATTEN SLIGHTLY

1/8 OR 3/16 TUBING

FORD OVERDRIVE

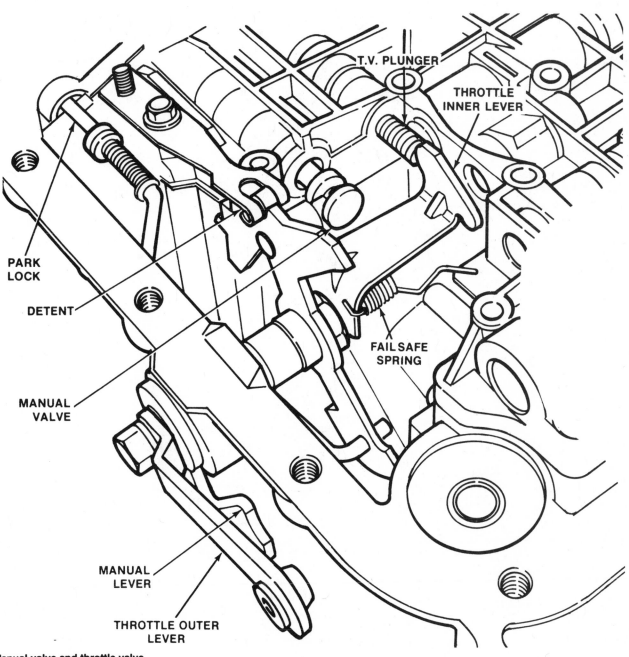

T.V. PLUNGER

THROTTLE
INNER LEVER

PARK
LOCK

DETENT

FAIL SAFE
SPRING

MANUAL
VALVE

MANUAL
LEVER

THROTTLE OUTER
LEVER

Manual valve and throttle valve

FORD OVERDRIVE

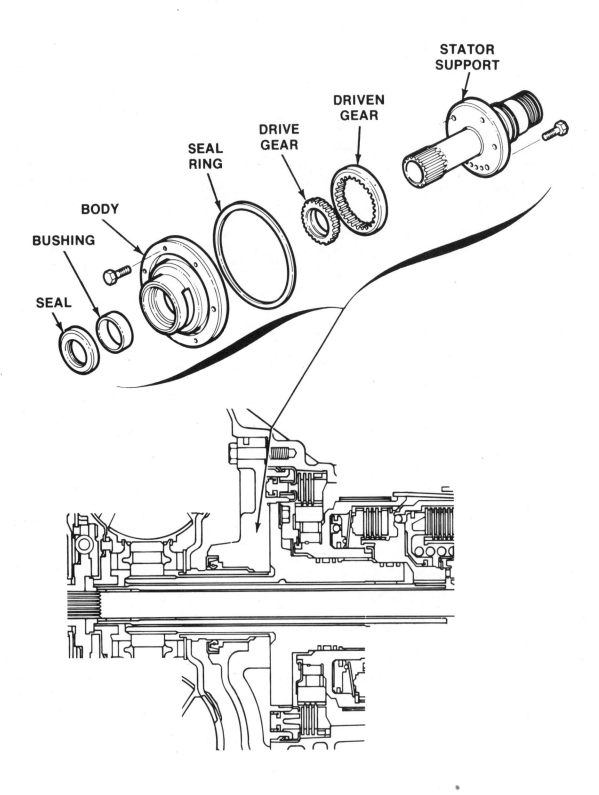

STATOR
SUPPORT

DRIVEN
GEAR

DRIVE
GEAR

SEAL
RING

BODY

BUSHING

SEAL

Oil pump location and components. Pump construction, operation and maintenance are similar to the Ford C6 transmission. Remove the pump attaching bolts and pry the gear train forward to remove the pump

FORD OVERDRIVE

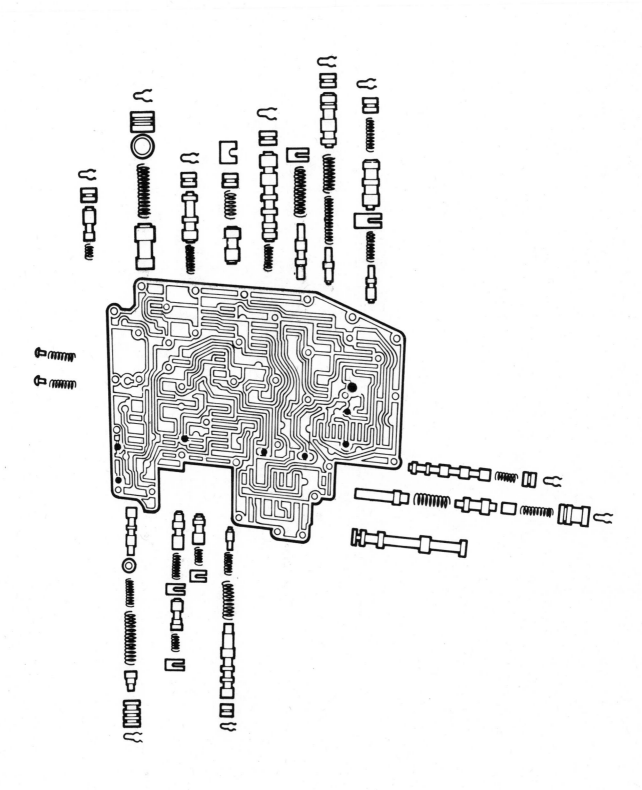

Valve body and components. Dots indicate the eight ball check valve locations

FORD OVERDRIVE

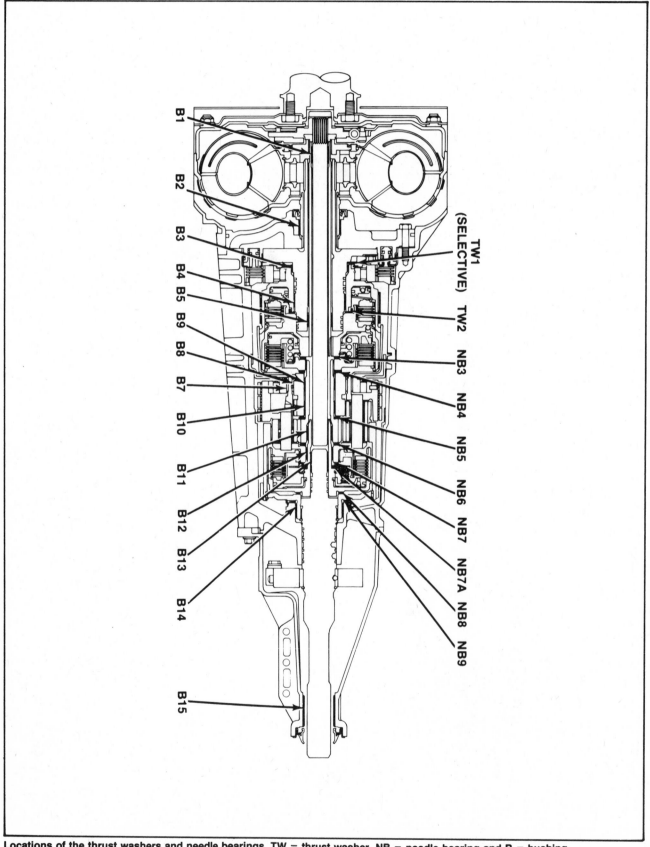

Locations of the thrust washers and needle bearings. TW = thrust washer, NB = needle bearing and B = bushing

FORD OVERDRIVE

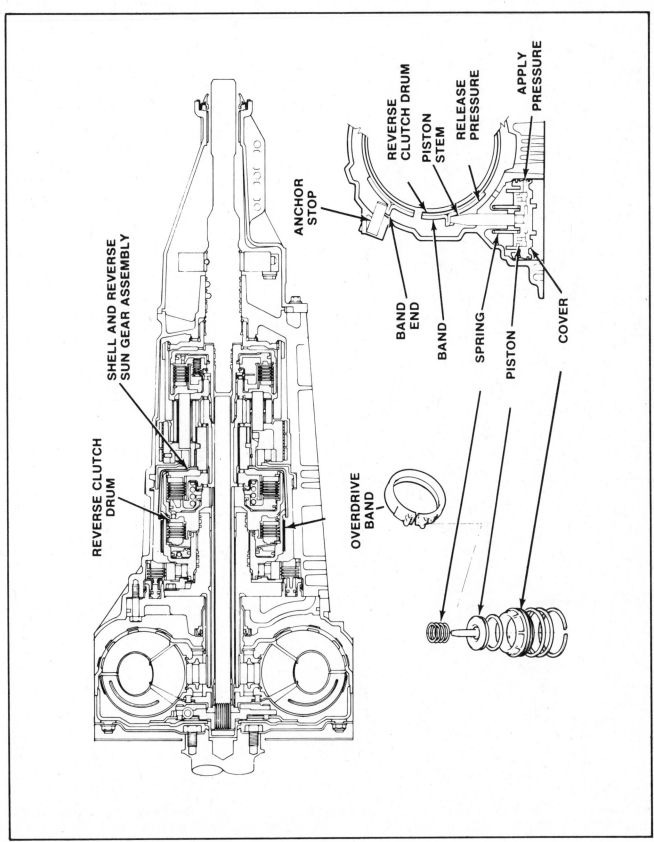

REVERSE CLUTCH DRUM

PISTON STEM

RELEASE PRESSURE

APPLY PRESSURE

ANCHOR STOP

SHELL AND REVERSE SUN GEAR ASSEMBLY

REVERSE CLUTCH DRUM

OVERDRIVE BAND

BAND END

BAND

SPRING

PISTON

COVER

Overdrive band and overdrive servo locations and components

Component Removal and Overhaul

FORD OVERDRIVE

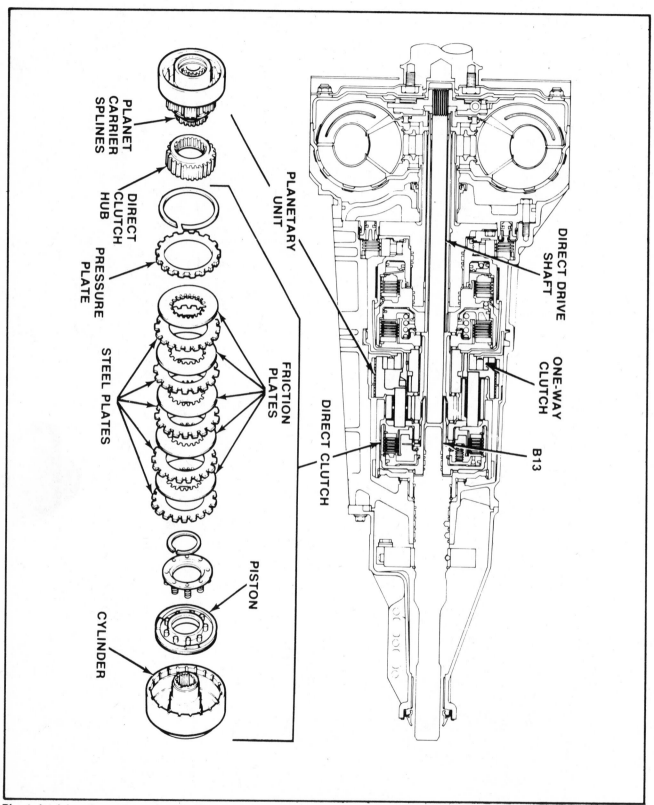

Direct clutch location and components. The direct clutch is driven mechanically through the direct drive shaft. B13 is the direct clutch cylinder support bushing

FORD OVERDRIVE

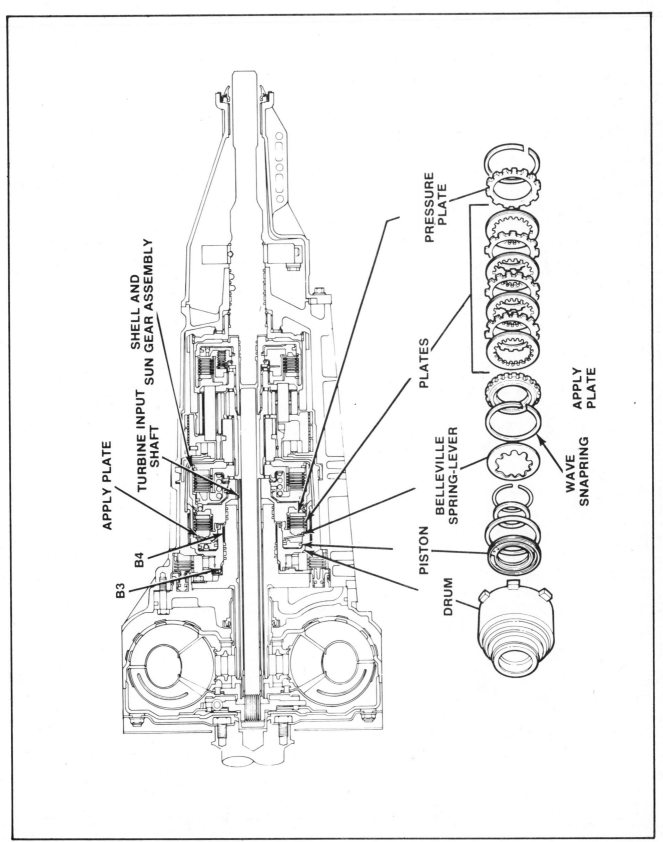

Reverse clutch assembly location and components

FORD OVERDRIVE

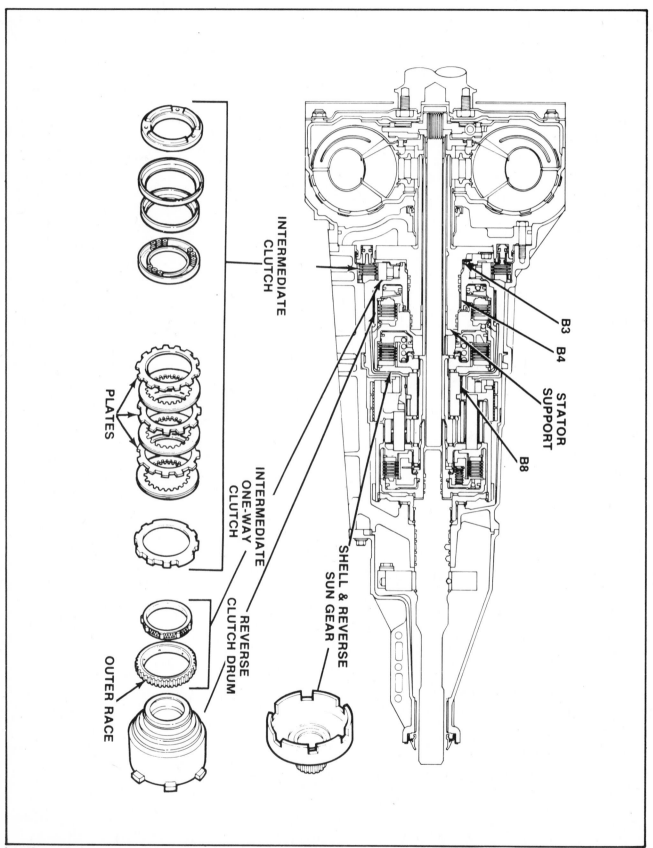

Location and components of the intermediate clutch, intermediate one-way clutch, reverse clutch drum and shell and reverse sun gear. B3, B4 and B8 are support bushings

FORD OVERDRIVE

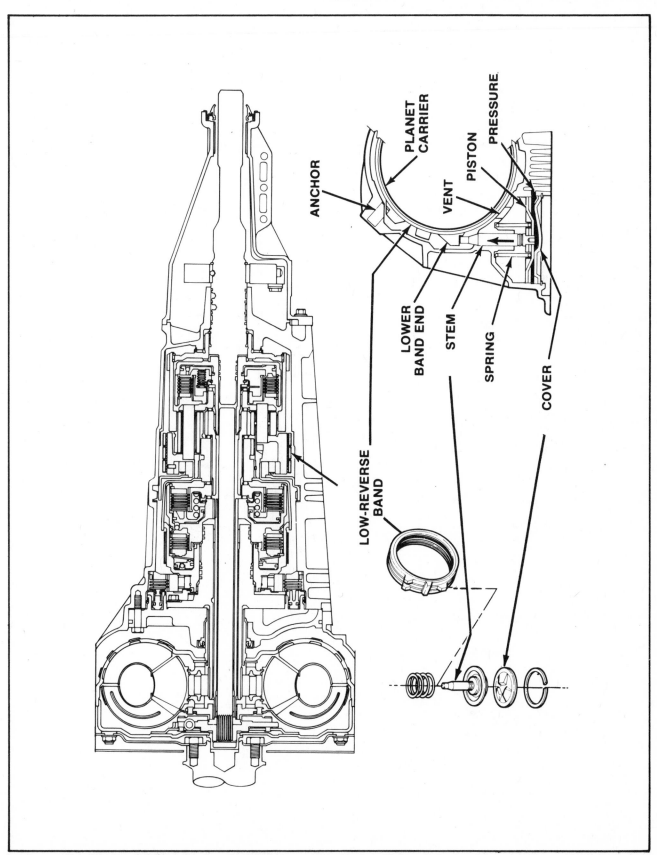

Low/reverse band and low/reverse servo locations and components. The band is installed around the outer drum surface of the planet carrier and the spring-loaded servo piston is located in the case

FORD OVERDRIVE

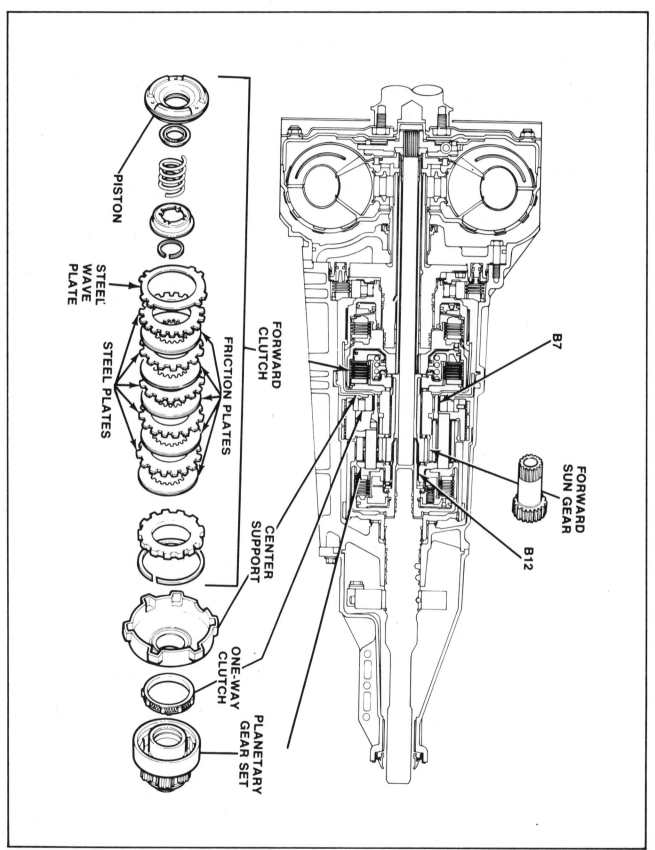

PISTON

STEEL WAVE PLATE

STEEL PLATES

FRICTION PLATES

FORWARD CLUTCH

CENTER SUPPORT

ONE-WAY CLUTCH

PLANETARY GEAR SET

B7

FORWARD SUN GEAR

B12

Locations and components of the forward clutch, forward sun gear, center support, one-way planetary clutch and planetary gearset. B7 and B12 are planet carrier support bushings

FORD OVERDRIVE

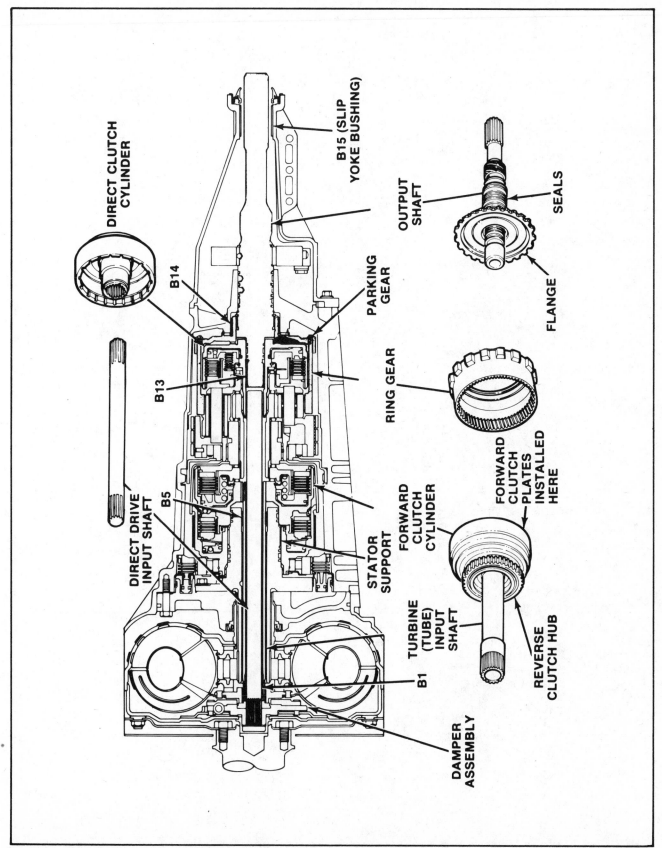

Input and output shafts and supports. B13 is a bushing on the output shaft which supports the Direct clutch cylinder. B1 and B5 are turbine input shaft support bushings. B14 and B15 are output shaft support bushings

FORD OVERDRIVE

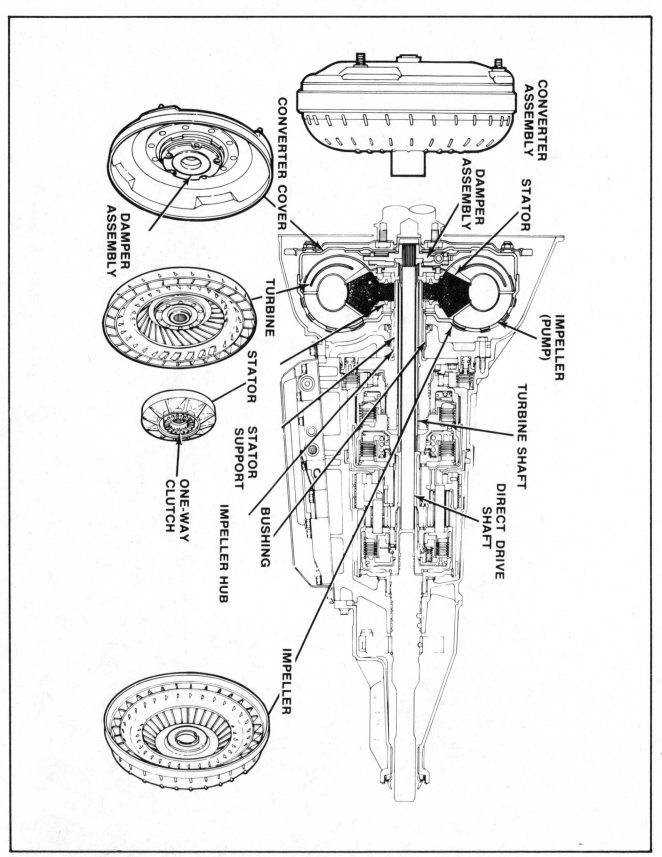

Torque converter assembly and attaching components. As a fluid coupler and torque multiplier; this torque converter is basically the same as in the C6 transmission

GM 125 TRANSAXLE

BAND AND CLUTCH APPLICATION CHART

Gear	Intermediate Friction Clutch	Intermediate Roller Clutch	Overdrive Band	Reverse Clutch	Forward Clutch	Planetary (Low) One-Way Clutch	Low and Reverse Band	Direct Clutch	Gear Ratio①
1 Range Low					APPLIED	HOLDING	APPLIED		2.4–to–1
O/D and 3 Range									
Low					APPLIED	HOLDING			2.4–to–1
Second									
(Intermediate)	APPLIED	HOLDING			APPLIED	OVERRUNS			1.467–to–1
Third (Direct)	APPLIED	OVERRUNS			APPLIED	OVERRUNS		APPLIED	1–to–1
Fourth (Overdrive)	APPLIED		APPLIED			OVERRUNS		APPLIED	0.667–to–1
Reverse				APPLIED			APPLIED		2–to–1

① Not including torque converter reduction in 1st, Second and Reverse.

GM 125 AUTOMATIC TRANSAXLE
TROUBLE SHOOTING

PROBLEM	POSSIBLE CAUSE	POSSIBLE CONDITION
NO DRIVE IN DRIVE RANGE (CHECK OIL PRESSURES)	Low Oil Level	Incorrect level. External leaks.
	Manual Linkage	Misadjusted.
	Low Oil Pressure	Plugged or restricted oil strainer. Strainer "O" ring seal cut, missing. Pressure regulator valve stuck. Pump rotor splines damaged by shaft. Porosity in intake bore.
	Case Cover	Manual valve disconnected. Gaskets mispositioned.
	Forward Clutch	Forward clutch does not apply—piston cracked, seals missing, damaged; clutch plates burned; snap ring out of groove. Forward clutch oil seal rings missing or damaged on input shaft; leak in feed circuits. Clutch housing ball check stuck, missing, damaged. Wrong number of clutch plates. Feed orifice plugged in input shaft.
	Roller Clutch	Springs missing. Rollers galled or missing.
HIGH OR LOW OIL PRESSURES	Throttle Valve Cable	Misadjusted, binding, unhooked, or broken.
	Throttle Lever and Bracket Ass'y	Bent, binding, broken, unhooked or mispositioned.
	Throttle Valve or Plunger.	Binding.
	Shift T.V. Valve	Binding.
	Line Boost Valve	Binding. Bore plug installed wrong.
	T.V. Boost Valve & Reverse Boost Valve	Binding. Plugged orifices in bushings.
	Pressure Regulator Valve & Spring	Binding.

GM 125 TRANSAXLE

GM 125 AUTOMATIC TRANSAXLE
TROUBLE SHOOTING

PROBLEM	POSSIBLE CAUSE	POSSIBLE CONDITION
	Pressure Relief Valve	Ball missing. Spring damaged.
	Manual Valve	Unhooked.
	Pump	Pump slide stuck. Pump slide seal damaged, missing.
1–2 SHIFT—FULL THROTTLE ONLY	Throttle Valve Cable	Binding, unhooked, or broken. Misadjusted.
	Throttle Lever and Bracket Ass'y	Binding, unhooked, or mispositioned. NOTICE: Allowing line boost valve to seat causes full T.V. pressure, regardless of throttle valve position.
	Throttle Valve and plunger	Binding.
	Pump and Control Valve Ass'y	Gaskets or spacer plate leaking, damaged, incorrectly installed.
	Case Assembly	Porosity.
FIRST SPEED ONLY, NO 1–2 UPSHIFT	Governor and Governor Feed Passages	Plugged governor oil feed orifice in spacer plate. Governor ball or balls missing in assembly. Governor cover rubber "O" ring seal missing or leaking. NOTICE: If governor cover "O" ring seal leaks, an external leak will be present, along with no upshifts. Governor shaft seal missing or damaged. Governor driven gear stripped. Governor weights binding on pin. Governor assembly missing. Governor drive gear loose or missing. Governor springs damaged or missing. Governor oil pipe plugged or mislocated.
	1–2 Shift Train	1–2 shift valve or 1–2 throttle valve stuck in downshift position.
	Case and Case Cover	Porosity in case channels or undrilled 2nd speed feed holes. Excessive leakage between case bore and intermediate band apply pin. Broken or missing band.
	Intermediate Servo Ass'y	Servo oil seal ring missing or damaged. Porosity in servo cover or piston. Wrong intermediate apply pin.
FIRST AND SECOND SPEEDS ONLY, NO 2–3 SHIFT	Pump and Control Valve Ass'y	Direct clutch feed orifice in spacer plate plugged. Spacer plate or gaskets leaking, damaged, or incorrectly installed. Check ball #5 missing or mislocated.
	2–3 Valve Train	2–3 Shift valve or 2–3 throttle valve stuck in downshift position.
	Case and Case Cover	Porosity in case channels. 3rd oil cup plug missing. Direct clutch accumulator check valve missing. Servo bleed orifice cup plug missing.
	Driven Sprocket Support	Oil seals cut, missing, damaged. Feed passages blocked.
	Direct Clutch	Oil seals missing or damaged on piston. Direct clutch piston or housing cracked. Direct clutch plates damaged or missing. Direct clutch backing plate snap ring out of groove.
	Intermediate Servo Ass'y	Servo oil seal ring broken or missing on piston.
	Governor Ass'y	Shaft seal missing or damaged.

GM 125 AUTOMATIC TRANSAXLE
TROUBLE SHOOTING

PROBLEM	POSSIBLE CAUSE	POSSIBLE CONDITION
THIRD SPEED ONLY	2–3 Shift Valve	Stuck in upshift position.
	Governor Ass'y	Governor oil pipe plugged or mislocated. Governor feed passages plugged.
DRIVE IN NEUTRAL	Manual Linkage	Misadjusted or disconnected.
	Forward Clutch	Clutch does not release; exhaust check ball sticking, plates burned together.
	Case and Case Cover	Cross leakage to forward clutch passages.
NO DRIVE IN REVERSE OR SLIPS IN REVERSE (REFER TO OIL PRESSURE CHECKS)	Throttle Valve Cable	Binding or misadjusted.
	Manual Linkage	Misadjusted.
	Throttle Valve	Binding.
	Shift T.V. Valve	Binding.
	Reverse Boost Valve	Binding. Bushing orifices blocked.
	Lo and Reverse Clutch Ass'y	Piston—cracked, broken or missing seals. Clutch plates burned, wrong number.
	Case and Case Cover	Porosity in passages. Check ball #4 missing, mislocated. Lo and reverse cup plug missing. Reverse oil pipe plugged, mislocated. Reverse oil pipe "O" ring seal cut, missing.
	Direct Clutch	Sprocket cover oil seal rings damaged or missing. Piston or housing cracked. Piston seals cut or missing. Housing ball check stuck, leaking or missing. Plates burned. Orifice plugged in spacer plate.
SLIPS ON 1–2 SHIFT	Low Oil Level	Oil will aerate-correct oil level.
	Spacer Plate and Gaskets	Second speed feed orifice partially blocked. Gaskets damaged or mispositioned.
	1–2 Accumulator Valve	Valve sticking in valve body causing low 1–2 accumulator pressure. Weak or missing spring.
	1–2 Accumulator Piston	Seal leaking, spring broken or missing. Leak between piston and pin.
	Intermediate Band Apply Pin	Wrong apply pin. Excessive leakage between apply pin and case.
	Intermediate Servo Assembly	Porosity in piston. Servo oil seal ring damaged or missing.
	Throttle Valve Cable	Not adjusted properly.
	Throttle Valve	Binding, causing low T.V. pressure.
	Shift T.V. Valve	Binding.
	Intermediate Band	Worn or burned.
	Case	Porosity in passages.
ROUGH 1–2 SHIFT	Throttle Valve Cable	Not adjusted properly. Binding.
	Throttle Valve and T.V. Plunger	Binding.
	Shift T.V. Valve	Binding.
	1–2 Accumulator Valve	Binding.

Component Removal and Overhaul

GM 125 TRANSAXLE

GM 125 AUTOMATIC TRANSAXLE
TROUBLE SHOOTING

PROBLEM	POSSIBLE CAUSE	POSSIBLE CONDITION
ROUGH 1–2 SHIFT	1–2 Accumulator	Oil ring damaged. Piston stuck. Broken or missing spring. Bore damaged.
	Intermediate Servo Ass'y	Wrong pin. Servo piston oil seal ring damaged or missing.
SLIPS 2–3 SHIFT	Oil Level Low	Correct oil level.
	Throttle Valve Cable	Not adjusted properly.
	Throttle Valve	Binding.
	Spacer Plates & Gaskets	Direct clutch orifice partially blocked in spacer plate. Gaskets mispositioned or damaged.
	Intermediate Servo Ass'y	Servo oil seal ring damaged.
	Direct Clutch	Case porosity in direct clutch feed channels. Driven sprocket channels cross feeding, leaking or restricted. Driven sprocket oil seal rings damaged or missing. Direct clutch piston or housing cracked. Piston seals cut or missing. Direct clutch plates burned. Housing check ball damaged, missing, binding.
ROUGH 2–3 SHIFT	Throttle Valve Cable	Not adjusted properly. Binding.
	Throttle Valve and Plunger	Binding.
	Shift T.V. Valve	Binding.
NO ENGINE BRAKING—INTERMEDIATE RANGE 2ND GEAR (CHECK OIL PRESSURES)	Intermediate Servo Ass'y	Servo oil seal ring missing or damaged.
	Intermediate Band	Mispositioned. Broken or burned.
NO ENGINE BRAKING—LO RANGE—1ST GEAR (CHECK OIL PRESSURES)	Lo-Reverse Clutch Ass'y	NOTICE: No reverse should also be a complaint with any of the (4) following conditions: Piston seals broken or missing. Porosity in piston or housing. Clutch housing snap ring out of case. Cup plug seal missing or damaged between case and lo and reverse clutch housing.
NO PART THROTTLE OR DETENT DOWNSHIFTS (CHECK OIL PRESSURES)	Throttle Plunger Bushing	Passages not open.
	2–3 Throttle Valve Bushing	Passages not open.
	Valve Body Gaskets	Mispositioned or damaged.
	Spacer Plate	Hole plugged or undrilled.
	Throttle Valve Cable	Misadjusted.
	Shift T.V. Valve	Binding.
	Throttle Valve	Binding.
LOW OR HIGH SHIFT POINTS (CHECK OIL PRESSURES)	Throttle Valve Cable	Binding or disconnected.
	Throttle Valve	Binding.
	Shift T.V. Valve	Binding.
	Line Boost Valve	Missing or binding.

216
footer

GM 125 AUTOMATIC TRANSAXLE
TROUBLE SHOOTING

PROBLEM	POSSIBLE CAUSE	POSSIBLE CONDITION
LOW OR HIGH SHIFT POINTS (CHECK OIL PRESSURES)	Throttle Valve Plunger	Binding.
	1–2 or 2–3 Throttle Valve	Binding.
	Valve Body Spacer Plate or Gaskets	Damaged, mispositioned.
	Throttle Lever and Bracket Ass'y	Binding, unhooked or loose. Not positioned at the throttle valve plunger bushing pin locator.
	Governor Shaft Seal Ring	Broken or missing.
	Governor Cover "O" Ring	Broken or missing. NOTICE: The ring will leak externally.
	Case	Porosity.
WON'T HOLD IN PARK	Manual Linkage	Misadjusted.
	Internal Linkage	Park pawl binding in case. Actuator rod or plunger damaged. Parking pawl broken. Parking bracket, loose or damaged.
	Inside Detent Lever and Pin Ass'y	Nut loose. Hole in lever worn or damaged.
	Manual Detent Roller and Spring Ass'y	Bolt loose that holds roller assembly to valve body. Pin or roller damaged, mispositioned, or missing.
TRANSMISSION NOISY	Pump Noise	Oil level setting low. Cavitation due to plugged strainer, cut or missing "O" ring, porosity in intake circuit, or water in oil.
	Gear Noise	Transaxle grounded to body. Roller bearings worn or damaged. If noisy in 3rd gear or on turns only—check differential and final drive unit.

GM 125 TRANSAXLE

GM 125 Automatic Transaxle

Applications

Buick—1980 Skylark
Chevrolet—1980 Citation
Oldsmobile—1980 Omega
Pontiac—1980 Phoenix

Description

This unit is fully automatic, and consists primarily of a three-element torque converter, compound planetary gear set and dual sprocket and drive link assembly. Additionally, the transaxle incorporates a differential and final drive gear set. Three multiple disc clutches, a roller clutch and a band provide the friction elements required to obtain the desired function of the planetary gear sets, and a hydraulic system pressurized by a vane-type pump provides the working pressure required to operate the friction elements and automatic controls.

Approximately four quarts of fluid are required to refill the transaxle after the pan has been drained. The total capacity of the 125 transaxle and converter assembly is approximately nine quarts, but correct level is determined by mark on the dipstick.

Trouble Diagnosis

ROAD TEST

1 Position selector lever in DRIVE RANGE and accelerate the vehicle. A 1-2 shift and 2-3 shift should occur at throttle openings. These shift points will vary with throttle openings.
2 In DRIVE RANGE check part throttle 3-2 downshift at 30 m.p.h. by quickly opening throttle approximately three-fourths, and the transaxle should downshift at 50 m.p.h. by fully depressing the accelerator.
3 Position the selector lever in INTERMEDIATE RANGE and accelerate the vehicle. A 1-2 shift should occur at

all throttle openings. The shift point will vary with throttle opening and will be firmer than in DRIVE RANGE. (A 2-3 shift cannot be obtained in this range.)
4 In INTERMEDIATE RANGE check for detent 2-1 downshift at 20 m.p.h.
5 In LO RANGE accelerate the vehicle. No shift should occur.
6 In DRIVE RANGE with vehicle speed at approximately 50 m.p.h., close the throttle and move the selector lever to INTERMEDIATE RANGE. The transaxle should downshift to *second*. Engine speed should increase, and there should be noticeable engine braking.
7 In INTERMEDIATE RANGE with vehicle speed at approximately 40 m.p.h., close the throttle and move the selector lever to LO. The transaxle should downshift to *first* between 40 and 25 m.p.h. depending on valve body calibration. This downshift

GM 125 TRANSAXLE

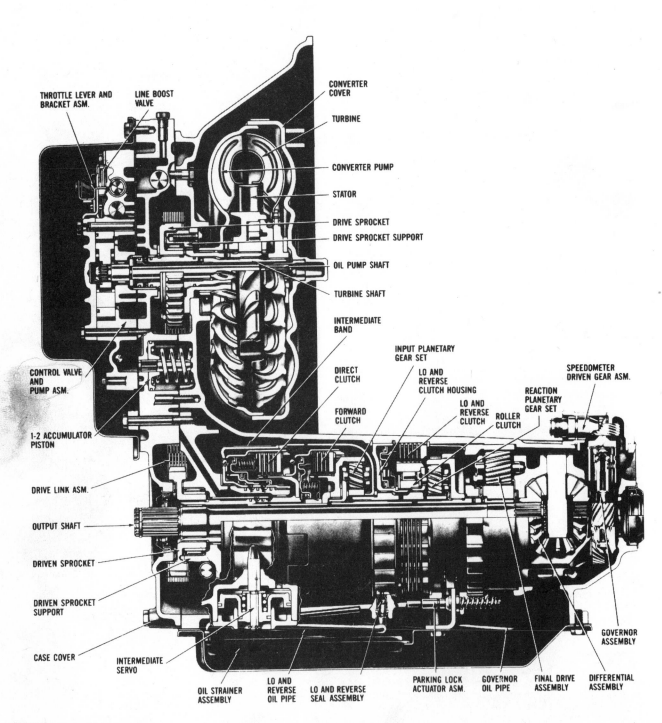

GM 125 automatic transaxle

THROTTLE LEVER AND BRACKET ASM.

LINE BOOST VALVE

CONVERTER COVER

TURBINE

CONVERTER PUMP

STATOR

DRIVE SPROCKET

DRIVE SPROCKET SUPPORT

OIL PUMP SHAFT

TURBINE SHAFT

INTERMEDIATE BAND

INPUT PLANETARY GEAR SET

DIRECT CLUTCH

LO AND REVERSE CLUTCH HOUSING

LO AND REVERSE CLUTCH

REACTION PLANETARY GEAR SET

SPEEDOMETER DRIVEN GEAR ASM.

FORWARD CLUTCH

ROLLER CLUTCH

CONTROL VALVE AND PUMP ASM.

1-2 ACCUMULATOR PISTON

DRIVE LINK ASM.

OUTPUT SHAFT

DRIVEN SPROCKET

DRIVEN SPROCKET SUPPORT

CASE COVER

INTERMEDIATE SERVO

OIL STRAINER ASSEMBLY

LO AND REVERSE OIL PIPE

LO AND REVERSE SEAL ASSEMBLY

PARKING LOCK ACTUATOR ASM.

GOVERNOR OIL PIPE

FINAL DRIVE ASSEMBLY

DIFFERENTIAL ASSEMBLY

GOVERNOR ASSEMBLY

GM 125 TRANSAXLE

PARK

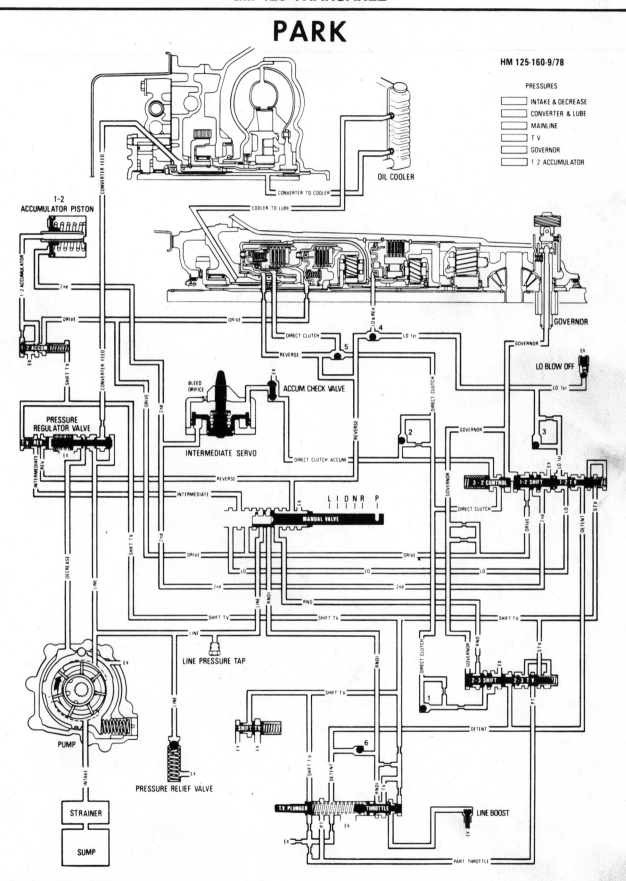

Complete hydraulic schematic

should be accompanied by increased engine speed and noticeable engine braking.

LOW OIL PRESSURE

Insufficient hydraulic pressure can be attributed to one or more of the following:
1 Low oil level
2 T.V. system (pressure low in Neutral or Drive; low to normal in Intermediate and Reverse)
   a. T.V. cable misadjusted or sticking
   b. T.V. linkage—binding, incorrect cable, incorrect link
   c. Throttle valve stuck
   d. Shift T.V. valve stuck
3 Oil screen plugged or "O" ring seal damaged
4 Oil screen "O" ring seal leaking
5 Control valve assembly bolts loose
6 Pressure regulator stuck or wrong size
7 Control valve assembly check balls 4, 5 or 6 missing or out of location
8 Control valve assembly shift T.V. valve or reverse boost valve stuck
9 Control valve assembly 1-2 accumulator piston missing or seal cut or leaking
10 Control valve assembly internal leaks
11 Lo/Reverse Clutch housing-to-case seal and cup plug leaking. (Reverse only)
12 Pump gears broken
13 Shift T.V. passage blocked

HIGH OIL PRESSURE

Too high oil pressure can be attributed to one or more of the following:
1 T.V. system (pressure high in Neutral, Drive; normal to high in Intermediate and Reverse)
   a. T.V. cable misadjusted or sticking
   b. T.V. linkage—binding, incorrect cable, incorrect link
   c. Throttle valve stuck
   d. Shift T.V. valve stuck
2 Pressure regulator valve stuck or wrong size
3 Control valve pump assembly
   a. Valves stuck
   b. T.V. valve and plunger
   c. Shift T.V. valve
   d. Pump slide stuck
   e. Reverse boost valve
4 Reverse boost orifice in spacer plate plugged (Reverse only)
5 Internal pump or case leaks

DIRECT CLUTCH

A burned direct clutch can be caused by one or more of the following:
1 Driven sprocket support
   a. Leaking (damaged) seal rings on driven sprocket support
   b. Driven sprocket support sleeve loose or mispositioned
2 Numbers 4 or 5 check balls missing or off location in case
3 Low hydraulic pressure
4 Channels blocked or interconnected
5 Accumulator check valve missing

6 Direct clutch assembly
   a. Seals cut, missing or rolled out of groove
   b. Apply ring, release spring, assembly or wrong number of clutch plates
   c. Exhaust ball capsule in housing damaged and not sealing
   d. Piston or housing damaged
7 Intermediate servo assembly
   a. Seals missing or damaged
   b. Servo bore scored or damaged
   c. Servo orifice bleed plug missing
   d. Accumulator check valve not sealing
8 Control valve assembly
   a. Control valve assembly-to-case bolts loose
   b. Sealing surface on control valve assembly, spacer plate, case and/or gaskets damaged and leaking
   c. Porosity in control valve assembly and/or case channels

FORWARD CLUTCH

A burned forward clutch may be caused by one or more of the following:
1 Driven sprocket support
   a. Leaking (damaged) seal rings on driven sprocket support
   b. Driven sprocket support sleeve loose or mispositioned
2 Numbers 4 or 5 check balls missing or off location in case
3 Low hydraulic pressure
4 Channels blocked or interconnected
5 Forward clutch assembly
   a. Seal rings on input shaft damaged or missing
   b. Input shaft feed passage or orifice restricted
   c. Housing exhaust ball capsule damaged or missing
   d. Housing or shaft seal surface damaged
   e. Piston seals missing or damaged
   f. Apply ring missing, wrong apply ring or wrong number of clutch plates
   g. Piston damaged, leaking
6 Control valve pump assembly and case
   a. Control valve assembly-to-case bolts loose
   b. Sealing surface on control valve assembly, spacer plate, case or gaskets damaged or leaking
   c. Porosity between channels in control valve assembly or case

LO-REVERSE CLUTCH

A burned lo-reverse clutch may be caused by one or more of the following:
1 Driven sprocket support
   a. Leaking (damaged) seal rings on driven sprocket support
   b. Driven sprocket support sleeve loose or mispositioned
2 Lo-reverse clutch assembly
   a. Housing seal area damaged
   b. Piston or seals damaged

   c. Apply ring missing, wrong apply ring or wrong number of clutch plates
3 Control valve pump assembly
   a. Reverse boost valve sticking
   b. Check balls missing or out of location
4 Case assembly
   a. Lo-reverse clutch housing-to-case cup plug assembly hole restricted, damaged or not correctly seated
   b. Number 4 check ball missing or off location in case
   c. Reverse feed passage restricted or leaking
   d. Low hydraulic pressure

INTERMEDIATE BAND ASSEMBLY

A burned intermediate band assembly may be caused by one or more of the following:
1 Band anchor pin missing or not engaged in the band
2 Band not properly aligned or apply pin not engaged
3 Intermediate servo assembly
   a. Seals missing or damaged
   b. Wrong band apply pin
4 Leak in clutch apply stream
5 Control valve assembly
   a. 1-2 accumulator piston missing or seal leaking
   b. 1-2 accumulator valve sticking
6 Case
   a. Accumulator check valve missing or not seating correctly

NOISY TRANSAXLE

Transaxle noise may be caused by one or more of the following:
1 Pump
   a. Oil level setting low
   b. Cavitation due to plugged screen, porosity intake circuit, water in oil
   c. Damaged pump vanes
2 Gears
   a. Transaxle grounded to body
   b. Roller bearings worn or damaged

THUMPING OR CLUNKING SOUND AT 1–5 M.P.H. IN DRIVE OR INTERMEDIATE RANGE

1 Governor assembly
   a. Governor springs damaged, mispositioned or tilted

CONVERTER STATOR FREEWHEELS BOTH DIRECTIONS

If the stator roller clutch becomes ineffective, the stator assembly freewheels at all times in both directions. With this condition, the vehicle will accelerate poorly. At speeds above 30–35 m.p.h. the vehicle may respond normally.

CONVERTER STATOR LOCKED UP

If the stator assembly remains locked up at all times, engine and vehicle speed will be limited or restricted at high speed.

Performance when accelerating from a standstill will be normal. Engine overheating may be noted, and visual inspection of the converter may reveal an overheated blue cover.

On Car Serviceable Components

TV CABLE

INSPECTION

To check T.V. cable for freeness, *cap liber* pull out on the upper end of the cable. The cable should travel a short distance with light spring resistance. This light resistance is due to the small coiled return spring on the T.V. lever and bracket that returns the lever to the zero T.V. or closed throttle position. Pulling the cable farther out moves the lever to contact the T.V. plunger which compresses the T.V. spring which has more resistance. By releasing the upper end of the T.V. cable, it should return to the zero T.V. position. This checks the cable in the housing, the T.V. lever and bracket, and the T.V. plunger in its bushing for freeness.

ADJUSTMENT

1 Install line pressure gauge.
2 Check line pressure with selector lever in PARK and engine at 1000 rpm.
3 Check line pressure in NEUTRAL at 1000 rpm. Pressure should be the same to no more than 5 psi higher than in PARK.
4 Increase engine rpm to 1200, and line pressure should increase.

Transaxle Disassembly

EXTERNAL COMPONENTS

1 Remove speedometer driven gear and sleeve assembly.
2 Remove governor cover and "O" ring.
3 Remove governor cover-to-speedometer drive gear thrust washer and speedometer drive gear assembly.
4 Remove oil pan and oil strainer.
5 Remove reverse oil pipe retaining brackets and intermediate servo cover.
6 Remove intermediate servo assembly.
7 Check for correct intermediate band apply pin as follows:
 a. Install intermediate band apply pin gauge, Tool J-28535
 b. Remove the band apply pin from the intermediate servo assembly
 c. Install Tool J-28535 on to band apply pin
 d. Install Tool J-28535 and band apply pin into gauge
 e. Apply 100 in./lbs. torque to compress the band
 f. If any part of the white line appears in the window, the pin is the correct length. If the white line cannot be seen, change the band apply pin and recheck
8 Remove 3rd accumulator check valve.

NOTE: Early model transaxles were built with a cup plug to retain the 3rd accumulator check valve. To remove this cup plug, use a #6 Easyout with ¼ inch cut from it.

9 Remove reverse oil pipe and seal backup ring.
10 Using a #4 Easyout shortened by ¾ inch, remove lo-reverse cup plug seal. Remove lo-reverse cup plug.
11 Remove dipstick stop and parking lock bracket.

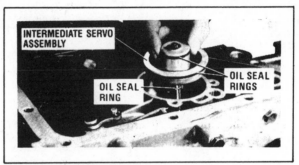

Intermediate servo assembly removal

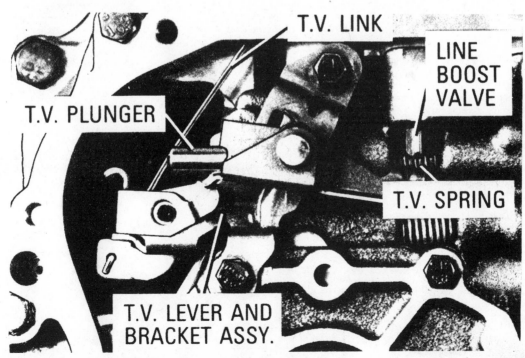

Throttle valve cable mechanism and component identification

12 Turn the final drive unit until the "C" ring is visible. Position the "C" ring so the open side is facing the access window. Tool J-28583 will push the "C" ring partially off the output shaft. Carefully turn the output shaft so the "C" ring is up, and remove the "C" ring.

13 Remove the output shaft.

VALVE BODY, CASE COVER & SPROCKET-LINK ASSEMBLY

1 Remove valve body cover.
2 Remove throttle lever and bracket assembly with T.V. cable link.
3 Remove pump cover screws except one screw shown in illustration. Loosen this screw.
4 Remove control valve and pump assembly.
5 Remove the #1 check ball from the spacer plate.
6 Remove the pump shaft.
7 Remove the spacer plate and five check balls located in the case cover.
8 Check input shaft to case cover end play for correct selective snap ring as follows:
  a. Install adaptor J-26958-10 in right-hand axle end
  b. Install J-26958, output shaft loading tool and bracket J-26958-11
  c. Adjust loading tool by turning the handle in until knob bottoms
  d. Install J-28544 in the input shaft bore, and tighten by turning handle
  e. Install J-25025-A dial indicator post and a dial indicator
  f. End play should be .004 to .033 inch. The selective snap ring which controls end play is located on the input shaft
9 Disconnect the manual valve rod from the manual valve.
10 Remove two screws (shown in illustration) from the converter housing side.
11 Remove remaining case cover screws.
12 Remove case cover, and lay case cover on bench with the 1-2 accumulator side up. The 1-2 accumulator pin may fall out.
13 Remove 1-2 accumulator spring and center case-to-case cover gasket.
14 Remove case cover-to-drive sprocket thrust washer and driven sprocket thrust bearing assembly. The case cover-to-drive sprocket thrust washer may be on the case cover.
15 Remove drive, driven sprockets and link assembly.
16 Remove drive and driven sprocket-to-support thrust washers. These washers may have come off with the sprockets.

INPUT UNIT COMPONENTS

1 Remove the detent lever-to-manual shaft pin using a ⅛ inch pin punch.
2 Remove manual shaft-to-case nail.

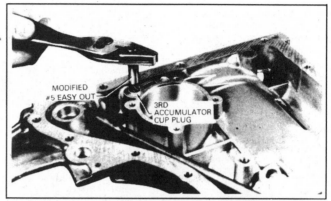

Removing 3rd accumulator cup plug (if so equipped)

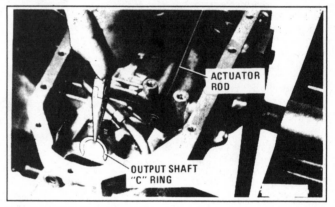

Output shaft "C" ring removal

3 Remove detent lever, manual shaft and parking lock actuator rod.
4 Remove driven sprocket support and driven sprocket support-to-direct clutch housing thrust washer. Thrust washer may come out with driven sprocket support.
5 Remove band anchor hole plug.

6 Remove intermediate band.
7 Holding the input shaft, remove the direct and forward clutch assemblies. Separate the direct and forward clutch assemblies.
8 Remove input internal gear-to-input shaft thrust washer. Washer may be located on end of input shaft.

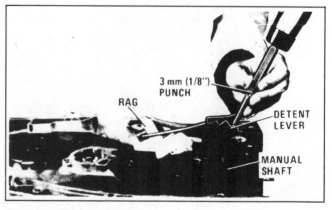

Manual shaft pin removal

GM 125 TRANSAXLE

Loosen, but do not remove, the indicated pump cover and control valve assembly bolt

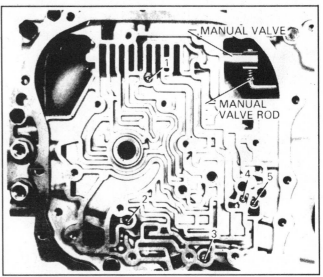

Five check valve locations

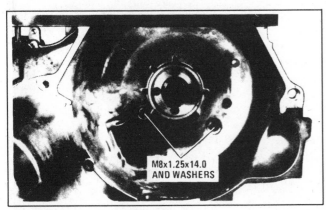

Two case cover bolts on converter side

Drive sprocket, driven sprocket and link assembly

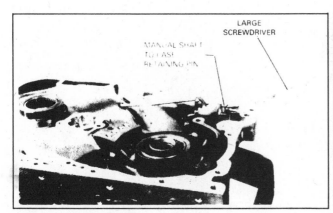

Manual shaft nail removal

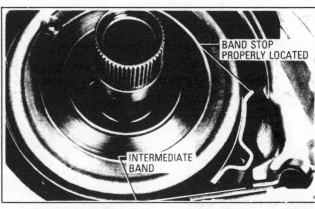

Intermediate band

GM 125 TRANSAXLE

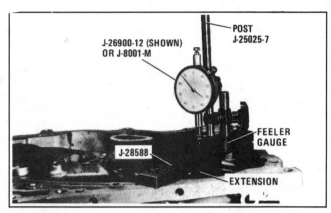

Measuring for correct reaction sun gear selective snap ring.

Measuring for correct lo roller clutch race selective thrust washer

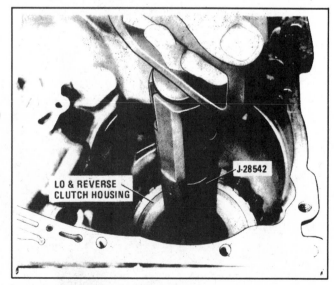

Lo and reverse clutch housing removal

9 Remove input internal gear.
10 Remove input carrier assembly, input carrier-to-input internal gear thrust washer and input carrier-to-input sun gear thrust washer.
11 Remove input sun gear.
12 Remove the input drum.

REACTION UNIT COMPONENTS

1 Measure the reaction sun gear-to-input drum selective snap ring as follows:
  a. Tool J-26958 must be in the loaded position
  b. Position J-28588 as shown in illustration using two case cover bolts. Locate the extension between the open ends of selective snap ring
  c. Press the reaction sun gear down to make sure it is seated
  d. Install Tool 25025-A and a dial indicator
  e. Position thickness gauge under extension shoulder and zero the dial indicator
  f. Rotate the selective snap ring under the extension and swing the thickness gauge away from the extension. Check the amount of dial indicator needle movement making sure the reading is between $+0.013$ to -0.0005 inch. The selective snap ring which controls this end play is located on the reaction sun gear shaft
2 Measure the lo-reverse clutch housing-to-lo roller clutch race selective thrust washer as follows:
  a. Zero dial indicator and leave J-26958 in the loaded position
  b. Place a screwdriver through the parking pawl case opening next to parking pawl and lift the reaction internal gear to check lo-reverse clutch selective end play
  c. End play should be 0.003–0.046 inch. The selective washer which controls this end play is located between the lo and reverse clutch housing and the lo roller clutch assembly
3 Remove reaction sun gear.
4 Remove lo and reverse clutch housing-to-case snap ring. The snap ring is 0.092 inch thick.
5 Using Tool J-28542 and adaptor, remove the lo-reverse clutch housing.
6 Remove lo-reverse clutch housing-to-case spacer ring. The spacer ring is 0.042 inch thick.
7 Grasp the final drive sun gear shaft and pull out the reaction gear set.
8 Roller clutch and reaction carrier assembly:
  a. Remove roller clutch and reaction carrier assembly from final drive sun gear shaft
  b. Remove four-tanged reaction carrier-to-reaction internal gear thrust washer off the end of the reaction carrier, or inside the reaction internal gear

9 Remove lo and reverse clutch plates off final drive sun gear shaft.

10 Reaction internal gear:
   a. Remove reaction internal gear-to-reaction sun gear thrust bearing assembly off reaction gear
   b. Remove reaction internal gear off final drive sun gear shaft

FINAL DRIVE COMPONENTS

1 Check final drive-to-case end play to determine the proper selection as follows:
   a. Remove load from J-26958
   b. Remove J-26958 and J-26958-11, output shaft loading tool. Leave J-26958-10, axle adaptor in place
   c. Install J-25025-A into one of the engine mount bolt holes
   d. Install dial indicator; positioning the dial indicator stem on adaptor J-26958-10
   e. Zero the dial indicator while pushing down on the adaptor J-26958-10
   f. Place a large screwdriver in the governor bore and lift on the governor drive gear and measure the final drive-to-case end play. End play should be 0.005–0.032 inch
   g. The selective washer which controls this end play is located between the differential carrier and the differential carrier case thrust bearing assembly

2 Remove final drive internal gear spacer-to-case snap ring. This snap ring is 0.092 inch thick.

3 Remove final drive internal gear spacer.

4 Remove governor if it has not otherwise been removed.

5 Remove final drive unit by installing Tool J-28545 into final drive assembly. Pull the unit out.

6 Remove final drive differential-to-case thrust selective washer and differential carrier-to-case thrust roller bearing assembly from the final drive assembly.

Transaxle Assembly & Component Inspection

CASE ASSEMBLY

1 Inspect case assembly for damage, cracks, porosity or interconnected oil passages.

2 Inspect exhaust vents to make sure they are open.

3 Inspect case lugs, intermediate servo bore and snap ring grooves for damage.

4 Inspect for damaged or stripped bolt holes.

5 Inspect case bushing for damage or scoring.

6 Inspect converter oil seal.

7 Inspect drive sprocket support roller bearing assembly for damage. If replacement is necessary, do the following:
   a. Using Tool J-26941 and a slide hammer, remove roller bearing assembly
   b. Inspect drive sprocket support roller bearing bore for damage
   c. Use Tool J-28677 and install a new roller bearing assembly. *Bearing identification must be installed up*
   d. If a new roller bearing is installed, the drive sprocket race must be checked for damage or wear. If necessary to replace, the drive sprocket must be replaced

8 Inspect drive sprocket support for damaged journals or splines. If removal is necessary:
   a. Remove converter oil seal
   b. Place hand under sprocket support, and remove case bolts using Tool J-25359-5
   c. Remove drive sprocket support
   d. Install drive sprocket support
   e. Install a new converter oil seal

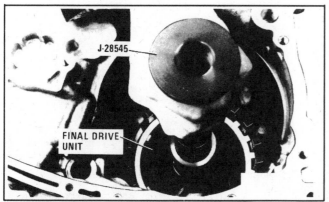

Use pulling tool J-28545 to remove final drive unit

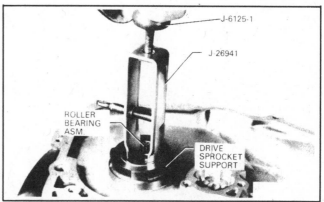

Removing drive sprocket support roller bearing

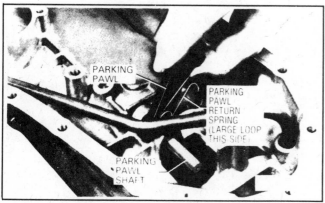

Parking pawl placement

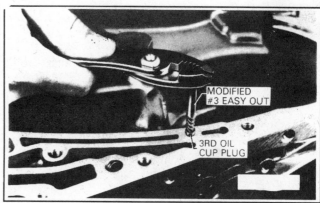

Removing 3rd oil cup plug

GM 125 TRANSAXLE

9 Inspect parking pawl shaft cup plug for damage. If replacement is necessary, remove and install using a ⅜ inch pin punch.

10 If parking pawl needs replacement, do as follows:
  a. Remove parking pawl shaft retainer
  b. Remove parking pawl shaft, parking pawl and parking return spring

11 Install parking pawl as follows:
  a. Place the return spring on the parking pawl and place in case
  b. Install parking pawl shaft
  c. Install parking shaft retainer

12 Inspect governor pipe for damage, cracks or possible leak points. If necessary to replace, do as follows:
  a. Remove governor pipe retaining clamp and screw
  b. Remove the right-hand axle end first. Great effort may be needed to break the pipe loose
  c. Install governor pipe using a liquid thread lock
  d. Install governor pipe retainer strap and screw

13 If necessary to remove the 3rd oil cup plug, do the following:

  a. Using a #3 Easyout cut off ½ inch, remove the 3rd oil cup plug
  b. Install the 3rd oil cup plug using a ¼ inch rod until it seats in the bore

14 Inspect manual shaft oil seal for damage. If replacement is necessary:
  a. Remove manual shaft oil seal
  b. Install a new manual shaft oil seal using a 9/16 inch socket

15 Inspect axle oil seal for damage.

DIFFERENTIAL AND FINAL DRIVE

1 Remove final drive internal gear.
2 Remove final drive internal gear thrust bearing. Thrust bearing may be located on final drive sun gear.
3 Remove final drive sun gear.
4 Remove final drive sun gear-to-differential carrier thrust bearing. Thrust bearing may be located on the carrier.
5 Inspect governor drive gear.
6 If governor drive gear needs replacement, proceed as follows:
  a. Using J-8433 or a similar puller and a thick flat washer, remove the governor drive gear
  b. Install governor drive gear. Using a soft hammer, drive the gear until

it is seated on the carrier

7 Inspect differential side gears and pinions for damage or excessive wear.

8 If replacement of differential side gears or pinions is required:
  a. Remove differential shaft retaining pin using a 3/16 inch drift. The pin can be removed only in the direction shown in the illustration
  b. Remove the differential shaft
  c. Remove the differential pinion gears and thrust washers. This is done by rotating the pinion gears until they are in the differential carrier window. When in this position, they can be removed. The pinion thrust washers should stay with the pinions; if not, remove the dished pinion thrust washers
  d. Remove the differential side gears. These gears are removed by sliding out. Make sure the thrust washers are removed
  e. Inspect the side gear thrust washers and the pinion thrust washers for scoring or other damage
  f. Install side gear thrust washer on the side gears and install into the differential·

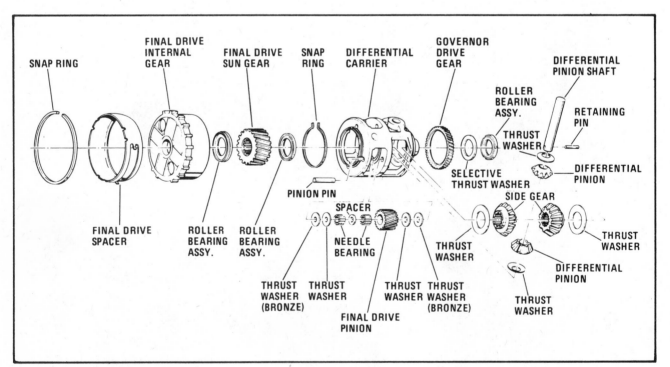

Differential and final drive components

g. Install dished thrust washers on the pinion gears. Retain with petroleum jelly
h. Install the pinion gears in each window
i. Align the pinion gears by sliding the differential pinion shaft through both pinion gears. Remove differential pinion shaft
j. Rotate the pinions into place by pushing on the pinions. Both pinions will rotate into place. Align the pinion differential shaft holes with differential pinions shaft hole in the center
k. Install the differential pinion shaft
l. Install the differential shaft retaining pin
9 Check final drive sun gear-to-differential roller bearing thrust washer for damage.
10 Inspect the final drive pinions for excessive wear or damage.
11 Check pinion end play with a thickness gauge. End play should be 0.009–0.025 inch.
12 If final drive pinions need replacement, do as follows:
a. Remove pinion pin snap ring
b. Remove pinion pins
c. Carefully remove pinions and thrust washers together to prevent dropping needle roller bearings
d. Remove both sets of needle roller bearings and the needle bearing spacer from each pinion
e. Inspect the needle roller bearings and spacer for damage
f. Inspect pinion for damage or wear
g. Install a steel pinion thrust washer on pinion pin and retain with petroleum jelly
h. Install needle roller bearings on the bottom half of pinion pins. Retain with petroleum jelly
i. Install the needle bearing spacer
j. Install the needle roller bearings in the top half of the pinion pin, and retain with petroleum jelly
k. Push the needle roller bearings, assembled on the pinion pin, into the pinion
l. Install a steel pinion thrust washer on the side of the pinion that does not have one
m. Install bronze pinion thrust washers on each side of the pinion on top of the steel thrust washers, and retain with petroleum jelly
n. Remove the pinion pin from the pinion assembly
o. Install the pinion assembly into the final drive carrier
p. Install the pinion pin, stepped end last. Align the step so the step is to the outside
q. Install the pinion pin snap ring after all four pinion pins are installed
13 Inspect final drive sun gear-to-differential carrier thrust bearing.
14 Install final drive sun gear-to-differen-

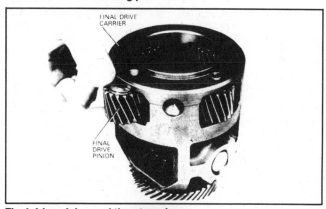
Differential shaft retaining pin removal

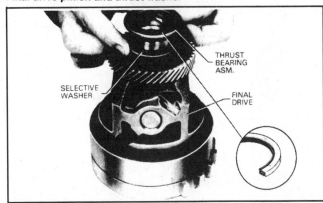
Final drive pinion and thrust washer

Final drive thrust washer and roller thrust bearing assembly

tial carrier thrust bearing outer race against carrier.
15 Inspect final drive sun gear splines and gear teeth for damage or excessive wear.
16 Install final drive sun gear with the step side up.
17 Inspect final drive internal gear for cracks, damage and worn gear teeth.
18 Inspect final drive internal gear bushing.
19 Inspect final drive sun gear-to-final drive internal gear thrust bearing.
20 Install final drive sun gear to final drive internal gear thrust bearing; cupped race side on to final drive internal gear.
21 Install final drive internal gear on final drive carrier.
22 Inspect case-to-differential selective thrust washer for pitted, rough or scored surfaces.
23 Install differential selective thrust washer on differential carrier and retain with petroleum jelly.
24 Inspect case-to-differential roller bearing thrust washer for damage.
25 Install case-to-differential roller bearing thrust washer with the inner race against the differential selective washer. Retain with petroleum jelly.
26 Install differential and final drive using Tool J-28545.
27 Inspect the final drive internal gear

GM 125 TRANSAXLE

spacer for damage.

28 Install final drive internal gear spacer with the cupped side against the final drive internal gear.

NOTE: The parking pawl opening in the spacer must be aligned with the parking pawl opening in the case. Check to see that the parking pawl passes through the spacer freely.

29 Inspect spacer-to-case snap ring for damage.

30 Install the spacer-to-case snap ring with the ring gap away from the parking pawl opening. The case snap ring is 0.092 inch thick.

31 Check final drive-to-case end play as follows:
   a. Install J-26958-10 adaptor, and install J-25025-7 into one of the engine mount bolt holes
   b. Install dial indicator; positioning the stem on adaptor J-26958-10
   c. Zero the dial indicator while pushing down on the adaptor
   d. Place a large screwdriver in the governor bore and lift on the governor drive gear and take the final drive-to-case end play. Measurement should be 0.004–0.032 inch
   e. Selective washer which controls this end play is located between the differential carrier and the differential carrier-to-case thrust bearing assembly
   f. Recheck if final drive-to-case selective thrust washer is changed

32 Install loading tool J-26958 and bracket J-26958-11 and turn the loading screw in until knob bottoms.

REACTION GEAR SET

1 Inspect final drive sun gear shaft and journals for damage or excessive wear. Inspect splines for damage.
2 Inspect the reaction internal gear splines, gear teeth and bearing surface for wear, cracks, or damage.
3 Install reaction internal gear on the final drive sun gear shaft with the gear teeth up.

4 Inspect the reaction internal gear-to-reaction sun gear roller thrust bearing for damage.
5 Install reaction internal gear-to-reaction sun gear roller thrust bearing with inner race against reaction internal gear.
6 Remove reverse clutch housing-to-Lo race selective washer. Inspect for damage.
7 Remove Lo roller clutch race. Inspect race and spline for scoring, cracks or wear.
8 Remove Lo roller clutch assembly. Inspect roller, springs and cage for damage or wear.
9 Remove reaction carrier-to-Lo race thrust washer. Inspect for damage.
10 Inspect the four-tanged reaction carrier-to-internal gear thrust washer for scoring or distorted tangs.
11 Inspect Lo roller clutch cam for damage.
12 Inspect reaction carrier bushing.
13 Inspect pinions for damage, rough bearings or tilt.
14 Check pinion pins for tightness. Pinion pins must not rotate.
15 Check pinion end play with a thickness gauge. End play should be 0.009″–0.027″ inch.
16 Install reaction carrier-to-Lo roller clutch race thrust washer.
17 Install any rollers that may have come out of the Lo roller clutch cage by compressing the spring with a finger and inserting the roller.
18 Install the Lo roller clutch assembly.
19 Install Lo roller clutch race splined side up. Rotate the race clockwise until it drops into position.
20 Install the four-tanged reaction carrier to reaction internal gear thrust washer. Retain the petroleum jelly.
21 Install reaction carrier and Lo roller clutch assembly into the reaction internal gear.
22 Install Lo and reverse clutch housing-to-Lo roller clutch race selective washer.
23 Install reaction gear set into the transaxle.

LO AND REVERSE CLUTCH

1 Inspect backing plate for cracks or other damage.
2 Install Lo and reverse clutch backing plate with the stepped side down.
3 Inspect the Lo and reverse clutch composition and steel clutch plates for wear or burning.
4 Oil the composition plates. Install clutch plates starting with one composition plate, then a steel clutch plate alternating until five flat steel and five composition plates are installed.
5 Install Lo and reverse clutch housing-to-case spacer ring. The case spacer ring is 0.0042 inch thick.
6 Compress Lo and reverse clutch spring retainer, remove snap ring and retainer.
7 Remove waved spring and inspect for damage.
8 Remove Lo and reverse clutch piston.
9 Remove inner and outer piston seals. Inspect for nicked, cut or hard seals.
10 Remove clutch apply ring. Measure the width of the apply ring. The apply ring should be 0.61 inch.
11 Inspect the Lo and reverse clutch housing for damage or plugged feed hole.
12 Inspect Lo and reverse clutch housing bushing for damage, cracks or scoring.
13 Inspect Lo and reverse clutch housing splines and snap ring groove for damage or burrs. Remove any burrs.
14 Inspect Lo and reverse clutch piston and apply ring for distortion, cracks or damage.
15 Install the apply ring on the Lo and reverse clutch piston.
16 Install inner and outer piston seals with the lips facing away from apply ring side.
17 Oil piston seals, and install clutch piston into Lo and reverse clutch housing. Use seal installation tool on the inner seal first. Then work tool around the outer seal until the piston is seated.
18 Install waved spring.
19 Install retainer, cupped side down.

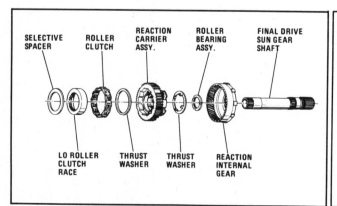

Reaction gear set components

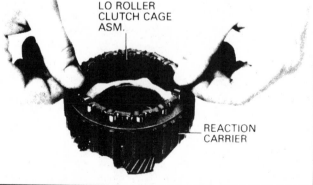

Lo roller clutch assembly

GM 125 TRANSAXLE

20 Compress retainer and install snap ring.

21 Using J-28542, install Lo and reverse clutch housing. Line up the Lo and reverse clutch feed hole with the feed hole in case. If the Lo and reverse clutch housing does not seat past the snap ring groove, proceed as follows:
a. Remove tool J-28542
b. Install the reaction sun gear
c. Rotate the reaction sun gear while pushing down on the Lo and reverse clutch housing until the Lo and reverse clutch housing drops below the snap ring groove. It may be necessary to remove the load from tool J-26958 to allow the housing to seat

22 Install Lo and reverse clutch housing-to-case snap ring. The case snap ring is 0.092 inch thick.

CHECKING REACTION SUN GEAR AND LO RACE END PLAY

1 Inspect reaction sun gear for cracks, splits, damaged spline, worn gear or journals and plugged lubrication holes.

2 Install reaction sun gear and selective snap ring assembly.

3 Take measurement of the reaction sun gear to input drum selective snap ring as follows:
a. Tool J-26958 must be in the loaded position
b. Position J-28588 as shown using two case cover bolts. Locate the extension between the open ends of selective snap ring
c. Press the reaction sun gear down to make sure it is seated
d. Install dial indicator
e. Position feeler gage under extension shoulder and set dial indicator to zero
f. Rotate the selective washer under the extension and swing the feeler gage away from the extension dial. Check the amount of dial indicator needle movement making sure the indicator reading is between +0.-013 to −0.005 inch from the zero reference point

NOTE: The selective snap ring which controls this end play is located on the reaction sun gear shaft.

4 Take measurement of the Lo and reverse clutch housing-to-Lo roller clutch race selective thrust washer as follows:
a. Leave dial indicator in place and zero dial indicator. Leave J-26958 in the loaded position
b. Place a screwdriver through the parking pawl case opening next to parking pawl and lift the reaction internal gear to check Lo and reverse clutch selective end play
c. End play should be 0.003–0.046 inch. The selective washer which controls this end play is located between the Lo and reverse clutch housing and Lo roller clutch assembly

INPUT UNIT COMPONENTS

1 Inspect input drum for damage. Angle of roll pins is normal, do not attempt to straighten roll pins.

2 Install input drum on the reaction sun gear.

3 Inspect the input sun gear splines and gear for damage or wear.

4 Inspect for pitting, scoring or damage.

5 Install input sun gear.

6 Inspect input carrier for damage.

7 Check pinion pins for tightness. Pinion pins must not rotate.

8 Inspect pinions for damage, rough operation or tilt.

9 Check pinion end play with a thickness gauge. End play should be 0.009–0.027 inch.

10 Inspect input sun gear-to-input internal gear four-tanged thrust washer for damage, scoring of distorted tangs.

11 Inspect the input carrier-to-input sun gear four-tanged thrust washer for damage, scoring or distorted tangs.

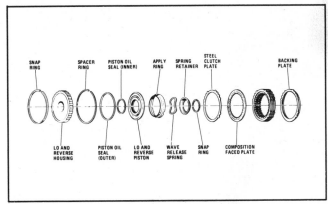

Lo/reverse clutch components

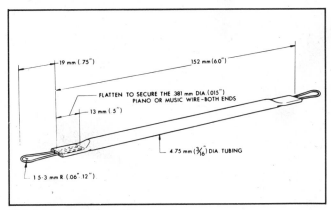

Clutch piston seal installation tool

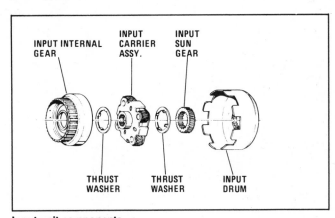

Input unit components

GM 125 TRANSAXLE

12 Install the input carrier-to-input sun gear four-tanged thrust washer; this is the smaller tanged thrust washer on the pinion side of the input carrier. Retain with petroleum jelly.

13 Install the input carrier-to-input internal gear four-tanged thrust washer. This is the larger diameter thrust washer on the input carrier. Retain with petroleum jelly.

14 Install input carrier.

15 Inspect input internal gear for damage or wear of gear teeth. Inspect input internal gear hub splines for damage.

16 Inspect input internal gear bushing for scoring or wear.

17 Install input internal gear.

FORWARD CLUTCH DISASSEMBLY, INSPECTION AND REASSEMBLY

1 Place forward clutch input shaft through hole in work bench.

2 Remove clutch pack snap ring.

3 Remove backing plate, composition plates and steel plates.

4 Using tools J-25018 A, J-23327 and J-23456, compress the spring pack, then remove the snap ring.

5 Remove tools J-25018 A and J-23327.

6 Remove the retainer and spring assembly.

7 Remove the forward clutch piston.

8 Remove inner and outer piston seals and inspect for nicks.

9 Inspect forward clutch housing for cracks, broken welds, damaged check ball capsule and for free operation of check ball. If the check ball capsule is damaged, replace as follows:

a. Set the forward clutch housing on the bench with the clutch side down

b. Using a ⅜ inch diameter drift, drive out the check ball capsule

c. To install new capsule, place the input shaft through the hole in the bench

10 Inspect input shaft splines and journals for damage.

11 Inspect input shaft sleeve for damage. The sleeve must not turn. The slot in the sleeve must be aligned with the hole in the input shaft.

12 Inspect input shaft teflon seal rings for damage. Do not remove unless replacing. If new seal rings are installed, make sure the cut ends are installed in the same way they are cut.

13 Inspect forward clutch piston and apply ring for damage or cracks. Measure the apply ring width; it should be 0.470 inch.

14 Install inner and outer seals on piston with lips facing away from the apply ring side.

NOTE: Use care when installing forward clutch piston past the large forward clutch snap ring groove. This groove could cut the seal. Oiling the seal will aid installation.

15 Install clutch apply ring on piston.

16 Install piston using tool around inner seal then outer seal.

17 Inspect spring guide.

18 Install spring guide.

19 Inspect forward clutch spring and retainer assembly for collapsed springs or bent retainer.

20 Install retainer and spring assembly.

21 Using J-25018 A and J-233274, compress the retainer spring assembly past the snap ring groove.

22 Install snap ring.

23 Remove J-25018 A and J-23327.

24 Inspect backing plate for damage or cracks.

25 Inspect composition and steel plates for wear or burning.

26 Oil the composition plates.

27 Install the clutch plates, starting with a waved steel plate, then a composition plate, alternating until four composition and three flat steel plates are installed.

28 Install backing plate, flat side up.

29 Install snap ring.

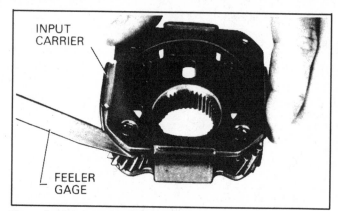

Measuring input carrier pinion end play

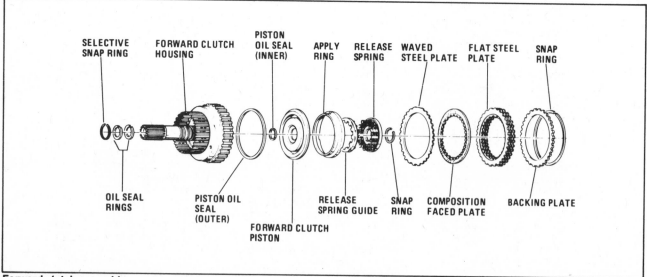

Forward clutch assembly components

GM 125 TRANSAXLE

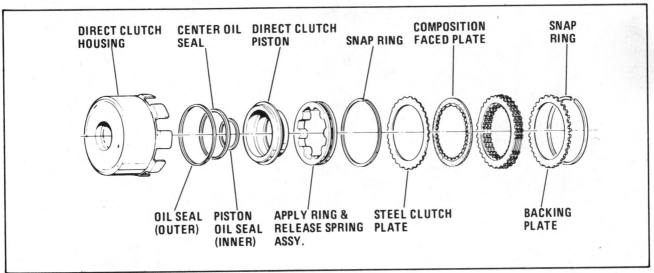

DIRECT CLUTCH HOUSING   CENTER OIL SEAL   DIRECT CLUTCH PISTON   SNAP RING   COMPOSITION FACED PLATE   SNAP RING

OIL SEAL (OUTER)   PISTON OIL SEAL (INNER)   APPLY RING & RELEASE SPRING ASSY.   STEEL CLUTCH PLATE   BACKING PLATE

Direct clutch assembly components

NOTE: Make sure the composition plates turn freely.

30 If removed, install new input shaft seal rings. Make sure cut ends are installed in the same way as they are cut and that the rings are seated in their groove.

31 Inspect input shaft-to-driven sprocket selective snap ring for damage or wear. Do not remove unless replacing.

DIRECT CLUTCH DISASSEMBLY, INSPECTION AND REASSEMBLY

1 Remove snap ring.
2 Remove backing plate, composition and steel plates.
3 Remove snap ring holding apply ring and release spring assembly.
4 Remove apply ring and release spring assembly.
5 Remove direct clutch piston.
6 Remove inner and outer piston seals.
7 Remove center seal from housing.
8 Inspect direct clutch housing for excessive wear. Inspect check ball capsule for free operation or any damage to check ball seat. If check ball replacement is necessary, do as follows:
   a. Set the forward clutch housing on the bench with the forward clutch side down
   b. Using a ⅜ inch diameter drift, drive out the check ball capsule.
   c. To install new capsule, place the input shaft through the hole in the bench
   d. Insert a new capsule from the forward clutch side
   e. Using the ⅜ inch diameter drift, set the check ball capsule
9 Inspect direct clutch housing bushings for damage, cracks or scoring.
10 Inspect piston for damage or cracks.
11 Inspect seals for burrs, nicks or brittleness.

12 Install inner and outer piston seals with the lips facing away from the clutch apply ring side.
13 Install center seal on housing with the lip facing up.

NOTE: Use care when installing direct clutch piston past snap ring grooves. Grooves could cut outer seal on the piston.

14 Oil seals and install piston. To make piston easier to install, use seal installation tool. Rotate piston once it is in place to seat seals.
15 Inspect apply ring, retainer and release spring assembly for damage, collapsed springs and proper apply ring height.
16 Install apply ring and release spring assembly.

17 Install snap ring.
18 Inspect backing plate for cracks or damage.
19 Inspect composition and steel plates for wear or burning.
20 Oil and install the clutch plates, starting with a steel plate, then a composition plate, alternating until four steel and four composition plates are installed.
21 Install backing plate; flat side up.
22 Install snap ring.

FORWARD AND DIRECT CLUTCH

1 Place the forward clutch assembly on the bench with input shaft up.
2 Install direct clutch assembly over the input shaft onto the forward clutch housing. When the forward clutch is seated, it will be about 1 7/32 inch

DIRECT CLUTCH HOUSING

21 MM

FORWARD CLUTCH HOUSING

Measuring for correct seating of forward clutch

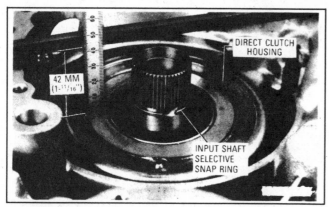

Measuring for correct installation of direct and forward clutch assemblies

Driven sprocket support installation

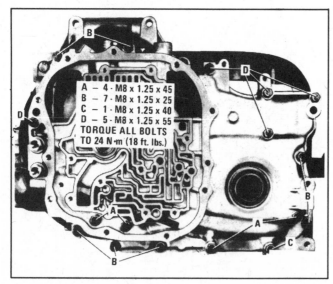

Case cover bolt location

from the tang end of the direct clutch housing to the end of the forward clutch drum.

3 Inspect input shaft-to-input internal gear thrust washer for damage.

4 Install the input shaft-to-input internal gear thrust washer rounded side against the input shaft, stepped side facing out. Retain with petroleum jelly.

5 Install direct and forward assemblies. The unit may install easier by rotating unit without pushing down.

6 When correctly installed the case face-to-direct clutch housing measurement will be approximately 1 1/16 inch.

INTERMEDIATE BAND ASSEMBLY

1 Inspect band for burns, flaking or damage.

2 Install intermediate band, locating the eye end into the case and aligning the lugged end with the apply pin bore.

NOTE: If the lug is not properly located, the band will be inoperative.

3 Install band anchor hole plug.

DRIVEN SPROCKET SUPPORT

1 Inspect driven sprocket support for cracks, burrs, or damage.

2 Inspect driven sprocket support sleeve for damage. It must be tight in its bore and align with holes in the support.

3 Inspect driven sprocket support bushing for damage or excessive wear.

4 Inspect driven sprocket bearings assembly for damage. If removal is necessary, do as follows:
   a. Using J-26941 and a slide hammer, remove the bearing
   b. Using J-28677 and J-8092, install a new bearing assembly

NOTE: The manufacturing identification must be installed facing up.

5 Inspect seal rings for nicks and cuts.

NOTE: If the driven sprocket bearing assembly is replaced, the race on the driven sprocket must be checked. If necessary to replace the race, the driven sprocket assembly must be replaced. Do not remove seal rings unless replacing.

6 If removed, install driven sprocket support seal rings. Make sure cut ends are assembled in the same way as cut.

7 Install direct clutch drum-to-support thrust washer. Retain with petroleum jelly.

8 Install driven sprocket support.

MANUAL SHAFT

1 Inspect manual valve rod and retainer for damage.

2 Inspect detent lever for damage.

3 Inspect manual shaft and retainer for damaged threads and the flats for raised edges. File down any raised edges.

GM 125 TRANSAXLE

4 Inspect parking lock actuator rod for damage, or broken retainer lugs.
5 Inspect parking lock actuator spring for damage.
6 If removed, install the parking lock actuator rod into the manual shaft lever.

NOTE: The manual shaft and detent lever assembly are made as a matched set. Any replacement must be with a matched set.

7 Install manual shaft and parking lock actuator rod into case and push through the driven sprocket support.
8 If removed, install manual valve rod into manual detent lever.
9 Install detent lever on the manual shaft, hub side away from the driven sprocket support. Push the manual shaft into place.
10 Install detent lever and hub assembly to manual shaft retaining pin.
11 Install manual shaft to case nail.

DRIVE LINK ASSEMBLY
1 Inspect drive sprocket teeth for nicks, burrs, scoring or excessive wear.
2 Inspect the turbine shaft seal ring grooves and seal ring for damage. Do not remove the seal ring unless replacing.
3 If necessary to replace either the turbine shaft or the drive sprocket, do as follows:
   a. Remove snap ring
   b. Separate the turbine shaft and drive sprocket
   c. Install drive sprocket, flat side up, onto the turbine shaft
   d. Install the snap ring
   e. If removed, install turbine shaft seal rings, making sure cut ends are assembled in the same relationship as cut and the rings are seated in their groove
4 Inspect link assembly for damage or loose links.
5 Inspect driven sprocket teeth for nicks, burrs, scoring or excessive wear.
6 Inspect internal splines for nicks, burrs, excessive wear.
7 Inspect driven gear thrust bearing race. If the race is to be replaced, the drive sprocket must be replaced. Also, the drive support bearing assembly must be replaced.
8 Install drive and driven sprocket thrust washers. Retain with petroleum jelly.

NOTE: Sprockets and link assemblies are different in some models. Refer to parts manual for correct usage.

9 Install drive and driven sprockets into link assembly; colored guide link, which has numerals, facing case cover.
10 Install drive link assembly into transaxle.

11 Install driven sprocket roller bearing thrust washer with outer race against sprocket.

CASE COVER
1 Inspect case cover for damage, cracks, porosity or interconnected passages.
2 Repair any damaged threads in the case cover with thread repair kit.
3 Inspect vent assembly for damage.
4 Inspect cooler line connectors for damage. If replacement is necessary:
   a. Remove connector(s)
   b. Coat the connector(s) with a thread sealer
   c. Install the connectors into case cover
5 Inspect manual detent spring and roller assembly for damage. If removal is necessary, do as follows:
   a. Remove detent spring retaining screw
   b. Remove detent spring and roller assembly
   c. Install detent spring and roller assembly
   d. Install detent spring retaining screw
6 Inspect axle oil seal for damage. If guard is damaged, the seal must be

replaced. If replacement is necessary:
   a. Remove axle oil seal
   b. Use J-26938 and install new axle oil seal
7 Inspect the case cover sleeve. Make sure the hole in the sleeve aligns with the case cover passages that intersect the case cover (pump shaft) bore.
8 Inspect manual valve for damage. If replacement is necessary:
   a. Remove manual valve cup plug
   b. Remove manual valve
   c. Install manual valve, small diameter first
   d. Coat cup plug with good grade commercial sealant and install until it is just below the surface
9 Remove 1-2 accumulator piston and 1-2 accumulator pin.
10 Inspect 1-2 accumulator piston for damage.
11 Inspect 1-2 accumulator piston oil seal ring for damage and free fit in the ring groove. Do not remove the seal ring unless replacing. If the seal ring is replaced, install the ends in the same relationship as they are cut.
12 Install 1-2 accumulator piston; flat side down.
13 Inspect the thermostatic valve. If repair is necessary do as follows:

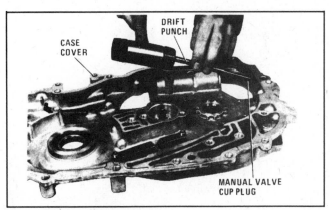

Manual valve cup plug removal

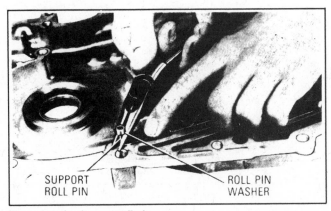

Thermostatic element roll pin

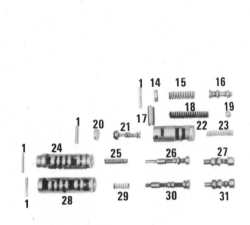

1. RETAINING COILED PIN
2. LINE BOOST VALVE PLUG
3. LINE BOOST VALVE
4. THROTTLE VALVE
5. THROTTLE VALVE SPRING
6. THROTTLE VALVE PLUNGER
7. T.V. PLUNGER BUSHING
8. PRESSURE REGULATOR VALVE
9. PRESSURE REGULATOR SPRING
10. REVERSE BOOST VALVE
11. REVERSE BOOST VALVE BUSHING
12. T.V. BOOST VALVE
13. T.V. BOOST VALVE BUSHING
14. VALVE BORE PLUG
15. SHIFT T.V. SPRING
16. SHIFT T.V. VALVE
17. SPRING RETAINING SLEEVE
18. PRESSURE RELIEF SPRING
19. PRESSURE RELIEF BALL

20. VALVE BORE PLUG
21. 1-2 ACCUMULATOR VALVE
22. 1-2 ACCUMULATOR BUSHING
23. 1-2 ACCUMULATOR SPRING
24. 2-3 THROTTLE VALVE BUSHING
25. 2-3 THROTTLE VALVE SPRING
26. 2-3 THROTTLE VALVE
27. 2-3 SHIFT VALVE
28. 1-2 THROTTLE VALVE BUSHING
29. 1-2 THROTTLE VALVE SPRING
30. 1-2 THROTTLE VALVE
31. 1-2 SHIFT VLAVE
32. 3-2 CONTROL VALVE
33. 3-2 VALVE SPRING
34. LO BLOW OFF BALL
35. LO BLOW OFF SPRING AND PLUG ASSEMBLY
36. LO BLOW OFF VALVE PLUG
37. CHECK BALL #1

Control valve components

a. Remove thermostatic element spring pin washer
b. Remove the thermostatic element
c. Remove thermostatic element roll pin washer
d. Remove the thermostatic element plate
e. Remove the roll pins if necessary
f. If removed, install the roll pins
g. If removed, install capped roll pin; setting height with J-29023
h. Inspect the thermostatic element plate for damage or distortion
i. Install thermostatic element plate
j. Install thermostatic element
k. Install roll pin washer using J-29203 to set the washer height
l. Install roll pin washer. Using J-

29023 set washer height
14 Install driven sprocket tanged thrust washer onto case cover, and retain with petroleum jelly.
15 Install two dowel pins in the case.
16 Install 1-2 accumulator spring in case.
17 Install center case-to-case cover gasket. Retain with petroleum jelly.
18 Install case-to-case cover gasket.
19 Install case cover.
20 Install case cover screws.
21 Install screws and new washers located on the converter housing side.
22 Clip manual valve rod to manual valve. Needle nose pliers may be used to pull manual rod up.
23 Install 1-2 accumulator pin chamfered end first.

24 Check input shaft end play. To check for proper selective washer do as follows:
  a. Install J-28544 in the input shaft bore; tighten by turning handle to secure lifter
  b. Install dial indicator post J-25025 A
  c. Install dial indicator, placing the dial indicator extension onto the end of the end play tool
  d. Press down on the lifting tool, then zero the dial indicator
  e. Pull up on the end play tool and take end play. End play should be .004–.033 inch. The selective snap ring which controls end play is located on the input shaft

CONTROL VALVE PUMPS DISASSEMBLY, INSPECTION AND REASSEMBLY

NOTE: As each valve train is removed, place the individual valve train in the order that it is removed and in a separate location relative to its position in the valve body. None of the valves, bushings or springs are interchangeable; some roll pins are interchangeable. Remove all roll pins by pushing through from the rough cast surface side of the control valve pump assembly, except for the blind hole roll pins.

1 Lay the control valve pump assembly machined face up, with the line boost valve at the top.
2 Check the operation of the line boost valve and if replacement is necessary do as follows:
  a. Grind a taper to one end of a #49 drill
  b. Lightly tap the tapered end into the roll pin
  c. Pull the drill and coil pin out
  d. Remove the line boost valve and plug by pushing the line boost valve out of the top of the valve body
3 Check the operation of the throttle valve by moving the valve against the spring. If it is necessary to remove the valve, do as follows:
  a. Remove the roll pin holding the T.V. plunger bushing
  b. Pull out the T.V. plunger and T.V. plunger bushing
  c. Remove the throttle valve spring
  d. Remove the throttle valve roll pin. Use a #49 drill tapered on one end.
  e. Remove the throttle valve
4 Remove the roll pin from next bore down.
5 Remove T.V. boost bushing and valve assembly.
6 Remove the reverse boost valve.
7 Remove the pressure regulator spring and valve.
8 Remove the roll pin from the next bore down, left side of control valve pump assembly.

NOTE: There is a pressure on this roll pin. When removing the pin, use a rag to prevent the bore plug and spring from being lost.

9 Remove the bore plug, shift T.V. spring and valve.
10 From the next bore down remove the spring retaining sleeve.

NOTE: This spring is under a load.

11 Remove pressure relief spring and check ball.
12 From the next bore down, remove the roll pin.
13 Remove the bore plug.
14 Remove the 1-2 accumulator bushing, valve and spring.
15 From the next bore down, remove roll pin.

16 Remove 2-3 throttle valve bushing and 2-3 throttle valve.
17 Remove the 2-3 shift valve.
18 From the next bore down, remove the roll pin.
19 Remove the 1-2 throttle bushing, valve and spring.
20 Remove the 1-2 shift valve.
21 Remove the spring retaining sleeve from the valve bore in the lower left hand corner.

NOTE: This sleeve is under load. Cover the open end of the bore to prevent loss of spring.

22 Remove the 3-2 control spring and valve.
23 Using a ¼ inch punch, remove the lo blowoff valve plug.
24 Remove the lo blowoff valve spring, spring plug and ball and discard.

NOTE: The lo blowoff assembly must be removed and replaced if the control valve pump body is washed in solvent.

25 If service is required on the pump assembly, do as follows:
  a. Remove roll pin from pump priming spring bore

NOTE: The pin is under spring load. Cover the open end with the bore to prevent loss of parts.

  b. Remove the priming spring cup plug and priming spring
  c. Remove pump cover screw and pump cover

  d. Remove pump slide, rotor, seven vanes and two vane rings
26 Wash control valve pump body, valves, springs, other valve train parts, pump cover, pump slide, pump rotor, vanes and vane rings. Do not put the pump seals into the solvent.
27 Inspect control valve pump body for cracks, damage or scoring of valve bores and pump pocket and pump cover.
28 Inspect pump shaft bearing assembly. If replacement is necessary, do as follows:
  a. Remove bearing with J-28698 and J-7079-2. Drive out toward pump pocket.
  b. Install new bearing assembly using tool J-28698 and J-7079-2. Install from pump pocket side. Install until bearing cup is 0.017–0.005 inch below the pump pocket face.
29 Inspect the valve bushings for cracks or scored bores.
30 Inspect springs for distortion or collapsed coils.
31 Inspect bore plugs for damage.
32 Inspect pump slide for damage, cracks or wear.
33 Inspect pump rotor for damage, cracks or wear.
34 Inspect pump vanes and vane rings for damage, cracks or wear.
35 Turn the valve assembly so the pump pocket side is up.
36 Install the pump slide into the pump pocket.

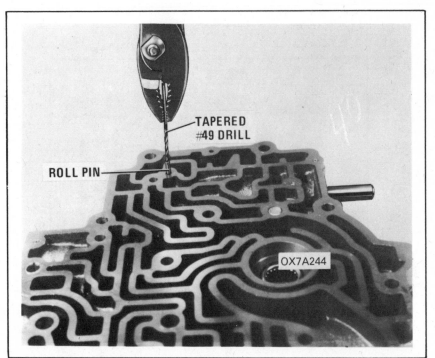

Removing blind hole roll pin

GM 125 TRANSAXLE

37 Install the pump slide seal support and pump slide seal. Retain with petroleum jelly. Make sure the slide seal is located properly.

38 Position slide with the pump slide pivot hole. Install the pump slide pivot pin.

39 Install a vane ring in the pump pocket.

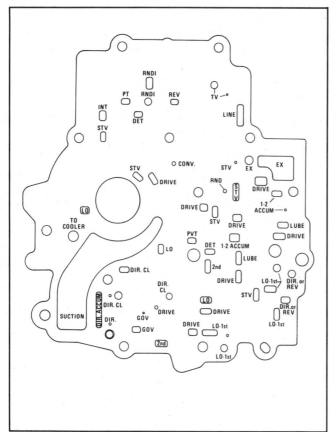

Spacer plate orifice identification

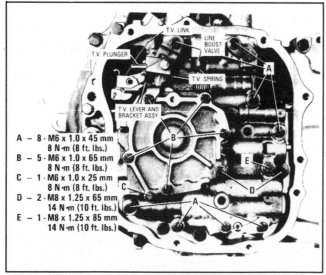

A – 8 - M6 x 1.0 x 45 mm
  8 N·m (8 ft. lbs.)
B – 5 - M6 x 1.0 x 65 mm
  8 N·m (8 ft. lbs.)
C – 1 - M6 x 1.0 x 25 mm
  8 N·m (8 ft. lbs.)
D – 2 - M8 x 1.25 x 65 mm
  14 N·m (10 ft. lbs.)
E – 1 - M8 x 1.25 x 85 mm
  14 N·m (10 ft. lbs.)

Control valve assembly bolt identification

40 Install pump rotor into the pump pocket.

41 Install seven vanes into the pump. Make sure the vane wear pattern is installed against the centering ring. Make sure each vane is seated flush with the rotor.

42 Install the top vane ring.

43 Install the slide "O" ring seal in the pump slide.

44 Install the pump slide to cover oil seal ring.

45 Check the pump cover sleeve for damage.

46 Install the pump cover.

47 Install and torque the pump cover screw.

48 Install pump primary spring and cup plug, flat side out.

49 Compress cup plug and spring and install roll pin.

50 Position control valve pump body with machined face up and the pump pocket on the right hand side.

51 In the lower right hand corner install the 3-2 control valve, 3-2 control valve spring, and the spring retaining sleeve from the machined face. Make sure the sleeve is level with or below the machined surface.

52 In the lower left hand bore install the 1-2 shift valve stem end last. Install 1-2 T.V. spring into the 1-2 T.V. bushing, next install 1-2 T.V. valve and bushing assembly into the 1-2 bore making sure the pin slot is aligned with the pin hole in the control valve pump body. Install roll pin.

53 In the next bore up install the 2-3 shift valve stem end last. Install the 2-3 T.V. spring into the 2-3 T.V. bushing. Next, install the 2-3 T.V. valve, stem end first, into the 2-3 bushing. Install the 2-3 T.V. valve and bushing assembly. Make sure the pin slot is aligned with the pin hole in the control valve pump body. Install roll pin.

54 In the next bore up install the 1-2 accumulator spring. Next install the 1-2 accumulator valve, small end first into the 1-2 accumulator bushing. Next, install the bore plug, flat side first. Install the 1-2 accumulator valve and bushing assembly into control valve pump body, aligning the pin slot in line with the pin hole in the control valve pump body. Push the bore plug in and install the roll pin.

55 In the next bore up install the pressure relief ball, the pressure relief spring and the spring retaining sleeve.

56 In the next bore up install the shift T.V. valve, small end first. Install the shift T.V. spring and bore plug, flat side first. Compress the spring and install the roll spring.

57 Install the pressure regulator, valve stem end last, in the next bore up—right hand side of control valve pump assembly. Next install the pressure regulator spring. Install the reverse boost valve into the reverse boost

GM 125 TRANSAXLE

bushing, stem end first into the large opening. Install the reverse bushing and valve with large end of valve first. Next install the T.V. boost valve, stem end first into the T.V. boost bushing. Install the T.V. boost bushing and valve assembly open end first. Be sure to align the pin slot in the bushing with the pin hole in the body. Install roll pin.

58 In the next bore up, install the throttle valve stem end last. Next install the throttle valve spring. Install the T.V. plunger into the T.V. bushing, small end first. Install the T.V. bushing and plunger assembly, large open end first. Align the T.V. plunger bushing pin slot with the pin hole in the valve body. Install roll pin. Install a roll pin in the throttle valve pin hole.

59 Install the line boost valve stem end first into the bore on the top of the valve body. Install the line boost bore plug ring end first. Install the roll pin. Make sure line boost valve moves freely.

60 Install lo blowoff ball in the lower right hand corner, rough casting side of valve body. Next, assemble the lo blowoff spring and spring plug. Install spring and plug assembly plug end first. Next, install the lo blowoff plug, cupped end first, using a rubber or plastic hammer until it seats in its bore.

CONTROL VALVE PUMP INSTALLATION

1 Install five check balls in the locations shown in the illustration.
2 Install one control valve pump body gasket.
3 Install the spacer plate.
4 Install one control valve pump body gasket.
5 Install #2 check ball.
6 Inspect pump shaft splines and journals for excessive wear.
7 Install pump shaft.
8 Install two 6 mm guide pins.
9 Install the control valve pump assembly.
10 Install control valve pump assembly bolts and torque to specifications.
11 Inspect throttle lever and bracket assembly for damage.
12 Install throttle lever link to T.V. bracket.
13 Install T.V. bracket on valve body.
14 Remove guide pins, install bolts and torque to specifications.
15 Clean and inspect valve body cover for damage.
16 Install new valve body cover gasket.
17 Install valve body cover and retaining screws.

OUTPUT SHAFT INSTALLATION

1 Remove output shaft loading tool bracket and adapter.
2 Install output shaft.
3 Turn the final drive so that "C" ring

groove on the output shaft is visible through the case access window.
4 Install new "C" ring in output shaft groove.
5 Complete installation of new "C" ring with J-28583.

REVERSE PIPE AND PARKING BRACKET

1 Inspect parking lock bracket for damage.

NOTE: The actuator rod must be positioned over the parking pawl and out of PARK range.

2 Install parking lock bracket and dip stick stop.
3 Install two m8.0 × 20 mm screws.
4 Check parking lock assembly for proper operation.
5 Use a ⅜ inch rod and install new Lo and reverse seal assembly.
6 Install "O" ring back up washer.
7 Install "O" ring seal on reverse oil pipe.
8 Install reverse oil pipe. Put the plain end in first, then put the "O" ring end into the Lo and reverse seal bore.
9 Install reverse oil pipe retainer bracket and bolt.

INTERMEDIATE SERVO INSPECTION

1 Using a flat blade screwdriver, remove the "E" ring from apply pin.
2 Separate the intermediate servo piston, cushion spring, spring retainer and apply pin.
3 Inspect orfice cup plug in intermediate bore.
4 Inspect apply pin for damage and free fit in pin bore.
5 Inspect apply pin seal ring and both piston seal rings for damage and free fit in their grooves. Do not remove seal rings unless replacing.
6 Inspect cushion spring and retainer for damage.
7 Inspect servo cover and piston assembly for porosity or damage.
8 Check for proper intermediate band

apply pin as follows:
a. Install J-28535, intermediate band apply pin gauge. Hold in place with two intermediate servo cover screws
b. Remove the band apply pin from the intermediate servo assembly.
c. Install J-28535 onto band apply pin
d. Install J-28535 and band apply pin into gauge
e. Apply 100 in./lbs. torque to compress the band
f. If any part of the white line appears in the window, the pin is the correct length. If the white line cannot be seen, change the band apply pin and recheck
g. Remove J-28535-4

INTERMEDIATE SERVO ASSEMBLY

1 Install the spring retainer, onto the small diameter end of the apply pin.
2 Install spring on retainer.
3 Install piston larger diameter first.
4 Compress the spring and install the "E" ring.
5 If removed, install new intermediate servo piston seal rings. Make sure the cut ends are installed in the same rela-

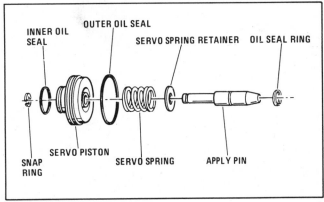

Intermediate servo components

tionship as the cut. Make sure the rings are seated in their grooves.
6 If removed, install intermediate band apply pin seal ring. Make sure the seal ring is seated in its groove.
7 Install 3rd accumulator check valve.
8 Install intermediate servo piston assembly.

NOTE: If the transaxle was equipped with a 3rd accumulator cup plug, and it was removed, do not reinstall. Replace servo cover and gasket with new version that takes the place of the cup plug.

9 Install intermediate servo cover gasket.
10 Inspect intermediate servo cover for damage or porosity.
11 Install intermediate servo cover and three servo cover screws.

GM 125 TRANSAXLE

12 Install the reverse oil pipe retainer bracket and screw.
13 Torque servo cover screws to correct specifications.

OIL STRAINER AND PAN
1 Install new strainer and "O" ring seal locating strainer against dipstick stop.
2 Clean and inspect transaxle oil pan for damage.
3 Install transaxle oil pan gasket on case.
4 Install transaxle oil pan and retaining bolts.

GOVERNOR AND SPEEDOMETER GEAR ASSEMBLY
1 Inspect for plugged oil passage.
2 Wash in cleaning solvent and blow out oil passage.
3 Inspect governor driven gear for nicks and damage.
4 Inspect governor shaft seal ring for cuts, damage and free fit in groove.
5 Inspect for free operation of governor weights. Weights must operate freely and independently of each other.
6 Inspect governor spring for damage and correct installation.
7 Inspect for presence of two check balls.
8 Inspect shaft for damage.
9 If damaged, cut seal ring off governor shaft.
10 If removed, install new seal ring; lubricate with petroleum jelly.
11 Install speedometer drive gear and thrust washer onto governor assembly.

12 Install governor assembly into transaxle.
13 Inspect governor cover for damage.
14 Install new governor cover "O" ring.
15 Install governor cover and retaining bolts.

NOTE: Make sure the governor shaft is piloted in the governor cover before torquing the retaining bolts.

16 Install speedometer driven gear and retainer.
17 Install converter. Make sure all shafts are fully engaged.

GM 125 AUTOMATIC TRANSAXLE OIL PRESSURE SPECIFICATIONS

Range	Normal Oil Pressure at Minimum T.V. P.S.I.	Normal Oil Pressure at Full T.V. P.S.I.
Park at 1,000 RPM	65–75	No T.V. pressure in Park Line pressure is same as Park at minimum T.V.
Reverse at 1,000 RPM	120–130 120–130	185–210 185–265
Neutral at 1,000 RPM	65–75 65–75	120–130 140–150
Drive at 1,000 RPM	65–75 65–75	120–130 140–150
Intermediate at 1,000 RPM	115–130	115–130
Low at 1,000 RPM	115–103	No T.V. pressure in Lo, Line pressure is the same as intermediate at minimum T.V.

CLUTCH PLATE & APPLY RING USAGE CHART

DIRECT CLUTCH						FORWARD CLUTCH						LOW & REV. CLUTCH				
Flat Steel Plate		Comp. Faced Plate	Apply Ring			Waved Plate		Flat Steel Plate		Comp. Faced Plate	Apply Ring		Comp. Flat Steel	Faced Plate	Apply Ring	
No.	Thickness	No.	I.D.*	Width	No.	Thickness	No.	Thickness	No.	I.D.*	Width	No.	No.	I.D.*	Width	
4	2.3 mm (0.09")	4	7	19 mm (0.75")	1	1.6 mm (0.06")	3	1.9 mm (0.08")	4	0	12 mm (0.47")	5	5	0	15.4 mm (0.61")	

The direct and forward clutch flat steel clutch plates and the forward clutch waved steel plate should be identified by their thickness.
The direct and forward production installed composition-faced clutch plates must not be interchanged. For service, direct and forward clutch use the same composition-faced plates.
*Measure the width of the clutch apply ring for positive identification.

GM 125 TRANSAXLE

TORQUE SPECIFICATIONS

Description of Usage	Quantity	Size	Torque Assembly
Pump Cover-to-Case			
Cover	5	M6 × 1.0 × 65.0	11 N·m (8 ft. lbs.)
Pump Cover-to-Case Cover	1	M8 × 1.25 × 65.0	14 N·m (10 ft. lbs.)
Pump Cover-to-Valve Body	1	M6 × 1.0 × 25.0	11 N·m (8 ft. lbs.)
Valve Body-to-Case Cover	8	M6 × 1.0 × 45.0	11 N·m (8 ft. lbs.)
Valve Body-to-Case	1	M8 × 1.25 × 85.0	14 N·m (10 ft. lbs.)
Valve Body-to-Driven			
Sprocket Support	1	M8 × 1.25 × 65.0	14 N·m (10 ft. lbs.)
Case Cover-to-Case	4	M8 × 1.25 × 45.0	24 N·m (18 ft. lbs.)
Case Cover-to-Case	5	M8 × 1.25 × 55.0	24 N·m (18 ft. lbs.)
Case Cover-to-Case	1	M8 × 1.25 × 40.0	24 N·m (18 ft. lbs.)
Case Cover-to-Case	7	M8 × 1.25 × 25.0	24 N·m (18 ft. lbs.)
Case-to-Case Cover	2	M8 × 1.25 × 35.0	24 N·m (18 ft. lbs.)
Case-to-Drive Sprocket Support	4	M8 × 1.25 × 23.5	24 N·m (18 ft. lbs.)
Oil Pan and Valve Body Cover	27	M8 × 1.25 × 16.0	16 N·m (12 ft. lbs.)
Manual Detent Spring			
Assembly-to-Case	1	M6 × 1.0 × 10.0	11 N·m (8 ft. lbs.)
Cooler Connector	2	¼-18 NPSF	38 N·m (23 ft. lbs.)
Line Pressure			
Take-off	1	⅛-27 NPTF	11 N·m (8 ft. lbs.)
Intermediate Servo Cover	4	M6 × 1.0 × 20.0	11 N·m (8 ft. lbs.)
Parking Lock Bracket-to-Case	2	M8 × 1.25 × 20.0	24 N·m (18 ft. lbs.)
Pipe Retainer-to-Case	2	M8 × 1.25 × 14.0	24 N·m (18 ft. lbs.)
Governor Cover-to-Case	2	M6 × 1.0 × 25.0	11 N·m (8 ft. lbs.)
Speedometer Driven			
Gear-to-Governor Cover	1	M6 × 1.0 × 16.0	9 N·m (75 in. lbs.)
T.V. Cable-to-Case	1	M6 × 1.0 × 16.0	9 N·m (75 in. lbs.)

FINAL DRIVE-TO-CASE END PLAY SELECTIVE THRUST WASHER IDENTIFICATION

Identification	Thickness
2	1.60–1.70 mm (0.062″–0.066″
3	1.70–1.80 mm (0.066″–0.070″)
4	1.80–1.90 mm (0.070″–).074″)
5	1.90–2.00 mm (0.074″–0.078″)
6	2.00–2.10 mm (0.078″–0.082″)
7	2.10–2.20 mm (0.082″–0.086″)

REACTION SUN GEAR-TO-INPUT DRUM SELECTIVE SNAP RING IDENTIFICATION

Identification/Color	Thickness
Pink	2.27–2.37 mm (0.089″–0.093″)
Brown	2.44–2.54 mm (0.096″–0.100″)
Lt. Blue	2.61–2.71 mm (0.103″–0.107″)
White	2.78–2.88 mm (0.109″–0.113″)
Yellow	2.95–3.05 mm (0.116″–0.120″)
Lt. Green	3.12–3.22 mm (0.123″–0.127″)
Orange	3.29–3.39 mm (0.129″–0.133″)
No Color	3.46–3.56 mm (0.136″–0.140″)

REVERSE CLUTCH HOUSING-TO-LO RACE SELECTIVE THRUST WASHER IDENTIFICATION

Identification	Thickness
1	1.00–2.20 mm (0.039″–0.043″)
2	1.42–1.52 mm (0.056″–0.060″)
3	1.84–1.94 mm (0.072″–0.076″)
4	2.26–2.36 mm (0.089″–0.093″)
5	2.68–2.78 mm (0.105″–0.109″)
6	3.10–3.20 mm (0.122″–0.126″)

INTERMEDIATE BAND APPLY PIN IDENTIFICATION CHART

Identification	Length
2 Grooves	Short
1 Groove	Medium
No Groove	Long

INPUT SHAFT SELECTIVE SNAP RING IDENTIFICATION

Identification/Color	Thickness
White	1.83–1.93 mm (0.071″–0.076″)
Blue	2.03–2.13 mm (0.078″–0.084″)
Brown	2.23–2.33 mm (0.088″–0.092″)
Yellow	2.43–2.53 mm (0.095″–0.099″)
Green	2.63–2.73 mm (0.103″–0.107″)

GM 180

GM 125 AUTOMATIC TRANSAXLE RANGE REFERENCE CHART

Range	Gear	Direct Clutch	Int. Band	Forward Clutch	Roller Clutch	Lo-Reverse Clutch
Park-Neut.						
	first			applied	holding	
Drive	second		applied	applied		
	third	applied		applied		
Int.	first			applied	holding	
	second		apllied	applied		
Lo	first			applied	holding	applied
	second		applied	applied		
Rev.		applied				applied

TROUBLESHOOTING
GM 180 AUTOMATIC
TRANSMISSION

OIL RELATED CONDITIONS

POSSIBLE CAUSE	POSSIBLE CONDITION
Low Oil Level	Oil coming out of oil filler tube. External oil leak. Failed vacuum modulator.
Oil Coming Out of Oil Filler Tube	Oil level too high. Coolant in transmission oil. External vent clogged with mud. Leak in oil pump suction circuit.
External Oil Leaks in the Area of the Torque Converter Housing	Leaking torque converter. Converter housing seal. Converter housing to case seal. Loose attaching bolts on front of transmission.
External Oil Leaks in the Area of Transmission Case and Extension	Shifter shaft seal. Extension seal. Oil pan gasket. Extension to case gasket. Vacuum modulator gasket. Drain plug gasket. Cooler line fittings. Oil filler tube seal ring. Detent cable seal ring. Line pressure gauge connection.
Low Oil Pressure	Low oil level. Clogged suction screen. Leak in oil pump suction circuit. Leak in oil pressure circuit. Pressure regulator valve malfunction.

STARTING RELATED CONDITIONS

POSSIBLE CAUSE	POSSIBLE CONDITION
No Starting in Any Drive Range	Low oil level. Clogged suction screen. Manual valve linkage or inner transmission selector level disconnected. Input shaft broken. Pressure regulator valve stuck in open position. Failed oil pump.
No Starting in Any Drive Range for a Time. Driving Possible Only After Repeatedly Moving Selector Lever To and Fro	Manual valve position does not coincide with valve body channels: Selector lever shaft retaining pin dropped out. Connecting rod to manual valve shifting. Selector lever shaft nut loose.
No Starting After Shifting Lever From "P" to "D", "L$_2$" or "L$_1$" (inadequate engine acceleration)	Parking pawl does not disengage.
Sudden Starting Only After Increase of Engine RPM	Band servo piston jamming. Low oil level. Oil pump defective. Oil screen missing. Sealing ball in valve body dropped out.
Heavy Jerking When Starting	Low oil pressure. Wrong modulator valve. Pressure regulator valve stuck. Sealing ball in valve body dropped out.

GM 180

TROUBLESHOOTING
GM 180 AUTOMATIC
TRANSMISSION

OIL RELATED CONDITIONS

POSSIBLE CAUSE	POSSIBLE CONDITION
Low Oil Level	Sealing ball in valve body dropped out.
High Oil Pressure	Modulator vacuum line leaky or interrupted. Failed vacuum modulator. Leak in any part of engine or accessory vacuum system. Pressure regulator valve malfunction.
Excessive Smoke Coming from Exhaust	Failed vacuum modulator. Oil from vent valve or leak on hot exhaust pipe.

STARTING RELATED CONDITIONS

POSSIBLE CAUSE	POSSIBLE CONDITION
No Starting in "D" or "L$_2$" Range, But in "L$_1$" and "R" Range	Input sprag installed backwards. Input sprag failure.
No Starting in "D" or "L$_2$" and "L$_1$" (proper driving in "R"; see also point 9)	Band worn, does not grip. Band servo piston jamming. Excessive leak in band servo. Parking pawl does not disengage.
No Starting in "R" Range (proper driving in all other ranges)	Reverse clutch failure.
Drive in Selector Lever Position "N"	Inadequate selector lever linkage. Planetary gear set broken. improper adjustment of band.

SHIFTING RELATED CONDITIONS

POSSIBLE CAUSE	POSSIBLE CONDITION
No 1–2 Upshift in "D" and "L$_2$" (transmission remains in 1st gear at all speeds)	Governor valves stuck. 1–2 shift valve stuck in 1st gear position. Seal rings (oil pump hub) leaky. Large leak in governor pressure circuit. Governor screen clogged.
No 2–3 Upshift in "D" (transmission remains in 2nd gear at all speeds)	2–3 shift valve stuck. Large leak in governor pressure circuit.
Upshifts in "D" and "L$_2$" Only at Full Throttle	Failed vacuum modulator. Modulator vacuum line leaky or interrupted. Leak in any part of engine or accessory vacuum system. Detent valve or cable stuck.
Upshifts in "D" and "L$_2$" Only at Part Throttle (no detent upshift)	Detent pressure regulator valve stuck. Detent cable broken or misadjusted.
Driving Only in 1st gear of "D" and "L$_2$" Range (transmission blocks in 2nd gear and "R")	"L$_1$" and "R" control valve stuck in "L$_1$" or "R" position.

SHIFTING RELATED CONDITIONS

POSSIBLE CAUSE	POSSIBLE CONDITION
No Part Throttle 3–2 Downshift at Low Vehicle Speeds	3–2 downshift control valve stuck.
No Forced Downshift	Detent cable broken or improperly adjusted. Detent pressure regulator valve stuck.
After Full Throttle Upshifting Transmission Shifts Immediately into Lower Gear Upon Easing Off Accelerator Pedal	Detent valve stuck in open position. Detent cable stuck. Modulator vacuum line interrupted.
At Higher Speeds, Transmission Shifts into Lower Gear	Retaining pin of selector lever shaft in transmission dropped out. Loose connection of selector lever linkage to manual valve. Pressure loss at governor.
Hard Disengagement of Selector Lever from "P" Position	Steel guide bushing of parking pawl actuating rod missing. Manual selector lever stuck.
Slipping 1–2 Upshifts (engine flares)	Low oil pressure. Sealing ball in valve body

---

GM 180

TROUBLESHOOTING
GM 180 AUTOMATIC
TRANSMISSION

SHIFTING RELATED CONDITIONS

POSSIBLE CAUSE	POSSIBLE CONDITION
	dropped out. Second clutch piston seals leaking. Second clutch piston centrifugal ball stuck open. Second clutch piston cracked or broken. Second clutch plates worn. Seal rings of oil pump hub leaky.
Slipping 2–3 Upshifts (engine flares)	Low oil pressure. Band adjustment loose. Third clutch piston seals leaking. Third clutch piston centrifugal ball stuck open. Third clutch piston cracked or broken. Wear of input shaft bushing. Sealing ball in valve body dropped out.
Abrupt 1–2 Upshift	High oil pressure. 1–2 accumulator valve stuck. Spring cushion of second clutch broken. Second gear ball valve missing.
Abrupt 2–3 Upshift	High oil pressure. Incorrect band adjustment.
Abrupt 3–2 Detent Downshift at High Speed	High speed downshift valve stuck open. Band adjustment.
Abrupt 3–2 Coast Downshift. Low Speed Downshift Timing Valve Stuck Open	
Flare on High Speed Forced Downshift	Low oil pressure. Band adjustment loose.
Flare on Low Speed Forced Downshift	Low oil pressure. Band adjustment loose. High speed downshift timing valve stuck in closed position. Sprag race does not grip on 3–1 down shifting.

NOISE RELATED CONDITIONS

POSSIBLE CAUSE	POSSIBLE CONDITION
Excessive Noises in All Drive Ranges	Too much backlash between sun gear and planetary gears. Lock plate on planetary carrier loose. Thrust bearing defective. Bearing bushings worn. Excessive transmission axial play. Unhooked parking pawl spring contacts governor hub. Converter balancing weights loose. Converter housing attaching bolt loose and contacting converter.
Screeching Noise When Starting	Converter failure.
Short Vibrating, Hissing Noise Shortly Before 1–2 Upshift	Dampening cushion of reverse clutch wearing into transmission case.

ENGINE BRAKING

POSSIBLE CAUSE	POSSIBLE CONDITION
No Engine Braking in "L_1" Range	Selector lever linkage improperly adjusted. Manual low control valve stuck.
No Engine Braking in "L_2" Range	Selector lever linkage improperly adjusted.
No Park	Selector lever linkage improperly adjusted. Parking lock actuator spring. Parking pawl. Governor hub.

GM180

GM 180 Automatic Transmission

Application

Chevrolet Chevette

Disassembly of Major Units
VALVE-BODY

1 Remove the manual detent roller and spring, retained by two bolts.
2 Remove the attaching bolts holding the strainer assembly to the valve body and remove. Discard gasket.
3 Remove the bolts from transfer plate reinforcement and remove reinforcement.
4 Remove the servo cover attaching bolts and remove servo cover and gasket.
5 Remove remaining bolts attaching valve body to case. Carefully remove valve body with gasket, and transfer plate. Care must be taken so that manual valve and manual valve link are not damaged or lost during removal of valve body from the case.
6 Remove two bolts holding transfer plate to valve body. Remove plate and gasket.
7 Remove the two check balls located in the oil passages in the transmission case. Location of these check balls must be noted so that they are installed correctly.

Manual detent spring removal/installation

SERVO PISTON

1 Compress Servo piston using compressor tool J-23075.
2 Using pliers, remove servo piston snap ring.
3 Loosen tool J-23075 slowly, as servo is under high spring tension. Remove tool and servo piston assembly.

NOTE: If necessary, remove burr from O.D. of piston bore to prevent damage to snap ring during reinstallation.

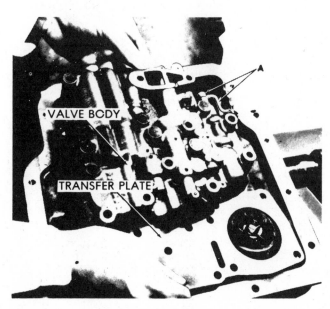

Valve body removal/installation

Compressing servo piston for removal

Component Removal and Overhaul

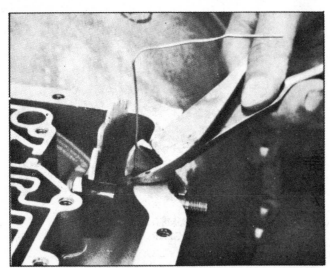

Removing selector lever shaft spring pin. Insert wire into middle of spring pin to prevent collapse

Vacuum modulator removal/installation

SELECTOR LEVER & SHAFT

1 Remove selector inner lever locknut from selector lever shaft.
2 Remove selector inner lever from selector lever shaft.
3 Remove selector lever shaft spring pin by pulling upwards with small pliers. To prevent collapse of spring pin, insert a wire into the middle of the spring pin.
4 Remove selector lever shaft.
5 Remove selector lever shaft oil seal and discard.

MODULATOR ASSEMBLY

1 Remove vacuum modulator from transmission case using tool J-23100. Care should be taken not to lose the modulator plunger.
2 Remove modulator valve and sleeve from transmission case. Remove O-ring seal by using internal snap ring pliers.

DETENT VALVE ASSEMBLY

1 Remove spring pin by pulling upward with pliers. To prevent collapse of spring pin, insert a wire into the middle of the spring pin.
2 Push on spring seat of detent valve assembly from front of case and remove detent valve, sleeve, spring, and spring seat from rear of case.

EXTENSION HOUSING & SPEEDOMETER GEAR

1 Remove bolt holding speedometer driven gear housing retainer and carefully remove retainer and pull speedometer driven gear assembly from extension housing to case.
2 Remove the attaching bolts from extension housing to case.
3 Remove extension housing and gasket. Note the position of the parking pawl in the extension housing.
4 Remove parking pawl actuator lever and actuator rod from extension housing.

SPEEDOMETER DRIVE GEAR, GOVERNOR BODY AND GOVERNOR HUB

1 Depress speedometer drive gear retaining clip and remove speedometer drive gear by sliding off output shaft.
2 Remove the four (4) attaching bolts from the governor body and remove governor and gasket.
3 Remove governor hub from output shaft using pliers such as J-8059.
4 Slide governor hub off the output shaft.

CONVERTER HOUSING, OIL PUMP, REVERSE CLUTCH & SECOND CLUTCH ASSEMBLY

1 Remove the seven converter housing attaching bolts which are the outer bolts in the housing. Loosen, but do not remove the five inner bolts.

GM 180

2 Remove converter housing with oil pump, reverse clutch and second clutch assemblies. Do not lose selective thrust washer, located between oil pump hub and second clutch drum. If the second clutch assembly remains in the case, remove it with the reverse clutch plates.

NOTE: Observe the position of the square-cut oil seal. Be sure to position the seal in the same location upon reassembly.

3 Remove reverse clutch plates and aluminum pressure plate from transmission case.

THIRD CLUTCH, PLANETARY CARRIER, REACTION SUN GEAR & DRUM AND LOW BAND

1 Remove third clutch assembly and input shaft.
2 Remove planetary carrier and output shaft by sliding out through front of case. Care should be taken not to lose the two torrington bearings and one thrust washer from the planetary carrier assembly.
3 Remove reaction sun gear-and-drum from case by pulling straight out.
4 Remove torrington bearing from rear of case.
5 Remove low band by slightly compressing band and pulling straight out.
6 If necessary to remove case vent, install new case vent. Do not attempt to reinstall old vent.

Component Disassembly, Inspection & Reassembly

CONVERTER HOUSING, OIL PUMP & REVERSE CLUTCH

1 Remove second clutch assembly from oil pump shaft.
2 Remove selective washer from oil pump shaft.
3 Remove the oil pump outer oil seal and discard.
4 Remove the five bolts holding the converter housing to the oil pump.
5 Separate the converter housing from the oil pump.
6 Remove oil pump wear plate "A."

7 Inspect front face of converter housing for oil leak. If problem is diagnosed as a front seal leak (presence of red oil coming from bottom of converter housing), remove front oil seal.
8 Inspect converter housing bushing. If worn, remove bushing using tool J-21465-17 with drive handle J-8092 from converter side of housing.
9 Clean converter housing thoroughly. Install converter housing bushing flush with front face of housing using tool J-21465-17 with driver handle J-8092.
10 Install new converter housing oil seal using tool J-21359.

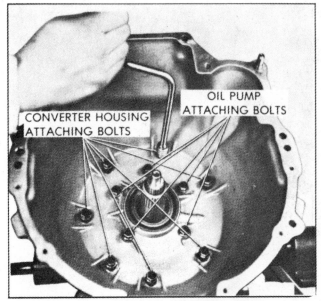

Converter housing and oil pump mount bolt locations

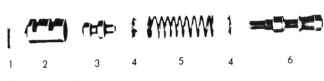

1. RETAINING PIN
2. PRESSURE REGULATOR BOOST VALVE SLEEVE
3. PRESSURE REGULATOR BOOST VALVE
4. PRESSURE REGULATOR VALVE SPRING SEAT
5. PRESSURE REGULATOR VALVE SPRING
6. PRESSURE REGULATOR VALVE
7. OIL PUMP BODY

Oil pump and pressure regulator components

GM 180

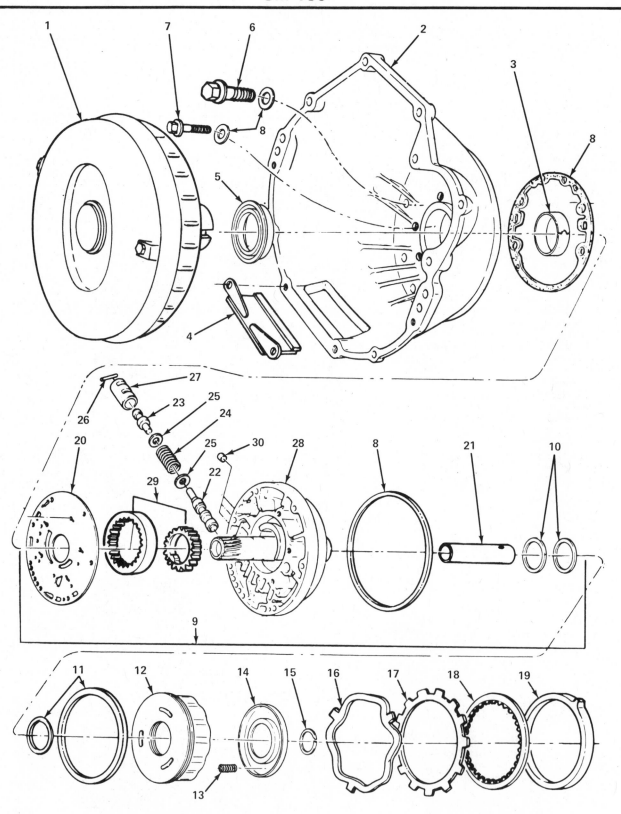

Converter, oil pump and reverse clutch components

1. Converter—Transmission
2. Housing—Transmission Converter
   w/Bushing

3. Bushing—Converter Housing
4. Plate—Converter Housing Protection
   (Lower)

5. Seal—Converter Hub to Housing
6. Screw—Converter Housing
7. Bolt—Converter Housing

GM 180

Spring Seat Retainer

8. Seal Kit—Converter Housing w/Gasket
9. Pump Package—Transmission Oil
10. Seal—Second Speed Clutch to Oil Pump
11. Seal Kit—Reverse Clutch Piston Inner & Outer
12. Piston—Reverse Clutch
13. Spring Kit—Reverse Clutch Piston Return (24 Spring)
14. Seat—Reverse Clutch Piston Retaining Spring
15. Ring—Reverse Clutch Piston Return
16. Spring—Reverse Clutch Cushion w/3 Prongs
17. Plate—Reverse Clutch Driven
18. Plate—Reverse Clutch Drive
19. Plate—Reverse Clutch Pressure
20. Plate—Transmission Oil Pump Wear
21. Bushing—Transmission Oil Pump Housing
22. Valve—Oil Pump Pressure Regulator
23. Valve—Oil Pump Pressure Regulator Boost
24. Spring—Oil Pump Pressure Regulator Valve
25. Seat—Pressure Regulator Valve Spring
26. Pin—Pressure Regulator Valve Retainer
27. Sleeve—Oil Pump Pressure Regulator Valve
28. Pump—Oil w/Gasket & Wear Plate
29. Gear—Oil Pump
30. Plug—Oil Pump

11 Check converter pump hub for nicks, burrs or damage which could have caused oil seal to leak or have worn bushing. Remove nicks and burrs.

12 Mark relative location of oil pump gears and remove oil pump gears.

13 Using compressor tool J-23327 on reverse clutch retaining seat, compress clutch return springs.

14 Remove snap ring using pliers such as J-8059.

15 Loosen compressor tool J-23327 and remove reverse clutch retaining ring and 24 reverse clutch springs.

16 Remove reverse clutch piston.

17 The pressure regulator and boost valve may be removed by using a pair of wire cutters to remove the retaining pin. However, it is not recommended that the pressure regulator valve and boost valve be disassembled during overhaul, unless it was determined by oil pressure checks to have been malfunctioning.

18 Remove pressure regulator boost valve sleeve, spring, pressure regulator valve and two washers.

19 Inspect, and if necessary replace the oil pump hub bushing by threading a ¾" standard pipe tap such as tool J-23130-5 into bushing.

20 Using a drift on tap, press out oil pump bushing with arbor press. Use rag or cloth to protect oil pump face.

21 Install new oil pump hub bushing with arbor press, using tool J-23130-1.

22 Clean pump body, including all holes and pockets thoroughly. With oil pump shaft hole "A" facing downward, scribe an aligning mark on oil pump shaft inner diameter at the center of the oil groove to the right of hole "A." Scribe mark on outer edge of bushing through the centers of the small and large drilled holes "B." Place bushing into oil pump shaft with small hole up, and align scribe marks on bushing with those made in oil pump shaft. Use arbor press to drive bushing into oil pump shaft until seated in the bore. Care must be taken so that bushing is pressed in straight, using the scribe marks as a guide until firmly seated.

23 Inspect and thoroughly clean the pressure regulator valve. Immerse in transmission fluid before installing in bores.

24 Install pressure regulator valve in oil pump body bore.

25 Install pressure regulator valve, two spring seats, spring, boost valve and sleeve in oil pump body bore.

26 Depress pressure regulator boost valve sleeve until back end lines up with pin hole and insert pin to secure.

27 Inspect oil pump hub oil seal rings. Replace if damaged or if side wear is noted.

28 Inspect reverse clutch piston for damage. Replace if necessary.

29 Install two new oil seals on reverse clutch piston.

30 Install reverse clutch piston onto rear face of oil pump using liberal amount of transmission fluid. J-28241 may be used to protect I.D. of seal.

31 Inspect reverse clutch piston springs. Evidence of extreme heat or buring in the area of the clutch may have caused the springs to take a heat set and would justify replacement of the springs.

32 Install the twenty-four reverse clutch piston return springs.

33 Install retaining seat.

34 Compress return springs using second and reverse clutch piston spring compressor tool J-23327. Care should be taken not to damage retainer should retainer catch in snap ring groove.

35 Install snap ring using pliers such as J-8059. Do not air check reverse clutch as the clutch is not complete and damage to the return spring retaining seat may occur.

36 Turn oil pump and reverse clutch assembly so that oil pump face is facing up.

37 Install oil pump gears using the location mark made before disassembly.

38 Check the end clearance of both gears to the oil pump face. (Be sure to measure between the face of the gears and the pump face, not between the crescent and the pump face.) Use a straight edge and feeler gage. Clearance should be between .0005 to .0035 inch. Install J-23082 onto oil pump drive gear. Measure clearance between drive gear and crescent segment while rotating gears one complete revolution. Clearance should be .005 to .009 inch. Next, measure clearance between outside of driven gear and pump housing. Rotate one complete revolution. Clearance should be between .003 and .007 inch. Now mea-

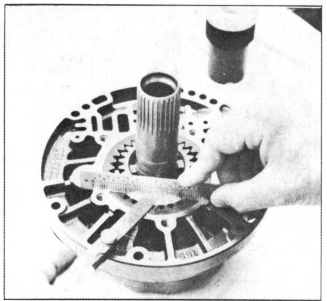

Measuring oil pump face-to-gears end clearance

Measuring oil pump operating clearances

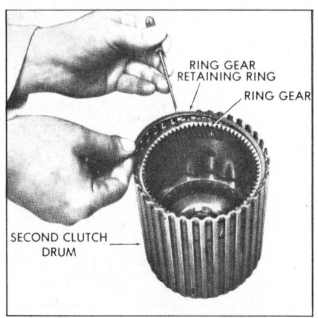

RING GEAR
RETAINING RING

RING GEAR

SECOND CLUTCH
DRUM

Removing second clutch drum ring gear retaining ring

sure between inside of driven gear and crescent segment. Rotate gear one revolution. Clearance should be a minimum of .005 inch.

39 Install oil pump wear plate onto oil pump.

40 Insert guide pin into oil pump for alignment of converter housing and lower housing onto pump.

41 Loosely install bolts into converter housing.

42 Use converter housing to oil pump aligning tool J-23082 to align converter housing to pump. Tool should bottom on oil pump gear.

43 Tighten the five inner bolts, to half-torque, then alternately to 13–17 lbs./ft. and remove aligning tool J-23082.

44 Install new converter housing-to-case rubber oil seal.

SECOND CLUTCH

1 Remove ring gear retaining ring from second clutch drum.

2 Remove ring gear.

3 Remove second clutch spacer plate retaining ring.

4 Remove second clutch spacer ring.

5 Remove second clutch steel and composition plates. The plates should be kept in the same sequence as they were installed in the clutch.

6 Remove second clutch assembly to third clutch assembly thrust washer.

7 Install second and reverse clutch spring compressor tool J-23327 on second clutch piston return spring retainer and compress second clutch piston return springs.

8 Remove snap ring using pliers such as J-8059.

9 Remove second clutch retaining seat and 22 return springs.

10 Remove second clutch piston.

11 Inspect second clutch piston. If piston is damaged or if check ball falls out upon inspection, replace piston. Install two new piston lip seals.

12 Inspect the piston return springs. Evidence of extreme heat or buring in the area of the clutch may have caused the springs to take a heat set and would justify replacement of the springs.

13 Inspect second clutch hub bushing for scoring or wear.

14 If necessary, remove second clutch hub bushing using remover and installer J-23130-6 with driver handle J-8092.

15 Clean in solvent to remove any foreign matter. Install new second clutch hub bushing using tool J-23130-6 and driver handle J-8092. Bushing must be driven in until tool bottoms on bench.

16 To install second clutch piston into second clutch drum, use tool J-23080 so as not to damage outer lip seal. Use liberal amount of transmission fluid for ease of installation and to prevent seal damage.

NOTE: Slide tool J-23080 over the piston to compress the outer lip seal.

17 Remove tool J-23080.
18 Install 22 springs and retaining seat on second clutch piston.
19 Using spring compressor tool J-23327 on retaining seat, compress second clutch piston return springs. Care should be taken so that retainer does not catch in snap ring groove and damage retainer.
20 Install snap ring with pliers such as J-8059.
21 Install second-to-third clutch thrust washer so that the tang seats in the slot of the second clutch hub. Secure with petroleum jelly (unmedicated).
22 Inspect condition of composition and steel plates.

NOTE: Do not diagnose a composition drive plate by color.

23 Install second clutch plates into second clutch drum with cushion plate (wave washer) first, then steel plate, composition plate, steel plate, etc. Use liberal amount of transmission fluid.
24 Install second clutch spacer plate into second clutch drum. If necessary, expand spacer plate with screw driver until ends of spacer are evenly butted together seating tightly into drum.
25 Install second clutch spacer retaining ring.
26 Install ring gear into second clutch drum, with the grooved edge facing upward.
27 Install ring gear retaining ring.

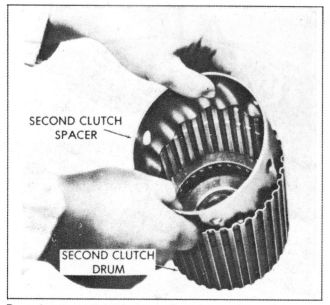

Removing second clutch spacer ring

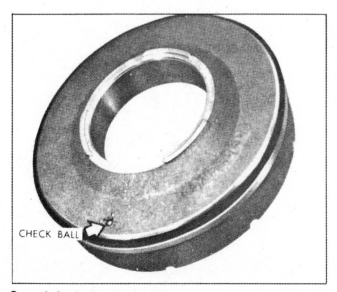

Second clutch piston check ball

Installing piston into second clutch drum

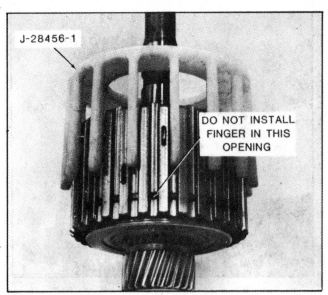

Placing Tool J-28456-1, retaining ring compressor, on the third clutch assembly

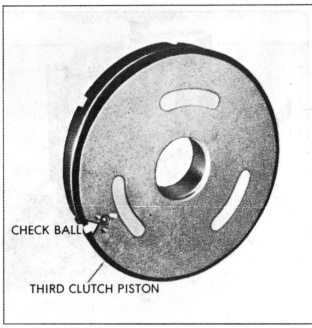

Third clutch piston check ball

THIRD CLUTCH, SPRAG & INPUT SUN GEAR

1 Install the fingers of J-28456-1 onto the third clutch.

NOTE: Do not install a finger in the opening of the retaining ring.

2 Compress the retaining ring by sliding J-28456-2 against the fingers.
3 After the retaining ring has been pressed out of the clutch drum groove, remove the sprag assembly.
4 Remove input shaft-to-input sun gear thrust washer and torrington bearing.
5 Remove third clutch plates from third clutch drum. The plates should be kept in the same sequence as they were installed in the clutch.
6 Remove input sprag race and retainer assembly from third clutch hub and input sun gear assembly.
7 Push sprag assembly and retaining rings from sprag race and retainer.
8 Using compressor tool J-23075 on third clutch piston retaining seat, compress third clutch piston return springs using arbor press.
9 Remove snap ring using pliers such as J-8059.
10 Remove the retaining seat and 12 return springs.
11 Remove third clutch piston from third clutch drum.
12 Inspect third clutch piston return springs. Evidence of extreme heat or burning in the area of this clutch may have caused the springs to take a heat set and would justify replacement of the springs.
13 Inspect check ball in third clutch piston. Shake the piston and listen for movement of check ball, which will indicate proper operation. If ball is missing or falls out upon inspection or piston is damaged, replace piston. Install new lip seal on piston.
14 Carefully install new oil lip seal on input shaft inside of third clutch drum, with the lip pointing downward.
15 Inspect steel thrust washer on front face of third clutch drum. Replace if scored or damaged.
16 Install third clutch piston into third clutch drum, using a liberal amount of transmission fluid, so that lip seal is not damaged upon installation.
17 Install the 12 third clutch piston return springs onto piston.
18 Install retaining seat.
19 Using compressor tool J-23075 on retaining seat, compress piston return springs. Care must be taken so that retaining seat does not catch in snap ring groove and damage retainer.
20 Install snap ring using pliers such as J-8059.
21 Inspect condition of the third clutch composition and steel plates.

GM 180

NOTE: Do not diagnose a composition drive plate by color.

22 Install third clutch plates into third clutch drum beginning with the conical plate toward the piston, with the outer edge against the piston, and the inner (high) edge toward the plates. Then install a steel plate, then a composition plate, steel plate, composition plate, etc. Use a liberal amount of transmission fluid.

23 Inspect the thick thrust washer and torrington bearing for damage. Replace if necessary.

24 Install thrust washer and bearing onto input shaft. The inner lips of the bearing should be toward the thrust washer. Secure with petroleum jelly (unmedicated).

25 Inspect sprag assembly for wear, damage or sprags that freely fall out of cage. Inspect input sun gear for chipped or nicked teeth or abnormal wear. Replace part if necessary.

26 Install sprag onto third clutch hub with the flare shoulder on sprag cage outer diameter toward input sun gear.

27 Install sprag race and retainer assembly over sprag assembly. Holding input sun gear with left hand, sprag race and retainer assembly should "lock up" when turned with right hand in a clockwise direction and should rotate freely when turned counterclockwise.

28 By slightly turning the clutch hub and sprag assembly, engage it to clutch plate splines until sprag race rests on the clutch drum.

29 Align teeth of sprag race with splines of the clutch drum. Using a small screwdriver, press retaining ring all around into the retaining ring groove, at the same time applying pressure on the sprag race.

30 Slide sprag into the third cluch drum until the retaining ring snaps into the clutch drum groove.

PLANETARY CARRIER

1 Inspect the planetary carrier and output shaft for distortion or damage.
2 Inspect the planetary pinions for excessive wear or damage, such as chipped teeth.
3 Check the end clearance of all planetary pinions with feeler gauge at points "A" and "B". Clearance should be between .005″ and .035″.
4 Replace entire assembly if damage or excessive wear is noted.
5 Tighten planetary carrier lock plate retaining screws to 20–35 lbs./in.

REACTION SUN GEAR AND DRUM

1 Inspect reaction sun gear for chipped or nicked teeth and inspect sun gear for scoring. If necessary, replace entire assembly.
2 Inspect reaction sun gear, drum, and bushing.
3 If necessary to replace bushing, use a

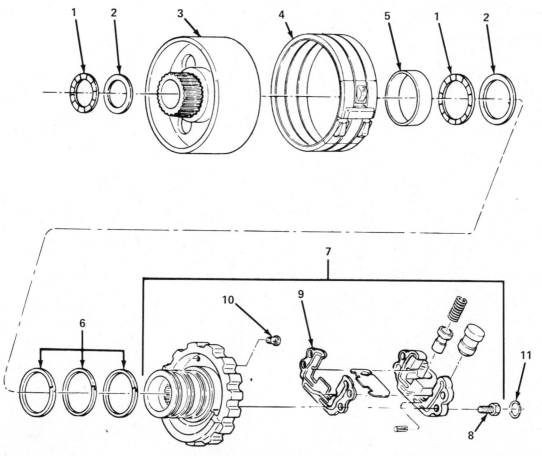

1. Bearing—Sun Gear Reaction
2. Race—Sun Gear Reaction
3. Gear—Reaction Sun w/Drum
4. Band—Low Brake
5. Bushing—Input Sun Gear

6. Ring—Governor Hub-Seal
7. Governor—Transmission w/Hub
8. Screw—Hex Head Body on Governor Hub
9. Gasket—Governor

10. Screen—Oil Pump Governor
11. Ring—Governor Hub to Input Shaft-Snap

Reaction sun gear and governor components

GM 180

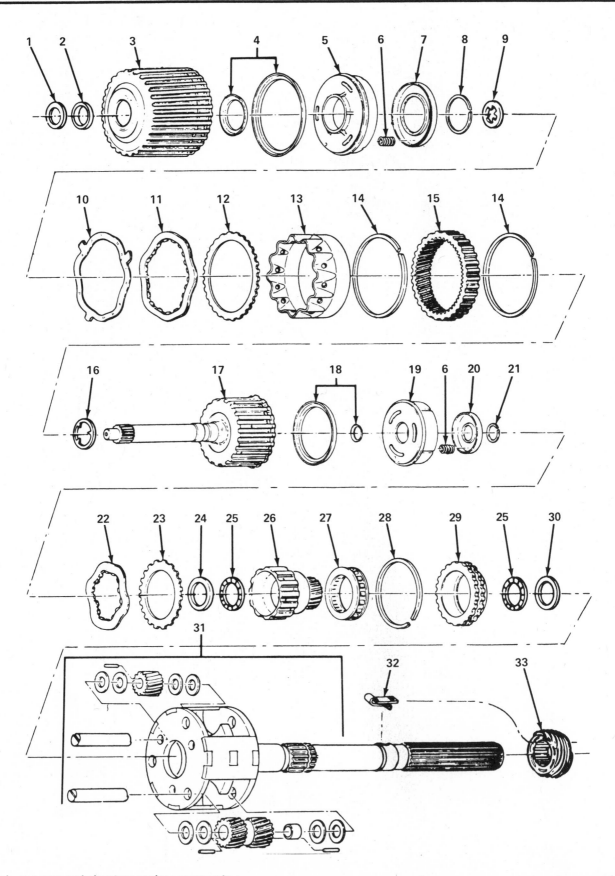

Clutch, sun gear and planetary carrier components

1. Thrust Washer—Between Oil Pump & Second Speed Clutch
2. Bushing—Clutch Drum Second Speed
3. Clutch Drum—Second Speed Outer (Includes Bushing)
4. Seal Ring Set—Clutch Piston Second Speed (Inner & Outer)
5. Piston—Clutch Second Speed (w/$^5/_{32}$" Ball)
6. Spring Kit—Second Speed Clutch Piston Return (24 Springs)
7. Seat—Clutch Piston Return Spring Second Speed
8. Ring—Clutch Piston Return Spring Seat Retainer Second Speed-Snap
9. Washer—Second & Third Speed Clutch Connection-Thrust (Bronze)
10. Plate—Clutch Cushion-Second Speed
11. Plate—Clutch Second Speed-Drive
12. Plate—Clutch Second Speed-Driven
13. Spacer—Clutch Plate Second Speed
14. Ring—Retaining Ring Gear
15. Gear—Ring
16. Washer—Second & Third Speed Clutch Connection-Thrust (Steel)
17. Drum—Clutch Third Speed w/Input Shaft-Inner
18. Seal Set—Third Speed Clutch Piston-Inner & Outer
19. Piston—Clutch Third Speed w/$^5/_{32}$" Ball
20. Seat—Clutch Piston Return Spring Third Speed
21. Ring—Clutch Piston Return Spring Seat Retainer Third Speed-Snap
22. Plate—Clutch Third Speed-Drive
23. Plate—Clutch Third Speed-Driven
24. Race—Input Sun Gear to Input Shaft
25. Bearing—Input Sun Gear
26. Gear—Input Sun
27. Sprag—Third Speed Clutch Input
28. Ring—Clutch Input Sprag Race Retaining Third Speed
29. Race & Retainer—Third Speed Clutch Input Sprag
30. Race—Input Sun Gear to Planetary Carrier
31. Carrier—Planetary
32. Clip—Speedometer Drive Gear
33. Gear—Speedometer Drive

chisel such as tool J-8400-1. Remove bushing from sun gear drum at bushing joint.

4 Thoroughly clean drum. Install new bushing using installer tool J-23130-2 with driver handle J-8092. Bushing should be installed flush with rear face of sun gear drum hub.

GOVERNOR BODY

1 Depress secondary valve spring with small screwdriver and remove secondary valve spring retainer.
2 Remove secondary valve spring, secondary valve, primary valve, and roll pin from governor body.
3 Inspect the primary and secondary valve for nicks, burrs, etc. If necessary, use crocus cloth to remove small burrs. Do not remove the sharp edges of the valve since these edges perform a cleaning action with the valve bore.
4 Inspect the secondary valve spring for distortion or breakage.
5 Clean in solvent, air clean, and blow out all oil passages. Inspect all oil passages, valve bores for nicks, burrs or varnish in governor body. Replace if necessary.
6 Install roll pin flush to .010" below the front face.
7 Install primary valve in governor placing the small portion of the valve in first. Use liberal amount of transmission fluid. There is no spring for the primary valve.
8 Install secondary valve with small spool portion of valve in first.
9 Install secondary valve spring.
10 Depress secondary valve spring with small screwdriver and install retainer.

GOVERNOR HUB

1 Inspect the three oil seal rings.
2 Remove governor hub oil screen. Use care not to damage or lose the oil screen. Inspect screen and clean with solvent and air dry. Replace if necessary.

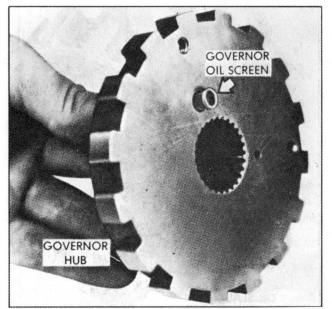

Governor oil screen location

3 Install oil screen flush to governor hub.
4 Inspect governor hub splines for cracks or chipped teeth in splines. Replace governor hub if required.

EXTENSION HOUSING

1 Inspect extension housing for damage. Replace housing if necessary.
2 Inspect parking pawl and spring for damage. Replace if necessary.
3 If lubricant leakage was noted prior to disconnecting propshaft from extension housing, extension housing rear seal should be replaced.
4 Use screwdriver to pry out extension housing seal.
5 Inspect extension housing bushing. If worn, scored or damaged, bushing can be removed with tool J-21424-9 used with driver handle J-8092.
6 Clean extension housing of dirt and foreign matter. Install new extension housing bushing using J-21424-9 with driver handle J-8092. Bushing must be installed flush to shoulder of extension housing.
7 Install new extension housing seal using installer tool J-21426.

SERVO PISTON

1 Remove servo piston apply rod.
2 Holding servo piston sleeve at flat portion of sleeve with wrench, loosen the adjusting bolt lock nut and remove.
3 Depress servo piston sleeve and remove piston sleeve retaining ring.
4 Push sleeve through piston and remove cushion spring and spring retainer.
5 Remove servo piston ring.
6 Inspect cushion spring, adjusting bolt, and piston sleeve for damage. Inspect piston for damage and piston ring for side wear, replace if necessary.
7 Reassemble servo piston, reversing disassembly procedure.

GM 180

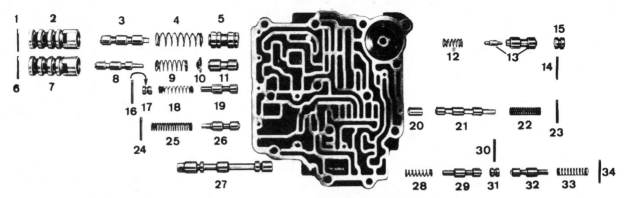

1. RETAINING PIN
2. 1-2 SHIFT CONTROL VALVE SLEEVE
3. T-2 SHIFT CONTROL VALVE
4. 1-2 SHIFT CONTROL VALVE SPRING HT. 2 7/16" DIA. 3/4"
5. 1-2 SHIFT VALVE
6. RETAINING PIN
7. 2-3 SHIFT CONTROL VALVE SLEEVE
8. 2-3 SHIFT CONTROL VALVE
9. 2-3 SHIFT CONTROL VALVE SPRING HT. 1 3/4" DIA. 11/16"
10. 2-3 SHIFT CONTROL VALVE SPRING SEAT
11. 2-3 SHIFT VALVE
12. 1-2 ACCUMULATOR VALVE SPRING HT. 1 1/16" DIA. 1/2"
13. 1-2 ACCUMULATOR VALVE
14. RETAINING PIN
15. 1-2 ACCUMULATOR VALVE PLUG
16. RETAINING PIN
17. 3-2 CONTROL VALVE PLUG
18. 3-2 CONTROL VALVE SPRING HT. 1 3/4" DIA. 7/16"

19. 3-2 CONTROL VALVE
20. REVERSE CONTROL VALVE
21. MANUAL LOW CONTROL VALVE
22. MANUAL LOW CONTROL VALVE SPRING HT. 1 5/16" DIA. 7/16"
23. RETAINING PIN
24. RETAINING PIN
25. DETENT PRESSURE REGULATOR VALVE SPRING HT. 1 5/8" DIA. 1/2"
26. DETENT PRESSURE REGULATOR VALVE
27. MANUAL VALVE
28. LOW SPEED DOWNSHIFT TIMING VALVE SPRING HT. 1 3/8" DIA. 7/16"
29. LOW SPEED DOWNSHIFT TIMING VALVE
30. RETAINING PIN
31. TIMING AND CONTROL VALVE PLUG
32. HIGH SPEED DOWNSHIFT TIMING VALVE
33. HIGH SPEED DOWNSHIFT TIMING VALVE SPRING HT. 1 5/16" DIA. 7/16"
34. RETAINING PIN

Valve body components

3 Check for good retention of band anchor pins.
4 Inspect all threaded holes for thread damage.
5 Inspect detent valve and modulator valve bores for scratches or scoring.
6 Inspect detent valve and modulator valve bores for scratches or scoring.
7 Inspect case bushing inside of case at rear. If damaged, remove bushing with remover and installer tool J-23130-3 and driver handle J-8092.
8 Inspect reaction sun gear drum bushing sleeve inside case at rear for scoring. If necessary, replace sleeve before installing rear case bushing.
9 Remove sleeve by grinding, or with a small chisel. Care must be used in order that aluminum case is not damaged when grinding sleeve.
10 Install new sleeve using installer tool J-23130-7 and driver handle J-8092.

NOTE: The case bushing must be removed before installing the sleeve, since J-23130-7 pilots in the bushing bore, not the bushing itself.

11 Install new case bushing using tool J-23130-3 and driver handle J-8092. Bushing should be installed flush with case at rear.

CONVERTER INSPECTION & TESTING
12 Drain converter. If clutch disc material or foreign matter has been found while draining converter, replace entire converter assembly as it can not be cleaned properly.
13 Air check converter for leaks using

VALVE BODY
1 Remove the manual valve and manual valve link from the valve body.
2 Turn the valve body so that the transfer plate is facing upward and remove the two bolts holding the transfer plate to the valve body.
3 Remove transfer plate and gasket.
4 Using small C-clamp on valve body, compress accumulator piston.
5 Remove the accumulator piston retaining ring with screwdriver.
6 Carefully loosen C-clamp as accumulator piston is under spring tension.
7 Remove accumulator piston, oil ring, and spring.
8 Inspect accumulator oil ring for damage or edge wear and piston for damage. Replace if necessary.
9 Remove 1–2 Shift control valve retaining pin, 1–2 shift control valve sleeve, control valve, 1–2 shift valve spring

and valve. It may be necessary to remove burrs in valve body bore made by retaining pin prior to removal of the sleeves and valves.
10 Remove the 2–3 shift control valve retaining pin and sleeve. Also, remove the 2–3 shift control valve, spring seat, spring and 2–3 shift valve.
11 Remove the 3–2 control valve retaining pin and plug. Remove 3-2 control valve spring and control valve.
12 Remove the detent pressure regulator valve retaining pin, spring, and detent pressure regulator valve.
13 Remove the high speed downshift timing valve retaining pin and spring and remove valve.
14 Remove the downshift timing valve plug retaining pin and remove downshift timing valve plug. Remove the low speed downshift timing valve and spring.
15 Remove the manual low and reverse control valve retaining pin. Remove the spring and the manual low control valve and the reverse control valve.
16 Remove the 1-2 accumulator valve retaining pin and remove the 1-2 accumulator valve plug, 1-2 accumulator valve and spring.
17 A clean work area which is free of dirt and dust should be used to inspect, clean and install the valves in the

valve body. Handle valve components with clean hands and tools. Since many valve failures are caused initially by dirt or other foreign matter preventing a valve from functioning properly, a thorough cleaning of all the components with a cleaning solvent is essential. Do not use paraffin to clean out the valve body passages and valve bore. Compressed air may be used to blow out the passages.

18 Inspect each valve for free movement in its respective bore in the valve body. If necessary, use crocus cloth to remove small burrs on a valve.

NOTE: Do not remove the sharp edges of the valves as these edges perform a cleaning action within the bore.

19 Inspect the valve springs for distortion or collapsed coils. Replace the entire valve body assembly if any parts are damaged.

20 Inspect the transfer plate for dents or distortion. Replace transfer plate if necessary.

21 Reassemble the valves, springs, plugs and retaining pins in their proper location and order into the valve body using a liberal amount of transmission fluid.

22 Install spring and accumulator piston in valve body.

23 Compress accumulator piston with C-clamp and install retaining ring.

24 Install new valve body gasket.

25 Bolt the transfer plate and gasket to the valve body. Torque to 6–8 lbs./ft.

CASE DISASSEMBLY & REASSEMBLY

1 Inspect case for damage.

2 Inspect and clean oil passages with cleaning solvent and air.
converter checking tool J-21369. Install tool and tighten. Apply 80 psi air pressure to tool.

14 Submerge in water and check for leaks.

15 Check converter hub surfaces for scoring or wear.

Assembly of Major Units

SELECTOR LEVER & SHAFT

1 Install new selector lever shaft oil seal in case, with the grooved end (with metric threads) outside of case. Insert selector lever shaft through case from outside. Case should be exercised so that oil seal is not damaged.

2 Insert spring pin in case to secure selector lever shaft.

3 Guide selector lever over shaft and secure with lock nut.

4 Insert parking pawl actuator rod from front of the case and through hole in case at rear.

5 Install parking pawl actuator rod retaining ring.

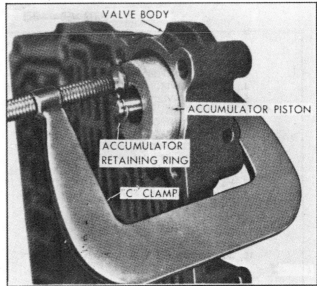

Compressing accumulator piston for retaining ring removal

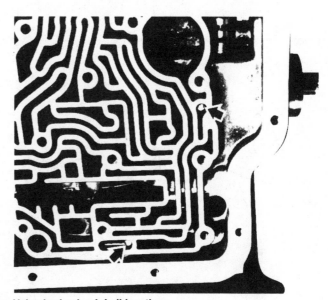

Valve body check ball locations

GM 180

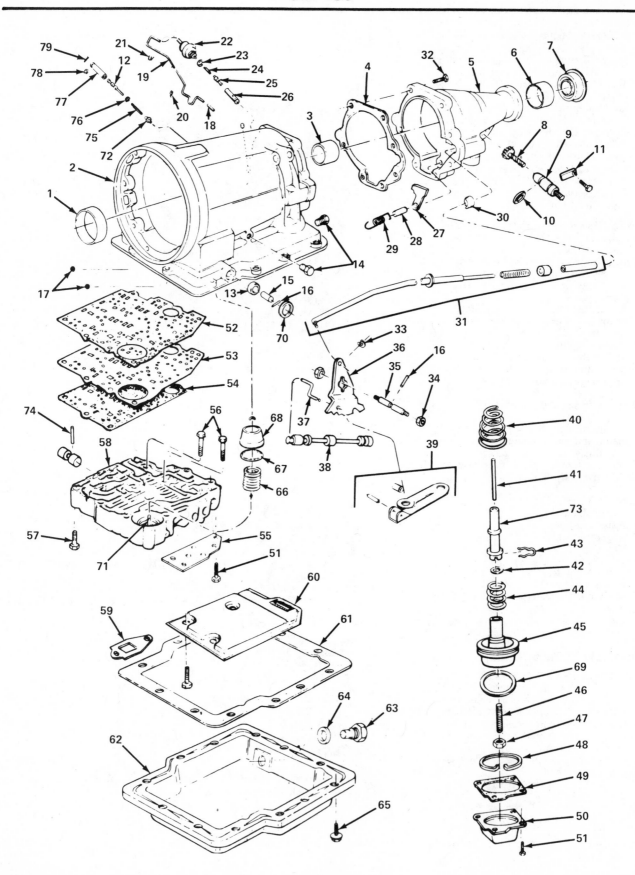

Case, valve body and low servo piston components

GM 180

1. Sleeve—Reaction Sun Gear Drum Bearing
2. Case—Transmission w/Bushing
3. Bushing—Transmission Case
4. Gasket—Extension to Transmission Case
5. Extension—Transmission w/Bushing
6. Bushing—Transmission Case Extension
7. Seal—Transmission Case Extension
8. Gear—Speedometer Driven
9. Guide—Speedometer Drive
10. Ring—Speedometer Drive Gear Seal
11. Bracket—Speedometer & Guide on Transmission
12. Valve—Transmission Detent
13. Seal—Transmission Range Selector Shaft
14. Plug—Pressure Tap
15. Plug—Manual Shaft Transmission Case Side
16. Pin—Manual Shaft Plug Lock Transmission Case Side
17. Ball—Transmission Valve Body
18. Grommet—Vacuum Line Hose
19. Line—Vacuum Modulator
20. Clamp—Spring Vacuum Line to Engine
21. Clamp—Spring Vacuum Line to Transmission
22. Modulator—Vacuum
23. Gasket—Vacuum Modulator
24. Sleeve—Vacuum Modulator
25. Valve—Vacuum Modulator
26. Sleeve—Vacuum Modulator Valve
27. Pawl—Transmission Parking Lock
28. Shaft—Parking Lock Pawl in Extension
29. Spring—Transmission Parking Lock Pawl Disengaging
30. Sleeve—Parking Lock Pawl Actuator (in Extension Housing)
31. Actuator—Parking Lock
32. Bolt—Transmission Extension to Case
33. Ring—Parking Lock Actuator on Lever-Snap
34. Nut—Parking Lock Lever
35. Shaft—Parking Lock & Range Selector
36. Lever—Transmission Parking Lock
37. Link—Transmission Manual Valve Lever
38. Valve—Transmission Manual
39. Spring—Detent w/Roller
40. Spring—Low Servo Piston Return
41. Rod—Low Servo Piston Apply
42. Seat—Low Servo Piston Cushion Spring
43. Ring—Low Servo Piston Retaining-Inner
44. Spring—Low Servo Piston Cushion
45. Piston—Low Servo
46. Stud—Low Servo Piston Adjusting
47. Nut—Hex Head Low Servo Piston Adjusting
48. Ring—Low Servo Piston Retaining-Outer
49. Gasket—Low Servo Cover on Transmission Case
50. Cover—Low Servo on Transmission Case
51. Bolt—Low Servo Cover on Transmission Case
52. Gasket—Transfer Plate to Transmission Case
53. Plate—Valve Body Transfer (See Parts Catalog)
54. Gasket—Transfer Plate to Valve Body
55. Plate—Reinforcement Oil Pump Suction Transfer
56. Bolt—Valve Body
57. Bolt Valve Body & Oil Pump Screen on Transmission Case
58. Body—Transmission Valve Complete
59. Gasket—Oil Pump Suction Screen
60. Screen—Oil Pump Suction
61. Gasket—Transmission Oil Pan
62. Pan—Transmission Oil
63. Screw—Oil Pan Drain Plug-Hex Head
64. Gasket—Oil Pan Drain Screw
65. Bolt—Transmission Oil Pan-Hex Head
66. Spring—Accumulator Piston Thrust
67. Ring—Low Servo Piston Oil Seal
68. Piston—Transmission Accumulator
69. Ring—Low Servo Piston Oil Seal
70. Seal—Manual Shaft Transmission Case Side
71. Pin—Front Accumulator Piston
72. Screw—Detent Valve Plug
73. Sleeve—Low Servo Piston Adjusting
74. Pin—Valves on Valve Body
75. Spring—Detent Valve
76. Seat—Detent Valve Spring
77. Sleeve—Detent Valve
78. Seal—Detent Valve
79. Pin—Detent Valve

Bolt—Oil Pan Suction Reinforcement Plate

LOW BAND

1 Turn transmission case so that front of case is upward.
2 Inspect band for cracks, flaking, burring or looseness. Replace if required.
3 Place band in case and locate band onto the anchor pins in case.

REACTION SUN GEAR AND DRUM

1 Place torrington bearing into case. Secure with petroleum jelly (unmedicated). The case bushing acts as a guide to center the bearing.
2 Insert reaction sun gear facing upward.
3 Place torrington bearing onto sun gear. Secure with petroleum jelly (unmedicated).

OUTPUT SHAFT & PLANETARY CARRIER

1 Install thrust washer, then torrington bearing into planetary carrier. Secure with petroleum jelly (unmedicated).
2 Insert output shaft and planetary carrier assembly from front of case to spline with reaction sun gear.

SECOND & THIRD CLUTCH ASSEMBLIES

1 With second clutch assembly on bench, align second clutch drive plates in second clutch drum.
2 Insert third clutch drum and input shaft through top of second clutch drum, seating third clutch drum splines into the second clutch plate splines.
3 Holding second and third clutch assemblies by the input shaft, lower into transmission case, indexing ring gear in second clutch drum with long planetary pinion gear teeth.

REVERSE CLUTCH PLATES

1 Inspect condition of the composition and steel plates.

NOTE: Do not diagnose a composition drive plate by color.

2 Dry composition plates with compressed air and inspect the composition surface for pitting and flaking, wear, glazing, cracking, charring, chips or metal particles imbedded in lining. If a composition drive plate exhibits any of the above conditions, replacement may be required. Wipe steel plates dry and check for heat discoloration. If the surface is smooth and an even color smear is indicated, the plates should be reused. If severe heat spot discoloration or surface scuffing is indicated, the plates must be replaced.

3 Install the aluminum pressure plate into the case, with the flat side up. Make sure that the lug on the pressure plate engages with one of the narrow notches in the case.
4 Install reverse clutch steel plate, composition plate, steel plate, composition plate, etc., into case. Use a liberal amount of transmission fluid.
5 Install reverse clutch cushion plate (wave washer) into case, so that all three of its lugs are engaged into narrow notches in the case.

SELECTIVE WASHER THICKNESS

1 Place gauging tool J-23085 on case flange and against input shaft.
2 Loosen thumb screw on J-23085 to allow inner shaft of tool to drop onto the second clutch drum hub.
3 Tighten thumb screw, then remove tool J-23085.
4 Compare the thickness of the selective washer removed earlier from transmission to the protruding portion of the inner shaft of tool J-23085. The selective washer used in reassembly should be the thickest washer available without exceeding the dimension of the shaft protruding from Tool J-23085. The dimension of the washer selected should be equal to or slightly

GM 180

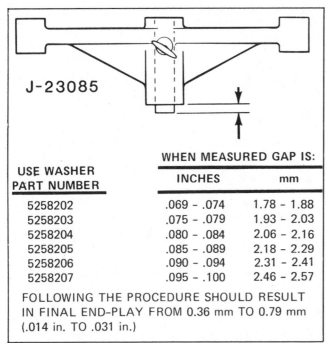

USE WASHER PART NUMBER	WHEN MEASURED GAP IS:	
	INCHES	mm
5258202	.069 – .074	1.78 – 1.88
5258203	.075 – .079	1.93 – 2.03
5258204	.080 – .084	2.06 – 2.16
5258205	.085 – .089	2.18 – 2.29
5258206	.090 – .094	2.31 – 2.41
5258207	.095 – .100	2.46 – 2.57

FOLLOWING THE PROCEDURE SHOULD RESULT IN FINAL END–PLAY FROM 0.36 mm TO 0.79 mm (.014 in. TO .031 in.)

Selective thrust washer chart for oil pump hub-to-second clutch

Aligning converter housing to oil pump

less than the inner shaft for correct end play in transmission.

CONVERTER HOUSING, OIL PUMP & REVERSE CLUTCH

1 Install new pump flange gasket.
2 Place the proper selective washer, as previously determined, onto the oil pump shaft and retain with petroleum jelly (unmedicated).
3 Install two guide pins in case and lower converter housing and oil pump into case.
4 Bolt converter housing to case. Torque to 22–26 lbs./ft.
5 Check for correct assembly by turning input shaft by hand.

GOVERNOR ASSEMBLY

1 Turn case so that bottom of transmission is facing upward.
2 Slide governor hub along output shaft and seat into case. Use liberal amount of transmission fluid on oil seal rings.
3 Install snap ring over output shaft using pliers such as J-8059.
4 Install new governor body gasket.
5 Bolt governor body to governor hub. Torque to 6–8 lbs./ft. The two governor valves should move freely after governor body is torqued.
6 Install speedometer drive gear retaining clip into output shaft.
7 While depressing retaining clip, slide speedometer gear over output shaft and install gear and retaining clip.

EXTENSION HOUSING

1 Install new extension housing gasket.
2 Slide extension housing over output shaft and align holes.
3 Align parking pawl shaft into extension housing.
4 Bolt extension housing to rear of case. Torque to 20–30 lbs./ft.

SPEEDOMOETER DRIVEN GEAR

1 Install speedometer driven gear and housing into extension housing.
2 Install speedometer driven gear housing retainer into slot provided in speedometer driven gear housing. Bolt retainer to extension housing. Torque to 6–8 lbs./ft.

DETENT VALVE, MODULATOR VALVE & MODULATOR ASSEMBLY

1 Inspect detent valve sleeve oil seal and replace if necessary.
2 Install detent valve, sleeve, spring, and spring seat into case bore using liberal amount of transmission fluid.
3 Depress detent valve spring and insert spring pin to secure detent valve assembly. Detent valve sleeve must be installed with slots facing oil pan. Care should be taken so that spring pin is inserted into the groove provided in sleeve and not into one of the oil passage slots in the sleeve.
4 Install modulator valve and sleeve into case with small end of modulator valve first.
5 Using new modulator assembly O-ring, install plunger and thread modulator into case and tighten to 38 lbs./ft, using tool J-23100.

SERVO

1 Install servo apply rod, spring and piston into case, using liberal amount of transmission fluid.
2 Compress servo piston spring using tool J-23075, lightly tapping servo piston while compressing until piston is seated to avoid damage to the oil seal ring.
3 Install servo retaining ring. Remove tool J-23075.
4 Using 3/16" hex head wrench on servo adjusting bolt, adjust servo apply rod by tightening adjusting bolt to 40 lbs./in. Be certain that lock nut remains loose. Back off bolt five (5) turns exactly.
5 Tighten lock nut, holding adjusting bolt and sleeve firmly.

VALVE BODY

1 Install steel balls in oil passages in case.
2 Install new case-to-transfer plate gasket.
3 Locate guide pins in transmission case for correct alignment of valve body and transfer plate. Install bolts holding transfer plate to valve body.
4 Install manual valve into valve body bore using liberal amount of transmission fluid.
5 Install long side of manual valve link pin into manual valve.
6 Install short end of manual valve link "A" into selector lever and install valve body and transfer plate assembly over guide pins.
7 Install selector lever roller spring and retainer. Torque to 13–15 lbs./ft. The valve body bolts should be torqued starting in the center of the valve body and working outward. Torque to 13–15 lbs./ft.
8 Install reinforcement plate bolts to case. Torque to 13–15 lbs./ft.
9 Inspect oil strainer. If foreign matter is present, install new strainer.

10 Install oil strainer assembly using new gasket. Torque to 13–15 lbs./ft.
11 Install new servo cover gasket.
12 Install servo cover. Torque to 17–19 lbs./ft.

TORQUE CONVERTER

1 Place transmission on portable jack.
2 Slide torque converter over stator shaft and input shaft.
3 Be sure that converter pump hub keyway is seated into oil pump drive lugs and the distance "A" is .20" to .28".
4 Rotate converter to check for free movement.

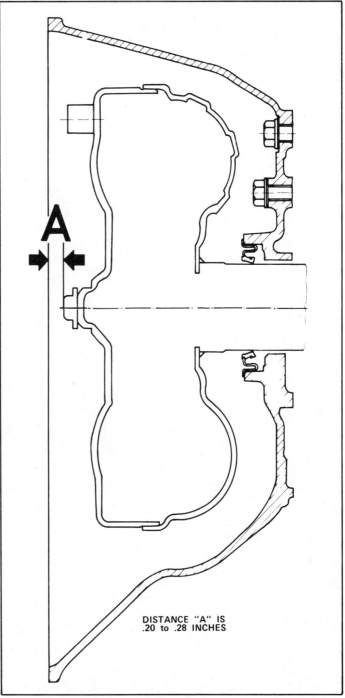

A

DISTANCE "A" IS .20 to .28 INCHES

Torque converter installed end clearance

Component Removal and Overhaul

GM 180

PRESSURE TAP CAN BE MADE ACCESSIBLE BY:

1. Provide proper transmission support.

2. Remove rear transmission crossmember side bolts.

3. Lower transmission enough to remove pressure tap plug which is located on the left side of the transmission.

NOTE: ON REPLACING PLUG, TORQUE TO 5-7 LB-FT.

4. After pressure gage and hose is installed, replace rear crossmember and side bolts and proceed with pressure checking procedure.

NOTE: PRESSURES ARE OFF OF THE SERVO APPLY.

Hydraulic pressure check specifications

Car Coasting @ 30 MPH; Vacuum Line Connected; Foot Off Throttle	Zero Output Shaft Speed; Vacuum Line Disc from the Mod; Engine RPM at 1500
Min. PSI	Max. PSI
D _____ 65	D _____ 118
L_2 _____ 65	L_2 _____ 118
L_1 _____ 80	L_1 _____ 160

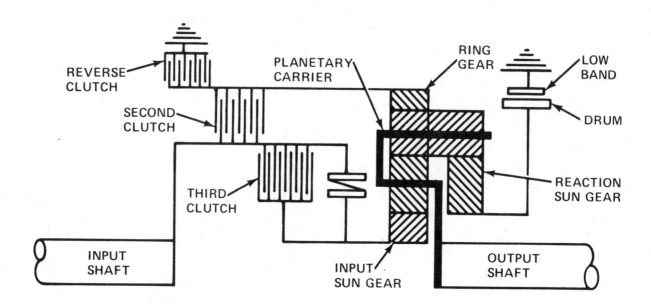

		REVERSE CLUTCH	SECOND CLUTCH	THIRD CLUTCH	LOW BAND	SPRAG
A.	NEUTRAL/PARK	Released	Released	Released	Released	Locked
B.	DRIVE RANGE, First Gear	Released	Released	Released	Applied	Locked
C.	L_1 RANGE	Released	Released	Applied	Applied	Locked
D.	DRIVE RANGE, Second Gear	Released	Applied	Released	Applied	Overrunning
E.	L_2 RANGE	Released	Applied	Released	Applied	Overrunning
F.	DRIVE RANGE, Third Gear	Released	Applied	Applied	Released	Locked
G.	REVERSE RANGE	Applied	Released	Applied	Released	Locked

Clutch application chart

GM 180

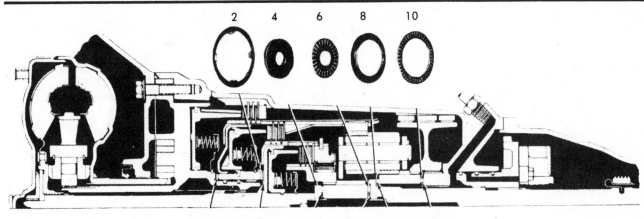

1. SELECTIVE THRUST WASHER
   (OIL PUMP HUB-TO-2ND CLUTCH)
2. BRONZE THRUST WASHER
   (2ND CLUTCH-TO-3RD CLUTCH)
3. STEEL THRUST WASHER
   (2ND CLUTCH-TO-3RD CLUTCH)
4. THRUST WASHER
   (INPUT SHAFT-TO-SUN GEAR)
5. TORRINGTON BEARING
   (INPUT SHAFT-TO-SUN GEAR)

6. TORRINGTON BEARING
   (SUN GEAR-TO-OUTPUT SHAFT)
7. THRUST WASHER
   (SUN GEAR-TO-OUTPUT SHAFT)
8. THRUST WASHER
   (OUTPUT SHAFT-TO-OUTPUT SUN GEAR)
9. TORRINGTON BEARING
   (OUTPUT SHAFT-TO-OUTPUT SUN GEAR)
10. TORRINGTON BEARING
    (REACTION SUN GEAR-TO-CASE)
11. THRUST WASHER (REACTION SUN GEAR-TO-CASE)

Bearing and thrust washer locations and identifications

TORQUE SPECIFICATIONS

ITEMS	TORQUE	
	N·m	Ft/lbs
Oil Pan-to-Case	9.8–12.8	7–10
Modulator Assembly	52	38
Extension Housing-to-Case	27.5–34.3	20–25
Oil Pressure Check Plug	6.4–9.8	5–7
Converter Housing-to-Cylinder Block	27–41	20–30
Transmission Support-to-Extension	39–48	29–36
Shift Lever-to-Selector Lever Shaft	20–34	15–25
Detent Cable, Retainer-to-Case	7–10	5–7
Oil Cooler Fittings-to-Case	11–16	8–12
Oil Cooler Fittings-to-Radiator	20–34	15–25
Oil Cooler Hose Clamps-to-Cooler Lines	0.8–1.2	.6–.9
Shifter Ass'y-to-Console	8.5–11	6–8
Neutral Safety Switch-to-Bracket	1.6–2.2	1.2–1.6
Lower Cover-to-Converter Housing	18–22	13–16
Flexplate-to-Converter	41–54	30–40
Transfer Plate-to-Valve Body	7.8–10.8	6–8
Reinforcement Plate-to-Case	17.7–20.6	13–15
Valve Body-to-Case	17.7–20.6	13–15
Servo Cover-to-Case	22.6–25.5	17–19
Converter Housing-to-Oil Pump	17.7–20.6	13–15
Converter Housing-to-Case	32.4–35.3	24–26
Selector Lever Locknut	10.8–14.7	8–11
Governor Body-to-Governor Hub	7.8–9.8	6–7
Servo Adjusting Bolt Locknut	16.7–20.6	12–15
Planetary Carrier Lock Plate	27–48	20–35

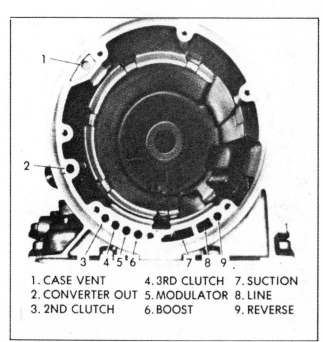

1. CASE VENT   4. 3RD CLUTCH   7. SUCTION
2. CONVERTER OUT   5. MODULATOR   8. LINE
3. 2ND CLUTCH   6. BOOST   9. REVERSE

Oil passage identification

GM TURBO HYDRA-MATIC 200

TROUBLESHOOTING
GM TURBO HYDRA-MATIC 200

THE PROBLEM	THE POSSIBLE CAUSE	THE POSSIBLE CONDITION
NO DRIVE IN DRIVE RANGE. **(CHECK OIL PRESSURE SPECS.)**	A. Low Oil Level	1. Incorrect level. 2. External leaks.
	B. Manual Linkage	Maladjusted.
	C. Low Oil Pressure	1. Plugged or restricted oil screen. 2. Oil screen gasket off location. 3. Pump assembly—pressure regulator. 4. Pump drive gear—tangs damaged by converter. 5. Case—porosity in intake bore.
	D. Valve Body Assembly	Manual valve disconnected from manual lever pin.
	E. Forward Clutch	1. Forward clutch does not apply—piston cracked, seals missing, damaged; clutch plates burned; snap ring out of groove. 2. Forward clutch oil seal rings missing or damaged on turbine shaft; leak in feed circuits; pump to case gasket mispositioned or damaged. 3. Clutch housing ball check stuck or missing. 4. Cup plug leaking or missing in the rear of the turbine shaft in the clutch apply passage. 5. Wrong forward clutch piston assy. or wrong number of clutch plates. 6. Feed orifice plugged in turbine shaft.
	F. Roller Clutch Assembly	1. Springs missing in the roller clutch. 2. Roller galled, or missing.
HIGH OR LOW OIL PRESSURES. **(SEE SPECS.)**	A. Throttle Valve Cable	Misadjusted, binding, unhooked, or broken.
	B. Throttle Valve Ass'y. Or #1 Ball Check	1. Throttle lever and bracket assembly binding, unhooked or mispositioned. 2. Throttle valve or plunger valve binding. 3. Shift TV valve binding. 4. #1 check ball missing or leaking.
	C. Pressure Regulator Valve & Spring	1. Valve binding. 2. Wrong spring. 3. Oil pressure control orifice in pump cover plugged, causing high oil pressure. 4. Pressure regulator bore plug leaking.
	D. Manual Valve	Unhooked manual valve.
	E. Intermediate Boost Valve	1. Valve binding—pressures will be incorrect in intermediate and low ranges only. 2. Orifice in spacer plate at end of valve plugged.
	F. Reverse Boost Valve	1. Valve binding—pressures will be incorrect in reverse only. 2. Orifice in spacer plate at end of valve plugged.
	G. Pump Gears & Body	Low oil pressures.

TROUBLESHOOTING
GM TURBO HYDRA-MATIC 200

THE PROBLEM	THE POSSIBLE CAUSE	THE POSSIBLE CONDITION
1-2 SHIFT—FULL THROTTLE ONLY.	A. Throttle Valve Cable	1. Binding, unhooked or broken. 2. Misadjusted.
	B. Throttle Lever & Bracket Assembly	Binding or unhooked.
	C. T.V. Exhaust Ball Lifter or #5 Ball	Binding, mispositioned or unhooked. NOTE: Allowing #5 ball to seat causes full TV pressure regardless of throttle valve position.
	D. Throttle Valve & Plunger	Binding.
	E. Control Valve Ass'y.	Valve body gaskets—leaking, damaged, incorrectly installed.
	F. Case Assembly	Porosity.
FIRST SPEED ONLY, NO 1-2 UPSHIFT.	A. Governor & Governor Feed Passages	1. Plugged governor oil feed orifice in spacer plate. 2. Plugged orifice in spacer plate that feeds governor oil to the shift valves. 3. Governor ball or balls missing in governor assembly. 4. Inner governor cover rubber "O" ring seal missing or leaking. NOTE: (If the outer governor cover "O" ring seal leaks, an external leak will be present along with no upshifts.) 5. Governor shaft seal missing or damaged. 6. Governor driven gear stripped. 7. Governor weights binding on pin. 8. Governor assembly missing.
	B. Control Valve Ass'y.	1-2 shift valve or 1-2 throttle valve stuck in downshift position.
	C. Case	1. Porosity in case channels or undrilled 2nd speed feed holes. 2. Excessive leakage between case bore and intermediate band apply ring. 3. Intermediate band anchor pin missing or unhooked from band. 4. Broken or missing band.
	D. Intermediate Servo Assembly	1. Servo to cover oil seal ring missing or damaged. 2. Porosity in servo cover or piston. 3. Wrong intermediate band apply pin. 4. Incorrect usage of cover and piston.
FIRST AND SECOND SPEEDS ONLY, NO 2-3 SHIFT.	A. Control Valve Assembly & Spacer Plate	1. 2-3 shift valve or 2-3 throttle valve stuck in the downshift position. 2. Direct clutch feed orifice in spacer plate plugged. 3. Valve body gaskets—leaking, damaged or incorrectly installed.
	B. Case	Porosity in case channels.

GM TURBO HYDRA-MATIC 200

TROUBLESHOOTING
GM TURBO HYDRA-MATIC 200

THE PROBLEM	THE POSSIBLE CAUSE	THE POSSIBLE CONDITION
	C. Pump	1. Channels in pump plugged or leaking. 2. Pump to case gasket off location. 3. Rear oil seal ring on pump cover leaking, or missing.
	D. Direct Clutch	1. Oil seals missing or damaged on piston. 2. Direct clutch piston or housing cracked. 3. Direct clutch plates damaged or missing. 4. Direct clutch backing plate snap ring out of groove.
	E. Intermediate Servo Ass'y. (Direct Clutch Accumulator Oil Passages).	1. Servo to case oil seal ring broken or missing on intermediate servo piston. 2. Exhaust hole in case between servo piston seal rings plugged or undrilled.
DRIVE IN NEUTRAL.	A. Manual Linkage	Misadjusted or disconnected.
	B. Forward Clutch	Clutch does not release.
	C. Pump	Cross leakage in pump passages.
	D. Case	Cross leakage to forward clutch passages.
NO DRIVE IN REVERSE OR SLIPS IN REVERSE. (REFER TO OIL PRESSURE CHECKS.)	A. Throttle Valve Cable	Binding or misadjusted.
	B. Manual Linkage	Misadjusted.
	C. Throttle Valve	Throttle valve binding.
	D. Shift T.V. Valve	Binding in valve body bore.
	E. Reverse Boost Valve	Binding in valve body bore.
	F. Low Overrun Clutch Valve	Binding in valve body bore. NOTE: (Line pressure readings will be normal).
	G. Reverse Clutch	1. Piston—cracked broken or missing seals. Clutch plates burned. 2. Wrong selective spacer ring.
	H. Direct Clutch Passages	1. Porosity in case passages. 2. Pump—case to pump gasket mispositioned or damaged. 3. Pump channels cross feeding, leaking or restricted. 4. Pump cover oil seal rings damaged or missing. 5. Piston or housing cracked. 6. Piston seals cut or missing. 7. Housing ball check stuck, leaking or missing. 8. Plates burned. 9. Incorrect piston. 10. Orifices plugged in spacer plate (see spacer plate). 11. Intermediate servo to case oil seal ring cut or missing.
SLIPS ON 1-2 SHIFT.	A. Low Oil Level	Oil will aerate. Correct oil level.
	B. Spacer Plate & Gaskets	1. Second speed feed orifice partially blocked. 2. Gaskets damaged or mispositioned.

GM TURBO HYDRA-MATIC 200

TROUBLESHOOTING
GM TURBO HYDRA-MATIC 200

THE PROBLEM	THE POSSIBLE CAUSE	THE POSSIBLE CONDITION
	C. 1-2 Accumulator Valve	1. Valve sticking in valve body causing low 1-2 accumulator pressure. 2. Weak or missing spring.
	D. 1-2 Accumulator Piston	1. Seal leaking, spring broken or missing. 2. Leak between piston and pin.
	E. Intermediate Band Apply Pin	1. Wrong selection pin. 2. Excessive leakage between apply pin and case.
	F. Intermediate Servo Assembly	1. Porosity in piston. 2. Cover to servo oil seal ring damaged or missing. 3. Incorrect usage of cover and piston.
	G. Throttle Valve Cable	Not adjusted properly.
	H. Throttle Valve	Binding, cause low TV pressure.
	I. Shift T.V. Valve	Binding.
	J. Intermediate Band	Worn or burned.
	K. Case	Porosity in 2nd clutch passages.
ROUGH 1-2 SHIFT.	A. Throttle Valve Cable	1. Not adjusted properly. 2. Binding.
	B. Throttle Valve	1. TV plunger binding. 2. Throttle valve binding.
	C. Shift T.V. Valve	1. Binding.
	D. 1-2 Accumulator Valve	1. Binding.
	E. Intermediate Servo Assembly	1. Wrong pin. 2. Servo piston to case oil seal ring damaged or missing.
	F. 1-2 Accumulator	1. Oil ring damaged. 2. Piston stuck. 3. Broken or missing spring. 4. Bore damaged.
SLIPS 2-3 SHIFT.	A. Oil Level Low	Correct oil level.
	B. Throttle Valve Cable	Not adjusted properly.
	C. Throttle Valve	Binding.
	D. Spacer Plate & Gaskets	1. Direct clutch orifice partially blocked in spacer plate. 2. Gaskets mispositioned or damaged.
	E. Intermediate Servo Assembly	Servo to case oil seal ring damaged.
	F. Direct Clutch Feed	1. Porosity in direct clutch feed channels in case. 2. Pump to case gasket mispositioned or damaged. 3. Pump channels cross feedings, leaking or restricted. 4. Pump cover oil seal rings damaged or missing. 5. Direct clutch piston or housing cracked. 6. Piston seals cut or missing. 7. Direct clutch plates burned.

GM TURBO HYDRA-MATIC 200

TROUBLESHOOTING
GM TURBO HYDRA-MATIC 200

THE PROBLEM	THE POSSIBLE CAUSE	THE POSSIBLE CONDITION
ROUGH 2-3 SHIFT.	A. Throttle Valve Cable	1. Not adjusted properly. 2. Binding.
	B. Throttle Valve and Plunger	1. TV plunger binding. 2. Throttle valve binding.
	C. Shift T.V. Valve	Binding.
	D. Intermediate Servo Assembly	Exhaust hole undrilled or plugged between intermediate servo piston seals, not allowing intermediate servo piston to complete its stroke.
	E. Direct Clutch Exhaust Valve Ball Check #4	Missing or mispositioned.
NO ENGINE BRAKING— INTERMEDIATE RANGE— 2nd GEAR (CHECK OIL PRESSURE SPECS.)	A. Intermediate Boost Valve	Binding in valve body.
	B. Intermediate—Rev. Ball Check #3 Ball	Mispositioned or missing.
	C. Shift T.V. Ball Check #1 Ball	Mispositioned or missing.
	D. Intermediate Servo Assembly	Servo to cover oil seal ring missing or damaged.
	E. Intermediate Band	1. Off anchor pin. 2. Broken or burned.
NO ENGINE BRAKING IN LOW RANGE 1st GEAR (CHECK OIL PRESSURE SPECS.)	A. Low Overrun Clutch Valve	Binding in valve body.
	B. Lo-Reverse Clutch Assembly	NOTE: (No reverse should also be a complaint with any of the (4) following conditions). 1. Piston seals broken or missing. 2. Porosity in piston or housing. 3. Clutch housing snap ring out of case. 4. Cup plug or rubber seal missing or damaged between case and low reverse clutch housing.
NO PART THROTTLE OR DETENT DOWNSHIFTS (CHECK OIL PRESSURE SPECS.)	A. Throttle Plunger Bushing	Passages not open.
	B. 2-3 Throttle Valve Bushing	Passages not open.
	C. Valve Body Gaskets	Mispositioned or damaged.
	D. Spacer Plate	Hole plugged or undrilled.
	E. Throttle Valve Cable	Improperly set.
	F. Shift T.V. Valve	Binding.
	G. Throttle Valve	Binding.
LOW OR HIGH SHIFT POINTS (CHECK OIL PRESSURE SPECS.)	A. Throttle Valve Cable	Binding or disconnected.
	B. Throttle Valve	Binding.
	C. Shift T.V. Valve	Binding.
	D. T.V. Shift Ball, #1 Ball	Missing or mispositioned.
	E. Throttle Valve Plunger	Binding.
	F. 1-2 or 2-3 Throttle Valves	Binding in bushings.
	G. Valve Body Gaskets	Mispositioned or damaged.
	H. Pressure Regulator Valve	Binding.
	I. T.V. Exhaust Ball #5 and Lifter	1. Mispositioned or unhooked. 2. Missing.

GM TURBO HYDRA-MATIC 200

TROUBLESHOOTING
GM TURBO HYDRA-MATIC 200

THE PROBLEM	THE POSSIBLE CAUSE	THE POSSIBLE CONDITION
	J. Throttle Lever and Bracket Assembly	1. Binding, unhooked or loose at mounting valve body bolt. 2. Not positioned at the throttle valve plunger bushing pin locator.
	K. Governor Shaft To Cover Seal Ring	Broken or missing.
	L. Governor Cover "O" Rings	Broken or missing. NOTE: The outer ring will leak externally and the inner ring internally.
	M. Case	Porosity.
WON'T HOLD IN PARK.	A. Manual Linkage	Misadjusted.
	B. Internal Linkage	1. Park pawl binding in case. 2. Actuator rod or plunger damaged. 3. Parking pawl broken. 4. Parking bracket, loose or damaged.
	C. Inside Detent Lever and Pin Assembly	1. Nut loose. 2. Hole in lever worn or damaged.
	D. Manual Detent Roller and Spring Assembly	1. Bolt loose that holds roller assembly to valve body. 2. Pin or roller damaged or mispositioned, or missing.
TRANSMISSION NOISY.	A. Pump Noise	1. Oil level setting low. 2. Cavitation due to plugged screen, porosity in intake circuit, water in oil. 3. Pump gears—damaged.
	B. Gear Noise	1. Transmission grounded to body. 2. Roller bearings worn or damaged,

GM TURBO HYDRA-MATIC 200

GM Turbo Hydra-matic 200

Applications

Chevrolet—Chevette, Nova, Monza
Oldsmobile—Omega (through '79), Starfire
Pontiac—Sunbird, Ventura, Phoenix (through '79)
Buick—Skyhawk, Skylark (through '79)

Throttle Valve Cable Check in-Car

To check the throttle valve (TV) cable for freeness, pull out on the upper end of the cable. The cable should travel a short distance with light spring resistance. This light resistance is due to the small coiled return spring on the TV lever and bracket that returns the lever to the zero TV or closed thröttle position. Pulling the cable farther out moves the lever to contact the TV plunger which compresses the TV spring which has more resistance. By releasing the upper end of the TV cable, it should return to the zero TV position.

This checks the cable in the housing, the TV lever and bracket, and the TV plunger in its bushing for freeness.

Checking Transmission Fluid Level

Proceed by having:
1 Engine running
2 Vehicle on level surface
3 Brakes applied
4 Lever run through all ranges
5 Transmission in park

Now check oil level. If oil is low, check for possible causes. See Oil Leak data which follows later.

The oil level, at room temperature of approximately 21° C (70° F.), should be approximately 6mm (¼") below the ADD mark.

The oil level should be between the ADD and FULL marks at normal operating temperature 90° C (200° F.). This temperature is obtained after at least 24 km (15 miles) of expressway driving or equivalent city driving. Also, at normal operating temperature, the oil will heat the gauge end of the dip stick to a degree where the average person cannot grasp it

firmly with his bare hand without discomfort.

CHILTON CAUTION: *Do not overfill transmission, as this may cause foaming and loss of oil through the vent pipe.*

Outside Manual Linkage Adjustment

The transmission manual linkage must be adjusted so that the indicator quadrant and stops correspond with the transmission detents. If the linkage is not adjusted properly, an internal leak could occur which could cause a clutch or band to slip.

Refer to the car division shop manual for manual linkage adjustment procedure.

NOTE: If a manual linkage adjustment is made, the neutral safety switch should be adjusted, if necessary. The neutral safety switch should be adjusted so that the engine will start only in park and neutral positions. With the selector lever in the park position, the parking pawl should freely engage and prevent the vehicle from rolling.

GM TURBO HYDRA-MATIC 200

GM TURBO HYDRA-MATIC 200 OIL PRESSURE CHECKS

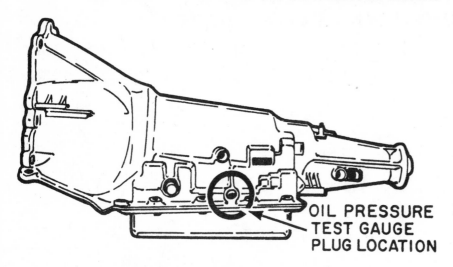

OIL PRESSURE
TEST GAUGE
PLUG LOCATION

PRELIMINARY CHECKING PROCEDURE

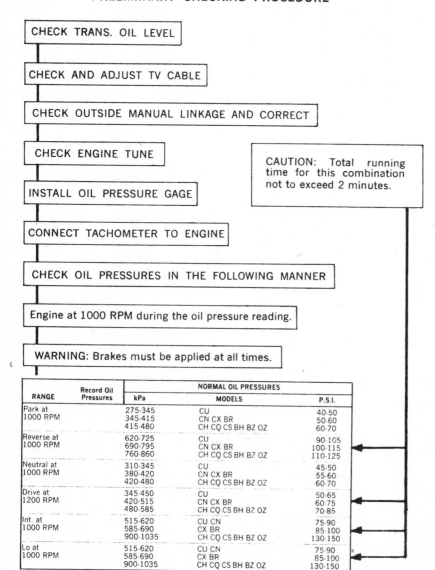

CHECK TRANS. OIL LEVEL

CHECK AND ADJUST TV CABLE

CHECK OUTSIDE MANUAL LINKAGE AND CORRECT

CHECK ENGINE TUNE

INSTALL OIL PRESSURE GAGE

CONNECT TACHOMETER TO ENGINE

CHECK OIL PRESSURES IN THE FOLLOWING MANNER

Engine at 1000 RPM during the oil pressure reading.

WARNING: Brakes must be applied at all times.

CAUTION: Total running time for this combination not to exceed 2 minutes.

RANGE	Record Oil Pressures	NORMAL OIL PRESSURES		
		kPa	MODELS	P.S.I.
Park at 1000 RPM		275-345	CU	40-50
		345-415	CN CX BR	50-60
		415-480	CH CQ CS BH BZ OZ	60-70
Reverse at 1000 RPM		620-725	CU	90-105
		690-795	CN CX BR	100-115
		760-860	CH CQ CS BH BZ OZ	110-125
Neutral at 1000 RPM		310-345	CU	45-50
		380-420	CN CX BR	55-60
		420-480	CH CQ CS BH BZ OZ	60-70
Drive at 1200 RPM		345-450	CU	50-65
		420-515	CN CX BR	60-75
		480-585	CH CQ CS BH BZ OZ	70-85
Int. at 1000 RPM		515-620	CU CN	75-90
		585-690	CX BR	85-100
		900-1035	CH CQ CS BH BZ OZ	130-150
Lo at 1000 RPM		515-620	CU CN	75-90
		585-690	CX BR	85-100
		900-1035	CH CQ CS BH BZ OZ	130-150

Road Test Procedures

DRIVE RANGE

Position selector lever in DRIVE RANGE and accelerate the vehicle. A 1–2 and 2–3 shift should occur at all throttle openings. (The shift points will vary with the throttle openings). Check part throttle 3–2 downshift at 50 km/h (30 mph) by quickly opening throttle approximately three-fourths, the transmission should downshift 3–2. Check for 3–2 downshifts at 80 km/h (50 mph), by depressing the accelerator fully.

INTERMEDIATE RANGE

Position the selector lever in INTERMEDIATE RANGE and accelerate the vehicle. A 1–2 shift should occur at all throttle openings. (No 2–3 shift can be obtained in this range). The 1–2 shift point will vary with throttle opening. Check detent 2–1 downshift at 32 km/h (20 mph) The transmission should downshift 2–1.

NOTE: The 1–2 shift in INTERMEDIATE RANGE is somewhat firmer than in DRIVE RANGE. This is normal.

LO RANGE

Position the selector lever in Lo RANGE and accelerate the vehicle. No upshift should occur in this range, except possibly in some vehicles which have a high numerical axle ratio and/or engine (r.p.m.).

INTERMEDIATE RANGE OVERRUN BRAKING

Position the selector lever in DRIVE RANGE, and with the vehicle speed at approximately 80 km/h (50 mph) closed or 0 throttle, move the selector lever to INTERMEDIATE RANGE. The transmission should downshift to 2nd. An increase in engine r. p. m. and an engine braking effect should be noticed.

LO RANGE OVERRUN BRAKING:

At 64 km/h (40 mph), with throttle closed, move the selector lever to Lo. A 2–1 downshift should occur in the speed range of approximately 64 to 40 km/h (40 to 25 mph), depending on axle ratio and valve body calibration. The 2–1 downshift at closed throttle will be accompanied by increased engine rpm and an engine braking effect should be noticed. Stop vehicle.

REVERSE RANGE

Position the selector lever in REVERSE POSITION and check for reverse operation.

Converter Stator Operation Diagnosis

The torque converter stator assembly and its related roller clutch can possibly have one of two different type malfunctions. They are:

A. The stator assembly freewheels in

GM TURBO HYDRA-MATIC 200

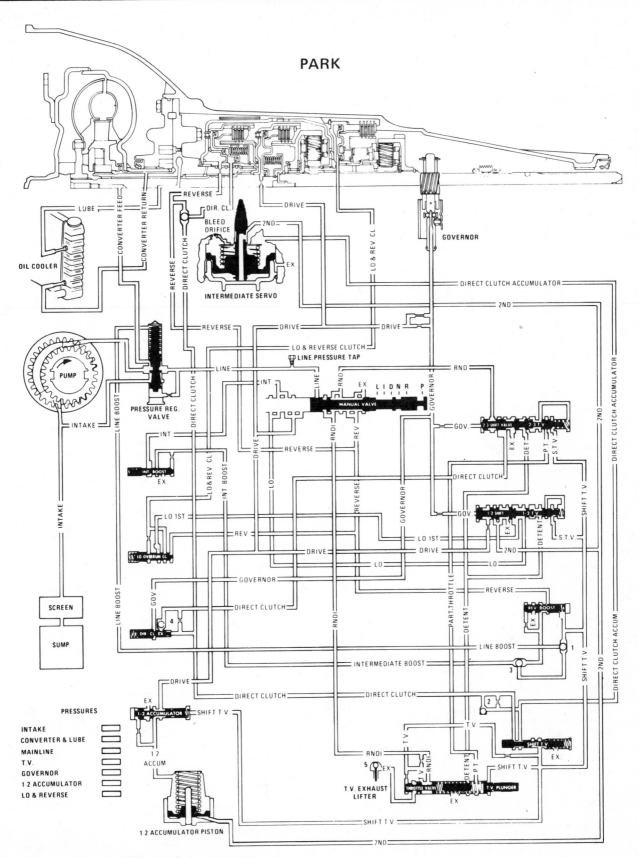

Hydraulic schematic GM Turbo Hydra-matic 200

GM TURBO HYDRA-MATIC 200

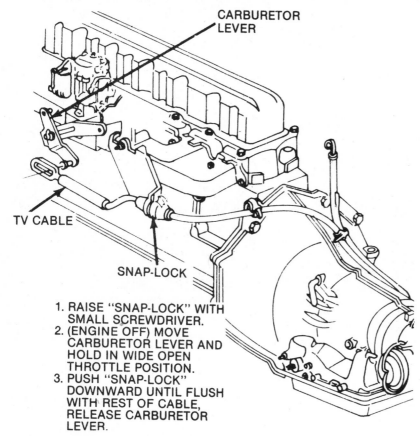

1. RAISE "SNAP-LOCK" WITH SMALL SCREWDRIVER.
2. (ENGINE OFF) MOVE CARBURETOR LEVER AND HOLD IN WIDE OPEN THROTTLE POSITION.
3. PUSH "SNAP-LOCK" DOWNWARD UNTIL FLUSH WITH REST OF CABLE, RELEASE CARBURETOR LEVER.

Throttle valve (T.V) cable adjustment

both directions
B. The stator assembly remains locked up at all times

CONDITION A

If the stator roller clutch becomes ineffective, the stator assembly freewheels at all times in both directions. With this condition, the vehicle will tend to have poor acceleration from a standstill. At speeds above 30–35 mph, the vehicle may act normal. If poor acceleration problems are noted, it should first be determined that the exhaust system is not blocked, the engine is in good tune and the transmission is in first (1st) gear when starting out.

If the engine will freely accelerate to high r. p. m. in Neutral (N), it can be assumed that the engine and exhaust system are normal. Driving the vehicle in Reverse (R) and checking for poor performance will help determine if the stator is freewheeling at all times.

CONDITION B

If the stator assembly remains locked up at all times, the engine r. p. m. and vehicle speed will tend to be limited or restricted at high speeds. The vehicle performance when accelerating from a standstill will be normal. Engine over heating may be noted. Visual examination of the

converter may reveal a blue color from the over heating that will result.

Under conditions A or B above, if the converter has been removed from the transmission, the stator roller clutch can be checked by inserting a finger* into the splined inner race of the roller clutch and trying to turn the race in both directions. The inner race should turn freely in the clockwise direction, but not turn or be very difficult to turn in the counterclockwise direction.

*NOTE: Do not use such items as the pump cover or stator shaft to turn the race as the results may be misleading.

External Oil Leaks

Before attempting to correct an oil leak, the actual source of the leak must be determined. In many cases, the source of the leak can be deceiving due to wind flow around the engine and transmission.

The suspected area should be wiped clean of all oil before inspecting for the source of the leak. Red dye is used in the transmission oil at the assembly plant and will indicate if the oil leak is from the transmission.

Oil leaks around the engine and transmission are generally carried toward the rear of the car by the air stream. For example, a transmission oil filler tube to case leak will sometimes appear as a leak at the rear of the transmission. In determining

the source of an oil leak, two checks should be made.
1 With the engine running, check for external oil pressure leaks.
2 With the engine off, check for oil leaks due to the raised oil level caused by drain-back of converter oil into the transmission.

POSSIBLE POINTS OF OIL LEAKS
1 Transmission Oil Pan Leak
   a. Attaching bolts not correctly torqued.
   b. Improperly installed or damaged pan gasket.
   c. Oil pan gasket mounting face not flat.
2 Oil Comes Out Vent Pipe
   a. Transmission over-filled or underfilled.
   b. Water in oil.
   c. Screen gasket damaged or improperly assembled, causing oil to foam.
3 Case Leak
   a. Filler pipe "O" ring seal damaged or missing; misposition of filler pipe bracket to engine—"Loading" one side of the "O" ring.
   b. T. V. cable "O" ring seal missing, damaged or improperly installed.
   c. Rear seal assembly—damaged or improperly installed.
   d. Governor cover and "O" ring—damaged or missing.
   e. Speedo gear "O" ring damaged.
   f. Manual shaft seal—damaged improperly installed.
   g. Line pressure tap plug.
   h. Porous case.
4 Front End Leak
   a. Front seal—damaged check converter neck for nicks, etc., also for pump bushing moved forward) garter spring missing.
   b. Pump attaching bolts and seals—damaged, missing, bolts loose.
   c. Converter—leak in weld.
   d. Pump "O" ring seal—damaged. (Also check pump oil ring groove and case bore).
   e. Porous casting (pump or case).

Disassembly of Major Units

Removal of Governor, Oil Pan and Screen

1 Using small screwdriver, remove governor cover retaining ring.
2 Using pliers, remove governor cover and discard two (2) seal rings; seal rings may be located in case.

NOTE: Governor assembly and governor to case washer may come out with cover.

CHILTON CAUTION: *Do not use any type of pliers to remove governor assembly.*

3 Remove governor assembly and governor to case washer from case. It may be necessary to rotate output shaft counter-clockwise while removing the governor.

GM TURBO HYDRA-MATIC 200

NOTE: If the governor to case washer falls off the governor, use a small magnet to remove washer from case.

4 Remove oil pan and discard gasket.
5 Remove oil screen and discard gasket.

Removal of Valve Body and Linkage

1 Remove throttle lever and bracket assembly. Do not bend throttle lever link.

NOTE: Throttle lever plunger and spring may separate from throttle lever and bracket assembly.

2 Remove manual detent roller and spring assembly
3 Remove remaining control valve assembly attaching bolts.

NOTE: Manual valve can fall out of either end of its bore.

4 Holding manual valve with finger, remove control valve assembly, spacer plate, and gaskets together, to prevent the dropping of four (4) check balls, located in the control valve assembly.
5 Lay control valve assembly down with spacer plate side up and discard gaskets.

NOTE: Some spacer plates may be of a rubberized type which do not require gaskets and are re-used if not damaged.

6 Remove 1–2 accumulator spring.
7 Remove fifth (5th) check ball located in case.

Removal of Speedometer Driven Gear and Intermediate Servo

1 Remove speedometer driven gear bolt, washer and retainer.
2 Remove speedometer driven gear.
3 Using small screwdriver, remove intermediate servo cover retaining ring.
4 Using pliers, remove intermediate servo cover and discard seal rings; seal ring may be located in case.
5 Remove intermediate servo piston and band apply pin assembly.

NOTE: If intermediate servo cover and seal assembly will not remove, place shop towels and hand over cover and case.

Apply air pressure into the direct clutch accumulator port.

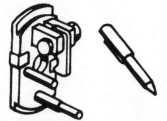

J25014 Intermediate band apply pin guage

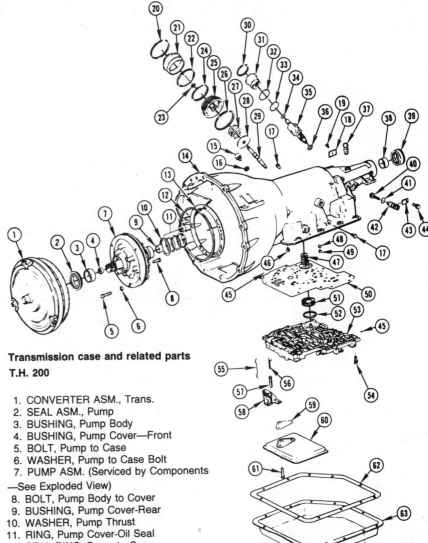

Transmission case and related parts
T.H. 200

1. CONVERTER ASM., Trans.
2. SEAL ASM., Pump
3. BUSHING, Pump Body
4. BUSHING, Pump Cover—Front
5. BOLT, Pump to Case
6. WASHER, Pump to Case Bolt
7. PUMP ASM. (Serviced by Components —See Exploded View)
8. BOLT, Pump Body to Cover
9. BUSHING, Pump Cover-Rear
10. WASHER, Pump Thrust
11. RING, Pump Cover-Oil Seal
12. SEAL RING, Pump to Case
13. GASKET, Pump to Case
14. CASE ASM., Trans.
15. CONNECTOR, Cooler Pipe (Inverted Flare)
16. GASKET, Oil Cooler Fitting
17. PLUG, Case (⅛" Pipe—Hex. Hd.)
18. PLATE, Trans. Name
19. SCREW, Nameplate Attaching
20. RING, Servo Cover Retaining
21. COVER, Intermediate Servo
22. SEAL, Cover "O" Ring
23. RING, Retainer
24. RING, Piston Oil Seal—Inner
25. PISTON, Intermediate Servo
26. RING, Piston Oil Seal—Outer
27. SPRING, Intermediate Servo Cushion
28. WASHER, Intermediate Servo Piston
29. PIN, Intermediate Band Apply
30. RING, Governor Cover Retaining
31. COVER, Governor
32. SEAL, Governor Cover—Inner
33. SEAL, Governor Cover—Outer
34. RING, Governor Oil Seal
35. GOVERNOR ASM., Trans.
36. WASHER, Governor Driven Gear to Case
37. VENT ASM., Trans.
38. BUSHING, Trans. Case—Rear
39. SEAL ASM., Rear Oil
40. GEAR, Speedo. Driven
41. SEAL, Speedo. Fitting "O" Ring
42. FITTING ASM., Speedo. Driven Gear
43. RETAINER, Speedo. Gear
44. BOLT, Retainer to Case
45. BALL, Valve Body Check
46. PIN, Band Anchor
47. SPRING, 1-2 Accumulator
48. SEAL, Housing to Case (Part of #49)
49. PLUG, Case & Housing Cup
50. PLATE, Valve Body Spacer
51. PISTON, 1-2 Accumulator
52. SEAL, Accumulator Piston (Part of #51)
53. VALVE ASM., Control
54. BOLT, Valve Body to Case
55. LINK, Throttle Lever to Cable
56. LIFTER, T.V. Exhaust Valve
57. SPRING, T.V. Exhaust Valve Lifter
58. LEVER & BRACKET ASM., Throttle
59. GASKET, Valve Body Screen
60. SCREEN ASM., Trans. Oil
61. BOLT, Valve Body to Case
62. GASKET, Oil Pan
63. PAN, Trans. Oil
64. SCREW, Pan to Case

6 Using J-25014, intermediate band apply pin gauge kit, check for proper intermediate band apply pin as follows:

a. Install J-25014-2 in intermediate servo bore and retain with intermediate servo cover retaining ring, aligning ring with gap at case slot.

b. Install pin J-25014-1 into J-25014-2.

c. Install a dial indicator and position dial indicator point on top of J-25014-2 zero post and set dial indicator to zero.

d. Align stepped side of pin J-25014-1 with torquing arm of J-25014-2. Arm must stop against step of pin J-25014-1.

NOTE: If band selection pin does not register between the high and low limits, look for possible problem with the intermediate band, direct clutch or case.

e. Make sure band anchor pin is located in the case.

f. Apply 12 Nm (100″ lbs.) of torque to hex nut on side of gage. Slide dial indicator over pin J-25014-1. Read dial indicator and see chart below for proper size.

INTERMEDIATE BAND APPLY PIN CHART

Dial Indicator Reading	Apply Pin Identification
.0 - .72mm (.0 -.029″)	1 Ring
.72-1.44mm (.029-.057″)	2 Rings
1.44-2.16mm (.057-.086″)	3 Rings
2.16-2.88mm (.086-.114″)	Wide Band

NOTE: Dial indicator travel is reversed, making the indicator readings backwards. On an indicator that ranges from 0–100, a .5 mm (.020″) travel will appear to read 2 mm (.080″), a 1.5 mm (.060″) travel will read 1 mm (.040″). The identification ring is located on the band end of the pin.

7 After check is completed, remove retaining ring, Tool J-25014-2 and Tool J-25014-1.

Removal of Oil Pump and Front Unit Components

1 Using special tools J-24773 oil pump remover and end play checking fixture, J-25013 output shaft and rear unit support fixture and J-25022 end play checking fixture adapter, check front unit end play as follows:

a. Install J-25013-1 sleeve on output shaft first, then bolt J-25013-5 on end of case.

b. Turn transmission to vertical position, pump side up.

c. Remove pump to case bolt and washer and install 278 mm (11″) long bolt and locking nut.

d. Push turbine shaft downward.

e. Install J-25022 on J-24773 tool and secure on end of turbine shaft.

f. Mount dial indicator and clamp assembly on bolt, positioning indica-

J24773 Oil pump remover and end play checking fixture

J25022 End-play checking fixture adapter

tor point against cap nut of J-24773.

g. Move output shaft upward by turning the adjusting screw on J-25013-5 until the red line on sleeve J-25013-1 begins to disappear, then set dial indicator to zero.

h. Pull J-24773 on turbine shaft upward and read end play. Front unit end play should be 0.56 mm-1.30 mm (.022″-.051″).

i. Remove dial indicator, clamp assembly, J-24773 and J-25022.

j. Do not remove J-25013-5 or J-25013-1.

2 If necessary, remove pump oil seal and discard.

3 Remove remaining pump to case bolts and washers; discard washers.

4 Using J-24773 tool, remove pump assembly, pump to case gasket and discard gasket.

5 Grasp turbine shaft and remove direct and forward clutch assemblies.

NOTE: Selective washer controlling end play is located between the output shaft and turbine shaft. If more or less washer thickness is required to bring end play within specifications, select proper washer from the following chart:

FRONT UNIT END PLAY CHART

Thickness	Identification Number and/or Color
1.66-1.77mm (.065-.070″)	1
1.79-1.90mm (.070-.075″)	2
1.92-2.03mm (.076-.080″)	3 Black
2.05-2.16mm (.081-.085″)	4 Light Green
2.18-2.29mm (.086-.090″)	5 Scarlet
2.31-2.24mm (.091-.095″)	6 Purple
2.44-2.55mm (.096-.100″)	7 Cocoa Brown
2.57-2.68mm (.101-.106″)	8 Orange
2.70-2.81mm (.106-.111″)	9 Yellow
2.83-2.94mm (.111-.116″)	10 Light Blue
2.96-3.07mm (.117-.121″)	11
3.09-3.20mm (.122-.126″)	12

6 Lift direct clutch assembly off forward clutch assembly.

NOTE: Do not inter-mix the forward and direct clutch steel plates. The direct clutch steel plates have four (4) teeth

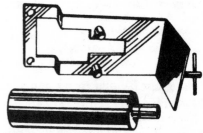

J25013 Output shaft and rear unit support fixture

omitted, two (2) directly opposite the other.

7 Remove intermediate band assembly.

8 Remove band anchor pin.

9 Remove output shaft to turbine shaft front selective washer.

NOTE: This washer may be stuck to the end of the turbine shaft.

10 Check rear unit end play as follows:

a. Loosen J-25013-5 adjusting screw on output shaft and push output shaft downward.

CHILTON CAUTION: Do not install clamp assembly on any machined case surfaces.

b. Install gage clamp on case as shown in figure 31a.

c. Install dial indicator gage and extension. Position indicator point against end of output shaft and set dial indicator to zero.

d. Move output shaft upward by turning adjusting screw on J-25013-5, until red line on sleeve J-25013-1 begins to disappear; then read end play (rear unit end play should be 0.10-0.64mm (.004-.025″).

e. Remove dial indicator and clamp assembly. Do not remove J-25013 tools.

NOTE: It may be necessary to tighten J-25013 adjusting screw on output shaft to remove snap ring.

11 Using snap ring pliers, remove output shaft to selective washer snap ring.

NOTE: Selective washer controlling end play is located between the front internal gear thrust washer and output shaft snap ring. If more or less washer thickness is required to bring end play within pseicifications, select proper washer from the following chart:

REAR UNIT END PLAY CHART

Thickness	Identification Number and/or Color
2.90-3.01mm (.114-.119″)	1 Orange
3.08-3.19mm (.121-.126″)	2 White
3.26-3.37mm (.128-.133″)	3 Yellow
3.44-3.55mm (.135-.140″)	4 Blue
3.62-3.73mm (.143-.147″)	5 Red
3.80-3.91mm (.150-.154″)	6 Brown
3.98-4.09mm (.157-.161″)	7 Green
4.16-4.27mm (.164-.168″)	8 Black
4.34-4.45mm (.171-.175″)	9 Purple

GM TURBO HYDRA-MATIC 200

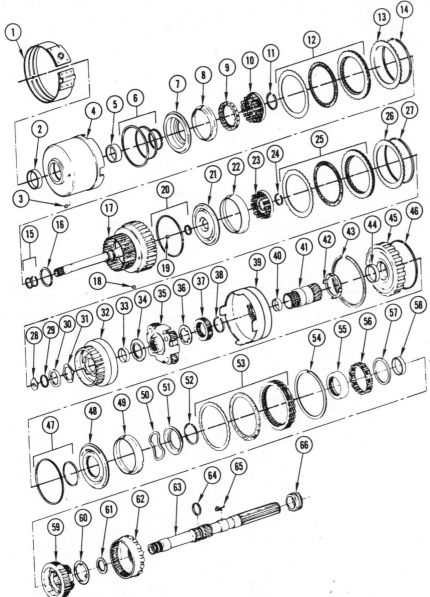

13. PLATE, Clutch Backing
14. RING, Clutch Snap
15. SEAL, Turbine Shaft Ring
16. WASHER, Direct to Forward Clutch Thrust
17. HOUSING ASM., Forward Clutch
18. RETAINER & BALL ASM., Check Valve
19. PLUG, Clutch Housing Cup
20. SEAL PKG., Forward Clutch Piston
21. PISTON ASM., Forward Clutch
22. RING, Forward Clutch Apply
23. RETAINER & SPRING ASM.,
    Forward Clutch
24. RING, Clutch Snap
25. PLATE, Forward Clutch
26. PLATE, Clutch Backing
27. RING, Clutch Snap
28. WASHER, Thrust (Selective—Front)
29. RING, Output Shaft Snap
30. WASHER, Thrust (Selective—Rear)
31. WASHER, Intermediate Gear Thrust
32. GEAR & BUSHING ASM., Front Int.
33. BUSHING, Front Int. Gear
34. BEARING ASM., Int. Gear to Carrier Roller
35. CARRIER ASM., Front—Complete
36. BEARING ASM., Front Carrier to
    Sun Gear Thrust
37. GEAR, Front Sun
38. RING, Input Drum to Rear Sun Gear Snap
39. DRUM, Input
40. BUSHING, Rear Sun Gear
41. GEAR & BUSHING ASM., Rear Sun
42. WASHER, Drum to Housing Thrust
43. RING, Housing to Case Snap
44. BUSHING, Lo & Reverse Clutch Housing
45. HOUSING ASM., Lo & Reverse Clutch
46. SPACER, Housing to Case
47. SEAL PKG., Lo & Reverse Piston
48. PISTON ASM., Lo & Reverse Clutch
49. RING, Clutch Apply
50. SPRING, Reverse Clutch Release Wave
51. RETAINER, Lo & Reverse Clutch Spring
52. RING, Retainer to Housing Snap
53. PLATE, Lo & Reverse Clutch
54. WASHER, Lo & Reverse Clutch—Selective
55. RACE, Lo Clutch Roller
56. ROLLER ASM., Lo Clutch
57. WASHER, Rear Carrier to Lo Race Thrust
58. BUSHING, Rear Carrier
59. CARRIER ASM., Rear—Complete
60. WASHER, Rear Carrier to Lo Race Thrust
61. BEARING ASM., Rear Sun Gear to Rear
    Internal Gear Roller Thrust
62. GEAR, Rear Internal
63. SHAFT, Output
64. RING, Rear Internal Gear to Output
    Shaft Snap
65. CLIP, Speedo. Drive Gear
66. GEAR, Speedo. Drive

Internal transmission and components T.H. 200

1. BAND ASM., Intermediate
2. BUSHING, Direct Clutch—Front
3. RETAINER & BALL ASM., Check Valve
4. HOUSING & DRUM ASM., Direct Clutch
5. BUSHING, Direct Clutch—Rear
6. SEAL PKG., Direct Clutch Piston

7. PISTON ASM., Direct Clutch
8. RING, Direct Clutch Apply
9. GUIDE, Release Spring
10. RETAINER & SPRING ASM., Direct Clutch
11. RING, Spring Retainer Snap
12. PLATE, Direct Clutch

Removal of Front Internal Gear

1 Remove front internal gear, rear selective washer and tanged thrust washer.
2 Remove rear selective washer and tanged thrust washer from front internal gear.
3 Remove front carrier assembly and the front internal gear to front carrier roller bearing assembly.

NOTE: The front sun gear to front carrier roller thrust bearing and race may

come out as the front carrier is removed. The CU and CN models have a two pinion carrier, with the pinions directly opposite each other.

4 Remove front sun gear and front sun gear to front carrier roller thrust bearing and thrust race.

NOTE: This thrust bearing requires only one thrust race or could use the optional bearing and race assembly.

Removal of Input Drum and Rear Sun Gear Assembly

1 Loosen J-25013-5 adjusting screw and remove input drum and rear sun gear.
2 Remove the four (4) tanged input drum to reverse clutch housing thrust washer from rear of input drum or from reverse clutch housing.

If lo and reverse clutch housing assembly has to be removed the lo and reverse clutch piston travel must be checked.

Component Removal and Overhaul

GM TURBO HYDRA-MATIC 200

1 Using a #14 sheet metal screw, remove housing to case cup plug and seal by turning screw 2 or 3 turns and pulling straight out. Discard cup plug and seal.

NOTE: If cup plug will not remove, grind approximately 20 mm (¾") from end of 6.3 mm (#4) easy out to remove cup plug. Then use #14 sheet metal screw to remove seal.

2 Remove lo and reverse clutch housing to case beveled snap ring.

NOTE: The flat side of the ring should have been against the lo and reverse clutch housing with beveled side up.

3 Using J-25012 reverse clutch housing installer and remover remove lo and reverse clutch housing assembly by moving J-25012 back and forth.
4 Remove lo and reverse clutch housing to case spacer ring.

Removal of Rear Gear Components

1 Grasp output shaft and lift out remainder of rear unit parts and lay down in horizontal position.
2 Remove lo and reverse clutch selective spacer. Remove roller clutch and rear carrier assembly off output shaft.

NOTE: The CU and CN models have a two pinion carrier, with the pinions directly opposite each other.

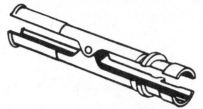

J25012 Reverse clutch housing installer and remover

3 Remove the four (4) tanged rear carrier to rear internal gear thrust washer off the end of the rear carrier, or inside the rear internal gear.
4 Remove lo and reverse clutch plates off output shaft.
5 Remove rear internal gear to rear sun gear roller thrust bearing assembly off rear internal gear.
6 Remove rear internal gear off output shaft.
7 Turn transmission to horizontal position and remove J-25013-5 and J-25013-1 tools from case. Turn transmission to vertical position with rear end up.
8 If necessary, remove rear oil seal.

Removal of Manual Shaft and Parking Pawl Components

1 Remove hex nut which holds inside detent lever to manual shaft.
2 Remove parking brake actuator rod and inside detent lever assembly.

3 Remove manual shaft retaining pin from case and slide manual shaft out.
4 Inspect manual shaft to case seal for damage. If necessary, pry out manual shaft seal using screwdriver.
5 Remove parking brake bracket.
6 Remove parking brake pawl shaft retaining pin.
7 Using 6.3 mm (#4) easy out, remove parking brake pawl cup plug and discard.

NOTE: Grind approximately 20 mm (¾") from end of 6.3 mm (#4) easy out to remove cup plug.

8 Using sheet metal screw or 4 mm (#3) easy out, remove parking brake pawl shaft.
9 Remove parking pawl and return spring.

Transmission Inspection and Reassembly

TEFLON SEALS

If any Teflon seal rings are damaged (distorted, cut, scored, etc.), or do not rotate freely in their groove, and replacement is necessary, do the following:

1 Remove and discard old scarf cut seal rings; full circle rings must be cut off.
2 Inspect seal ring groove for burrs or damage.
3 When installing scarf cut seal rings, do not overstretch. Make sure cut ends are in same relation as cut. Also, make sure rings are seated in the grooves to

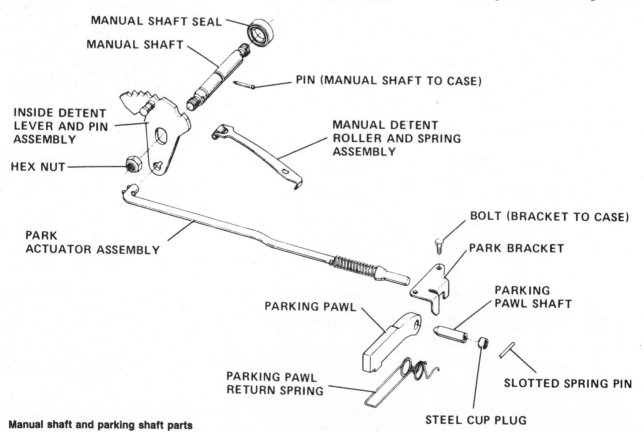

MANUAL SHAFT SEAL

MANUAL SHAFT

PIN (MANUAL SHAFT TO CASE)

INSIDE DETENT LEVER AND PIN ASSEMBLY

MANUAL DETENT ROLLER AND SPRING ASSEMBLY

HEX NUT

PARK ACTUATOR ASSEMBLY

BOLT (BRACKET TO CASE)

PARK BRACKET

PARKING PAWL SHAFT

PARKING PAWL

PARKING PAWL RETURN SPRING

SLOTTED SPRING PIN

STEEL CUP PLUG

Manual shaft and parking shaft parts

GM TURBO HYDRA-MATIC 200

prevent damage to the rings during re-assembly of mating part over rings. Retain with petrolatum.

4 When installing full circle seal rings, size and reshape seals, using proper tools.

5 New scarf cut or full circle teflon seal rings may appear to be distorted after being installed. Once exposed to normal transmission oil temperatures, the new seal rings will return to their normal shape and fit freely in their bores.

6 The teflon seal rings allow for a free fit in their bores after operation. The free fit of the rings in their bores does not indicate leakage during operation.

CASE INSPECTION

CU and CN models have a taped hole with pipe plug in it near the intermediate servo bore, except air conditioned vehicles which will have an A/C overdrive pressure switch. Do not remove this plug. This is direct clutch pressure.

1 Inspect case assembly for damaged, cracks, porosity, or interconnected oil passages. If case is porous, refer to porosity repair section.

2 Inspect the exhaust vents for being opened.

3 Inspect reverse clutch lugs, governor bore, intermediate servo bore, speedometer bore and snap ring grooves for damage.

4 Inspect reverse clutch seal and intermediate band anchor pin bores for damage.

5 Inspect vent assembly in case for damage. Do not remove unless replacement is required.

6 Turn transmission to vertical position, rear end up.

7 Inspect cooler line connectors for damage. Do not remove unless replacing.

8 Inspect case bushing for damage or scoring.

9 If removed, install a new oil seal, using J-21426 rear oil seal installer.

10 If removed, install vent assembly into case using rubber or plastic hammer.

11 If removed, install new cooler line washer(s) and connector(s); torque connector(s) to 11–16 Nm (8–10 ft./lbs.).

Manual Shaft and Parking Pawl Components

INSPECTION

1 Align actuator rod tangs with hole in inside detent lever and separate.

2 Inspect parking brake actuator rod for damage, or broken retainer lugs.

3 Inspect parking brake actuator spring for damage.

4 Inspect actuator for free fit on actuator rod.

5 Inspect parking pawl for cracks or damage.

6 Inspect parking pawl return spring for deformed end or coils.

7 Inspect parking pawl shaft for damage.

J21426 rear oil seal installer

8 Inspect parking brake bracket for cracks or wear.

9 Inspect inside detent lever for cracks or loose pin.

10 Inspect manual shaft for damaged threads and the flats for raised edges. File down any raised edges.

REASSEMBLY

1 Turn transmission to horizontal position, oil pan side up.

2 If removed, install new manual shaft seal with lip facing inward into transmission case using a 14 mm (9/16″) socket to seat seal.

3 Install parking pawl and return spring with tooth toward inside of case and parking pawl return spring under pawl tooth with spring ends toward inside of case. Make sure spring ends locate against case pad.

4 Align parking pawl and return spring with case shaft bore.

5 Install parking pawl shaft, tapered end first.

6 Using 10 mm (3/8″) rod, install new parking pawl shaft cup plug, open end out, past retaining pin hole.

7 Install parking pawl shaft retaining pin.

8 Install parking brake bracket with the parking pawl between the guides of the bracket and install two (2) attaching bolts. Torque bolts to 20–27 N-m (15–20 ft./lbs.).

9 Install parking brake actuator rod into inside detent lever on pin side, locating lever between actuator rod tangs.

10 Install parking brake actuator rod and inside detent lever with detent lever pin toward center of transmission and actuator plunger between parking pawl and parking brake bracket.

11 Install manual shaft, small identification ring groove first, through case, next; install manual shaft to case retaining pin, indexing with larger groove on manual shaft.

12 Aligning inside detent lever with flats on manual shaft, install inside detent lever on shaft.

13 Install hex nut on manual shaft and torque to 27–34 N-m (20–25 ft./lbs.).

Rear Gear Components

OUTPUT SHAFT

NOTE: There are two (2) types of speedometer drive gears. When replacing the first type drive gear and/or clip, the first type drive gear and/or clip must be used. When replacing the second type drive gear and/or clip, the second type drive gear and/or clip must be used.

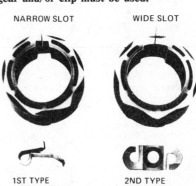

NARROW SLOT   WIDE SLOT

1ST TYPE SPEEDOMETER DRIVE GEAR AND CLIP   2ND TYPE SPEEDOMETER DRIVE GEAR AND CLIP

1 Inspect journals and snap ring grooves for wear or damage.

2 Inspect lubrication passages for being plugged or damaged.

3 Inspect splines for damage.

4 Inspect governor drive gear for rough or damaged teeth.

5 Inspect speedometer drive gear for rough or damaged teeth and also the clip for damage.

6 If necessary to replace speedometer drive gear, proceed as follows:

NOTE: When replacing an output shaft

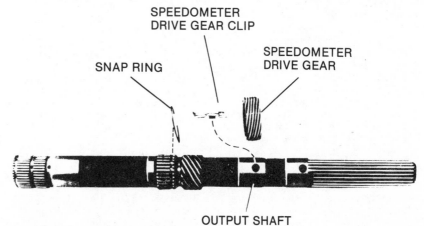

SPEEDOMETER DRIVE GEAR CLIP

SNAP RING

SPEEDOMETER DRIVE GEAR

OUTPUT SHAFT

Speedometer drive gear placement on output shaft

which uses the first type speedometer drive gear and clip, the speedometer drive gear and clip must be replaced with the second type drive gear and clip. The first type speedometer drive gear and clip will not fit on service replacement output shafts.

    a. Depress speedometer drive gear clip.
    b. Remove gear and clip, tapping gear lightly with plastic hammer.

7 Install speedometer drive gear clip, aligning tang in hole of output shaft and square end facing toward longer spline end of shaft.
8 Align slot in speedometer drive gear with clip, and install new gear.
9 If necessary, remove rear internal gear to output shaft snap ring and replace, with new snap ring if damaged.

REAR INTERNAL GEAR
1 Inspect rear internal gear, splines, teeth and bearing surface of wear, cracks or damage.
2 Inspect parking pawl lugs for cracks or damage.
3 Install rear internal gear, hub end first, on output shaft.
4 Thoroughly clean, air dry and inspect closely, the rear internal gear to rear sun gear roller thrust bearing assembly for pitted or rough conditions.
5 Install rear internal gear to rear sun gear roller thrust bearing assembly by placing the small diameter race over the output shaft.

NOTE: The optional roller thrust bearing and thrust races could be used. Install this type of thrust bearing and its races as follows:

    a. Place inside diameter flanged thrust race against rear internal gear.
    b. Place roller thrust bearing against inside diameter race.
    c. Place outside diameter thrust race on roller thrust bearing, with the cupped side over bearing.

ROLLER CLUTCH AND REAR CARRIER ASSEMBLY
INSPECTION
1 Remove roller clutch race. Inspect race and spline for scoring or wear.
2 Remove roller clutch assembly and inspect roller bearings, cage and springs for damage or wear.
3 Remove and inspect rear carrier to roller clutch thrust washer for signs of scoring or excessive wear.
4 Inspect four (4) tanged rear carrier to rear internal gear thrust washer for being scored or distorted tangs.
5 Inspect rear carrier for damage.
6 Inspect roller clutch cam ramps for damage.
7 Inspect bushing for damage or scoring.
8 Inspect planet pinions for damage, rough bearings or tilt.

NOTE: The CU and CN models have a two pinion carrier, with the pinions directly opposite each other.

9 Check pinion end play. Pinion end play should be 0.24–0.69mm (0.009″ -0.027″).

REASSEMBLY
1 Install roller clutch to rear carrier thrust washer.
2 Install rollers that may have come out of roller clutch cage, by compressing the energizing spring with forefinger and inserting roller from outer edge.
3 Install roller clutch assembly into roller clutch cam.
4 Install roller clutch race, spline side out and rotate clutch race counter-clockwise into position.
5 Install four (4) tanged rear carrier to rear internal gear thrust washer. Align tangs into slots of rear carrier and retain with petrolatum.
6 Install roller clutch and rear carrier assembly into rear internal gear.
7 Install J-25013-1, sleeve hole side first, into case; then bolt J-25013-5 on end of case.
8 Turn case to vertical position, pump end up.
9 Install rear unit parts into case and into J-25013-1 sleeve, indexing rear internal gear parking pawl lugs to pass by parking pawl tooth.
10 Using J-25013-5 adjusting screw and looking through parking pawl case slot, adjust the height of the rear internal gear parking pawl lugs to align flush with the parking pawl tooth.

NOTE: Make sure speedometer drive gear is visable through speedometer gear bore. If drive gear is not visable, it may be located on wrong journal of shaft.

Lo and Reverse Clutch
CHILTON CAUTION: *If Lo and Reverse clutch housing has been removed, the Lo and Reverse clutch piston travel must be checked.*

    Inspect lo and reverse clutch composition-faced and steel clutch plates for signs of wear or burning.

DISASSEMBLY
1 Compress lo and reverse clutch spring retainer, remove snap ring and inspect for damage or distortion.
2 Remove retainer and waved spring.
3 Remove lo and reverse clutch piston.
4 Remove inner and outer piston seals.
5 Remove clutch apply ring.

INSPECTION
1 Inspect lo and reverse clutch housing for damage, plugged feed hole.
2 Inspect lo and reverse clutch housing bushing for damage or scoring.
3 Inspect lo and reverse clutch housing bushing for damage or scoring.
4 Inspect lo and reverse clutch splines and snap ring groove for damage or

burrs. Remove any burrs on splines or snap ring groove.
5 Inspect lo and reverse clutch piston assembly for distortion, cracks or damage.
6 Inspect the clutch apply ring for damage. The apply ring is identified by the number (9) located on the outside of the apply ring.
7 Inspect lo and reverse clutch spring retainer for damage.
8 Inspect waved spring for damage.
9 Inspect lo and reverse clutch housing to case spacer ring for damage.

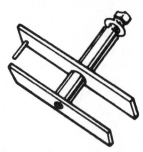

J25023 Reverse clutch selective shim gauge

REASSEMBLY AND INSTALLATION
1 Install clutch apply ring on lo and reverse clutch piston.
2 Install new outer and inner seals on piston with lips facing away from clutch apply ring side.
3 Install seal protector J-25011.

NOTE: Apply transmission fluid to all clutch seals before re-assembly.

CHILTON CAUTION: *Flat screw driver surface area must be smooth to prevent damaging outer seal.*

4 Using flat edged small screwdriver, install lo and reverse clutch piston, while rotating and pushing down into place.
5 Remove seal protector J-25011.
6 Install waved release spring.
7 Install retainer, cupped face down.
8 Compress retainer and install snap ring.
9 Using special tool J-25023, reverse clutch selective shim guage, check lo and reverse clutch piston travel as follows:
    a. Center the lo and reverse clutch housing and piston assembly on J-25023-1 plate.
    b. Center steel and composition clutch plates on clutch apply ring, then place J-25023-2 with gage pin toward housing.
    c. Install spring J-25023-7 over bolt of J-25023-1.
    d. Using nut and washer, compress spring J-25023-7 until washer bottoms against J-25023-1.
    e. Using feeler gage determine distance between gage pin on J-25023-2 and top edge of housing. See

GM TURBO HYDRA-MATIC 200

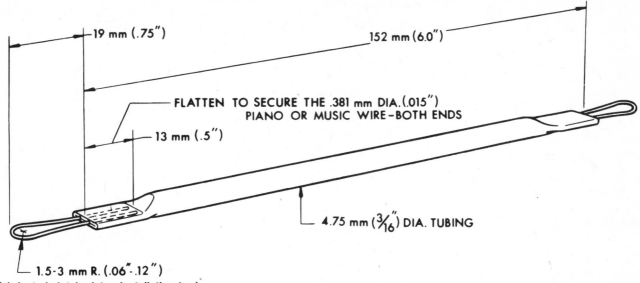

19 mm (.75″)

152 mm (6.0″)

FLATTEN TO SECURE THE .381 mm DIA.(.015″) PIANO OR MUSIC WIRE—BOTH ENDS

13 mm (.5″)

4.75 mm (³⁄₁₆″) DIA. TUBING

1.5-3 mm R. (.06″-.12″)

Fabricated clutch piston installation tool

chart below to determine proper selective washer.

NOTE: The lo and reverse clutch selective spacer identification number is also the last number of the part number.

LO AND REVERSE CLUTCH SELECTIVE SPACER THICKNESS

Feeler Stock Thickness	Identification Number
.0- .4mm (0-.016″)	6
.4- .8mm (.016-.032″)	5
.8-1.2mm (.032-.048″)	4
1.2-1.6mm (.048-.064″)	3
1.6-2.0mm (.064-.080″)	2
2.0-2.4mm (.080-.096″)	1

  f. Remove J-25023.
  g. Install proper lo and reverse clutch selective spacer in the case.
10 Oil and install lo and reverse clutch plates, starting with a steel clutch plate, aligning wide space in steel plate with wide tooth in case and alternating with composition-faced clutch plates.

NOTE: There are five (5) steel clutch plates and four (4) composition-faced clutch plates.

11 Install lo and reverse clutch housing to case spacer ring in case.
12 Install lo and reverse clutch housing assembly aligning reverse clutch housing feed hole to reverse clutch case feed passage, using J-25012. If the lo and reverse clutch housing does not seat past the case snap ring groove, proceed as follows:
  a. Remove tool J-25012.
  b. Using rear sun gear and input drum as a tool, install input drum and rear sun gear in case.
  c. Rotate rear sun gear back and forth, tapping lightly with input

drum, to align roller clutch race and lo and reverse clutch hub splines.
  d. Remove tool (input drum and rear sun gear).

NOTE: It may be necessary to loosen adjusting screw on J-25013-5 on output shaft to install snap ring.

  e. Repeat the above steps if lo and reverse clutch housing is not fully seated past case snap ring groove.
13 Install lo and reverse clutch housing to case snap ring, flat side against housing (beveled side up). Position snap ring gap opposite parking brake rod.

REAR SUN GEAR

INSPECTION
1 Inspect rear sun gear for cracks, splits, damage spline, worn gear or journals and plugged lubrication holes.
2 Inspect rear sun gear bushing for damage or scoring.
3 If necessary, remove input drum to rear sun gear snap ring and remove sun gear from input drum.
4 Inspect input drum for damage.
5 Inspect four (4) tanged input drum to low and reverse clutch housing thrust washer for scoring or distorted tange.
6 If damaged, replace rear sun gear to input drum snap ring.

REASSEMBLY
1 Install rear sun gear into input drum, spline side first, and retain with snap ring.
2 Install four (4) tanged thrust washer on input drum over sun gear end; align tangs into input drum and retain with petrolatum.
3 Install rear sun gear and input drum assembly.

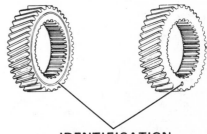

IDENTIFICATION
Front sun gear identification

FRONT SUN GEAR
1 Inspect front sun gear splines and teeth for damage or wear.
2 Inspect machined face for pitting, scoring or damage.
3 Install front sun gear face with the identification mark (a drill spot or groove) against input drum to rear sun gear snap ring.
4 Thoroughly clean, air dry and inspect front sun gear to front carrier thrust race and thrust bearing for pitted or rough conditions.
5 Install front sun gear to front carrier thrust bearing and race assembly with roller thrust bearing against the front sun gear.

NOTE: This thrust bearing requires only one (1) thrust race. An optional thrust bearing and race could be used. Install this type of roller thrust bearing and race as follows:

  a. Place roller thrust bearing on sun gear.
  b. Place O.D. flanged thrust race on bearing with cupped side over bearing.

FRONT CARRIER ASSEMBLY
1 Inspect front carrier for damage.
2 Inspect pinions for damage, rough bearings or tilt.

3 Check pinion end play. Pinion end play should be 0.24mm-0.69mm (0.-009"-0.27").

NOTE: The CU and CN models have a two pinion carrier, with the pinions directly opposite each other.

4 Thoroughly clean, air dry and inspect closely, front carrier to front internal gear roller thrust bearing assembly for pitted or rough conditions.
5 Install front carrier to front internal gear roller thrust bearing assembly by placing the smaller diameter race against carrier. Retain with petrolatum.

NOTE: The optional roller thrust bearing and thrust races could be used. Install this type of thrust bearing and its races as follows:

 a. Place inside diameter flanged thrust race against front carrier on the splined hub side.
 b. Place roller thrust bearing against inside diameter race.
 c. Place outside diameter thrust race on roller thrust bearing, with the cupped side over bearing.
6 Install front carrier.

FRONT INTERNAL GEAR
1 Inspect forward clutch hub for worn splines and for lubrication holes.
2 Inspect internal gear for cracks or damage.
3 Inspect gear teeth for excessive wear or damage.
4 Inspect bushing for damage or scoring.
5 Inspect tanged thrust washer, front internal gear to selective washer for scoring or damage.
6 Install tanged thrust washer on front internal gear and retain with petrolatum.
7 Install front internal gear.
8 Inspect selective washer for scoring or damage.
9 Install rear selective washer.
10 Inspect output shaft to selective thrust washer snap ring for damage or distortion.

NOTE: It may be necessary to move output shaft upward by turning the adjusting screw on J-25013-5 to install output shaft to selective washer snap ring.

11 Install snap ring on output shaft.

CHILTON CAUTION: *Make sure snap ring is fully seated in output shaft groove.*

12 Check rear unit end play as outlined in the disassembly section.
13 Loosen J-25013-5 adjusting screw on output shaft.
14 Inspect output shaft to turbine shaft front selective thrust washer for damage or scoring.
15 Install output shaft to turbine shaft front selective thrust washer, locating in output shaft and retain with petrolatum.

Front Unit Components
DIRECT CLUTCH HOUSING ASSEMBLY

DISASSEMBLY
1 Remove snap ring.
2 Remove the clutch backing plate and clutch plates from the direct clutch housing.
3 Inspect composition-faced plates and steel clutch plates for wear or burning.

CHILTON CAUTION: *Do not intermix the direct clutch composition clutch plates with the forward clutch composition plates.*

4 Inspect clutch backing plate for scoring or other damage.
5 Using J-23327-1, clutch spring compressor, compress retainer and spring assembly. Using arbor press or J-6129. remove snap ring and inspect for damage or distortion.
6 Remove J-23327-1 and/or J-6129.
7 Remove retainer and spring assembly from housing.
8 Inspect release spring retainer for distortion.
9 Inspect release springs for being collapsed.
10 Remove direct clutch piston.
11 Remove outer and inner seals from direct clutch piston and discard.
12 Remove clutch apply ring and inspect for damage. The apply ring is identified by the number 2, located on the outside of the apply ring.
13 Inspect direct clutch piston assembly for distortion, cracks or damage.
14 Remove center seal from direct clutch housing and discard.
15 Inspect direct clutch housing for cracks, wear and open oil passages.
16 Check for free operation of check ball.
17 Inspect direct clutch housing snap ring grooves for damage.
18 Inspect direct clutch bushings for damage or scoring.

REASSEMBLY
1 Install clutch apply ring on piston.
2 Install new inner and outer seals on piston with lips facing away from clutch apply ring side.
3 Install new center seal on direct clutch housing with lip facing up.
4 Install seal protector J-25010 over oil seals.

CHILTON CAUTION: *Use extreme care when installing direct clutch piston past larger direct clutch snap ring groove. Groove could cut outer seal on piston.*

5 Oil seals and install direct clutch piston.
6 Remove seal protector J-25010.
7 Install Spring Release Guide, retainer and spring assembly.

CHILTON CAUTION: *Retainer could locate in snap ring groove and forcing retainer to compress springs; could damage retainer plate.*

8 Using J-23327-1 compress retainer past snap ring groove using arbore press or J-6129 and install snap ring.
9 Remove tool J-23327-1 and/or J-6129 if used.
10 Oil and install clutch plates, starting with flat steel plate, then alternate composition-faced and flat steel plate.

NOTE: The direct clutch steel plates have four (4) teeth omitted, two (2) directly opposite each other. There are three (3) flat steel clutch plates and three (3) composition clutch plates.

11 Install backing plate, chamfered side up.
12 Install snap ring.

NOTE: Make sure composition clutch plates turn freely.

13 Set direct clutch assembly aside.

FORWARD CLUTCH HOUSING ASSEMBLY

DISASSEMBLY AND INSPECTION
1 Inspect teflon oil seals on turbine shaft for damage and free fit in grooves. **Do not remove unless replacing.**
2 Remove and inspect forward clutch to direct clutch thrust washer for damage.
3 Place forward clutch down with turbine shaft through hole in work bench.
4 Remove snap ring and inspect for damage.
5 Remove backing plate and clutch plates from the forward clutch housing.
6 Inspect composition-faced and steel clutch plates for signs of wear or burning.

CHILTON CAUTION: *Do not intermix the forward clutch composition plates with the direct clutch composition plates.*

7 Inspect backing plate for scratches or damage.
8 Using an arbor press compress retainer and spring assembly and remove snap ring and inspect for damage or distortion.

NOTE: Tool J-23327-1 with adapter J-25018-4 is available to compress retainer and spring assembly. Tool J-25018 is also available when using these tools.

9 Remove retainer and spring assembly from housing.
10 Inspect release spring retainer for distortion.
11 Inspect release springs for being collapsed.
12 Remove forward clutch piston.
13 Remove forward clutch outer and inner piston seals and discard.
14 Remove clutch apply ring and inspect for damage. The apply ring is identified by the number O located on the outside of the apply ring.

GM TURBO HYDRA-MATIC 200

15 Inspect forward clutch piston assembly for distortion or cracks.
16 Inspect forward clutch housing for cracks, opened oil passages or other damage.
17 Check for free operation of check ball.
18 Inspect forward clutch housing snap ring groove for damage or burrs.
19 Inspect turbine shaft for open oil passages on both ends of shaft and journals for damage.
20 Inspect cup plug for damage. If cup plug is damaged or missing, proceed as follows:
   a. Use 4 mm (#3) easy out (grind to fit) and remove cup plug.
   b. Install new cup plug to 1.0 mm (.039") below surface.

REASSEMBLY
1 Install clutch apply ring on piston.
2 Install new outer and inner seals on piston with lips facing away from clutch apply ring side.

CHILTON CAUTION: *Use extreme care when installing forward clutch piston past large forward clutch snap ring groove. Groove could cut outer seal on piston.*

3 Lubricate seals and install forward clutch piston.

NOTE: All models contain twelve (12) clutch release springs.

4 Install retainer and spring assembly.

CHILTON CAUTION: *Retainer could locate in snap ring groove and forcing retainer to compress springs, could damage retainer plate.*

5 Using an arbor press compress retainer past snap ring groove and install snap ring.
6 Oil and install waved steel clutch plate first (waved plate has three teeth missing) then alternate composition-faced and flat steel clutch plates.

NOTE: There are two (2) flat steel clutch plates, one (1) waved steel clutch plate and three (3) composition-faced clutch plates.

7 Install backing plate, chamfered side up.
8 Install snap ring.

NOTE: Make sure composition clutch plates turn freely.

9 Install forward to direct clutch thrust washer and retain with petrolatum.
10 If removed, install new turbine shaft seal rings.

INSTALLATION
1 Position direct clutch assembly, clutch plate end up, over hole in bench.

NOTE: Align direct clutch composition-faced clutch plate teeth one above the other to make the forward clutch assembly easier to install.

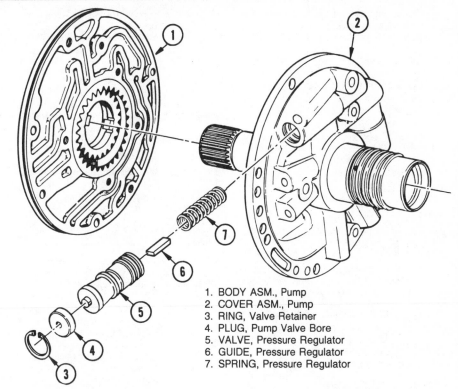

1. BODY ASM., Pump
2. COVER ASM., Pump
3. RING, Valve Retainer
4. PLUG, Pump Valve Bore
5. VALVE, Pressure Regulator
6. GUIDE, Pressure Regulator
7. SPRING, Pressure Regulator

Transmission pump T.H. 200

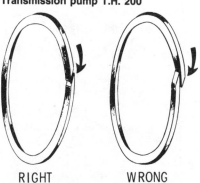

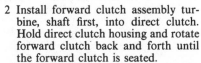

RIGHT   WRONG

Correct installation of Teflon oil seal rings

2 Install forward clutch assembly turbine, shaft first, into direct clutch. Hold direct clutch housing and rotate forward clutch back and forth until the forward clutch is seated.
3 When the forward clutch is seated, it will be approximately 19.05 mm (¾") from the tang end of the direct clutch housing to the end of the forward clutch drum.
4 Grasp direct and forward clutch assemblies to prevent their separation and position on bench, with the turbine shaft up.
5 Inspect the intermediate band for burning, flaking or damage.
6 Install intermediate band, locating band apply lug and anchor pin lug in case slots.
7 Install direct and forward clutch assemblies and rotate into position.

NOTE: The direct clutch housing will be

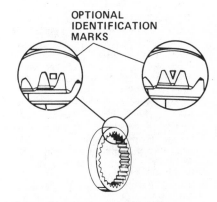

OPTIONAL IDENTIFICATION MARKS

Pump driven gear identification

approximately 33.34 mm (1–5/16") from the pump face in case when correctly seated.

OIL PUMP ASSEMBLY
1 Remove pump to case seal ring and inspect groove for damage.
2 Place pump over hole in bench with pump cover side up.
3 Remove pump to direct clutch thrust washer and inspect for damage or wear.

NOTE: If pump was difficult to lift from case, or off the direct clutch assembly, remove and discard the three (3) teflon oil ring seals.

4 Inspect three (3) teflon oil ring seals for damage and free fit in grooves. Do not remove unless replacing.
5 Using small screwdriver, push on bore plug, compressing pressure regulator

spring; and using snap ring pliers, remove retaining ring.
6 Release valve spring tension slowly and remove valve train.
7 Inspect pressure regulator valve for nicks or damage.
8 Inspect spring and guide for damage or distortion.
9 Inspect pressure regulator valve for free operation in bore.
10 Remove pump cover to pump body attaching bolts and separate pump cover from pump body.

J25015 Oil pump body & cover alignment band

REASSEMBLY

1 Place pump body over hole in bench.
2 Assemble pump cover to pump body with attaching bolts, finger tight.
3 Align pump cover and pump body using J-25015, oil pump body and cover alignment band, and place bolt or screwdriver through pump to case bolt hole and bench.
4 Torque pump cover attaching bolts to 20–27 Nm (15–20 ft./lbs.).
5 Remove J-25015.
6 Install pressure regulator spring first, spring guide, valve, stem end out and then bore plug hole side out.
7 Compress pressure regulator valve spring by pushing on bore plug with small screwdriver and install retaining ring.
8 If removed, install three (3) new oil seal rings, making sure cut ends are assembled in the same relationship as cut. Also, make sure rings are seated in the grooves to prevent damage to the rings during re-assembly of mating part over rings. Retain with petrolatum.
9 Install pump to case seal ring, chamfered side out, making sure the ring is not twisted.
10 Install pump to direct clutch thrust washer and retain with petrolatum.
11 Remove bolt or screwdriver.
12 Install new pump to case gasket on pump and retain with petrolatum.

NOTE: Make two (2) 8 mm 1.25 6 g thread alignment pins for installing the pump, 75 mm (3″) long.

13 Install two (2) pump to case alignment pins in case.

NOTE: Before installing pump, make sure intermediate band anchor pin lug is aligned with band anchor pin hole in case.

14 Install pump assembly and finger start pump to case bolts and new washers, except one bolt hole which will be used to make the front unit end play check.

NOTE: If turbine shaft cannot be rotated as pump is being pulled into place, the forward or direct clutch housings have not been installed properly to index with all the clutch plates. This condition must be corrected before pump is pulled fully into place.

15 Replace two (2) alignment pins with two (2) bolts and new washers.
16 Torque pump to case bolts to 20–27 Nm (15–20 ft./lbs.).

NOTE: Make sure turbine shaft rotates freely.

17 Check front unit end play as outlined in the disassembly section.
18 Remove front unit end play checking tools.
19 Install remaining pump to case bolt and new washer, torquing bolt to 20–27 Nm (15–20 ft./lbs.)
20 Remove J-25013-5 and J-25013-1 from rear end of transmission.
21 Turn transmission to horizontal position, oil pan side up.

External Components
GOVERNOR ASSEMBLY
1 Inspect governor cover for damage or plugged oil passage.
2 Wash in cleaning solvent and blow out oil passage.
3 Inspect governor driven gear for nicks or damage.
4 Inspect governor shaft seal ring for cuts, damage and free fit in groove.
5 Inspect for free operation of governor weights.
6 Inspect secondary spring for damage.
7 Inspect for presence of two (2) check balls.
8 Inspect shaft for damage.

NOTE: If governor assembly requires disassembly, proceed as follows:

9 Inspect governor washer for damage.
10 If damaged, cut seal ring off governor shaft.

CHILTON CAUTION: *Do not damage seal ring when removing seal.*

11 If removed, install new seal ring on shaft and place seal ring end into governor cover to size seal; lubricate with petrolatum.
12 Lubricate with petrolatum and install two (2) new seal rings on governor cover.

CHILTON CAUTION: *The governor cover seal rings must be well lubricated with petrolatum to prevent damage or cutting of the rings.*

13 Install governor assembly, seal ring end first, into cover.

CHILTON CAUTION: *Do not use any type of hammer to install governor assembly and cover into case. Damage to case, governor or cover could result.*

14 Install governor to case washer, against the governor driven gear and retain with petrolatum.
15 Install governor and cover assembly, aligning governor shaft with shaft hole in case. Rotate governor and cover assembly and output shafts slightly. The governor cover fits tight in the bore the last 1.5 mm (1/16″).

NOTE: Governor shaft is not aligned with case hole, if retaining ring cannot be installed.

16 Install governor retaining ring. Align ring gap with an end showing in case slot.

INTERMEDIATE BAND ANCHOR PIN AND INTERMEDIATE SERVO PISTON ASSEMBLY
1 Inspect anchor pin for damage and install stem end first, making sure stem locates in hole of intermediate band lug.
2 Inspect intermediate servo pin for damage and fit in case bore.
3 Inspect inner and outer piston seal rings for damage and free fit in ring grooves. Do not remove unless replacing.
4 Inspect spring.
5 Inspect intermediate servo cover and piston assembly for proper combination and usage.

NOTE: The cover is identified by the part number stamped on the cover. The piston is identified by a single groove located between the oil seal ring grooves. The CU and CN model has three grooves.

6 Check for proper intermediate band. Apply pin as outlined in the disassembly section.

J22269–01 Low servo cover remover and installer

NOTE: If new apply pin or replacement of piston is required proceed as follows:

a. Using special tool J-22269-01, low servo cover remover and installer, compress intermediate servo piston spring.
b. Using small flat edge screwdriver, remove intermediate pin to piston snap ring.
c. Remove J-22269-01 and separate band apply pin, spring and washer from intermediate servo piston.

GM TURBO HYDRA-MATIC 200

d. Install washer on snap ring end of band apply pin.
e. Install spring on washer.
f. Install band apply pin, spring end first through intermediate servo pistons.
g. Using J-22269-01, compress intermediate servo piston spring.
h. Using small pliers, install intermediate servo pin to piston snap ring. Remove J-22269-01.

7 If removed, install new intermediate servo piston, inner and outer seal rings. Make sure cut ends are assembled in the same relationship as cut and make sure rings are seated in the grooves to prevent damage to the rings. Retain with petrolatum.

8 Lubricate with petrolatum and install new seal ring on intermediate servo cover.

9 Install intermediate servo piston assembly into intermediate servo cover.

10 Install intermediate servo assembly into case tapping lightly with rubber hammer, if necessary.

11 Install servo retaining ring. Align ring gap with an end showing in case slot.

LO AND REVERSE CLUTCH HOUSING TO CASE CUP PLUG AND SEAL
1 Install new seal.

1. PIN, Valve Body—Coiled Spring
2. PLUG, T.V. Shift Valve Bore
3. PLUG, Reverse Boost Valve Bore
4. PIN, Valve Body—Coiled Spring
5. VALVE, Manual
6. PLUG, 1-2 Accumulator Valve Bore

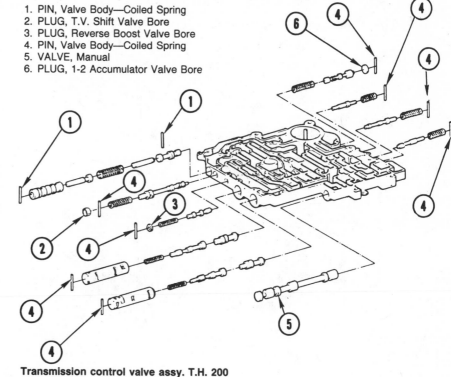

Transmission control valve assy. T.H. 200

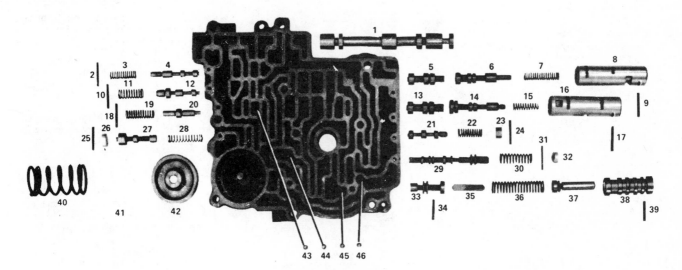

1. Manual valve
2. Retaining coiled pin
3. Intermediate boost spring
4. Intermediate boost valve
5. 2-3 shift valve
6. 2-3 throttle valve
7. 2-3 throttle valve spring
8. 2-3 throttle valve bushing
9. Retaining coiled pin
10. Retaining coiled pin
11. Lo overrun clutch spring
12. Lo overrun clutch valve
13. 1-2 shift valve
14. 1-2 throttle valve
15. 1-2 throttle valve spring
16. 1-2 throttle valve bushing

17. Retaining coiled pin
18. Retaining coiled pin
19. Direct clutch exhaust spring
20. Direct clutch exhaust valve
21. Reverse boost valve
22. Reverse boost spring
23. Reverse boost bore plug
24. Retaining coiled pin
25. Retaining coiled pin
26. 1-2 accumulator bore plug
27. 1-2 accumulator valve
28. 1-2 accumulator valve spring
29. Shift T.V. valve
30. Shift T.V. spring
31. Retaining coiled pin
32. Shift T.V. bore plug

33. Throttle valve
34. Retaining coiled pin
35. Detent pin
36. Throttle valve spring
37. Throttle valve plunger
38. Throttle valve plunger bushing
39. Retaining coiled pin
40. 1-2 accumulator spring
41. 1-2 accumulator piston seal
42. 1-2 accumulator piston
43. Check ball #4
44. Check ball #3
45. Check ball #2
46. Check ball #1
(5th check ball is in the case.)

Exploded view of valve body

GM TURBO HYDRA-MATIC 200

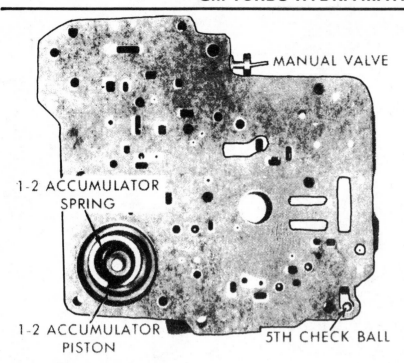

Valve body with fifth ball location

Typical valve body spacer plate

2 Install cap plug, large hole out, below rough casting surface.

CONTROL VALVE ASSEMBLY

DISASSEMBLY

As each valve train is removed, place the individual valve train in the order that it's removed and in a separate location relative to its position in the valve body. None of the valves, bushings, or springs are interchangeable; some coiled pins are interchangeable. Remove all coiled pins from machined face side, except the coiled pin which retains the throttle valve bushing, plunger, spring and detent pin. Remove that coiled pin from rough casting side.

1 Remove the four (4) check balls, and position the control valve assembly as illustrated with the 1–2 accumulator at lower left corner.
2 Remove 1–2 accumulator piston.
3 Remove manual valve from upper bore.

NOTE: Use a 1.5 mm (1/16″) allen wrench with two (2) ground sides to remove coiled pins that cannot be removed with needle nose pliers.

4 Remove coiled pin from upper right bore. Remove 2–3 throttle valve bushing, 2–3 throttle valve spring, 2–3 throttle valve and the 2–3 shift valve.

NOTE: The 2–3 throttle valve spring and 2–3 throttle valve may be inside the 2–3 throttle valve bushing.

5 From next bore down, remove coiled pin. Remove 1–2 throttle valve bushing, 1–2 throttle valve spring, 1–2 throttle valve and 1–2 shift valve.

NOTE: The 1–2 throttle valve spring and the 1–2 throttle valve may be inside the 1–2 throttle valve bushing.

6 From next bore down, remove coiled pin and bore plug. Remove reverse boost spring and reverse boost valve.
7 Check the operation of the shift T.V. valve in the next bore down, by moving the valve against the spring. If it is necessary to remove the valve, proceed as follows:
   a. Remove coiled pin and place valve body on shop towel with rough casting surface up.
   b. Using needle nose pliers, compress the shift T.V. spring by pushing on the shift T.V. valve, hold valve with small screwdriver.
   c. Place 6.3 mm (¼″) rod, 9.5 mm (⅜″) long against the end of the shift T.V. valve.
   d. Prying on end of rod with screwdriver, remove small screwdriver and remove shift T.V. plug, shift T.V. spring and shift T.V. valve.
   e. Discard shift T.V. plug and remove 6.3 mm (¼″) rod from shift T.V. bore.
8 From next bore down, first remove

outer coiled pin. Remove the throttle valve plunger, throttle valve plunger bushing, throttle valve spring and detent pin. Next, remove inner coiled pin and throttle valve.

9 From upper left bore, remove coiled pin, intermediate boost spring and intermediate boost valve.

10 From next bore down, remove coiled pin, lo overrun clutch spring and lo overrun clutch valve.

11 From next bore down, remove coiled pin, direct clutch exhaust spring and direct clutch exhaust valve.

12 From next bore down, remove coiled pin, valve bore plug, 1–2 accumulator valve and 1–2 accumulator valve spring.

INSPECTION

1 Wash control valve body, valves, springs and other parts in clean solvent and air dry.

2 Inspect 1–2 accumulator piston for damage.

3 Inspect 1–2 accumulator piston seal for damage and free fitting groove; do not remove seal unless replacing.

4 Inspect valves for scoring, cracks and free movement in their bores.

5 Inspect bushings for cracks or scored bores.

6 Inspect valve body for cracks, damage or scored bores.

7 Inspect springs for distortion or collapsed coils.

8 Inspect bore plugs for damage.

REASSEMBLY

Install all coiled pins from machined face side except the coiled pin retaining the throttle valve bushing, plunger, spring and detent pin. Install this coiled pin from rough casting side. Coiled pins do not fit flush on rough casting face.

1 For reassembly position control valve body with 1–2 accumulator at lower left corner.

2 Install into lower left bore, 1–2 accumulator spring, 1–2 accumulator valve, stem end out. Install bore plug, hole out, and coiled pin.

3 In next bore up, install direct clutch exhaust valve, longer stem end out, direct clutch exhaust spring and coiled pin.

4 In next bore up, install lo overrun clutch valve, longer stem end out, lo overrun clutch spring and coiled pin.

5 In next bore up, install intermediate boost valve, longer stem end out, intermediate boost spring and coiled pin.

6 In lower right bore, install throttle valve, smaller outside diameter land first, making sure valve is seated at the bottom of the bore. Install inner coiled pin between the lands of this valve. Install throttle valve plunger, stem end first, into throttle valve plunger bushing. Install throttle valve spring end first, into bore. Install outer coiled

pin from rough casting side, aligning pin with slot in bushing.

7 In next bore up, install shift T.V. valve, larger outside diameter stem end out, shift T.V. spring and coiled pin. Then using plastic hammer, install new shift T.V. plug flush with rough casting surface.

8 In next bore up, install reverse boost valve, stem end out, reverse boost spring. Install bore plug, hole side out and coiled pin.

9 In next bore up, install 1–2 shift valve, longer stem end out. Install 1–2 throttle valve spring into the 1–2 throttle valve bushing and 1–2 throttle valve, stem end first, into the bushing. Pushing on 1–2 throttle valve, compress 1–2 throttle valve spring and retain spring by installing small screwdriver in slot nearest closed end of bushing. Install 1–2 throttle bushing assembly, valve end first, into bore and remove screwdriver. Install coiled pin, aligning pin with slot in bushing.

10 In next bore up, install 2–3 shift valve, longer stem end out. Install 2–3 throttle valve spring into the 2–3 throttle valve bushing and 2–3 throttle valve, stem end first, into the bushing. Install 2–3 throttle bushing assembly, valve end first, into bore, install coiled pin, aligning pin with slot in bushing.

11 Install manual valve with the inside detent lever pin groove to the right.

12 If removed, install new seal ring on 1–2 accumulator piston.

13 Oil and install 1–2 accumulator piston, spring pocket side out, into 1–2 accumulator piston bore of valve body.

INSTALLATION

1 Inspect 1–2 accumulator spring for damage.

2 Install 1–2 accumulator spring into case.

NOTE: Make two (2) 6 mm 1–6 g thread alignment pins for installing the valve body.

3 Install control valve assembly to case guide pins as shown in illustration.

4 Inspect new control valve assembly gaskets and spacer plate for damage.

5 Install fifth (5th) check ball in case as shown in illustration.

6 Install control valve assembly spacer plate case gasket marked "C".

7 Install control valve assembly spacer plate.

NOTE: Some spacer plates may be of a rubberized type which do not require gaskets.

8 Install spacer plate to control valve assembly gasket marked "VB".

9 Install four (4) check balls into ball seat pockets in control valve assembly and retain with petrolatum.

10 Install control valve assembly, aligning manual valve with pin on inside detent lever.

NOTE: Make sure check balls, 1–2 accumulator piston and manual valve do not fall out.

11 Start control valve assembly to case attaching bolts, except the throttle lever and bracket assembly and the oil screen attaching bolts.

NOTE: The two oil screen attaching boltgs are longer than the rest of the control valve assembly attaching bolts.

12 Inspect inside manual detent roller and spring assembly for damage.

13 Remove guide pins and replace with bolts and inside manual detent roller and spring assembly, locating the tang in the control valve body, and the roller on the inside detent lever.

14 Inspect throttle lever and bracket assembly for damage.

15 If removed, install spring on top of lifter then lifter spring first into throttle bracket.

16 Install link on throttle lever making sure link is hooked as shown in illustration.

17 Install throttle lever and bracket assembly, locating slot in bracket with coiled pin, aligning lifter through valve body hole and link through T.V. linkage case bore. Retain with bolt.

18 Torque all control valve assembly to case attaching bolts to 8–14 Nm (6–10 ft./lbs.).

OIL SCREEN AND OIL PAN

1 Thoroughly, air dry, clean and inspect oil screen assembly.

2 Install new oil screen gasket on screen and retain with petrolatum.

3 Install oil screen assembly and attaching bolts. Torque bolts to 8–14 Nm (6–10 ft./lbs).

4 Clean and inspect oil pan for damage.

5 Install new oil pan to case gasket on case.

6 Install oil pan and retaining bolts. Torque bolts to 14–18 Nm (10–13 ft./lbs.).

SPEEDOMETER DRIVEN GEAR

1 Remove speedo driven gear from housing and inspect gear for damage.

2 Inspect housing for damage and "O" ring for damage or cuts.

3 If damaged, remove and discard "O" ring.

4 If removed, install new "O" ring on housing.

5 Install speedo driven gear into housing.

6 Install speedo driven gear assembly into case.

7 Install speedo retainer and attaching bolt, align slot in speedo driven gear housing with retainer. Torque bolt to 8–14 Nm (6–10 ft./lbs.).

CONVERTER
INSPECTION

1 Using J-21369 series special tools check converter for leaks as follows:
   a. Install J-21369-2 and J-21369-6

GM TURBO HYDRA-MATIC 200

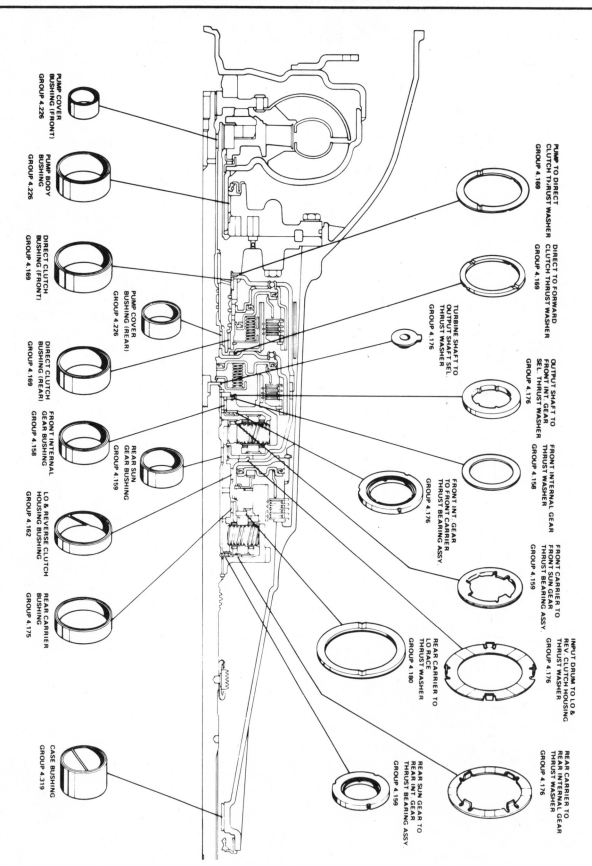

PUMP COVER BUSHING (FRONT)
GROUP 4.226

PUMP BODY BUSHING
GROUP 4.226

DIRECT CLUTCH BUSHING (FRONT)
GROUP 4.169

PUMP COVER BUSHING (REAR)
GROUP 4.226

DIRECT CLUTCH BUSHING (REAR)
GROUP 4.169

FRONT INTERNAL GEAR BUSHING
GROUP 4.158

REAR SUN GEAR BUSHING
GROUP 4.159

LO & REVERSE CLUTCH HOUSING BUSHING
GROUP 4.162

REAR CARRIER BUSHING
GROUP 4.175

CASE BUSHING
GROUP 4.319

PUMP TO DIRECT CLUTCH THRUST WASHER
GROUP 4.169

DIRECT TO FORWARD CLUTCH THRUST WASHER
GROUP 4.169

TURBINE SHAFT TO OUTPUT SHAFT SEL. THRUST WASHER
GROUP 4.176

OUTPUT SHAFT TO FRONT INT. GEAR SEL. THRUST WASHER
GROUP 4.176

FRONT INTERNAL GEAR TO FRONT SUN GEAR THRUST WASHER
GROUP 4.158

FRONT INT. GEAR TO FRONT CARRIER THRUST BEARING ASSY.
GROUP 4.176

FRONT CARRIER TO FRONT SUN GEAR THRUST BEARING ASSY.
GROUP 4.159

REAR CARRIER TO LO RACE THRUST WASHER
GROUP 4.180

INPUT DRUM TO LO & REV. CLUTCH HOUSING THRUST WASHER
GROUP 4.176

REAR CARRIER TO REAR INTERNAL GEAR THRUST WASHER
GROUP 4.176

REAR SUN GEAR TO REAR INT. GEAR THRUST BEARING ASSY.
GROUP 4.159

Bushing locations and identification

GM TURBO HYDRA-MATIC 200

and tighten hex nut on J-21369-2.
b. Fill converter with air; 551 kilopascals (80 p.s.i.).
c. Submerge in water and check for leaks.
d. Bleed air pressure from J-21369-2 and remove tools.
2 Check converter hub surfaces for signs of scoring or wear.

END CLEARANCE CHECK
NOTE: Both converter stator and turbine end-play checking fixtures J-25020 and J-21371 are required to perform this inspection.
1 Fully release collet end of tool J-25020.

2 Install collet end of tool J-25020 into converter hub until it bottoms; then tighten cap nut to 6.7 Nm (5 ft./lbs.).
3 Install tool J-21371-3 and tighten hex nut to 4.0 Nm (3 ft./lbs).
4 Install dial indicator and set it at zero while its plunger rests on the cap nut of tools J-25020.
5 Loosen hex nut while holding cap nut stationary. With hex nut loose and holding tool J-21371-3 firmly against the converter hub, the reading obtained on the dial indicator will be the converter end clearance. End should be less than 0–1.27 mm (0–.050″). If the end clearance is greater, the converter must be replaced.

INSTALLATION
1 With the transmission in cradle or portable jack, install the converter assembly into the pump assembly, making sure that the converter hub drive slots are fully engaged with the pump drive gear tangs and the converter installed fully towards the rear of the transmission.
2 The converter is correctly installed when the minimum distance between the engine mounting face of the case and the front face of the converter cover drive lugs is 25.4 MM (1.0 inch).

TURBO-HYDRAMATIC 200 TORQUE SPECIFICATIONS

	ft./lbs.
Pump Cover Bolts	18
Pump to Case Attaching Bolts	18
Parking Pawl Bracket Bolts	18
Control Valve Body Bolts	11
Oil Screen Retaining Bolts	11
Bottom Pan Attaching Bolts	12
Converter to Flywheel Bolts	35
Transmission to Engine Mounting Bolts	25
Converter Dust Shield Screws	8
Manual Shaft Nut	23
Speedometer Driven Gear Attaching Bolts	8
Detent Cable Attaching Screw	6
Oil Cooler Line to Transmission Connector	25
Oil Cooler Line to Radiator Connector	20
Linkage Swivel Clamp Nut	30
Shifter Assembly to Sheet Metal Screws	8
Converter Bracket to Adapter Nuts	13
Catalytic Converter to Rear Exhaust Pipe Nuts	17
Exhaust Pipe to Manifold Nuts	12
Rear Transmission Support Bolts	40
Mounting Assembly to Support Nuts	21
Mounting Assembly to Support Center Nut	33
Adapter to Transmission Bolts	33

GM TURBO HYDRA-MATIC 250

TROUBLESHOOTING
GM TURBO HYDRA-MATIC 250

	ITEMS TO CHECK	
THE PROBLEM	IN CAR	OUT OF CAR
SLIPS IN ALL RANGES	Low fluid level/water in fluid. Modulator and/or valve. Strainer and/or gasket leak. Valve body gasket/plate. Pressure regulator and/or boost valve. Ball No. 1 missing. Manual valve linkage. Porosity/cross leak—manual valve. Gasket screen-pressure.	Pump gears. Clutch sealing rings. Leaking or porous seal rings. Porosity or cross-leaking. 1-2 accumulator.
DRIVE SLIPS—NO 1st	Low fluid level/water in fluid. Modulator and/or valve. Strainer and/or gasket leak. Valve body gasket/plate. Pressure regulator and/or boost valve. Ball No. 1 missing. Manual valve linkage. Porosity/cross leak—manual valve. Gasket screen-pressure.	Pump gears. Clutch sealing rings. Leaking or porous seal rings. Porosity or cross-leaking. 1-2 accumulator. Intermediate servo. Forward clutch assembly. Low & Reverse clutch assembly.
OIL PRESSURE—ALL TOO LOW	Low fluid level/water in fluid. Modulator and/or valve. Strainer and/or gasket leak. Governor valve/screen. Valve body gasket/plate. Pressure regulator and/or boost valve Ball No. 1 missing. Manual valve linkage. Porosity/cross leak—manual valve. Gasket screen-pressure.	Pump gears. Clutch sealing rings. Leaking or porous seal rings. Porosity or cross-leaking. 1-2 accumulator. Intermediate servo.
OIL PRESSURE—ALL TOO HIGH	Vacuum leak. Modulator and/or valve. Governor valve/screen. Pressure regulator and/or boost valve. Detent valve and linkage. Porosity/cross leak—manual valve.	Porosity or cross-leaking.
1-2 INTERMEDIATE PRESSURE—HIGH	Vacuum leak. Modulator and/or valve. Pressure regulator and/or boost valve. Porosity/cross leak—manual valve.	Porosity or cross-leaking.
1-2 INTERMEDIATE PRESSURE—LOW	Low fluid level/water in fluid. Strainer and/or gasket leak. Valve body gasket/plate. Ball No. 1 missing. 1-2 shift valve. 2-3 accumulator. Porosity/cross leak—manual valve.	Pump gears. Clutch sealing rings. Leaking or porous seal rings. Porosity or cross-leaking. 1-2 accumulator. Intermediate servo.
2-3 DIRECT CLUTCH PRESSURE—HIGH	Vacuum leak. Pressure regulator and/or boost valve. Porosity/cross leak—manual valve.	Porosity or cross-leaking.
2-3 CIRCUIT CLUTCH PRESSURE—LOW	Low fluid level/water in fluid. Strainer and/or gasket leak. Valve body gasket/plate. Pressure regulator and/or boost valve. 2-3 shift valve. Porosity cross/leak—manual valve.	Pump gears. Clutch sealing rings. Leaking or porous seal rings. Porosity or cross-leaking. 1-2 accumulator. Forward clutch assembly.
NO 1-2 UPSHIFT	Vacuum leak. Governor valve/screen. Valve body gasket/plate. Pressure regulator and/or boost valve. 1-2 shift valve.	Clutch sealing rings. Leaking or porous seal rings. Porosity or cross-leaking. 1-2 accumulator. Intermediate band assembly.
1-2 UPSHIFT—EARLY OR LATE	Vacuum leak. Governor valve/screen. Valve body gasket/plate. Ball No. 1 missing. 1-2 shift valve. Porosity/cross leak—manual valve.	Porosity or cross-leaking.
1-2 UPSHIFT AT WIDE-OPEN THROTTLE ONLY	Vacuum leak. 2-3 shift valve. Detent regulator valve and linkage. Porosity/cross leak—manual valve.	Porosity or cross-leaking.
SLIPS ON 1-2 UPSHIFT	Low fluid level/water in fluid. Modulator and/or valve. Valve body gasket/plate. Pressure regulator and/or boost valve. Ball No. 1 missing. 1-2 shift valve. 2-3 accumulator. Porosity/cross leak—manual valve.	Pump gears. Clutch sealing rings. Leaking or porous seal rings. Porosity or cross-leaking. 1-2 accumulator. Intermediate servo. Intermediate band assembly.
ROUGH ON 1-2 UPSHIFT	Vacuum leak. Modulator and/or valve. Pressure regulator and/or boost valve. Porosity/cross leak—manual valve.	Leaking or porous seal rings. Porosity or cross-leaking. 1-2 accumulator.
NO 2-3 UPSHIFT	Valve body gasket/plate. 2-3 shift valve. Porosity/cross leak—manual valve.	Clutch sealing rings. Leaking or porous seal rings. Porosity or cross-leaking. Direct clutch assembly.
2-3 UPSHIFT—EARLY OR LATE	Vacuum leak. Modulator and/or valve. Governor valve/screen. Valve body gasket/plate. Ball No. 1 missing. 2-3 shift valve. Detent valve and linkage.	Porisity or cross-leaking. Direct clutch assembly.

GM TURBO HYDRA-MATIC 250

TROUBLESHOOTING
GM TURBO HYDRA-MATIC 250

	ITEMS TO CHECK	
THE PROBLEM	IN CAR	OUT OF CAR
SLIPS ON 2-3 UPSHIFT	Low fluid level/water in fluid. Vacuum leak. Valve body gasket/plate. Pressure regulator and/or boost valve. Ball No. 1 missing. 2-3 shift valve. Porosity/cross leak—manual valve.	Pump gears. Clutch sealing rings. Leaking or porous seal rings. Gasket screen-pressure. Porosity or cross leaking. Direct clutch assembly.
2-3 UPSHIFT WIDE-OPEN THROTTLE ONLY	Vacuum leak. Detent valve and linkage.	
ROUGH 2-3 UPSHIFT	Vacumm leak. Modulator and/or valve. Governor valve/screen. 2-3 shift valve. 2-3 accumulator. Porosity/cross leak—manual valve.	Porosity or cross-leaking.
NO FULL THROTTLE 1-2 UPSHIFT	Detent regulator valve. Porosity/cross leak—manual valve.	Leaking or porous seal rings.
NO PART-THROTTLE DOWNSHIFT	2-3 shift valve. Detent valve and linkage.	
NO FULL-THROTTLE DOWNSHIFT	Vacuum leak. Detent valve and linkage.	
HARSH DOWNSHIFT	Ball No. 1 missing.	
NO ENGINE BRAKING L1 RANGE	Pressure regulator and/or boost valve. 1-2 shift valve. Manual low control valve. Manual valve linkage. Porosity/cross leak—manual valve.	Clutch sealing rings. Porosity or cross-leaking. Forward clutch assembly. Low & Reverse clutch assembly.
NO ENGINE BRAKING L2 RANGE	Pressure regulator and/or boost valve. Manual valve linkage. Porosity/cross leak—manual valve.	Clutch sealing rings.
CAR DRIVES IN NEUTRAL		Forward clutch assembly.
NO REVERSE	Pressure regulator and/or boost valve. Ball No. 1 missing. Manual valve linkage. Porosity/cross leak—manual valve. Gasket screen-pressure.	Clutch sealing rings. Leaking or porous seal rings. Porosity or cross-leaking. Forward clutch assembly. Direct clutch assembly. Low & Reverse clutch assembly.
SLIPS IN REVERSE	Low fluid level/water in fluid. Modulator and/or valve. Strainer and/or gasket leak. Valve body gasket/plate. Pressure regulator and/or boost valve. Ball No. 1 missing. 1-2 shift valve. Manual valve linkage. Porosity/cross leak—manual valve. Gasket screen-pressure.	Clutch sealing rings. Leaking or porous seal rings. Porosity or cross-leaking. Direct clutch assembly. Low & Reverse clutch assembly.
NO PARK OR "RATCHETS"		Parking pawl/linkage.
NOISE ALL RANGES	Low fluid level/water in fluid. Strainer and/or gasket leak. Valve body gasket/plate. Gasket screen-pressure.	Pump gears. Converter assembly. Gearset and bearings.
1-2/2-3—SHIFTS NOISY	Low fluid level/water in fluid.	Direct clutch assembly. Intermediate band assembly. Gearset and bearings. Converter assembly
R, D, L1, L2—NOISY		Converter assembly.
SPEWS OIL OUT BREATHER	Low fluid level—water in fluid. Strainer and/or gasket leak.	Leaking or porous seal rings.
"HUNTS"—2-3/3-2	Detent valve and linkage.	

GM TURBO HYDRA-MATIC 250

GM Turbo Hydra-matic 250
Identification

On all Chevrolet vehicles, see the engine identification chart, and check engine I.D. tag for transmission identification.

Intermediate Band Adjustment

1 Raise the vehicle on a lift, and put selector in Neutral.

2 Position a special tool J-24367 over the adjusting screw and locknut on the right side of the transmission. Loosen the locknut ¼ turn and hold.

3 Attach a torque wrench onto the special tool, and torque adjusting screw to 30 in./lbs. Back the screw off exactly three turns, using the mark on the tool.

4 Hold screw in position, and torque the locknut to 15 ft./lbs.

Capacity and General Specifications

Fluid capacity (dry) . . . . . 21½ pints
Cooling . . . . . . Water (except Vega)
Fluid filter type
. . . . Bottom suction screen
Clutches 3
Roller clutches 1
Band adjustment . . . . . Intermediate

GM TURBO HYDRA-MATIC 250

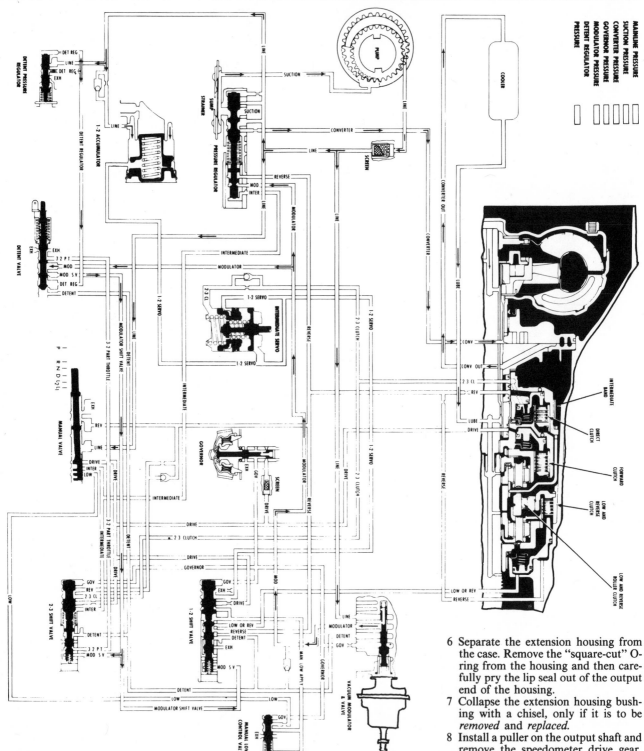

Hydraulic Schematic GM Turbo-Hydramatic 250

Transmission Disassembly

Clean the outside of the transmission thoroughly, before beginning disassembly.

1 Install the transmission in a stand with the torque converter facing up.
2 Lift the torque converter assembly out of the transmission.

3 Unfasten the vacuum modulator securing bolt and retainer.
4 Remove the vacuum modulator, O-ring, and modulator valve from the case. Throw the old O-ring away.
5 Unfasten the 4 bolts which attach the extension housing to the transmission case.

6 Separate the extension housing from the case. Remove the "square-cut" O-ring from the housing and then carefully pry the lip seal out of the output end of the housing.
7 Collapse the extension housing bushing with a chisel, only if it is to be *removed* and *replaced*.
8 Install a puller on the output shaft and remove the speedometer drive gear. Take off the retaining clip.
9 Carefully pry the governor retainer off with a large, flat screwdriver.
10 Work the screwdriver along the lip of the governor cover, while tapping it *gently* with a hammer. Remove the cover.

CHILTON CAUTION: *Do not pry the governor cover off the transmission case, as this could damage the case.*

GM TURBO HYDRA-MATIC 250

11 Remove and discard the O-ring seal.
12 Remove the governor assembly from the case. Inspect the governor boss and sleeve for scoring.
13 Unfasten the screws which secure the oil pan. Remove the oil pan and gasket. Throw the gasket away.
14 Unfasten the screws which secure the oil pump strainer to the valve body, and remove the strainer and gasket from the valve body.
15 Remove the detent spring and roller assembly from the valve body. Unfasten the bolts securing the valve body to the transmission case.
16 Remove the actuator pin from the detent actuator lever and unfasten the control wire, then separate the manual control valve link from the range selector inner lever. Remove the valve body from the case and remove the intermediate servo return spring.
17 Remove the bolts which attach the transfer support plate and then remove the plate.
18 Take off the top gasket, valve body spacer plate, and the spacer plate gasket.
19 Remove all 4 check balls from the passage in the face of the case.

NOTE: Be sure that the check balls are kept in their proper order for installation.

20 Remove the screen from the oil pump pressure hole in the transmission case. Remove the screen from the governor feed, as well.
21 Gently pry out the manual shift-to-case retainer with a screwdriver.
22 Loosen the nut which secures the range selector inner lever to the manual shaft. Separate the range selector lever from the manual shaft, by pulling on the shaft and removing the nut. Remove the manual shaft from the case.
23 Take the inner lever and parking pawl actuator out of the case. Separate the inner lever and pawl actuator rod.
24 If replacement is required, pry the manual shaft seal out of the case.
25 Remove the two special bolts which secure the parking lock and lock bracket; remove the lock and bracket.
26 Use a pair of needlenosed pliers to remove the parking pawl disengagement spring.
27 If they are to be replaced, remove the parking pawl shaft securing plug, pawl shaft, and parking pawl.

NOTE: The shaft retaining plug must be removed with a bolt extractor.

28 Unfasten the 8 pump retaining bolts and remove them, complete with sealing washers.
29 Install two threaded slide hammers into the two threaded holes in the pump body.
30 Loosen, but do not remove, the nut and intermediate band anchor bolt. Take out the intermediate band.

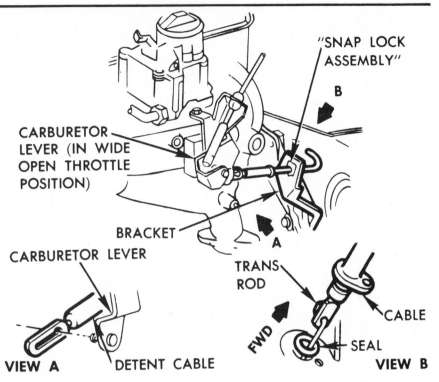

Dentent downshift cable–typical

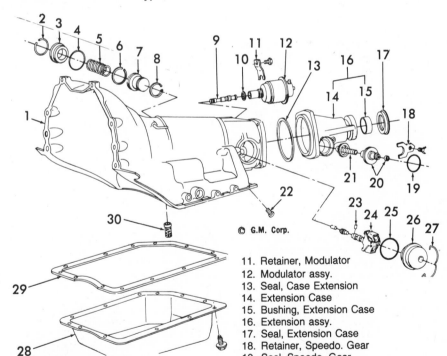

Case and parts 250 trans.

1. Case Assy.
2. Retainer, Acuumulator Piston Cover
3. Cover, Acum. Piston
4. Seal, Accum. Piston cover
5. Spring, Acum. Piston
6. Seal, Accum, Piston—Outer
7. Piston, Inter Clutch Accum.
8. Seal, Accum, Piston—Inner
9. Valve, Modulator
10. Seal, Modulator
11. Retainer, Modulator
12. Modulator assy.
13. Seal, Case Extension
14. Extension Case
15. Bushing, Extension Case
16. Extension assy.
17. Seal, Extension Case
18. Retainer, Speedo. Gear
19. Seal, Speedo. Gear
20. Sleeve, Speedo. Gear
21. Gear, Speedo. Driven
22. Screen, Governor Pressure
23. Pin, Governor
24. Governor Assy.
25. Seal, Governor
26. Cover, Governor
27. Retainer Gov. Cover
28. Pan Assy.
29. Gasket, Oil Pan
30. Screen Pump Pressure

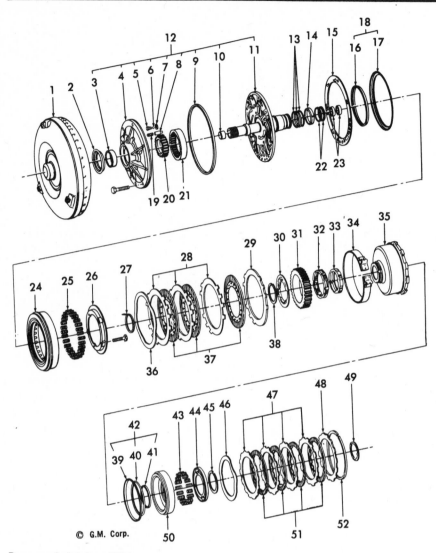

© G.M. Corp.

Pump and clutches 250 trans.

1. Converter Assy. (rebuilt)
2. Seal, Oil pump
3. Bushing, Oil pump
4. Body, Oil pump
5. Spring, Pump by-pass valve
6. Ball, pump by-pass valve
7. Seat, pump by-pass valve
8. Valve, pump priming
9. Seal, pump to case
10. Bushing, starter shaft—front
11. Cover, oil pump
12. Pump assy.
13. Ring, clutch drum
14. Bushing, clutch drum
15. Gasket, pump cover
16. Seal, Inter. Clutch—inner
17. Seal, Inter. Clutch—outer
18. Seal kit, inter. clutch
19. Spring pump priming valve
20. Gear, pump drive
21. Gear pump driven
22. Ring forward clutch
23. Bushing, starter shaft—rear
24. Piston, Inter clutch
25. Spring inter. clutch piston
26. Seat, inter. clutch piston

27. Washer, direct clutch
28. Plate, Inter. clutch reaction
29. Plate, Inter Clutch—pressure
30. Retainer, Inter clutch
31. Race, Inter clutch—outer
32. Clutch, Inter Overrun
33. Cam, Inter clutch
34. Band, Inter clutch
35. Drum, direct clutch
36. Spring, Inter clutch
37. Plate, Inter. clutch drive
38. Retainer ring, Inter clutch
39. Seal, Direct clutch piston, outer
40. Seal, direct clutch piston, center
41. Seal, direct clutch piston, inner
42. Seal kit, direct clutch piston
43. Spring, direct clutch piston, return
44. Seat, direct clutch piston
45. Ring, direct clutch piston
46. Spring, direct clutch piston
47. Plate, direct clutch driven
48. Plate, direct clutch—pressure
49. Bearing, direct clutch drum
50. Piston, direct clutch
51. Plate, direct clutch-drive
52. Retainer ring, direct clutch

31 Remove the intermediate servo and related components from the case.
32 Grasp the input shaft, and remove the direct and forward clutch assemblies from the case.
33 Straighten out the 3 tangs on the thrust washer and remove the input ring gear from it.
34 Remove the snap-ring which secures the input gear to the output shaft. Remove the ring gear.
35 Remove the output carrier thrust washer and the output carrier assembly.
36 Remove the sun gear drive shell.
37 Carefully pry out the Low/Reverse roller clutch support retaining ring. Remove the Low/Reverse clutch support and race assembly, and the "anti-clunk" spring.
38 Remove the Low/Reverse clutch pack.
39 Separate the reaction carrier from the ring gear/shaft assembly.
40 Remove the output ring gear and shaft, as an assembly, from the transmission case.
41 Remove the 3-tanged thrust washer from the front of the output ring gear.
42 Slide the needle bearing off the output ring gear.
43 Use a compressor to compress the Low/Reverse piston spring retainer. Remove the retaining ring and the spring retainer.
44 Take all 17 piston springs off the piston.
45 Apply compressed air to the passage illustrated, in order to remove the Low/Reverse clutch piston assembly. Slip the outer seal off the piston.
46 Remove the center and inner seals from the Low/Reverse clutch piston.
47 Compress the First/Second accumulator piston cover. Remove the retaining ring, piston cover and O-ring.
48 Remove the First/Second accumulator piston spring and piston assembly.
49 If necessary, remove the "hook-type" inner and outer oil seal rings.

Component Disassembly

Valve Body

DISASSEMBLY

1 Position the valve body assembly with the cored face up and the direct clutch accumulator piston pocket located as illustrated in the valve body assembly illustration.
2 Remove the manual valve from the lower left-hand bore, A.
3 From the lower right-hand bore, B, remove the pressure regulator valve train retaining pin, boost valve sleeve, intermediate boost valve, reverse and modulator boost valve, pressure regulator valve spring, and the pressure regulator valve.
4 From the next bore, C, remove the Second/Third shift valve train retain-

Component Removal and Overhaul

GM TURBO HYDRA-MATIC 250

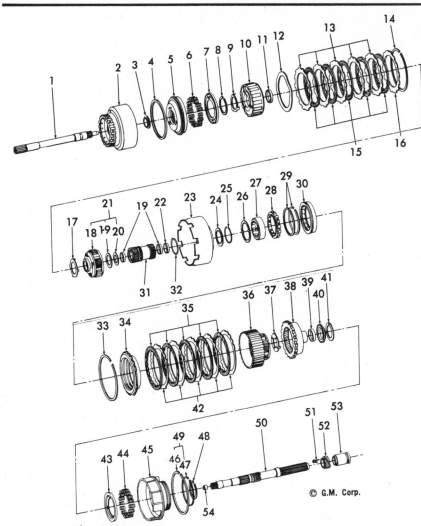

7. Seat, forward clutch
8. Retainer ring, forward clutch
9. Washer, input ring gear—front
10. Gear, input ring
11. Bushing, input ring gear
12. Spring, forward clutch, cushion
13. Plate, forward clutch, driven
14. Plate, forward clutch, drive
15. Plate, forward clutch, drive
16. Plate, forward clutch, pressure
17. Washer, input gear supt.
18. Carrier, output
19. Bearing, sun gear
20. Race, sun gear
21. Carrier assy., output
22. Bushing, sun gear
23. Shell, sun gear
24. Washer, sun gear
25. Retainer ring, sun gear shell
26. Washer, clutch tace and shell
27. Race, low and rev. clutch—inner
28. Clutch, low and reverse
29. Retainer ring, clutch to cam
30. Cam, low and rev. clutch
31. Gear, Sun
32. Retainer ring, sun gear shell
33. Retainer ring, clutch to cam
34. Support, low & rev. clutch (inc. cam)
35. Plate, rev. clutch—drive
36. Carrier, reaction planet
37. Washer, output ring gear, front
38. Gear, output ring
39. Bushing, case to shaft
40. Bearing, ring gear supt. to case
41. Ring, reverse clutch
42. Plate, low & rev. clutch reaction
43. Seat, reverse clutch
44. Spring, rev. clutch—return
45. Piston, low & rev. clutch
46. Seal, low & rev. clutch outer
47. Seal, low & rev. clutch, inner
48. Retainer ring, output carrier
49. Seal, kit, low & rev. clutch
50. Shaft assy., output
51. Clip, speedo. gear
52. Gear, speedo. drive
53. Bushing output shaft

Shafts and clutches 250 trans.

1. Shaft, input
2. Housing and shaft forward clutch
3. Seal, forward clutch—inner

4. Seal, forward clutch—outer
5. Piston, forward clutch
6. Spring, forward clutch—return

ing pin, sleeve, control valve spring, Second/Third shift control valve, shift valve spring, and the Second/Third shift valve.

5 From the next bore, D, remove the First/Second shift valve train retaining pin, sleeve, shift control valve spring, First/Second shift control valve, and the First/Second shift valve.

6 From the next bore, E, remove retaining pin, plug, manual low control valve spring, and the manual low control valve.

7 From the next bore, F, remove retaining pin, spring seat and the detent regulator valve.

8 Remove the following from the bore on the opposite side: detent actuating bracket bolt, bracket, stop, spring retainer, seat, outer spring, inner spring, washer, and detent valve.

INSPECTION

1 Wash all parts in solvent. Air dry.

Intermediate servo component identification

291

GM TURBO HYDRA-MATIC 250

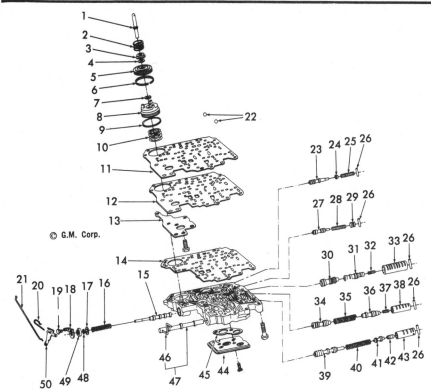

1. Rod inter. servo apply
2. Spring, inter servo piston
3. Seat, inter servo spring
4. Washer, inter servo piston
5. Piston, inter. servo
6. Seal, inter servo piston
7. Retainer ring, accum. piston
8. Piston, direct clutch accum.
9. Seal, direct clutch accum. piston
10. Spring, direct clutch accum, piston
11. Gasket, upper valve body
12. Plate, valve body spacer
13. Support, spacer plate
14. Gasket, lower valve body
15. Valve detent
16. Spring, detent valve
17. Seat, detent valve spring
18. Bracket, detent valve lever
19. Bolt, detent lever
20. Pin, detent lever
21. Wire, detent valve
22. Ball, check valve
23. Valve detent regulator
24. Seat, detent reg. valve spring
25. Spring, detent reg. valve
26. Pin, detent reg. valve
27. Screen, oil pump
28. Gasket, oil pump screen
29. Link, manual valve
30. Valve assy. manual
31. Lever detent valve

Valve body assy. 250 trans.

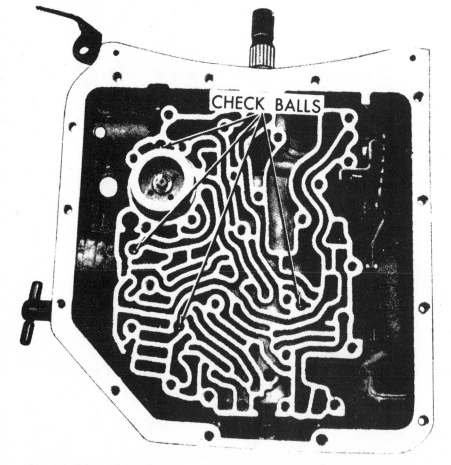

Check ball four (4) locations

Blow out all passages.
2 Inspect all valves for scoring, cracks, and free movement in their bores.
3 Inspect all sleeves for cracks, scratches, or distortion.
4 Inspect the valve body for cracks, scored bores, interconnected oil passages, and flatness of mounting face.
5 Check all springs for distortion or collapsed coils.

ASSEMBLY

The valve body is assembled in the reverse order of disassembly. Be careful not to mix the valve springs. Torque the detent valve actuating lever bracket bolt to 52 in./lbs.

Oil Pump

DISASSEMBLY

1 Remove the 5 pump cover-to-body securing bolts.
2 Remove the forward clutch Teflon® oil rings (two) and the direct pump-to-clutch hook-type oil seal rings (3).
3 Remove the direct drum-to-housing selective-fit thrust washer.
4 Separate the pump stator shaft/cover assembly from the pump body.
5 Withdraw the drive and driven gears from the pump.
6 Remove and discard the "square-cut" O-ring seal.

INSPECTION

1 Wash all parts in solvent. Air dry. Blow out all passages.
2 Inspect the drive and driven gears,

GM TURBO HYDRA-MATIC 250

gear pocket, and crescent for nicks, galling or other damage.

3 Inspect the pump body and cover for nicks or scoring.

4 Check the pump cover outer diameter for nicks or burrs.

5 Inspect the pump body bushing for galling or scoring. Check the clearance between the pump body bushing and the converter hub. It must be no more than .005 in. If the bushing is damaged, replace the pump body.

6 Install the pump gears in the body and check pump body face-to-gear face clearance. It should be from .0005–.0015 in.

7 Inspect the pump body-to-converter hub lip oil seal. Inspect the converter hub for nicks or burrs which might have damaged the pump lip oil seal or pump body bushing.

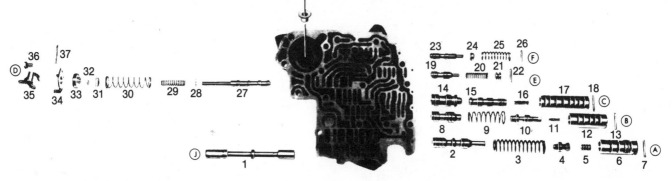

Valve body component identification

J— 1 MANUAL VALVE

A
   2 PRESSURE REGULATOR VALVE
   3 PRESSURE REGULATOR VALVE SPRING
   4 REVERSE AND MODULATOR BOOST VALVE
   5 INTERMEDIATE BOOST VALVE
   6 BOOST VALVE SLEEVE
   7 RETAINING PIN

B
   8 2-3 SHIFT VALVE
   9 2-3 SHIFT VALVE SPRING
  10 2-3 SHIFT CONTROL VALVE
  11 2-3 SHIFT CONTROL VALVE SPRING
  12 2-3 SHIFT CONTROL VALVE SLEEVE
  13 RETAINING PIN

C
  14 1-2 SHIFT VALVE
  15 1-2 SHIFT CONTROL VALVE
  16 1-2 SHIFT CONTROL VALVE SPRING
  17 1-2 SHIFT CONTROL VALVE SLEEVE
  18 RETAINING PIN

E
  19 MANUAL LOW CONTROL VALVE
  20 MANUAL LOW CONTROL VALVE SPRING
  21 PLUG
  22 RETAINING PIN

F
  23 DETENT REGULATOR VALVE
  24 DETENT REGULATOR VALVE SPRING SEAT
  25 DETENT REGULATOR VALVE SPRING
  26 RETAINING PIN

  27 DETENT VALVE
  28 WASHER
  29 DETENT VALVE INNER SPRING
  30 DETENT VALVE OUTER SPRING
  31 DETENT VALVE OUTER SPRING SEAT

D
  32 DETENT VALVE SPRING RETAINER
  33 DETENT VALVE STOP
  34 DETENT VALVE ACTUATING LEVER BRACKET
  35 DETENT VALVE ACTUATING LEVER
  36 RETAINING BOLT
  37 RETAINING PIN
  38 CAP

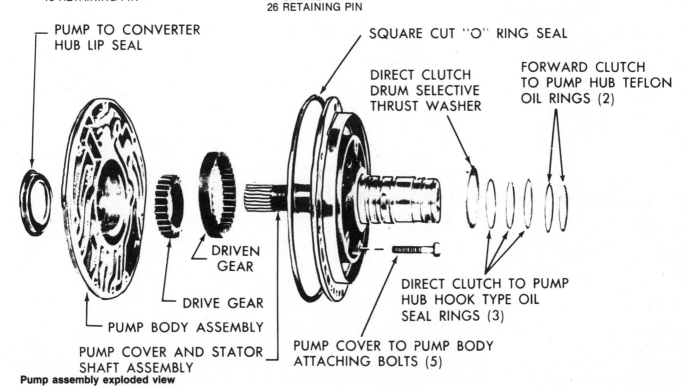

PUMP TO CONVERTER HUB LIP SEAL

SQUARE CUT "O" RING SEAL

DIRECT CLUTCH DRUM SELECTIVE THRUST WASHER

FORWARD CLUTCH TO PUMP HUB TEFLON OIL RINGS (2)

DRIVEN GEAR

DRIVE GEAR

PUMP BODY ASSEMBLY

PUMP COVER AND STATOR SHAFT ASSEMBLY

DIRECT CLUTCH TO PUMP HUB HOOK TYPE OIL SEAL RINGS (3)

PUMP COVER TO PUMP BODY ATTACHING BOLTS (5)

Pump assembly exploded view

GM TURBO HYDRA-MATIC 250

WHEN DIM. C IS	USE (OR EQUIVALENT)
.0160— .0520	6261072
.0520— .0830	6261349
.0830— .1218	6261350

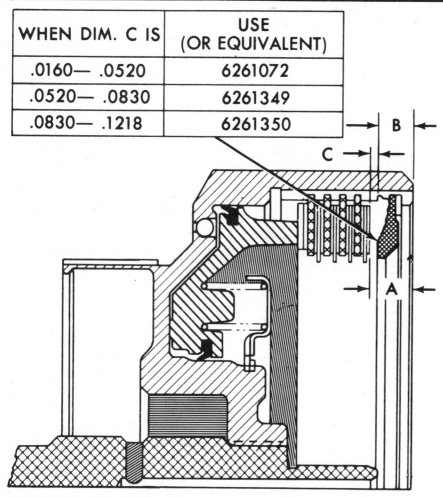

Determining selective fit for forward clutch pressure plate

ASSEMBLY

1 If the hub lip seal is defective, pry it out with a screwdriver. Place the pump body on wooden blocks first.
2 Coat the inside diameter of a new seal with a non-hardening sealer. With the pump still resting on the wooden blocks, seat the seal with a drift.
3 Install the pump drive and driven gears, so that the marks on them are aligned, as illustrated.
4 Fit the selective-fit thrust washer over the pump cover delivery sleeve.
5 Install the 3 direct pump-to-clutch hook-type oil seal rings and the two Teflon forward clutch oil seal rings.

NOTE: Inspect the pump cover and body oil passages to see that they are not blocked.

6 Fit the square-cut O-ring on the case.
7 Assemble the pump body to the cover and secure them with the 5 bolts; tighten the bolts to 18 ft./lbs.

Direct Clutch

DISASSEMBLY

1 Remove the retaining ring and pressure plate from the direct clutch drum.

2 Remove the following:
 a. 3 lined drive plates
 b. 3 steel driven plates
 c. Cushion spring
3 Remove the direct clutch piston return spring seat retaining ring and the 17 coil return springs, by using a spring compressor.
4 Remove the piston.
5 Remove the outer and inner direct clutch seals.
6 Extract the direct clutch piston center seal from the drum.

INSPECTION

1 Check the clutch piston for burring, scoring, or wear.
2 Examine each of the 17 coil springs for collapsed coils, cracks, or distortion.
3 Check the piston for cracks.
4 Examine the clutch housing for wear, scores, open oil passages, and free movement of the ball check.

ASSEMBLY

1 Install new direct clutch piston outer and inner seals.
2 Fit a new center seal on the drum, so that its lip is facing upward.
3 Fabricate a tool from a piece of 0.020

in. piano wire crimped into a copper tubing handle and use it to install the direct clutch piston into the housing.
4 Install all 17 clutch return springs.
5 Install the piston return spring seat and compress it. Fit the retaining ring with the spring seat compressed.
6 Lubricate the cushion spring, lined plates, and steel plates with transmission fluid. Start the installation with the cushion spring and then alternate steel and lined plates.
7 Install the direct clutch pressure plate and secure it with the retaining ring.

Forward Clutch

DISASSEMBLY

1 Remove the direct clutch drum-to-housing needle roller bearing. Use a screwdriver to pry out the retaining ring and remove the pressure plate.
2 Remove the forward clutch pressure plate.
3 Remove the forward clutch housing faced plates, steel plates and cushion spring.
4 Compress the springs. Remove the forward clutch piston return spring seat retaining ring and spring seat.
5 Remove all 21 clutch return springs.
6 Remove the forward clutch piston assembly.
7 Remove the forward clutch piston inner and outer seals.
8 Be sure that the ball check exhaust in the drum is operable and free of dirt.

INSPECTION

1 Wash all parts in solvent. Blow out all passages. Air dry.
2 Inspect the clutch plates for burning, scoring, or wear.
3 Check all springs for distortion or collapsed coils.
4 Inspect the piston for cracks.
5 Inspect the clutch drum for wear, scoring, cracks, proper opening of oil passages, and free operation of the ball check.
6 Check the input shaft for:
 a. Open lubrication passage at each end.
 b. Damage to the splines or shaft.
 c. Damage to the ground bushing journals.
 d. Cracks or distortion of the shaft.

ASSEMBLY

1 Install the forward clutch piston inner and outer seals.
2 Install the forward clutch piston with a loop of .020 in. wire crimped into a length of copper tubing.
3 Install the 21 forward clutch return springs. These springs are identical to those used in the direct clutch. Install the spring seat.
4 Compress the springs and replace the spring seat retaining ring.
5 Replace the forward clutch housing cushion spring. Replace the steel and faced plates alternately, starting with a steel plate.

GM TURBO HYDRA-MATIC 250

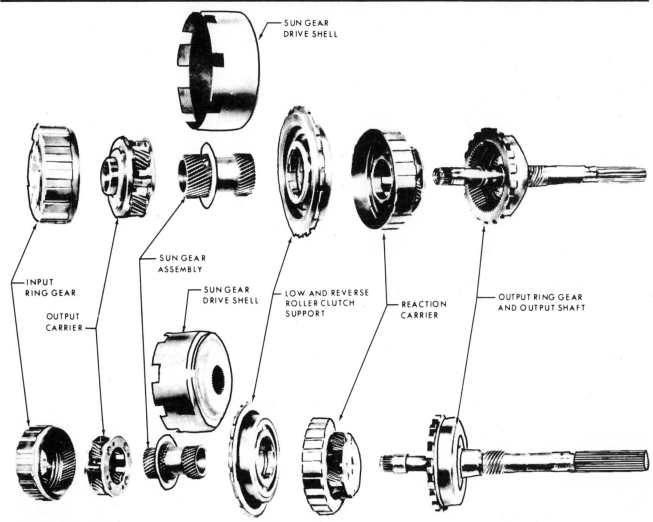

SUN GEAR
DRIVE SHELL

INPUT
RING GEAR

OUTPUT
CARRIER

SUN GEAR
ASSEMBLY

SUN GEAR
DRIVE SHELL

LOW AND REVERSE
ROLLER CLUTCH
SUPPORT

REACTION
CARRIER

OUTPUT RING GEAR
AND OUTPUT SHAFT

Planetary Gear Train

6 Use the following chart to determine the proper-selective-fit pressure plate, by measuring the distance from the top of the clutch pack to the top of the clutch drum and subtracting it from the distance measured between the edge of the notch on the inner surface of the drum to the end of the drum.

If the difference is (in.):	Use Part No.:
0.0160–0.0520	621072
0.0520–0.0830	621349
0.0830–0.1218	621350

7 Install the pressure plate selected above and secure it with its retaining ring.

8 Install the needle roller bearing.

Sun Gear and Sun Gear Drive Shell

DISASSEMBLY

1 Remove the sun gear-to-sun gear drive shell rear retaining ring.

2 Remove the sun gear-to-drive shell flat rear thrust washer.

3 Remove the sun gear and front retaining ring from the drive shell.

4 Remove the front retaining ring from the sun gear.

INSPECTION

1 Wash all parts in solvent. Air dry.

2 Inspect the sun gear and sun gear drive shell for wear or damage.

3 Inspect the sun gear bushings for galling or scoring. Drive out the damaged bushings. Install new bushings flush to 0.010 in. below the surface of the counterbore.

ASSEMBLY

1 Install the new front retaining ring on the sun gear. Be careful not to over-stress this ring.

2 Install the sun gear and retaining ring in the drive shell.

3 Install the sun gear-to-drive shell flat rear thrust washer.

4 Install the new sun gear-to-sun gear drive shell retaining ring.

Low and Reverse Roller Clutch Support

DISASSEMBLY

1 Remove the Low and Reverse clutch-to-sun gear shell thrust washer.

2 Remove the Low and Reverse overrun clutch inner race.

3 Remove the Low and Reverse roller clutch retaining ring.

4 Remove the Low and Reverse roller clutch assembly.

INSPECTION

1 Wash all parts in solvent. Air dry.

2 Inspect the roller clutch inner and outer races for scratches, wear, or indentations.

3 Inspect the roller clutch assembly rollers for wear and roller springs for distortion. If the rollers are removed from the assembly, install the rollers from the outside in to avoid bending the springs.

ASSEMBLY

1 Install the Low and Reverse roller clutch assembly.

2 Install the Low and Reverse roller clutch retaining ring.

3 Install the Low and Reverse overrun clutch inner race. The inner race must free-wheel in the clockwise direction only.

Transmission Assembly

NOTE: During assembly, use transmission fluid or petroleum jelly only, as lubricants to retain bearings and races. Be sure to lubricate all bearings, seal rings, and clutch plates, before assembly.

1 Fit the Low/Reverse clutch piston with its notch next to the parking pawl.

2 Install all 17 piston return coil springs.

3 Install the retainer and its retaining ring. Compress the retainer, prior to installing the ring.

4 Assemble the output ring gear to the output shaft.

5 Fit the reaction carrier front thrust washer (the one with 3 tangs) into the output ring gear support.

6 Place the output shaft assembly into the transmission case.

7 Put the reaction carrier assembly in the output ring gear/shaft assembly.

8 Lubricate the Low/Reverse clutch steel reaction and face plates. Install them starting with a steel reaction plate and then alternating them with faced plates. Install the anti-clunk spring.

NOTE: The notch on the steel separator should be installed toward the bottom of the case.

9 Fit the Low/Reverse clutch support assembly with the notch and anti-clunk spring positioned at 5 o'clock.

CHILTON CAUTION: *Be sure that the splines on the inner race of the roller clutch are aligned with the reaction carrier splines.*

10 Assemble the Low/Reverse roller clutch inner race to the sun gear thrust washer.

11 Fit the case-to-Low/Reverse clutch support snap-ring so that the anti-clunk spring is in the gap.

12 Install the rear thrust washer/sun gear drive shell assembly, the output carrier assembly, input ring gear rear thrust washer, and the input ring gear.

13 Use a new snap-ring to secure the ring gear to the output shaft. Do not overstress the snap-ring.

14 Install the front thrust washer on the input gear.

15 Assemble the direct clutch assembly and its special thrust washer to the forward clutch.

16 Install the combined direct clutch and forward clutch assemblies in the transmission case.

CHILTON CAUTION: *Be sure that the forward clutch face plates are positioned over the input ring gear and also, that the direct clutch housing tangs are engaged in the slots on the sun gear drive shell.*

17 Install the intermediate servo and the intermediate band. Be sure that the ends of the band are correctly positioned on the adjusting screw and servo rod ends. Turn the adjusting screw in, until the end of the screw has engaged the slot in the band lug.

18 The oil pump cover and the direct clutch assembly have a select-fit thrust washer between them; to determine the proper size washer to be used, proceed as follows:
   a. Temporarily install the thrust washer, oil pump gasket, and the oil pump. Use guide studs to install the pump. Secure the pump with its bolts.

b. Position the transmission with the output shaft pointing downward.

c. Attach a dial indicator so that it measures vertical output shaft movement. Set the indicator to zero.

d. Lift up on the output shaft as far as it will go and read the total movement registered on the indicator.

e. The indicator reading should fall between 0.032 and 0.064 in. If it does, go on to the next step. If the reading is not within specifications, use a thinner or thicker selective-fit thrust washer, as necessary. Thrust washers are available in the following sizes:

   0.065–0.067 in.
   0.082–0.084 in.
   0.099–0.101 in.

f. Repeat the checking procedure, to be sure that the proper size washer was installed.

19 Install both a *new* pump assembly gasket and a *new* square-cut oil seal ring.

20 Fit the guide pins into the case.

21 Fit the pump assembly into the case and secure it with its mounting bolts. Use new washer-type bolt seals.

NOTE: If the input shaft won't turn, as the pump is being pulled into the case, the faced plates in the direct and forward clutch housings are not indexed properly with their respective parts, because of incorrect housing installation. Go back and reinstall the housings, before completing pump installation.

22 Adjust the intermediate band once the pump has been installed:
   a. Tighten the adjusting screw to 30 in./lbs.
   b. Then back off on the adjusting screw 3 full turns.
   c. Hold the adjusting screw in this position and tighten the locknut to 30 ft./lbs.

23 Install the speedometer drive gear in the following order:
   a. Fit the speedometer drive gear retaining clip into the output shaft hole.
   b. Use a heat lamp to heat a new speedometer drive gear.
   c. Install the speedometer gear on the output shaft, by aligning the slot in the gear with the retaining clip.

24 Fit the square-cut O-ring seal on the extension housing.

25 Secure the extension housing to the transmission housing by tightening the securing bolts to 25 ft./lbs.

26 If the old extension housing rear seal was removed, install a new one, by using a drift to drive it into place.

27 If it was removed, install a new manual shaft-to-case lip seal, by driving it into place with a ¾ in. diameter rod.

28 Fit the parking pawl into the transmission case with its tooth facing the inside of the case. Place the parking pawl shaft into the transmission case and through the pawl.

29 Drive the parking pawl retaining plug

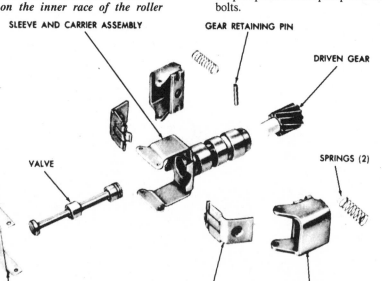

SLEEVE AND CARRIER ASSEMBLY   GEAR RETAINING PIN

DRIVEN GEAR

PINS   VALVE   SPRINGS (2)

THRUST CAP   SECONDARY WEIGHT (2)   PRIMARY WEIGHT (2)

Governer Assembly-exploded view (typical)

GM TURBO HYDRA-MATIC 250

into the case with a ⅜ in. diameter rod, so that the plug is 0.13–0.17 in. below the face of the case. Stake the plug in 3 places.

30 Hook the square end of the parking pawl disengaging spring into the pawl.

31 Secure the parking lock bracket with the two special bolts and torque the bolts to 29 ft./lbs.

CHILTON CAUTION: *Use only the specified bolts, GM 300M or equivalent, which have six marks on their heads.*

32 Connect the range selector lever to the parking pawl actuator rod.

33 Install the actuator rod beneath the parking lock bracket and pawl.

34 Fit the manual shaft through the transmission case and range selector lever. Torque the manual shaft securing nut to 30 ft./lbs.

35 Fit the spacer clip between the case and manual shaft.

36 Install the oil pump pressure and governor feed screens in their respective holes.

37 Insert each of the 4 check balls into the same holes from which they were removed. Do not mix the check balls.

38 Install the valve body transfer plate lower valve, lower gasket, and upper gasket.

39 Insert the intermediate servo return spring.

40 Install the valve body in the case. Assemble the manual control valve link to the range selector inner lever. Tighten the valve body bolts, randomly, to 130 in./lbs.

41 Secure the transfer support plate with its mounting bolts. Tighten the bolts to 130 in./lbs.

42 Attach the detent control valve wire to the detent valve operating lever, and then secure the lever to the valve body.

43 Attach the detent roller/spring assembly to the valve body.

44 Line up the suction screen lubrication holes with those in the valve body, and install it, complete with gasket.

NOTE: It is very important that the screen and valve body holes are aligned.

45 Use a new oil pan gasket, and install the oil pan. Tighten the oil pan bolts, in succession, until they will maintain 130 in./lbs. torque.

NOTE: When using a new oil pan, be sure to transfer the production code number from the old pan to the right-side of the new pan.

46 Install the governor assembly, cover, seal, and retaining wire.

47 Install the vacuum modulator valve, modulator, and retaining clip. Tighten the securing bolts to 130 in./lbs. Be sure that the tang on the clip is pointing toward the modulator.

48 Insert the First/Second accumulator piston assembly and spring.

49 Fit a new O-ring into the groove in the case; the cover and retaining ring.

GM Type 250 Torque Specifications

Component	Tightening Torque
Oil pan-to-transmission case	130 in/lbs
Pump assembly-to-transmission case	18.5 ft/lbs
Vacuum modulator retainer-to-case	130 in/lbs
Valve body-to-support plate	130 in/lbs
Cover-to-pump body	17 ft/lbs
Bracket—parking lock-to-case	28 ft/lbs
Extension housing-to-transmission case	25 ft/lbs
Oil cooler fitting-to-case	120 in/lbs
Oil pickup screen-to-valve body	40 in/lbs
Bracket—detent valve actuating lever-to-valve body	52 in/lbs
Transmission-to-engine	35 ft/lbs
Shaft-to-inner selector lever	25 ft/lbs
Lever-to-manual shaft	20 ft/lbs
Swivel-to-control rod	120 in/lbs
Control lever	25 in/lbs
Intermediate band adjustment locknut	15 ft/lbs

GM Type 250 Capacity & General Specifications

Fluid Capacity (dry)	21½ pints
Cooling	Water (except Vega)
Fluid filter type	bottom suction screen
Clutches	3
Roller clutches	1
Band adjustment	Intermediate

GM Type 250 Oil Pressure Checks With Vacuum Modulator

Model	Altitude (ft. above At sea level)	Oil Pressure (psi) in Gear Range*		
		D,N,P	L₁,L₂	R
Vega	0	55.0	81.9	88.4
(L4)	4,000	66.9	88.8	107.5
	4,000	81.9	97.6	131.5
	6,000	95.6	105.6	153.6
	8,000	107.9	112.8	173.4
All Models	0	55.0	79.7	83.4
(L6)	2,000	68.9	89.9	104.4
	4,000	86.3	102.8	130.4
	6,000	102.4	114.6	155.2
	8,000	116.7	125.2	176.9

* Made with vehicle stationary, brake on, engine rpm set to maintain 16 in. Hg manifold pressure, oil pressure gauge connected to line pressure tap, and vacuum modulator tube **connected.**

GM Type 250 Oil Pressure Checks Without Vacuum Modulator

Model	Oil Pressure (psi) in gear range @ 1200 rpm ▲		
	D	L₁, L₂	R
Vega (L4)	120.0	120.0	245.2
All Models (L6)	168.9	163.6	256.0

▲ Made with vehicle stationary, brake on, 1200 rpm engine speed, oil pressure gauge connected to line pressure tap and vacuum modulator tube **disconnected.**

GM POWERGLIDE

TROUBLESHOOTING
GM POWERGLIDE

	ITEMS TO CHECK	
THE PROBLEM	**IN CAR**	**OUT OF CAR**
CAR WILL NOT MOVE IN ANY SELECTOR POSITION	Oil level. Oil screen. Pressure regulator valve.	Front pump priming valve. Front pump. Input shaft.
ENGINE SPEED FLARES, AS WITH SLIPPING CLUTCH	Oil level. Oil screen. Band adjustment. Servo blocked. Servo seal.	Low band. Low band linkage. Converter stator.
ENGINE SPEED FLARES ON UPSHIFT	Oil level. Band adjustment. Oil screen. Vacuum modulator or line.	Clutch feed blocked. High clutch. Front clutch relief valve.
TRANSMISSION WILL NOT UPSHIFT	Governor. Throttle valve. Throttle linkage.	Low clutch valve stuck. Rear pump priming valve. Rear pump or drive.
HARSH UPSHIFT	Throttle linkage. Band adjustment. Vacuum modulator or line. Hydraulic modulator valve.	
HARSH DECELERATION DOWNSHIFT	Band adjustment. Too high idle speed. Valves malfunctioning. Make pressure tests. Hydraulic modulator valve.	
NO DOWNSHIFT	Valves malfunctioning. Governor. Throttle valve.	
CLUTCH FAILURE, BURNT PLATES	Band adjustment. Oil level. Governor. Driving too fast in low.	High clutch. Front clutch relief valve.
EXCESSIVE CREEP IN DRIVE	Throttle linkage. Too high idle speed.	
CAR CREEPS IN NEUTRAL	Manual valve lever adj.	High clutch. Low band.
NO DRIVE IN REVERSE	Manual valve lever adj.	Reverse clutch relief valve. Low clutch. Clutch feed blocked.
IMPROPER SHIFT POINTS	Throttle linkage. Vacuum modulator or lines. Throttle valve. Governor.	Rear pump priming valve.
UNABLE TO PUSH-START CAR		Rear pump or drive.
OIL LEAKS	Oil leaks at external points. Oil cooler or lines. Vacuum modulator or line.	Front pump attaching bolts. Front pump.
OIL FORCED OUT AT FILLER TUBE	Oil level. Oil cooler or lines.	Pump circuit leakage.

GM POWERGLIDE

GM Powerglide

Identification

CHEVROLET

On all Chevrolet vehicles, see the engine identification chart in the appropriate car section of this manual. Check the engine identification tag for transmission I.D.

PONTIAC

On Pontiac vehicles the transmission identification number is on the right front of the transmission.

Transmission Disassembly

EXTENSION, GOVERNOR AND GOVERNOR SUPPORT

1 Place the transmission in a holding fixture, if possible.
2 Remove the converter holding tool, then lift off the converter.
3 If replacement is necessary, remove the speedometer driven gear. Loosen the cap screw and retainer clip and remove the gear from the extension.
4 Remove the transmission extension by removing the five attaching bolts. Note the seal ring on the rear pump body.
5 Remove the speedometer drive gear from the output shaft.
6 Remove the C-clip from the governor shaft of the weight side of the governor, then remove the shaft and governor valve from the opposite side of the governor assembly and the two belleville springs.
7 Loosen the governor drive screw and slide the governor over the end of the output shaft.
8 Remove the four bolts which hold the governor support to the transmission case and remove the support body, gasket, and extension seal ring.

TRANSMISSION INTERNAL COMPONENTS

9 Rotate holding fixture, or turn the transmission, until the front end is pointing up. Then remove the seven front oil pump bolts. (The bolt holes are of unequal spacing to prevent incorrect location upon installation.)
10 Remove the front oil pump and stator shaft assembly and the selective fit thrust washer using an inertia puller or substitute.
11 Release tension on the low band adjustment, then with transmission horizontal, grasp the transmission input shaft and carefully work it and the clutch drum out of the case. Be careful not to lose the low sun gear bushing from the input shaft. The low sun gear thrust washer will probably remain in the planet carrier.
12 The low brake band and struts may now be removed.
13 Remove the planet carrier and the output shaft thrust caged bearing from the front of the transmission.
14 Remove reverse ring gear if it did not come out with the planet carrier.

GM POWERGLIDE

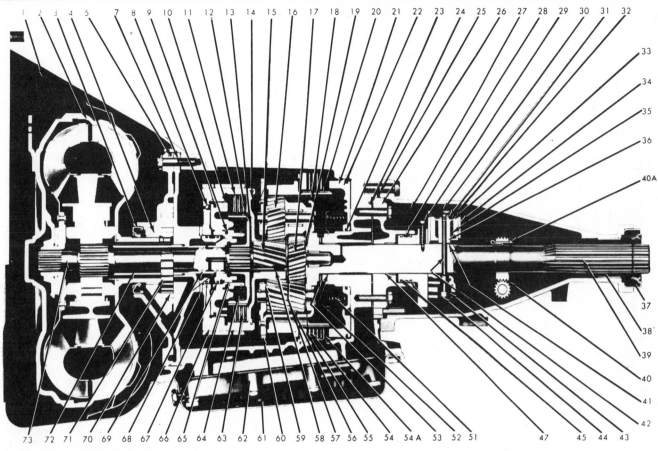

Cutaway view GM Powerglide Transmission

1. TRANSMISSION CASE
2. WELDED CONVERTER
3. OIL PUMP SEAL ASSEMBLY
4. OIL PUMP BODY
5. OIL PUMP BODY SQUARE RING SEAL
7. OIL PUMP COVER
8. CLUTCH RELIEF VALVE BALL
9. CLUTCH PISTON INNER AND OUTER SEAL
10. CLUTCH PISTON
11. CLUTCH DRUM
12. CLUTCH HUB
13. CLUTCH HUB THRUST WASHER
14. CLUTCH FLANGE RETAINER RING
15. LOW SUN GEAR AND CLUTCH FLANGE ASSEMBLY
16. PLANET SHORT PINION
17. PLANET INPUT SUN GEAR
18. PLANET CARRIER
19. PLANET INPUT SUN GEAR THRUST WASHER
20. RING GEAR
21. REVERSE PISTON
22. REVERSE PISTON OUTER SEAL
23. REVERSE PISTON INNER SEAL
24. GOVERNOR SUPPORT GASKET
25. EXTENSION SEAL RING
26. GOVERNOR SUPPORT
27. EXTENSION

28. GOVERNOR HUB
29. GOVERNOR HUB DRIVE SCREW
30. GOVERNOR BODY
31. GOVERNOR SHAFT RETAINER CLIP
32. GOVERNOUR OUTER WEIGHT RETAINER RING
33. GOVERNOR INNER WEIGHT RETAINER RING
34. GOVERNOR OUTER WEIGHT
35. GOVERNOR SPRING
36. GOVERNOR INNER WEIGHT
37. EXTENSION REAR OIL SEAL
38. EXTENSION REAR BUSHING
39. OUTPUT SHAFT
40. SPEEDOMETER DRIVE AND DRIVEN GEAR
40A. SPEEDOMETER DRIVEN GEAR RETAINING CLIP
41. GOVERNOR SHAFT URETHANE WASHER
42. GOVERNOR SHAFT
43. GOVERNOR VALVE
44. GOVERNOR VALVE RETAINING CLIP
45. GOVERNOR HUB SEAL RINGS
47. GOVERNOR SUPPORT BUSHING
51. REVERSE PISTON RETURN SPRINGS, RETAINER AND RETAINER RING

52. TRANSMISSION REAR CASE BUSHING
53. OUTPUT SHAFT THRUST BEARING
54. REVERSE CLUTCH PACK
54A. REVERSE CLUTCH CUSHION SPRING (WAVED)
55. PINION THRUST WASHER
56. PLANET LONG PINION
57. LOW SUN GEAR NEEDLE THRUST BEARING
58. LOW SUN GEAR BUSHING (SPLINED)
59. PINION THRUST WASHER
60. PARKING LOCK GEAR
61. TRANSMISSION OIL PAN
62. VALVE BODY
63. HIGH CLUTCH PACK
64. CLUTCH PISTON RETURN SPRING, RETAINER AND RETAINER RING
65. CLUTCH DRUM BUSHING
66. LOW BRAKE BAND
67. HIGH CLUTCH SEAL RINGS
68. CLUTCH DRUM THRUST WASHER (SELECTIVE)
69. TURBINE SHAFT SEAL RINGS
70. OIL PUMP DRIVEN GEAR
71. OIL PUMP DRIVE GEAR
72. STATOR SHAFT
73. INPUT SHAFT

15 With a large screwdriver, remove the reverse clutch pack retainer ring and lift out the reverse clutch plates and the cushion spring.

16 Install reverse piston spring compressor through rear bore of the case, with the flat plate on the rear face of the case, and turn down wing nut to compress the rear piston spring retainer and springs. Then remove the snap ring. A spring compressor may be made up from a suitable length bolt and large flat washers.

17 Remove the compression tool, the reverse piston spring retainer, and the

GM POWERGLIDE

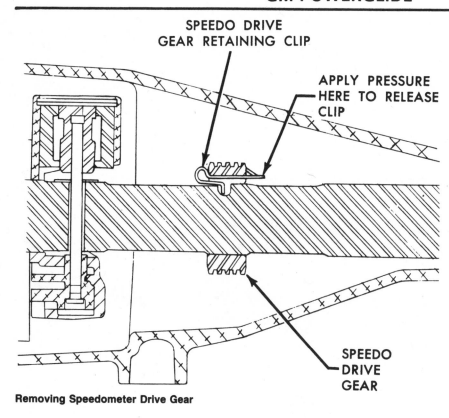

Removing Speedometer Drive Gear

SPEEDO DRIVE
GEAR RETAINING CLIP

APPLY PRESSURE
HERE TO RELEASE
CLIP

SPEEDO
DRIVE
GEAR

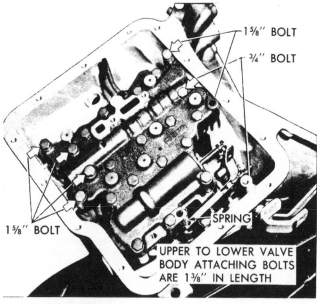

1⅝" BOLT

¾" BOLT

1⅝" BOLT

SPRING

UPPER TO LOWER VALVE
BODY ATTACHING BOLTS
ARE 1⅜" IN LENGTH

Valve Body Removal

17 piston return springs.
18 Remove the rear piston by applying air pressure to the reverse port in the rear of the transmission case. Remove inner and outer seals.

19 Remove the three servo cover bolts, servo cover, piston and spring.

OIL PAN AND VALVE BODY
NOTE: The oil pan and valve body may

be serviced without removing the extension, and internal components, covered in the preceding steps.

20 Rotate the transmission until the unit is upside down (oil pan on top). Remove oil pan attaching bolts, oil pan, and gasket. Remove the screen. Replace at assembly, using a new case screen.
21 Remove vacuum modulator and gasket, and vacuum modulator plunger, dampening spring, and valve.
22 Remove two bolts holding the detent guide plate to the valve body and the transmission case. Remove the guide plate and the range selector detent roller spring.
23 Remove the remaining valve body-to-transmission case attaching bolts and lift out the valve body and gasket. Disengage the servo apply tube from the transmission case as the valve body is removed. On Pontiac applications, remove the manual control valve from the valve body.
24 If necessary, the throttle valve, shift and parking actuator levers and the parking pawl and bracket may be removed, as follows:
   a. Loosen the allen head screw on the inner TV lever and remove the inner TV lever and shaft from the outer TV lever. Remove the O-ring seal and special washer from the outer TV shaft and remove the inner TV shaft from the case.
   b. Remove the selector outer lever and shaft from the selector inner lever. Remove the inner selector lever parking pawl actuator from the case. Separate the actuator from the inner lever.
   c. Remove the parking pawl bracket from the case. Remove the parking pawl spring and E-clip and drive the pawl shaft forward out of the case. Remove the parking pawl.

Unit Assembly Overhaul

CONVERTER AND STATOR
The converter is a welded assembly and no internal repairs are possible. Check the seams for stress or breaks and replace the converter if necessary.

Front Pump

SEAL REPLACEMENT
If the front pump seal requires replacement, remove the pump from the transmission, pry out and replace the seal. Drive new seal into place. Then, if no further work is needed on the front pump, install it in the case. (The outer edge of the seal should be coated with non-hardening sealer before installation.)

DISASSEMBLY
1 Remove pump cover-to-body attaching bolts and the cover.
2 On Pontiac models, remove the 2 oil seal rings from the hub of the pump

GM POWERGLIDE

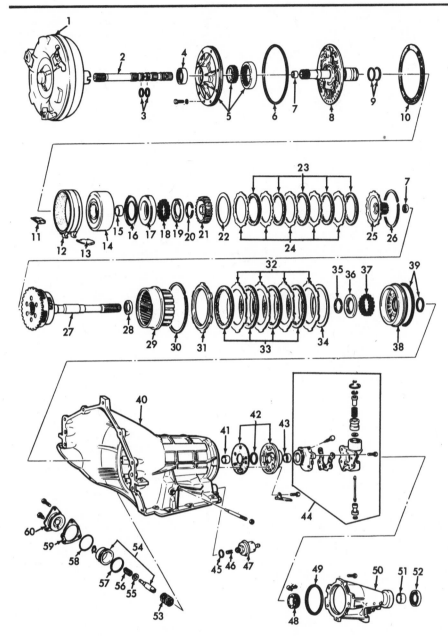

6. Seal (front pump)
7. Washer unit
8. Pump cover and stator shaft
9. Seal rings
10. Front pump gasket
11. Strut assembly
12. Band assembly (low)
13. Strut anchor
14. Clutch drum
15. Bushing (clutch drum
16. Piston seal unit
17. Piston (clutch)
18. Spring (Hi clutch
19. Seat (Clutch spring)
20. Snap ring
21. Clutch Hub
22. Spring (clutch cushion)
23. Plate assembly (drive)
24. Plate assembly (driven)
25. Sun bear and clutch flange
26. Ring (clutch flange)
27. Planet carrier assy.
28. Bearing assy. (outputshaft)
29. Gear (reverse ring)
30. Snap ring
31. Pressure plate assy.
32. Reaction plate assy.
33. Clutch driven plate
34. Spring (cushion)
35. Ring (Piston return)
36. Spring seat
37. Spring (Piston return)
38. Piston (reverse clutch)
39. Seal Unit (Piston)
40. Transmission case
41. Bushing (Trans. Case)
42. Body Assy. (Governor Support)
43. Bushing
44. Governor Assy.
45. Gasket (Vacuum Modulator)
46. Plunger (Vacuum Modulator)
47. Vacuum Modulator
48. Speedo Drive Gear
49. Seal (Extension Case)
50. Extension Case
51. Bushing (Extension)
52. Seal Assy. (Extension
53. Return Spring (Servo Low)
54. Rod assembly
55. Washer
56. Spring (Piston Apply)
57. Ring (Piston)
58. Seal (Low Servo Cover)
59. Gasket (Low Servo Cover)
60. Cover (Low Servo)

Shafts and clutches—powerglide trans.

1. Torque Converter
2. Input shaft assy.

cover. Remove the downshift timing valve from the pump cover.

3 Remove the oil seal ring, matchmark the gears for assembly, and remove the gears from the body.

CHILTON CAUTION: *Do not drop or nick the gears. They are not heat treated.*

INSPECTION

1 Wash all parts in solvent. Blow out all oil passages.
2 Inspect pump gears for nicks or damage.
3 Inspect body and cover faces for nicks or scoring. Inspect cover hub outside diameter for nicks or burrs.

3. Oil seal rings
4. Oil seal (inner)
5. Front oil pump

4 Check for free operation of priming valve. Replace if necessary.

5 Inspect body bushing for galling or scoring. Check clearance between body bushing and converter pump hub. Maximum clearance is .005 in. If the bushing is damaged, replace the pump body.

6 Inspect converter housing hub outside diameter for nicks or burrs. Repair or replace.

7 If oil seal is damaged or leaking, pry out and drive in a new seal.

8 Check condition of oil cooler bypass valve. Replace if leaking. An "Easy-Out" type remover may be used to

remove the valve. Tap new valve seat into place with a soft hammer or brass drift so that it is flush or up to .010 in. below the surface.

9 With all parts clean and dry, install pump gears and check:
a. clearance between outside diameter of driven gear and body should be .0035–.0065 in.
b. clearance between inside diameter of driven gear and crescent should be .003–.009 in.
c. gear end clearance should be .0005–.0015 in.

ASSEMBLY

1 Remove the input shaft, clutch drum,

low band and struts as outlined under "Transmission Disassembly."

2 Install downshift timing valve, conical end out, into place in the pump cover to a height of 17/32 in. measured from shoulder of valve assembly to face of pump cover.

3 Oil the drive and driven gears and install them into the pump body.

4 Set pump cover in place over the body and loosely install two attaching bolts. Assemble drive gear with recessed side of drive lug facing the converter.

5 Place pump assembly, less the rubber seal ring, upside down into the pump bore of the case. Install remaining attaching bolts and torque to 20 ft./lbs.

6 Remove pump assembly from case bore. On Pontiac models, install 2 high clutch oil seal rings. Replace the clutch drum and input shaft, low band and struts as described under "Transmission Assembly."

7 Renew rubber seal ring in its groove in the pump body and install the pump assembly in place in the case bore, using a new gasket. Be sure that the selective fit thrust washer is in place.

8 Install attaching bolts. (Use new bolt sealing washers if necessary.)

Clutch Drum

DISASSEMBLY

CHILTON CAUTION: *When working with the clutch drum, use extreme care that the machined face on the front of the drum not be scratched, scored, nicked, or otherwise damaged. This machined face must be protected whenever it is brought to bear on a press or tool of any sort.*

1 Remove retainer ring, low sun gear and clutch flange assembly from the clutch drum.

2 Remove the hub rear thrust washer.

3 Lift out clutch hub, then remove clutch pack and hub front thrust

washer. Note the number and sequence of plates.

4 Remove spring retainer. Compress the springs with a spring compressor or an arbor press enough to permit removal of the retainer snap-ring. Then, releasing pressure on the springs, remove retainer and the 24 springs.

5 Lift up on the piston with a twisting motion to remove it from the drum, then remove inner and outer seals.

INSPECTION

1 Wash all parts in solvent, blow out all passages, and air dry. Do not use rags to dry parts.

2 Check drum bushing for scoring or excessive wear.

3 Check steel ball relief valve in clutch drum. Be sure that it is free to move and that orifice in front face of drum is open. If ball is loose enough to come out, or not loose enough to rattle, replace drum. Do not attempt replacement or restaking of ball.

4 Check fit of low sun gear and clutch flange assembly in clutch drum slots. There should be no appreciable radial play.

5 Check low sun gear for nicks or burrs. Check gear bore for wear.

6 Check clutch plates for burring, wear, pitting, or metal pick-up. Faced plates should be a free fit over clutch hub; steel plates should be a free fit in clutch drum slots.

7 Check condition of clutch hub splines and mating splines of clutch faced plates.

8 Check clutch pistons for cracks or distortion.

CLUTCH DRUM BUSHING REPLACEMENT

If replacing drum bushing, carefully press out the old bushing. Then press (don't hammer) the new bushing into

place from the machined face side of the drum. Press only far enough to bring the bushing flush with the clutch drum. Do not force the tool against the clutch drum machined face.

ASSEMBLY

1 Install new piston inner seal into hub of clutch drum with seal lip toward front of transmission.

2 Install new piston seal into clutch piston. Seal lips must be pointed toward the clutch drum, (front of transmission). Lubricate the seals and install piston into clutch drum with a twisting motion.

3 Place 24 springs in position on the piston, then place the retainer on the springs.

4 Depress the retainer plate and springs far enough to permit installation of the spring retainer snap-ring into its groove on the clutch drum hub.

5 Install the hub front washer with its lip toward the clutch drum, then install the clutch hub.

6 Install the cushion spring if used. On Pontiac V8 and Chevrolet 350 V8 applications, the first steel drive plate to the rear of the cushion spring is a selective fit. To correctly install the clutch pack proceed as follows:

a. Before installing the clutch pack into the drum, stack the 5 steel driven plates (except selective fit) and 5 faced drive plates. Measure this stack height.

b. Using the measurement obtained, select the proper selective fit drive plate from the chart and install it on top of the cushion spring. Then install the faced drive plates and steel drive plates (see chart) alternately, beginning with a steel driven plate, for Chevrolet applications. On Pontiac applications, install the faced drive plates and steel

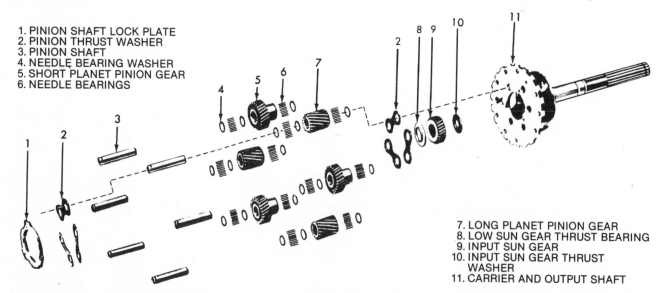

1. PINION SHAFT LOCK PLATE
2. PINION THRUST WASHER
3. PINION SHAFT
4. NEEDLE BEARING WASHER
5. SHORT PLANET PINION GEAR
6. NEEDLE BEARINGS

7. LONG PLANET PINION GEAR
8. LOW SUN GEAR THRUST BEARING
9. INPUT SUN GEAR
10. INPUT SUN GEAR THRUST WASHER
11. CARRIER AND OUTPUT SHAFT

Small Planet Carrier Assembly

driven plates, alternately, beginning with a faced drive plate.

c. On all six-cylinder models, install the drive plates and driven plates (see chart) alternately, beginning with a steel driven plate.

CLUTCH ASSEMBLIES

NOTE: The number and sequence of plates varies with the power and torque requirements of the car model involved. On some models, the first driven plate is a selective fit.

7 Install the rear hub thrust washer with its flange toward the low sun gear, then install the low sun gear and flange assembly and secure with retaining ring. When installed, the openings in the retainer ring should be adjacent to one of the lands of the clutch drum.

8 Check assembly for freedom of movement by turning the clutch hub.

Planet Assembly and Input Shaft

INSPECTION

1 Wash planet carrier and input shaft in cleaning solvent, blow out all passages, and air dry. Do not use rags to dry parts.

2 Inspect planet pinions for nicks or other tooth damage.

3 Check end clearance of planet gears. The clearance should be .006–.030 in.

4 Check input sun gear for tooth damage. Check thrust washer for damage.

5 Inspect output shaft bearing surface and input pilot bushing for nicks or scoring.

6 Inspect input shaft splines for nicks or damage. Check fit in clutch hub, input sun gear, and turbine hub.

7 Check oil seal rings for damage; rings must be free in input shaft ring grooves. Remove rings and insert in stator support bore. Check to see that hooked ring ends have clearance. Replace rings on shaft.

REPAIRS

NOTE: For carriers without flared pinion shafts, the following procedure may be used for overhaul.

1 Place the planet carrier assembly in a padded vise so that the front (parking lock gear end) of the assembly is up.

2 Using a prick punch, mark each pinion shaft and the carrier assembly so that, when reassembling, each shaft will be returned to its original location.

3 Remove pinion shaft lockplate screws and rotate plate counterclockwise far enough to remove it.

4 Starting with a short planet pinion, drive the lower end of the pinion shaft up until the shaft is above the press fit area of the output shaft flange. Feed a dummy shaft into the short planet pinion from the lower end, pushing

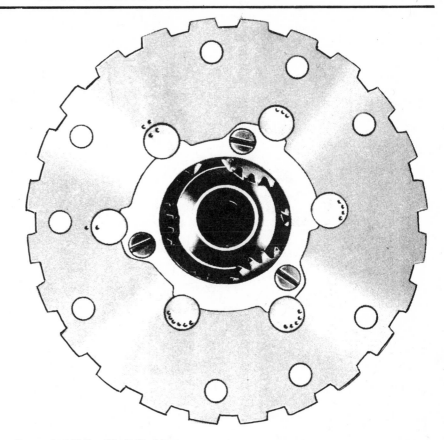

Suggested Pinion Shaft Markings

the planet pinion shaft ahead of it until the tool is centered in the pinion and the pinion shaft is removed.

5 Remove short planet pinion.

6 Remove dummy shaft, needle and bearing spacers from short pinion.

NOTE: Twenty needle bearings are used in each end of each gear and are separated by a bearing spacer in the center.

7 By following Steps 4, 5, and 6, remove the adjacent long planet pinion that was paired, by thrust washers, to the short pinion now removed.

8 Remove upper and lower thrust washers.

9 Remove and disassemble remaining planet pinions, in pairs, as above.

10 Remove low sun gear needle thrust bearing, input sun gear, and thrust washer.

11 Wash all parts in solvent and blow dry, then inspect.

12 Inspect input shaft bushing in base of output shaft. If damaged, it may be removed by using a slide hammer. New bearing can be installed by using pilot end of input shaft as press tool.

13 Using dummy shaft, assemble needle bearings and spacer (20 rollers in each end) in one of the long planet pinions. Use petroleum jelly to aid in holding the rollers in position.

14 Position long planet gear, with dummy shaft centered in the pinion and with thrust washers at each end,

in the planet carrier. Oil grooves on thrust washers must be toward the gears.

NOTE: Long pinions are located opposite the closed portions of the carrier and short pinions are located in the openings.

15 Feed a second dummy shaft in from the top, picking up the upper thrust washer and the pinion and pushing the already installed dummy shaft out the lower end. As the first dummy is pushed down, be sure that it picks up the lower thrust washer.

16 Select the correct pinion shaft, as marked in Step 2, lubricate the shaft and install it from the top, pushing the assembling tools (dummy) ahead of it.

17 Turn the pinion shaft so that the slot or groove at the upper end faces the center of the assembly.

18 With a brass drift, drive the shaft in until the lower end is flush with the lower face of the planet carrier.

19 Following the same procedure as outlined in Steps 13 through 18, assemble and install a short planet pinion into the planet carrier adjacent to the long pinion now installed.

NOTE: The thrust washers, already installed with the long planet pinion, also serve for this short planet pinion, because the two pinions are paired together on one set of thrust washers.

20 Install the input sun gear thrust

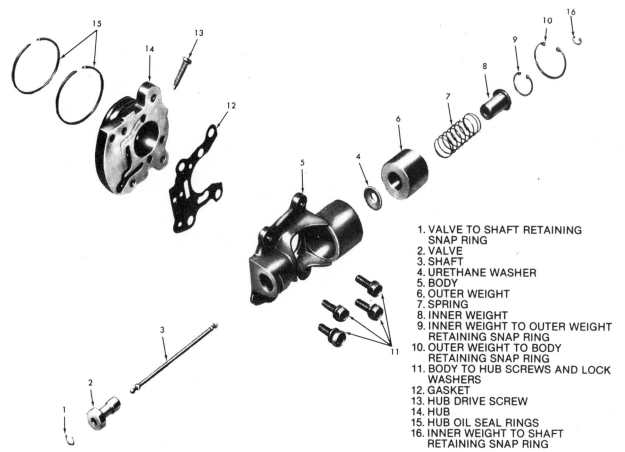

1. VALVE TO SHAFT RETAINING SNAP RING
2. VALVE
3. SHAFT
4. URETHANE WASHER
5. BODY
6. OUTER WEIGHT
7. SPRING
8. INNER WEIGHT
9. INNER WEIGHT TO OUTER WEIGHT RETAINING SNAP RING
10. OUTER WEIGHT TO BODY RETAINING SNAP RING
11. BODY TO HUB SCREWS AND LOCK WASHERS
12. GASKET
13. HUB DRIVE SCREW
14. HUB
15. HUB OIL SEAL RINGS
16. INNER WEIGHT TO SHAFT RETAINING SNAP RING

Governer Assembly

washer, input sun gear, and low sun gear needle thrust bearing.

21 Assemble and install the remaining planet pinions, in pairs, as previously explained.

22 Check end clearance of planet gears. This clearance should be .006–.030 in.

23 Place the shaft lockplate in position. Then, with the extended portions of the lockplate aligned with slots in the planet pinion shafts, rotate the lockplate clockwise until the three attaching screw holes are accessible.

24 Install lock plate attaching screws and torque to 2½ ft./lbs.

Governor

The governor assembly is a factory balanced unit. If body replacement is needed, the two sections must be replaced as a unit.

DISASSEMBLY
NOTE: The governor valve and shaft were removed in Step 6 of "Transmission Disassembly" procedures.

1 Remove the outer weight by sliding toward center of body.

2 Remove smaller inner weight retain-

ing snap-ring and remove inner weight and spring.

3 Remove the four body assembly bolts and separate the body, hub and gasket. Remove the two seal rings.

INSPECTION
1 Clean all parts in solvent and air dry.
2 Check all parts. Replace all bent, scored, or otherwise damaged parts. Body and hub must be replaced as a unit.

ASSEMBLY
1 Reassemble governor weights and install into body bore. Replace seal rings on hub.

2 Slide hub into place on output shaft and lock into place with drive screw. Install gasket and governor body over output shaft, install governor shaft, line up properly with output shaft and install body attaching bolts. Torque bolts to 6–8 ft./lbs. While tightening these bolts, engage the transmission selector lever in Park, to prevent the shaft from turning.

3 Check governor weight for free fit in body after the four attaching bolts are torqued. If the weight sticks or binds, loosen the bolts and retorque.

Valve Body

REMOVAL
Remove valve body, as described under "Transmission Disassembly". If performing the operation on the car, the vacuum modulator valve, oil pan and gasket, guide detent plate and range selector detent roller spring must be removed in order to remove the valve from the transmission.

DISASSEMBLY
1 Remove manual valve, suction screen and gasket. On Pontiac applications, remove the roller spring and E-clip from the range selector detent lever and remove the detent lever from side of valve body.

2 Remove cover bolts, then remove lower valve body and transfer plate from upper valve body. Discard gaskets.

3 From the upper valve body, remove the throttle valve and detent valve and the downshift timing valve as follows:
a. Throttle Valve and Detent Valve— Remove the retaining pin by wedging a thin screwdriver between its head and the valve body, then removing the detent valve assembly and throttle spring. Tilt valve

GM POWERGLIDE

body to allow the throttle valve to fall out. If necessary, remove the C-clip and disassemble the detent valve assembly.

NOTE: **Do not change adjustment of hex nut on the detent valve assembly. This is a factory setting and should not normally be changed. However, some adjustment is possible if desired. See "Throttle Valve Adjustment," in later text.**

    b. Downshift Timing Valve—Drive out the roll pin, remove valve spring and downshift timing valve.

4 From the lower valve body, remove the low-drive shift valve and the pressure regulator valve as follows:

    a. Low-Drive Shift Valve—Remove the snap-ring and tilt valve body to remove low-drive regulator valve sleeve and valve assembly, valve spring seat, valve springs and the shifter valve.

    b. Pressure Regulator Valve—Remove the snap-ring, then tilt valve body to remove the hydraulic modulator valve sleeve and valve, pressure regulator valve spring seat, spring, damper valve, spring seat and valve.

INSPECTION

1 Clean all parts in solvent. Air dry; use no rags.
2 Check all valves and valve bores for burrs or other deformities which could cause valve hang-up.

ASSEMBLY

1 Replace valve components in proper bores, reversing the steps of disassembly outlined above.
2 Install the gasket and transfer plate.
3 Install lower valve body and gasket and install attaching bolts. Torque to 15 ft./lbs.
4 Install valve body onto transmission, as outlined under "Transmission, Assembly" in later text.

Vacuum Modulator

The vacuum modulator is mounted on the left rear of the transmission and can be serviced from beneath the car.

REMOVAL

1 Remove vacuum line at the modulator.
2 Unscrew the modulator from the transmission with a thin 1 in. tappet-type wrench.
3 Remove vacuum modulator valve.

INSPECTION

1 Check the vacuum modulator plunger and valve for nicks and burrs. If such damage cannot be repaired with a stone, replace the part.
2 Check the vacuum modulator for leakage with a vacuum source. If the modulator leaks, replace the assembly.

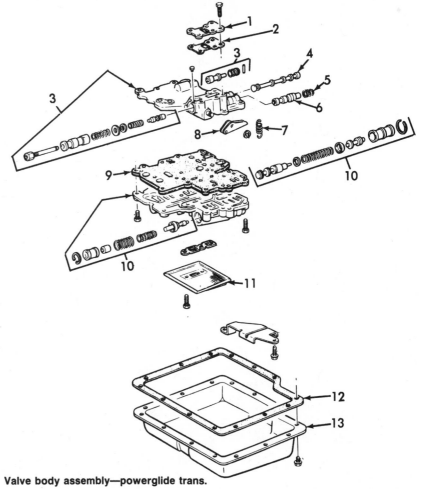

Valve body assembly—powerglide trans.

1. Plate assembly (upper)
2. Gasket (upper valve plate)
3. Body assy. (Main valve upper)
4. Valve (Manual)
5. Spring (Vac. Modulator)
6. Valve Assy. (Vac. Modulator)
7. Spring (range selector)
8. Lever assy. (range selector)
9. Plate assy. (w/gasket transfer)
10. Body assy. (main valve lower)
11. Screen assy. (w/gasket)
12. Oil pan gasket
13. Oil pan

INSTALLATION

Reverse removal procedure.

Transmission Assembly

NOTE: **If removed, assemble manual linkage to case, as described in Steps 1–7.**

1 Install parking lock pawl and shaft and insert a new E-ring retainer.
2 Install parking lock pawl pullback spring over its boss at rear of pawl. The short leg of the spring should locate in the hole in the pawl.
3 Install parking pawl reaction bracket with its two bolts.
4 Fit the actuator assembly between the parking pawl and the bracket.
5 Insert outer shift lever into the case. Pick up inner shift lever and parking lock assembly. Tighten Allen-head lock.
6 Insert outer throttle valve lever and shaft, special washer, and O-ring into case and pick up inner throttle valve lever. Tighten Allen-head lock.
7 Thread low band adjusting screw into case.

NOTE: **To prevent possible binding between throttle lever and range selector controls, allow .010–.020 in. clearance between inner throttle valve lever and inner shift lever.**

TRANSMISSION INTERNAL COMPONENTS

8 Install inner and outer rear piston seals onto reverse piston and, (with lubrication) install piston into the case. Thread low band adjusting screw part way into the case.
9 With transmission case facing up, install the 17 reverse piston springs and their retainer ring.
10 Install spring compressing tool. Compress the return springs, allowing the

Component Removal and Overhaul

GM POWERGLIDE

1. PARK LOCK AND RANGE SELECTOR OUTER LEVER AND SHAFT
2. THROTTLE VALVE CONTROL SHAFT OIL SEAL
3. THROTTLE VALVE CONTROL SHAFT WASHER
4. THROTTLE VALVE CONTROL LEVER AND SHAFT
5. THROTTLE VALVE CONTROL INNER LEVER TO CONTROL SHAFT ATTACHING SCREW AND NUT

6. THROTTLE VALVE CONTROL INNER LEVER
7. PARK LOCK AND RANGE SELECTOR INNER LEVER
8. PARK LOCK AND RANGE SELECTOR INNER LEVER ATTACHING SCREW AND NUT
9. PARK LOCK PAWL DISENGAGING SPRING
10. RANGE SELECTOR DETENT ROLLER SPRING
11. PARK LOCK ACTUATOR ASSEMBLY
12. RANGE SELECTOR DETENT ROLLER SPRING RETAINER
13. PARK LOCK PAWL SHAFT
14. PARK LOCK PAWL
15. PARK LOCK PAWL SHAFT RETAINING RING
16. PARK LOCK PAWL REACTION BRACKET
17. PARK LOCK PAWL REACTION BRACKET ATTACHING BOLTS
18. PARK LOCK ACTUATOR TO PARK LOCK AND RANGE SELECTOR INNER LEVER RETAINING CLIP

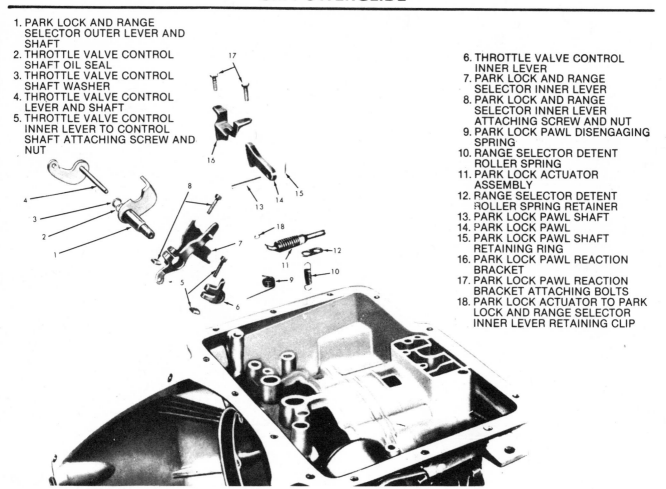

Manual Levers-Typical

retaining ring snap-ring to be installed. Remove the compressor.

11 Install the cushion spring.

12 Lubricate and install reverse clutch pack, beginning with a reaction spacer plate and alternating with the faced plates until all plates are installed.

NOTE: The number and sequence of plates varies with the power and torque requirements of the car model involved.

The notched lug on each reaction plate is installed in the groove at the seven o'clock position in the case. Then, install the thick pressure plate which has a dimple in one lug to align with the same slot in the case as the notched lugs on the other reaction plates.

13 Install clutch plate retainer ring.

14 Turn rear of transmission case down.

15 Align the internal lands and grooves of the reverse clutch pack faced plates, then engage the reverse ring gear with these plates. This engagement must be made by feel while turning the ring gear.

16 Place output shaft thrust bearing over the output shaft and install the planetary carrier and output shaft into the transmission case.

17 Move transmission to horizontal position.

18 The two input shaft seal rings should be in place on the shaft. Install clutch drum (machined face first) onto the input shaft and install the low sun gear bushing against shoulder.

19 Install clutch drum and input shaft assembly into case, aligning thrust washer on input shaft and indexing low sun gear with the short pinions on the planet carrier.

20 Remove rubber seal ring from the front pump body and install front pump, gasket and selective fit thrust washer into case. Install pump-to-case bolts.

21 To check for correct thickness of the selective fit thrust washer, move transmission so that output shaft points down and proceed as follows:
   a. Mount a dial indicator so that the indicator plunger is resting on the end of the input shaft. Zero the indicator.
   b. Push up on the output shaft and watch the total dial movement.
   c. The indicator should read .028–.059 in. If reading is not within specifications, remove front pump, change to a thicker or thinner se-

lective thrust washer. Repeat above checking procedure.

NOTE: Washers are available in thicknesses of .061, .078, .092 in. and .106 in.

22 Install servo piston, piston ring, and spring into the servo bore. Then, using a new gasket and O-ring, install the servo cover.

23 Remove front pump and selective fit washer from the case, and install the low brake band, anchor and apply struts into the case. Tighten the low band adjusting screw enough to prevent struts from falling out of case.

24 Place the seal ring in the groove around front pump body and the two seal rings on the pump cover extension. Install the pump, gasket and thrust washer into the case. Install all pump bolts. Torque bolts to 15 ft./lbs.

EXTENSION, GOVERNOR AND GOVERNOR SUPPORT

25 Turn transmission so that output shaft points upward.

26 Install governor support and gasket, drain back baffle, and support to case attaching bolts.

27 Install governor over output shaft. Install governor shaft and valve, ure-

GM POWERGLIDE

Low Band Adjustment

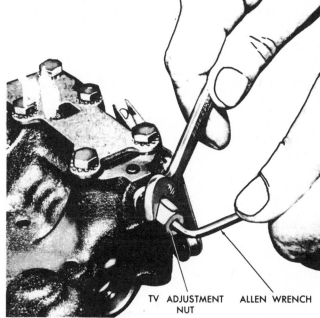

TV ADJUSTMENT NUT   ALLEN WRENCH

Throttle Valve Adjustment

thane washer, and retaining C clips. Center shaft in output shaft bore and tighten governor hub drive screw.

28 Install speedometer gear to output shaft.
29 Place extension seal ring over governor support. Install transmission extension and five retaining bolts.
30 Replace speedometer driven gear.

OIL PAN AND VALVE BODY

31 With transmission upside down, the manual linkage and the selector lever detent roller installed, install the valve body with a new gasket. (Carefully guide the servo apply line into its boss in the case as the valve body is set in place.) Position the manual valve actuating lever fully forward to more easily pick up the manual valve. Install six mounting bolts and the range selector detent roller spring. Install new gasket and suction screen to valve body.
32 Install the guide plate. Install attaching bolts.
33 Install vacuum modulator valve, the vacuum modulator and the gasket.
34 Install oil pan, using a new gasket, then the pan attaching bolts.
35 Install the converter and safety holding strap.

LOW BAND ADJUSTMENT

Tighten the low servo adjusting screw to 70 in./lbs. The input and output shaft must be rotated simultaneously to properly center the low band on the clutch drum. Then, back off four complete turns for a band which has been in use for 6,000 miles or more, or three turns for one in use less than 6,000 miles, and tighten the locknut.

CHILTON CAUTION: *The amount of back-off is very critical. Back off exactly three or four turns.*

THROTTLE VALVE (TV) ADJUSTMENT

No provision is made for checking TV pressures. However, if operation of the transmission is such that some adjustment of the TV is indicated, pressures may be raised or lowered by adjusting the position of the jam nut on the throttle valve assembly. To raise TV pressure 3 psi, back off the jam nut one full turn.

Conversely, tightening the jam nut one full turn, lowers TV pressure 3 psi. A difference of 3 psi in TV pressure will cause a change of about 2–3 mph in the wide open throttle upshift point. The end of the TV adjusting screw has an Allen head so

the screw may be held stationary while the jam nut is locked.

NOTE: Use care in changing this adjustment, as no pressure tap is provided to check TV pressure.

GM POWERGLIDE TORQUE SPECIFICATIONS

	Foot/Pounds
Transmission case to engine	35
Oil pan to case	7-10
Extension to case	20-30
Servo cover to case	20
Front pump to case	13-17
Front pump cover to body attaching bolts	20
Pinion shaft lock plate attaching screws	2-3
Governor body to hub	6-8
Governor hub drive screw	6-8
Rear pump or governor support to case	8-11
Valve body to case	15
Suction screen attaching screws	2-3
Upper valve body plate	5
Lower to upper valve body	15
Inner control lever allen head locks	2½
Low band adjusting locknut	13-17
Converter to engine	35

TROUBLESHOOTING
GM TURBO HYDRA-MATIC 350-375-M38

THE PROBLEM	ITEMS TO CHECK	
	IN CAR	OUT OF CAR
SLIPS IN ALL RANGES	Low oil level/water in oil. Modulator and/or valve. Strainer and/or gasket leak. Valve body gasket/plate. Pressure regulator and/or boost valve. Manual valve linkage. Gasket screen-pressure.	Pump gears.
DRIVE SLIPS—NO 1st GEAR	Low oil level/water in oil. Modulator and/or valve. Strainer and/or gasket leak. Valve body gasket/plate. Pressure regulator and/or boost valve. Manual valve linkage. Gasket screen-pressure.	Pump gears. Forward clutch assembly. Low & Reverse roller clutch.
NO 1-2 UPSHIFT	Vacuum leak. Governor valve/screen. Valve body gasket/plate. Pressure regulator and/or boost valve. 1-2 shift valve.	Intermediate clutch assembly. Intermediate roller clutch.
1-2 UPSHIFT, EARLY OR LATE	Vacuum leak. Governor valve/screen. Valve body gasket/plate. 1-2 shift valve.	
SLIPS, 1-2 UPSHIFT	Low oil level/water in oil. Modulator and/or valve. Valve body gasket/plate. Pressure regulator and/or boost valve. 1-2 shift valve. 2-3 accumulator.	Pump gears. Intermediate clutch assembly. Intermediate roller clutch.
HARSH 1-2 UPSHIFT	Vacuum leak. Modulator and/or valve. Pressure regulator and/or boost valve.	
NO 2-3 UPSHIFT	Valve body gasket/plate. 2-3 shift valve.	Direct clutch assembly.
2-3 UPSHIFT, EARLY/LATE	Vacuum leak. Modulator and/or valve. Governor valve/screen. Valve body gasket/plate. 2-3 shift valve. Detent valve and linkage.	
SLIPS, 2-3 UPSHIFT	Low oil level/water in oil. Modulator and/or valve. Valve body gasket/plate. Pressure regulator and/or boost valve. 2-3 accumulator.	Pump gears. Direct clutch assembly.
HARSH 2-3 UPSHIFT	Vacuum leak. Modulator and/or valve. Pressure regulator and/or boost valve. 2-3 shift valve. 2-3 accumulator.	
NO FULL THROTTLE DOWNSHIFT	Detent valve and linkage.	
2-3 UPSHIFT, WIDE OPEN THROTTLE ONLY	Vacuum leak. Detent valve and linkage.	
L1 GEAR—NO ENGINE BRAKING	Pressure regulator and/or boost valve. 1-2 shift valve. Manual low control valve. Manual valve linkage.	Band—intermediate overrun roller clutch. Forward clutch assembly. Intermediate roller clutch.
CAR DRIVES IN NEUTRAL	Detent valve and linkage.	Forward clutch assembly.
SLIPS IN REVERSE	Low oil level/water in oil. Modulator and/or valve. Strainer and/or gasket leak. Valve body gasket/plate. Pressure regulator and/or boost valve. 1-2 shift valve. Manual valve linkage. Gasket screen-pressure.	Direct clutch assembly. Low & Reverse clutch assembly.
1-2, 2-3 SHIFT NOISY	Low oil level/water in oil.	Direct clutch assembly. Intermediate clutch assembly. Gear set and bearings.
NOISY IN ALL RANGES	Low oil level/water in oil. Strainer and/or gasket leak Valve body gasket/plate. Gasket screen-pressure.	Pump gears. Converter assembly. Gear set and bearings. Parking pawl/linkage.
SPEWS OIL OUT OF BREATHER	Low oil level/water in oil. Strainer and/or gasket leak	

GM TURBO HYDRA-MATIC 350-375-M38

GM Turbo Hydra-matic 350 & 375
Identification
BUICK

On Buick vehicles, production numbers, transmission model, and model year numbers are stamped on the governor cover, which is located on the middle rear, left side of the transmission case.

Transmission Disassembly

Clean outside of transmission thoroughly to prevent dirt from entering the unit.

1 With transmission in a holding fixture, lift off torque converter assembly.
2 Remove vacuum modulator assembly attaching bolt and retainer.
3 Remove vacuum modulator assembly, O-ring seal and modulator valve from the case.
4 Remove four extension housing-to-case attaching bolts.
5 Remove extension housing and the square cut O-ring seal.
6 Remove extension housing lip seal from output end of housing, using a screwdriver.

GM TURBO HYDRA-MATIC 350–375–M38

7 Remove extension housing bushing, using chisel to collapse bushing.

8 Drive in new extension housing bushing.

9 Install extension housing lip seal.

10 Depress speedometer drive gear retaining clip, then slide speedometer drive gear off output shaft.

11 Remove governor cover retainer wire with a screwdriver.

12 Remove governor cover and O-ring seal from case, then remove O-ring seal from governor cover.

13 Remove governor assembly from case.

NOTE: Check governor bore and sleeve for scoring.

14 Remove oil pan attaching screws, pan, and gasket.

15 Remove two oil pump suction screen (strainer) to valve body attaching screws.

16 Remove oil pump screen (strainer) and gasket from valve body.

17 Remove detent roller and spring assembly from valve body. Remove valve body-to-case attaching bolts.

18 Remove manual control valve link from range selector inner lever. Remove valve body. Remove detent control valve link from detent actuating lever.

NOTE: Refer to later text for valve body disassembly and passage identification.

NOTE: At this time, when handling valve body assembly, do not touch sleeves, because retaining pins may fall into the transmission.

19 Remove valve body-to-spacer plate gasket.

20 Remove spacer support plate gasket. Remove spacer support plate.

21 Remove valve body spacer plate and plate-to-case gasket.

22 Remove four check balls from passages in case face.

NOTE: If, during assembly, any of the balls are omitted or installed in the wrong locations, transmission failure will result.

23 Remove oil pump pressure screen from oil pump pressure hole in the case.

24 Remove governor feed screen from governor feed passage in case. Remove the manual control valve link from the range selector inner lever.

25 Remove manual shaft to case retainer with a screwdriver.

26 Remove manual shaft to case retainer with a screwdriver.

27 Remove jam nut holding range selector inner lever to manual shaft. Remove manual shaft.

28 Disconnect parking pawl actuating rod from range selector inner lever. Remove both from case.

29 Remove manual shaft to case lip oil seal, using a screwdriver.

30 Remove parking lock bracket.

31 Disconnect and remove parking pawl

© G.M. Corp.

Case and parts—350, 375B trans.

1. Case Assy.
2. Retainer, Accumulator Piston Cover
3. Cover, Acum. Piston
4. Seal, Accum. Piston cover
5. Spring, Acum. Piston
6. Seal, Accum, Piston—Outer
7. Piston, Inter Clutch Accum.
8. Seal, Accum, Piston—Inner
9. Valve, Modulator
10. Seal, Modulator
11. Retainer, Modulator
12. Modulator assy.
13. Seal, Case Extension
14. Extension Case
15. Bushing, Extension Case
16. Extension assy.
17. Seal, Extension Case
18. Retainer, Speedo. Gear
19. Seal, Speedo. Gear
20. Sleeve, Speedo. Gear
21. Gear, Speedo. Driven
22. Screen, Governor Pressure
23. Pin, Governor
24. Governor Assy.
25. Seal, Governor
26. Cover, Governor
27. Retainer Gov. Cover
28. Pan Assy.
29. Gasket, Oil Pan
30. Screen Pump Pressure

disengaging spring.

32 Remove parking pawl shaft retaining plug with a bolt extractor. Cock parking pawl on shaft. Using drift and hammer, tap on drift to force the parking pawl shaft from the case. Remove the parking pawl.

33 Remove intermediate servo piston and metal oil seal ring. Remove washer, spring seat, and apply pin.

34 On Buick applications, check the band apply pin. Using band apply pin selection tool J-23071, and a straightedge, exert firm downward pressure on the selection pin. If the tool is below the straightedge surface, a long apply pin must be used. If the tool is above the straightedge surface, a short pin should be used. The long apply pin is identified by a groove, while the short pin has no groove. If a new pin is required, make note of the pin size and remove the selection tool. Selecting the proper pin is the equivalent of adjusting the band.

35 Remove eight pump attaching bolts with washer-type seals. Discard seals.

36 Install two threaded slide hammers into threaded holes in pump body. Tighten jam nuts and remove pump assembly from the case.

37 Remove pump assembly to case gasket and discard.

38 Remove intermediate clutch cushion spring.

39 Remove intermediate clutch faced plates and steel separator plates. Inspect lined plates for pitting, flaking, wear, glazing, cracking and chips or metal particles imbedded in lining. Replace any lined plates showing any of these conditions. Inspect steel plates for heat spot discoloration or surface scuffing. If the surface is smooth and has an even color smear, the steel plates may be re-used.

40 Remove intermediate clutch pressure plate.

41 Remove intermediate overrun brake band.

42 Remove direct and forward clutch assemblies.

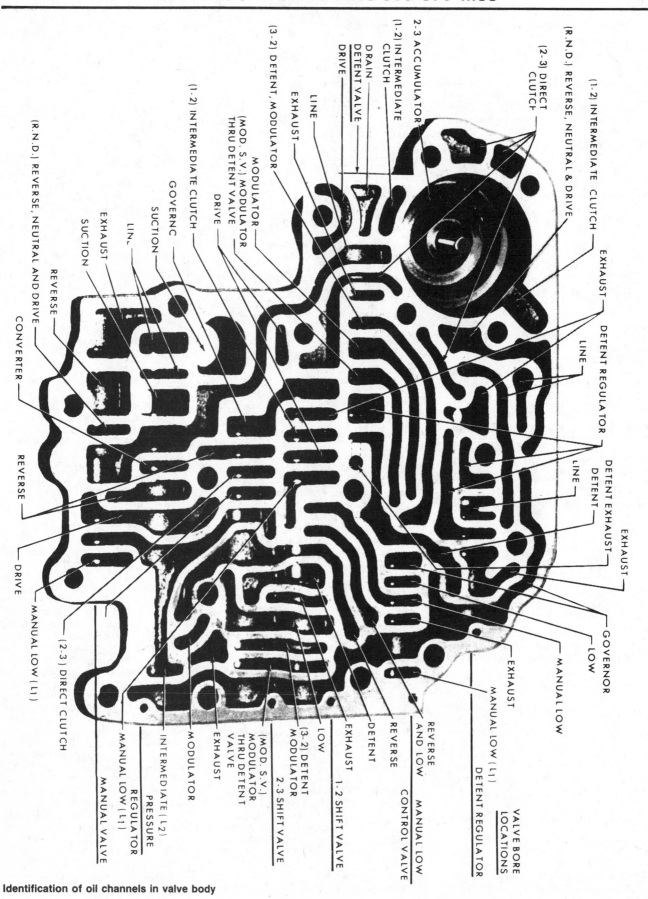

Identification of oil channels in valve body

GM TURBO HYDRA-MATIC 350–375–M38

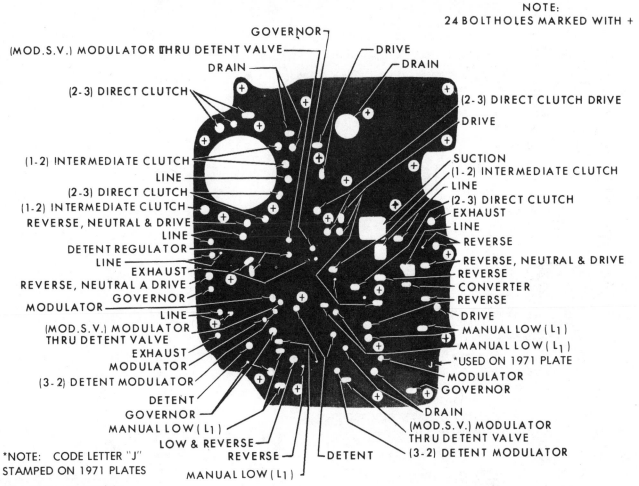

NOTE:
24 BOLT HOLES MARKED WITH +

GOVERNOR

(MOD.S.V.) MODULATOR THRU DETENT VALVE

DRIVE

DRAIN

DRAIN

(2-3) DIRECT CLUTCH

(2-3) DIRECT CLUTCH DRIVE

DRIVE

(1-2) INTERMEDIATE CLUTCH

LINE

(2-3) DIRECT CLUTCH

(1-2) INTERMEDIATE CLUTCH

REVERSE, NEUTRAL & DRIVE

LINE

DETENT REGULATOR

LINE

EXHAUST

REVERSE, NEUTRAL A DRIVE

GOVERNOR

MODULATOR

LINE

(MOD.S.V.) MODULATOR THRU DETENT VALVE

EXHAUST

MODULATOR

(3-2) DETENT MODULATOR

DETENT

GOVERNOR

MANUAL LOW (L₁)

LOW & REVERSE

REVERSE

DETENT

MANUAL LOW (L₁)

SUCTION

(1-2) INTERMEDIATE CLUTCH

LINE

(2-3) DIRECT CLUTCH

EXHAUST

LINE

REVERSE

REVERSE, NEUTRAL & DRIVE

REVERSE

CONVERTER

REVERSE

DRIVE

MANUAL LOW (L₁)

MANUAL LOW (L₁)

*USED ON 1971 PLATE

MODULATOR

GOVERNOR

DRAIN

(MOD.S.V.) MODULATOR THRU DETENT VALVE

(3-2) DETENT MODULATOR

*NOTE: CODE LETTER "J" STAMPED ON 1971 PLATES

Valve body spacer plate

43 Remove forward clutch housing-to-input ring gear front thrust washer.

NOTE: washer has three tangs.

44 Remove output carrier-to-output shaft snap-ring.
45 Remove input ring gear.
46 Remove input ring gear to output carrier thrust washer.
47 Remove output carrier assembly.
48 Remove sun gear drive shell assembly.
49 Remove low and reverse roller clutch support to case retaining ring.
50 Remove low and reverse roller clutch, support assembly, and retaining spring.
51 Remove low and reverse clutch faced plates and steel separator plates.
52 Remove reaction carrier assembly from output ring gear and shaft assembly.
53 Remove output ring gear and shaft assembly from case.
54 Remove reaction carrier to output ring gear tanged thrust washer. .
55 On Oldsmobile applications, remove output ring gear to output shaft snap-ring. Remove output ring gear from output shaft.

NOTE: Do not over-stress snap-ring upon reassembly. Use new snap-ring.

56 Remove output ring gear-to-case needle bearing assembly.
57 Compress low and reverse clutch piston spring retainer and remove piston retaining ring and spring retainer.
58 Remove 17 piston return coil springs from piston.
59 Remove low and reverse clutch piston by application of air pressure.
60 Remove low and reverse clutch piston outer seal.
61 Remove low and reverse clutch piston center and inner seal.
62 Install suitable tool to compress intermediate clutch accumulator cover and remove retaining ring.
63 Remove intermediate clutch accumulator piston cover. Remove cover O-ring seal from case.
64 Remove intermediate clutch accumulator piston spring.
65 Remove intermediate clutch accumulator piston assembly. Remove inner and outer hook-type oil seal rings if required.

Component Disassembly
Valve Body

DISASSEMBLY

1 Position valve body assembly with cored face up and direct clutch accumulator piston pocket located as illustrated in valve body assembly illustration.
2 Remove manual valve from lower left-hand bore, A.
3 From lower right-hand bore, B, remove the pressure regulator valve train retaining pin, boost valve sleeve, intermediate boost valve, reverse and modulator boost valve, pressure regulator valve spring, and the pressure regulator valve.
4 From the next bore, C, remove the second-third shift valve train retaining pin, sleeve, control valve spring, second-third shift control valve, shift valve spring, and the second-third shift valve.
5 From the next bore, D, remove the first-second shift valve train retaining pin, sleeve, shift control valve spring, first-second shift control valve, and the first-second shift valve.
6 From the next bore, E, remove retaining pin, plug, manual low control valve spring, and the manual low control valve.
7 From the next bore, F, remove retaining pin, spring seat and the detent regulator valve.

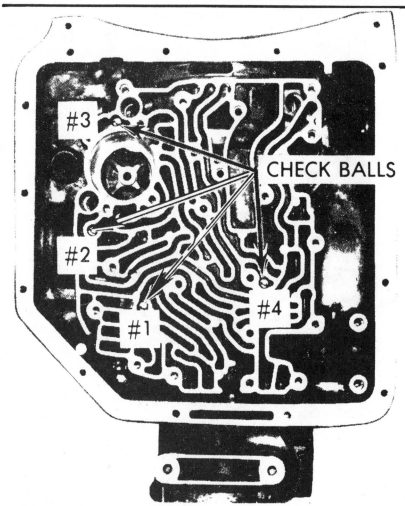

Control valve check ball locations

8 Install spring compressor onto direct clutch accumulator piston and remove retaining E-ring at H.

9 At location H, remove direct clutch accumulator piston, then metal oil seal ring and spring.

10 From the upper left hand bore, G, remove the detent actuating lever bracket bolt, bracket, actuating lever and retaining pin, stop, spring retainer, seat, outer spring, and the detent valve.

INSPECTION

1 Wash all parts in solvent. Air dry. Blow out all passages.

2 Inspect all valves for scoring, cracks, and free movement in their bores.

3 Inspect sleeves for cracks, scratches, or distortion.

4 Inspect valve body for cracks, scored bores, interconnected oil passages, and flatness of mounting face.

5 Check all springs for distortion or collapsed coils.

ASSEMBLY

1 Reverse disassembly procedures for assembly. Refer to Valve Body

Springs chart for identification of springs.

VALVE BODY SPRINGS

Valve	Free Length of Spring (in.)	Diameter (in.)
Detent regulator	1⅞	9/16
Manual low control valve	1½	7/16
1–2 shift control valve	1 15/16	¼
2–3 shift valve	2 1/16	⅞
2–3 shift control valve	11/16	3/16
Pressure regulator valve	1 11/16	17/32
Direct clutch accumulator	1¾	1½
Detent valve	1⅞	¾

Oil Pump

DISASSEMBLY

1 Remove five pump cover-to-body at-

taching bolts. Remove spring seat retainer.

2 Remove the intermediate clutch spring seat retainer, intermediate clutch return springs, and the intermediate clutch piston assembly.

3 Remove intermediate clutch piston inner and outer seals.

4 Remove two forward clutch-to-pump hub hook-type oil seal rings. Remove three direct clutch-to-pump hub hook type oil rings.

5 Remove pump cover-to-direct clutch drum selective thrust washer.

6 Remove pump cover and stator shaft assembly from pump body.

7 Remove pump drive gear and driven gear from pump body.

8 Remove pump outside diameter-to-case square cut O-ring seal.

9 Pry out pump body to converter hub lip seal, using a screwdriver. Place pump on wood blocks to prevent damage to surface finish.

10 Install pump-to-converter hub lip seal, using seal driver. Examine after installation to be sure the sealing surface is not damaged.

11 Remove cooler by-pass valve seat. Pack cooler by-pass passage with grease. Insert 5/16 in. dia. rod and tap with hammer to lift seat. Remove check ball and spring.

INSPECTION

1 Wash all parts in solvent. Air dry. Blow out all passages.

2 Inspect drive and driven gears, gear pocket, and crescent for nicks, galling or other damage.

3 Inspect pump body and cover for nicks or scoring.

4 Check pump cover outer diameter for nicks or burrs.

5 Inspect pump body bushing for galling or scoring. Check clearance between pump body bushing and converter hub. It must be no more than .005 in. If bushing is damaged, replace pump body.

6 Install pump gears in body and check pump body face to gear face clearance. It should be from .0008–.0035 in. (.0005–.0015 in. for Chevrolet).

7 Inspect pump body to converter hub lip oil seal. Inspect converter hub for nicks or burrs which might have damaged pump lip oil seal or pump body bushing.

8 Check priming valve for free operation. Replace if necessary.

9 Check condition of cooler bypass valve. Replace valve if it leaks excessively.

10 Check all springs for distortion or collapsed coils.

11 Check oil passages in pump body and in pump cover.

12 Inspect three pump cover stator shaft bushings for galling or scoring. If they are damaged, remove using a slide hammer. Drive the new bushings into

GM TURBO HYDRA-MATIC 350–375–M38

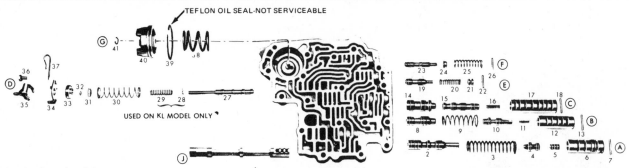

TEFLON OIL SEAL-NOT SERVICEABLE

USED ON KL MODEL ONLY

© G.M. Corp.

Valve body assembly

J— 1 MANUAL VALVE AND LINK ASSEMBLY
2 PRESSURE REGULATOR VALVE
3 PRESSURE REGULATOR VALVE SPRING
4 REVERSE AND MODULATOR BOOST VALVE
A 5 INTERMEDIATE BOOST VALVE
6 BOOST VALVE SLEEVE
7 RETAINING PIN
8 2-3 SHIFT VALVE
9 2-3 SHIFT VALVE SPRING
B 10 2-3 SHIFT CONTROL VALVE
11 2-3 SHIFT CONTROL VALVE SPRING
12 2-3 SHEET CONTROL VALVE SLEEVE
13 RETAINING PIN
14 1-2 SHIFT VALVE

15 1-2 SHIFT CONTROL VALVE
16 1-2 SHIFT CONTROL VALVE SPRING
C 17 1-2 SHIFT CONTROL VALVE SLEEVE
18 RETAINING PIN
19 MANUAL LOW CONTROL VALVE
20 MANUAL LOW CONTROL VALVE SPRING
E 21 PLUG
22 RETAINING PIN
23 DETENT REGULATOR VALVE
24 DETENT REGULATOR VALVE SPRING SEAT
F 25 DETENT REGULATOR VALVE SPRING
26 RETAINING PIN
27 DETENT VALVE
28 WASHER

29 DETENT VALVE INNER SPRING
30 DETENT VALVE OUTER SPRING
31 DETENT VALVE OUTER SPRING SEAT
D 32 DETENT VALVE SPRING RETAINER
33 DETENT VALVE STOP
34 DETENT VALVE ACTUATING LEVER BRACKET
35 DETENT VALVE ACTUATING LEVER
36 RETAINING BOLT
37 RETAINING PIN
38 DIRECT CLUTCH ACCUMULATOR SPRING
39 OIL SEAL RING
G 40 DIRECT CLUTCH ACCUMULATOR PISTON
41 RETAINER RING

1. Rod inter. servo apply
2. Spring, inter servo piston
3. Seat, inter servo spring
4. Washer, inter servo piston
5. Piston, inter. servo
6. Seal, inter servo piston
7. Retainer ring, accum. piston
8. Piston, direct clutch accum.
9. Seal, direct clutch accum. piston
10. Spring, direct clutch accum, piston
11. Gasket, upper valve body
12. Plate, valve body spacer
13. Support, spacer plate
14. Gasket, lower valve body
15. Valve detent
16. Spring, detent valve
17. Seat, detent valve spring
18. Bracket, detent valve lever
19. Bolt, detent lever
20. Pin, detent lever
21. Wire, detent valve
22. Ball, check valve
23. Valve detent regulator
24. Seat, detent reg. valve spring
25. Spring, detent reg. valve
26. Pin, detent reg. valve
27. Screen, oil pump
28. Gasket, oil pump screen
29. Link, manual valve
30. Valve assy. manual
31. Lever detent valve

Valve body assy.—350, 375 B trans.

place. The front stator shaft bushing must be .250 in. below the front face of the pump body. The center bushing should be 11/32 in. below the face of the pump cover hub. The rear bushing should be flush or up to .010 in. below the face of the pump cover hub.

13 Check three pump cover and hub lu-

brication holes to make certain they are not restricted.

ASSEMBLY

1 Install cooler by-pass valve spring,

GM TURBO HYDRA-MATIC 350–375–M38

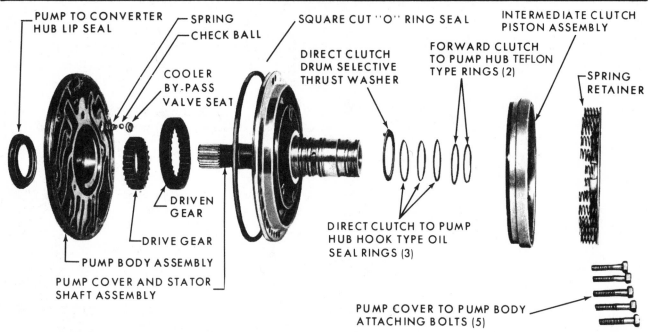

- PUMP TO CONVERTER HUB LIP SEAL
- SPRING
- CHECK BALL
- SQUARE CUT "O" RING SEAL
- INTERMEDIATE CLUTCH PISTON ASSEMBLY
- COOLER BY-PASS VALVE SEAT
- DIRECT CLUTCH DRUM SELECTIVE THRUST WASHER
- FORWARD CLUTCH TO PUMP HUB TEFLON TYPE RINGS (2)
- SPRING RETAINER
- DRIVEN GEAR
- DRIVE GEAR
- PUMP BODY ASSEMBLY
- PUMP COVER AND STATOR SHAFT ASSEMBLY
- DIRECT CLUTCH TO PUMP HUB HOOK TYPE OIL SEAL RINGS (3)
- PUMP COVER TO PUMP BODY ATTACHING BOLTS (5)

Oil pump assembly

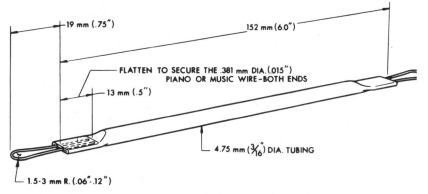

- 19 mm (.75")
- 152 mm (6.0")
- FLATTEN TO SECURE THE .381 mm DIA.(.015") PIANO OR MUSIC WIRE–BOTH ENDS
- 13 mm (.5")
- 4.75 mm (³⁄₁₆") DIA. TUBING
- 1.5-3 mm R. (.06"-.12")

Clutch piston installation tool fabricated from copper tubing and piano wire.

check ball and seat. Press seat into bore until top of seat is flush to .010 in. with face of pump body.

2 Install oil pump priming valve and spring. The priming valve is used on all pump bodies having a reamed hole in the priming valve area and on all replacement pump assemblies.

3 Install new pump outside diameter to case square cut O-ring seal.

4 Install pump drive and driven gears. If drive gear has offset tangs, assemble with tangs face up to prevent damage by converter.

5 Install selective thrust washer and five oil seal rings on pump cover hub.

6 Install intermediate clutch piston inner and outer seals.

7 Install pump cover on pump body.

8 Install intermediate clutch piston and clutch return springs.

9 Install spring retainer and pump cover bolts. Position aligning strap over pump body and cover. Tighten strap. Torque pump bolts to 18 ft./lbs.

Direct Clutch and Intermediate Overrun Roller Clutch

DISASSEMBLY

1 Remove intermediate overrun clutch front retainer ring and retainer.

2 Remove intermediate overrun clutch outer race.

NOTE: Before removal, check for correct assembly. The outer race should free wheel counterclockwise only.

3 Remove intermediate overrun roller clutch assembly.

4 Remove intermediate overrun roller clutch cam.

5 Remove direct clutch drum to forward clutch housing special needle bearing.

6 Remove direct clutch pressure plate to clutch drum retaining ring and pressure plate.

7 Remove direct clutch lined and steel plates.

8 Remove direct clutch piston return

spring seat retaining ring and spring seat.

9 Remove 17 clutch return coil springs and piston.

10 Remove direct clutch piston inner and outer seals.

11 Remove direct clutch piston center seal.

INSPECTION

1 Wash all parts in solvent. Blow out all passages. Air dry.

2 Inspect clutch plates for burning, scoring, or wear.

3 Check all springs for collapsed coils or distortion.

4 Inspect piston for cracks and free operation of ball check. Ball should be loose enough to rattle but not to fall out.

5 Check overrun clutch inner cam and outer race for scratches, wear, or indentations.

6 Inspect overrun roller clutch assembly rollers for wear. Check springs for distortion.

7 Inspect clutch drum for wear, scoring, cracks, proper opening of oil passages, wear on clutch plate drive lugs, and free operation of ball check.

8 Check direct clutch drum bushing for galling or scoring. When replacing bushing, drive the new bushing 9/32 in. below clutch plate side of hub face and .010 in. below slot in hub face.

ASSEMBLY

1 Install direct clutch drum center seal.

2 Install direct clutch piston inner and outer seals.

3 Install direct clutch piston into housing with a loop of .020 in. wire crimped into a length of copper tubing.

GM TURBO HYDRA-MATIC 350–375–M38

4 Install 17 clutch return springs on piston.

5 Install direct clutch piston return spring seat and retaining ring using snap-ring pliers and spring compressor.

6 Install direct clutch housing. Install steel and faced plates alternately, beginning with a steel plate.

7 Install direct clutch pressure plate in clutch drum. Install retaining ring.

8 Install intermediate overrun roller clutch inner cam on hub of direct clutch drum.

9 Replace intermediate overrun roller clutch assembly.

10 Install intermediate overrun clutch outer race. The outer race must free-wheel in the counter-clockwise direction only.

11 Replace intermediate overrun clutch retainer and retainer ring. If the retainer ring is dished, install the ring so that it compresses the retainer.

12 Install direct clutch drum to forward clutch housing needle thrust washer on hub of roller clutch inner race.

Forward Clutch

DISASSEMBLY

1 Remove forward clutch drum to pressure plate retaining ring.

2 Remove forward clutch pressure plate.

3 Remove forward clutch housing faced plates, steel plates and cushion spring.

4 Compress springs. Remove forward clutch piston return spring seat retaining ring and spring seat.

5 Remove 21 clutch return springs.

6 Remove forward clutch piston assembly.

7 Remove forward clutch piston inner and outer seals.

8 Be sure the ball check exhaust in the drum is operable and free of dirt.

INSPECTION

1 Wash all parts in solvent. Blow out all passages. Air dry.

2 Inspect clutch plates for burning, scoring, or wear.

3 Check all springs for distortion or collapsed coils.

4 Inspect piston for cracks.

5 Inspect clutch drum for wear, scoring, cracks, proper opening of oil passages, and free operation of ball check.

6 Check input shaft for:
   a. Open lubrication passages at each end.
   b. Damage to splines or shaft.
   c. Damage to ground bushing journals.
   d. Cracks or distortion of shaft.

ASSEMBLY

1 Install forward clutch piston inner and outer seals.

2 Install forward clutch piston with a loop of .020 in. wire crimped into a length of copper tubing.

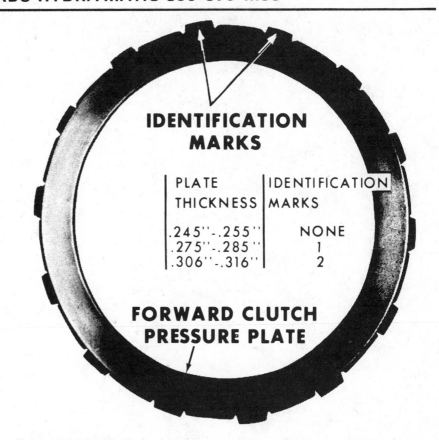

IDENTIFICATION MARKS

PLATE THICKNESS	IDENTIFICATION MARKS
.245''-.255''	NONE
.275''-.285''	1
.306''-.316''	2

FORWARD CLUTCH PRESSURE PLATE

Clutch plate identification

3 Install the 21 forward clutch return springs. These springs are identical to those used in the direct clutch. Install spring seat.

4 Compress springs and replace spring seat retaining ring.

5 Replace forward clutch housing cushion spring. Replace steel and faced plates alternately, starting with a steel plate.

6 Install forward clutch pressure plate and retaining ring.

7 Use a feeler gauge to check the clearance between the top faced plate and the pressure plate. The clearance should be .010–.080 in. for Oldsmobile, Buick and Pontiac applications. Selective pressure plates for Oldsmobile applications are available in .200 in., .230 in. and .260 in. thicknesses. Pontiac and Buick applications use selective pressure plates in .245–.255 in., .275–.285 in. and .306–316 in. thicknesses. Use the following chart to determine the proper plate for Chevrolet applications.

Subtract dim. B from C to obtain dim. A

When dim. A is:	*use plate number*
.016–.052 in.	6261072
.052–.083 in.	6261349
.083–.122 in.	6261350

8 Install direct clutch drum on input shaft and align faced plates with splines on forward clutch housing.

Sun Gear and Sun Gear Drive Shell

DISASSEMBLY

1 Remove sun gear to sun gear drive shell rear retaining ring.

2 Remove sun gear to drive shell flat rear thrust washer.

3 Remove sun gear and front retaining ring from drive shell.

4 Remove front retaining ring from sun gear.

INSPECTION

1 Wash all parts in solvent. Air dry.

2 Inspect sun gear and sun gear drive shell for wear or damage.

3 Inspect sun gear bushings for galling or scoring. Drive out damaged bushings. Install new bushings flush to .010 in. below surface of counterbore.

ASSEMBLY

1 Install new front retaining ring on sun gear. Be careful not to over-stress this ring.

2 Install sun gear and retaining ring in drive shell.

3 Install sun gear to drive shell flat rear thrust washer.

4 Install new sun gear to sun gear drive shell retaining ring.

GM TURBO HYDRA-MATIC 350–375–M38

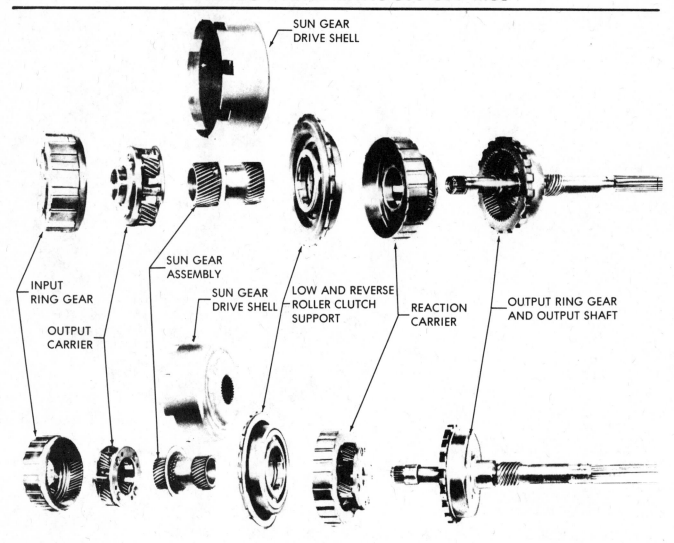

SUN GEAR
DRIVE SHELL

SUN GEAR
ASSEMBLY

SUN GEAR
DRIVE SHELL

LOW AND REVERSE
ROLLER CLUTCH
SUPPORT

REACTION
CARRIER

OUTPUT RING GEAR
AND OUTPUT SHAFT

INPUT
RING GEAR

OUTPUT
CARRIER

Planetary gear train

Low and Reverse Roller Clutch Support

DISASSEMBLY
1 Remove low and reverse clutch to sun gear shell thrust washer.
2 Remove low and reverse overrun clutch inner race.
3 Remove low and reverse roller clutch retaining ring.
4 Remove low and reverse roller clutch assembly.

INSPECTION
1 Wash all parts in solvent. Air dry.
2 Inspect roller clutch inner and outer races for scratches, wear, or indentations.
3 Inspect roller clutch assembly rollers for wear and roller springs for distortion. If rollers are removed from assembly, install rollers from outside in to avoid bending springs.

ASSEMBLY
1 Install low and reverse roller clutch assembly.

2 Install low and reverse roller clutch retaining ring.
3 Install low and reverse overrun clutch inner race. Inner race must free-wheel in the clockwise direction only.

Transmission Assembly
1 Install inner and outer hook type oil seal rings on accumulator piston. Install intermediate clutch accumulator piston assembly. Install O-ring seal against shoulder in case. Install accumulator piston spring and piston cover. Compress spring and cover as in disassembly and install retaining ring.
2 Install low and reverse clutch piston center, outer, and inner seal on piston. Install piston assembly. Notch in low and reverse clutch piston must be installed adjacent to parking pawl. Install 17 return springs in piston. Place spring seat on return springs. Compress springs as in disassembly and install retaining ring.
3 Install output ring gear to case needle bearing assembly, shoulder down. As-

semble output ring gear on output shaft. Use new snap-ring on shaft. Install reaction carrier to output ring gear tanged thrust washer. This washer has three narrow tangs on the outside circumference. Install output ring rear and shaft assembly in case.
4 Install reaction carrier assembly on output ring gear and shaft assembly.
5 Install low and reverse clutch plates. Start with a steel plate and alternate steel and faced plates. The notch in the steel plates must be toward the bottom of the case.
6 Install low and reverse roller clutch support retaining spring, then support assembly. Align tangs on inner race with slots in reaction carrier. Install retaining ring.
7 Install sun gear to roller clutch thrust washer on the roller clutch inner race. This washer has four tangs on the inside circumference. Install sun gear drive shell assembly.
8 Install output carrier assembly. Install new output carrier to output shaft snap-ring.

GM TURBO HYDRA-MATIC 350–375–M38

9 Install input ring gear to output carrier thrust washer. Install forward clutch housing to input ring gear front thrust washer. This washer has three wide tangs on the outside circumference.

10 Install direct and forward clutch assemblies in case. Align plates with splines and insert until tangs are 1/8 in. below notches in drive shell.

11 Install intermediate overrun brake band, intermediate clutch pressure plate, and steel and faced clutch plates. Insert a steel plate first, then alternate faced and steel plates. Install intermediate clutch cushion spring.

12 Lubricate case bore, then install new pump to case gasket in bore. Install the proper selective thrust washer to obtain .033–.064 in. shaft end-play with pump installed. Selective thrust washers are available in .066, .083 and .100 in. thicknesses.

13 Install pump into case. Drive in until seated. Install new washer type seals on pump attaching bolts. Torque to 15–25 ft./lbs. If the input shaft cannot be rotated clockwise as the pump is being bolted into place, the direct clutch and forward clutch housing has not been properly installed to index the faced plates with their respective parts. This condition must be corrected before the pump is bolted into place.

14 Invert transmission and install intermediate servo spring, spring seat, apply pin, washer, and piston assembly.

15 Install parking pawl tooth toward the inside of case. Install parking pawl shaft into case through disengaging spring. Install disengaging spring on parking pawl and slide shaft through pawl. Drive in pawl shaft retainer plug using a 3/8 in. drift. Drive plug

flush to .010 in. below face of case. Stake plug in three places. Install park lock bracket. Torque bolts to 29 ft./lbs.

16 Install range selector inner lever to parking pawl actuating rod. Install actuating rod under the park lock bracket and parking pawl.

17 Drive in new manual shaft lip seal using a 7/8 in. drift. Insert manual shaft through case and range selector inner lever. Torque manual shaft jam nut to 30 ft./lbs. Install manual shaft retainer.

18 Replace governor screens in case. Replace oil pump pressure screen in oil pump pressure hole in case. Ring end of screen goes in first. Replace four check balls in correct passages in case face. Replace valve body spacer plate to case gasket, spacer plate, spacer support plate, and seven bolts. Do not tighten bolts yet. Install valve body to

6. Spring, forward clutch—return
7. Seat, forward clutch
8. Retainer ring, forward clutch
9. Washer, input ring gear—front
10. Gear, input ring
11. Bushing, input ring gear
12. Spring, forward clutch, cushion
13. Plate, forward clutch, driven
14. Plate, forward clutch, drive
15. Plate, forward clutch, drive
16. Plate, forward clutch, pressure
17. Washer, input gear supt.
18. Carrier, output
19. Bearing, sun gear
20. Race, sun gear
21. Carrier assy., output
22. Bushing, sun gear
23. Shell, sun gear
24. Washer, sun gear
25. Retainer ring, sun gear shell
26. Washer, clutch tace and shell
27. Race, low and rev. clutch—inner
28. Clutch, low and reverse
29. Retainer ring, clutch to cam
30. Cam, low and rev. clutch
31. Gear, Sun
32. Retainer ring, sun gear shell
33. Retainer ring, clutch to cam
34. Support, low & rev. clutch (inc. cam)
35. Plate, rev. clutch—drive
36. Carrier, reaction planet
37. Washer, output ring gear, front
38. Gear, output ring
39. Bushing, case to shaft
40. Bearing, ring gear supt. to case
41. Ring, reverse clutch
42. Plate, low & rev. clutch reaction
43. Seat, reverse clutch
44. Spring, rev. clutch—return
45. Piston, low & rev. clutch
46. Seal, low & rev. clutch outer
47. Seal, low & rev. clutch, inner
48. Retainer ring, output carrier
49. Seal, kit, low & rev. clutch
50. Shaft assy., output
51. Clip, speedo. gear
52. Gear, speedo. drive
53. Bushing output shaft

Shafts and clutches—350, 375 B trans.

© G.M. Corp.

1. Shaft, input
2. Housing and shaft forward clutch

3. Seal, forward clutch—inner
4. Seal, forward clutch—outer
5. Piston, forward clutch

Component Removal and Overhaul

GM TURBO HYDRA-MATIC 350–375–M38

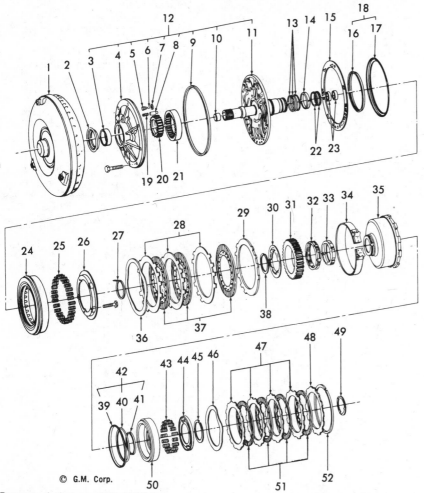

© G.M. Corp.

Pump and clutces—350, 375B trans.

1. Converter Assy. (rebuilt)
2. Seal, Oil pump
3. Bushing, Oil pump
4. Body, Oil pump
5. Spring, Pump by-pass valve
6. Ball, pump by-pass valve
7. Seat, pump by-pass valve
8. Valve, pump priming
9. Seal, pump to case
10. Bushing, starter shaft—front
11. Cover, oil pump
12. Pump assy.
13. Ring, clutch drum
14. Bushing, clutch drum
15. Gasket, pump cover
16. Seal, Inter. Clutch—inner
17. Seal, Inter. Clutch—outer
18. Seal kit, inter. clutch
19. Spring pump priming valve
20. Gear, pump drive
21. Gear pump driven
22. Ring forward clutch
23. Bushing, starter shaft—rear
24. Piston, Inter clutch
25. Spring inter. clutch piston
26. Seat, inter. clutch piston
27. Washer, direct clutch
28. Plate, Inter. clutch reaction
29. Plate, Inter Clutch—pressure
30. Retainer, Inter clutch
31. Race, Inter clutch—outer
32. Clutch, Inter Overrun
33. Cam, Inter clutch
34. Band, Inter clutch
35. Drum, direct clutch
36. Spring, Inter clutch
37. Plate, Inter. clutch drive
38. Retainer ring, Inter clutch
39. Seal, Direct clutch piston, outer
40. Seal, direct clutch piston, center
41. Seal, direct clutch piston, inner
42. Seal kit, direct clutch piston
43. Spring, direct clutch piston, return
44. Seat, direct clutch piston
45. Ring, direct clutch piston
46. Spring, direct clutch piston
47. Plate, direct clutch driven
48. Plate, direct clutch—pressure
49. Bearing, direct clutch drum
50. Piston, direct clutch
51. Plate, direct clutch-drive
52. Retainer ring, direct clutch

spacer plate gasket.

19 Install detent control valve link in detent actuating lever. Install valve body while installing manual control valve link in range selector inner lever. When handling valve body, be careful not to depress sleeves as retaining pins may fall out into the transmission.

20 Torque valve body attaching bolts to 13 ft./lbs. Install detent roller and spring to valve body.

21 Replace oil pump screen and gasket to valve body. Replace oil pump screen attaching screws.

22 Install pan and gasket. Torque bolts to 12 ft./lbs.

23 Check governor bore and sleeve for scoring. Install governor assembly into case. Install new O-ring on governor cover and install governor cover. Use a brass drift on the flange of the cover, but do not hammer on the cover. Replace governor cover retainer.

24 Place retainer clip on output shaft. Slide speedometer drive gear over output shaft until secured by clip.

25 Position yoke seal on output shaft. Tap seal into place with a suitable tool.

26 Install extension housing lip seal if not replaced on disassembly. Install extension housing square cut O-ring seal. Replace extension housing. Torque bolts to 35 ft./lbs.

27 Lubricate new modulator O-ring, then install on modulator. Replace vacuum modulator valve in case, bolting down retainer.

28 Install torque converter to transmission, engaging lugs, and install a holding tool to prevent converter from dropping off.

GM Type 350 Oil Pressure Checks Oldsmobile

Range	Normal Pressure (psi)
Drive—Brakes applied, engine @ 1,000 rpm	60-90
Super-low—Brakes applied, engine @ 1,000 rpm	85-110
Reverse—Brakes applied, engine @ 1,000 rpm	85-150
Neutral—Brakes applied, engine @ 1,000 rpm	55-70
Drive—Idle set to engine specifications	60-85
Drive—30 mph coast (closed throttle)*	55-70

*—Can be performed by running engine on lift at 2,000 rpm. Release throttle and read pressure before rpm drops below 1,200.

GM TURBO HYDRA-MATIC 350–375–M38

GM Type 350 Torque Specifications

Oil pan to transmission case	13 ft/lbs
Pump assembly to transmission case	20 ft/lbs
Vacuum modulator retainer to case	12 ft/lbs
Valve body assembly to case	13 ft/lbs
Oil channel support plate to case	13 ft/lbs
Pump body to pump cover	15-18 ft/lbs
Parking lock bracket to case	29 ft/lbs
Extension housing to case	35 ft/lbs
Inside shift nut	30 ft/lbs
External test plugs to case	8 ft/lbs
Cooler fitting to case	30 ft/lbs
Oil pickup screen to valve body	36 in/lbs
Detent valve actuating lever bracket to valve body	48 in/lbs
Detent valve actuating lever bracket to valve body	48-52 in/lbs

GM Type 350 Capacity General Specifications

Oil Capacity (dry)	20 Pints
Cooling	Water
Oil Filter Type	Bottom Suction Screen
Clutches	4
Roller Clutches	2
Band (non-adjustable)	1

GM Type 350 Oil Pressure Checks

	Oil Pressure (psi) @ Altitude (± 5 psi) ①								
	Sea Level			2,000 ft.			10,000 ft.		
Model	D, N, P	S, L	R	D, N, P	S, L	R	D, N, P	S, L	R
Buick, Pontiac	153	153	239	153	153	224	116	125	164
Chevrolet 307	60	87	92	69	93	104	N.A.	N.A.	N.A.
Chevrolet 350	60	87	92	69	93	104	N.A.	N.A.	N.A.

① Pressures are at 0 output speed, 1,200 engine rpm and vacuum line disconnected and plugged.

N.A.—Not Available

GM Type 350 Oil Pressure Checks

	Pressure (psi), with vehicle coasting @ 25 mph vacuum line connected, foot off throttle				
Model	Neutral ± 5 psi	Drive ± 5 psi	Super ± 5 psi	Low ± 5 psi	Reverse ± 5 psi
Buick, Pontiac	60	60	85	85	N.A.

GM TURBO HYDRA-MATIC 400–425–M40

TROUBLESHOOTING
GM TURBO HYDRA-MATIC 400-425-M40

THE PROBLEM	ITEMS TO CHECK	
	IN CAR	OUT OF CAR
NO DRIVE IN D RANGE	Oil level. Manual linkage (external). Check oil pressure. Manual control disconnected inside.	Front clutch. Clutch feed seals and gaskets. Low sprags.
NO DRIVE IN R or SLIPS IN REVERSE	Oil level. Manual linkage (external). Check oil pressure. Modulator and/or lines. Valves, body and/or leaks. Reverse feed passages. Valve check balls. Rear servo and accumulator. Pump regulator and boost valve.	Rear band. Direct clutch. Front clutch.
DRIVE IN NEUTRAL	Manual linkage (external).	Front clutch.
1st SPEED ONLY—NO 1-2 SHIFT	Governor and/or feed line seals. Valves, body and/or leaks.	Intermediate clutch.
1-2 SHIFT AT FULL THROTTLE ONLY	Detent solenoid. Detent switch. Valves, body and/or leaks.	
1st & 2nd SPEEDS ONLY—NO 2-3 SHIFT	Detent solenoid. Detent switch. Valves, body and/or leaks.	Direct clutch.
SLIPS IN ALL RANGES	Oil level. Check oil pressure. Modulator and/or lines. Clogged strainer or intake leaks. Valves body and/or leaks.	Front pump, Direct clutch. Front clutch. Pump-to-case gasket. Low sprags.
SLIPS ON 1-2 SHIFT	Oil level. Check oil pressure. Modulator and/or lines. Pump regulator and boost valve. Valves, body and/or leaks. Front servo and accumulator.	Intermediate clutch. Pump-to-case gasket.
ROUGH 1-2 SHIFT	Check oil pressure. Modulator and/or lines. Pump regulator and boost valve. Valves, body and/or leaks. Rear servo accumulator. Front servo and accumulator. Valves, body and/or leaks. Rear servo and accumulator.	Pump-to-case gasket. Intermediate check valve ball in case.
SLIPS ON 2-3 SHIFT	Oil level. Check oil pressure. Modulator and/or lines. Pump regulator and boost valve. Valves, body and/or leaks.	Direct clutch. Pump-to-case gasket.
ROUGH 2-3 SHIFT	Check oil pressure. Modulator and/or lines. Pump regulator and boost valve. Front servo and accumulator.	
SHIFTS OCCUR—TOO HIGH or TOO LOW CAR SPEED	Check oil pressure. Governor and/or feed line seals. Modulator and/or lines. Detent solenoid. Valves, body and/or leaks. Pump regulator and boost valve.	
NO DENTENT DOWNSHIFT	Detent switch. Detent solenoid. Valves, body and/or leaks.	
NO PART THROTTLE DOWNSHIFT	Check oil pressure. Modulator and/or lines. Valves, body and/or leaks.	
NO ENGINE BRAKING—SUPER RANGE—SECOND SPEED	Front servo and accumulator.	Front band.
NO ENGINE BRAKING—LOW RANGE—FIRST SPEED	Valves, body and/or leaks. Rear servo and accumlator.	Intermediate check valve ball in case. Rear band.
PARK WILL NOT HOLD	Manual linkage (external). Internal linkage. Parking pawl and/or link.	
POOR PERFORMANCE OR ROUGH IDLE—STATOR NOT FUNCTIONING	Stator switch. Valve body—stator section	Turbine shaft.
NOISY TRANSMISSION	Cooler or lines. Clogged strainer or intake leaks.	Front pump, Front clutch, Direct clutch. Intermediate clutch. Planetary assembly.

GM TURBO HYDRA-MATIC 400-425-M40

GM Turbo Hydra-Matic 400 & 425

Identification

CADILLAC

The Turbo Hydra-Matic 400 transmission unit number is located on a plate located on the right side of the case. This transmission is used in all rear wheel drive applications.

The Turbo Hydra-Matic 425 transmission unit number is located on the left side of the converter housing. This transmission is used with all front wheel drive applications.

CHEVROLET

On all Chevrolet vehicles, see the engine identification chart, and check engine identification tag for transmission identification.

OLDSMOBILE

On Oldsmobile vehicles, the serial plate

GM TURBO HYDRA-MATIC 400–425–M40

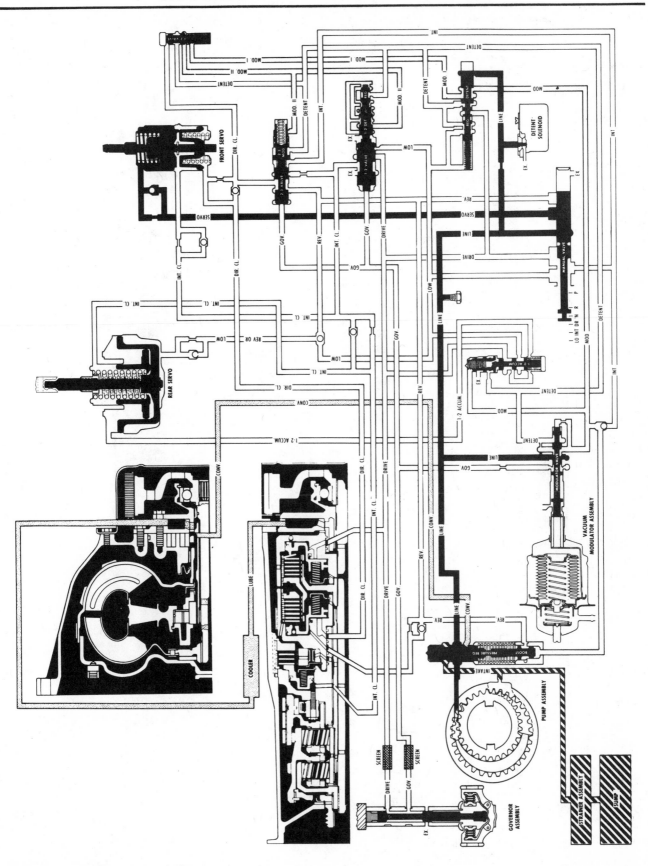

Hydraulic schematic 425 transmissions

GM TURBO HYDRA-MATIC 400–425–M40

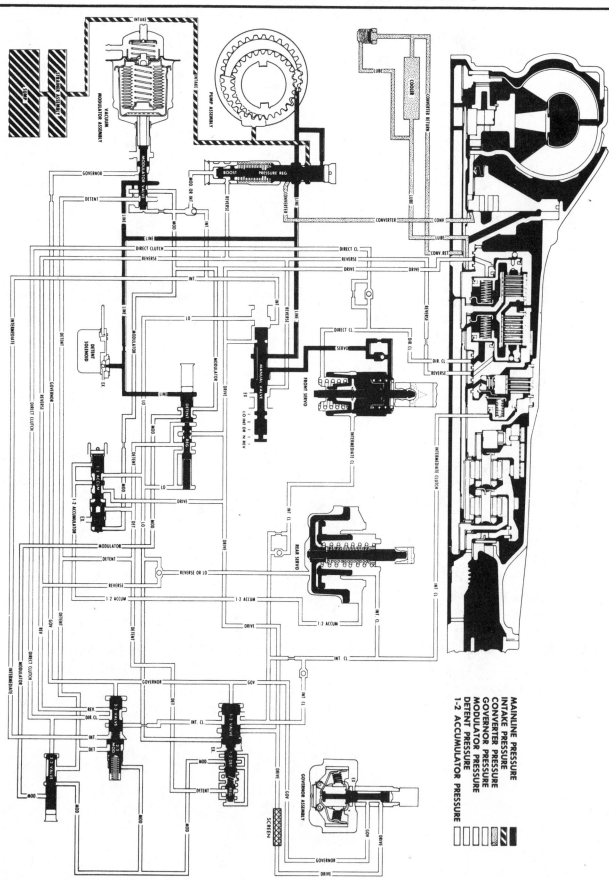

Hydraulic schematic 400 transmissions

GM TURBO HYDRA-MATIC 400–425–M40

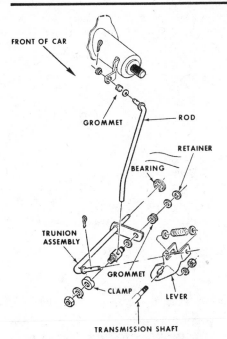

Typical external linkage 400 transmission

is located on the right side of the case, except on front wheel drive applications (Turbo Hydra-Matic 425), in which case it is located on the left side of the converter housing.

Transmission Disassembly

Clean outside of the unit thoroughly to prevent dirt from entering the unit.
1 With transmission in a work cradle or on a clean bench, lift the converter straight off the transmission input shaft.
2 With transmission bottom-up, remove modulator assembly attaching screw and retainer, then remove the modulator assembly and O-ring seal.

GOVERNOR, SPEEDOMETER DRIVEN GEAR, OIL PAN, STRAINER AND INTAKE PIPE REMOVAL

1 Remove attaching screws, governor cover and gasket, then withdraw governor from the case.
2 Remove speedometer driven gear attaching screw and retainer, then withdraw the driven gear assembly.
3 Remove oil pan attaching screws, then the oil pan. Discard gasket.
4 Remove pump intake pipe and strainer assembly, then the pipe-to-case O-ring seal.

CONTROL VALVE ASSEMBLY, GOVERNOR PIPES AND DETENT SPRING ASSEMBLY REMOVAL

1 Disconnect the lead wire from the pressure switch assembly. Remove the switch assembly from the valve body, if necessary. This switch is installed on 1972 models only. Pontiac model

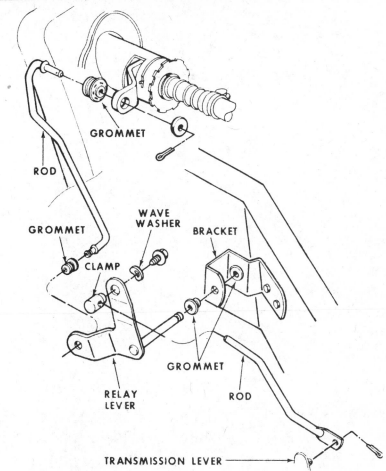

Typical external linkage 425 transmission

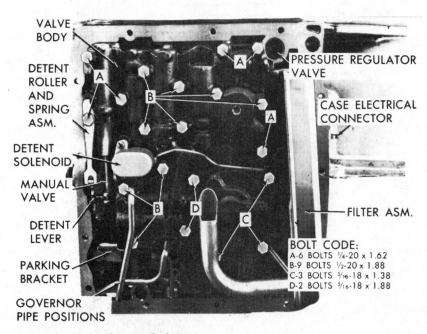

BOLT CODE:
A-6 BOLTS ¼-20 x 1.62
B-9 BOLTS ½-20 x 1.88
C-3 BOLTS ⁵⁄₁₆-18 x 1.38
D-2 BOLTS ⁵⁄₁₆-18 x 1.88

Oil pan removed 425 transmissions

PQ and Chevrolet model CM are not equipped with a pressure switch assembly.

1 Remove control valve body attaching screws and detent roller and spring assembly.

Component Removal and Overhaul

GM TURBO HYDRA-MATIC 400–425–M40

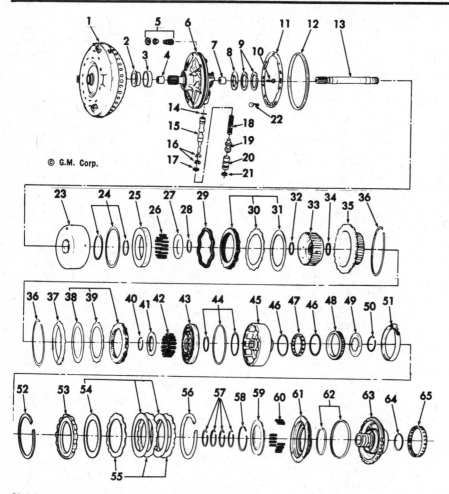

© G.M. Corp.

Clutch, pump and center support—400 and 425 trans.

1. Torque converter	22. Bolt	44. Seal kit
2. Oil seal	23. Forward clutch housing	45. Direct clutch housing
3. Bushing	24. Seal kit	46. Intermediate sprag
4. Bushing	25. Forward clutch piston	47. Intermediate sprag
5. Front oil pump	26. Release spring	48. Outer race
6. Front oil pump	27. Spring retainer	49. Clutch retainer
7. Bushing	28. Snap ring	50. Snap ring
8. Thrust washer	29. Wave plate, forward clutch	51. Front band
9. Seal ring	30. Clutch plate (driven)	52. Snap ring
10. Bolt	31. Clutch plate (drive)	53. Backing plate
11. Gasket	32. Thrust washer	54. Clutch plate (driving)
12. Seal	33. Forward clutch hub	55. Reaction plate
13. Turbine shaft	34. Thrust washer	56. Snap ring
14. Pin	35. Direct clutch hub	57. Seal ring
15. Pressure regulator valve	36. Snap ring	58. Snap ring
16. Spacer pressure regulator valve	37. Backing plate	59. Spring retainer
17. Retaining washer	38. Driving plate	60. Return spring
18. Regulator spring	39. Driven plate	61. Intermediate piston
19. Valve and bushing kit	40. Snap ring	62. Seal kit
20. Valve and bushing kit	41. Spring retainer	63. Center support
21. Retaining ring	42. Release spring	64. Thrust washer
	43. Direct clutch piston	65. Low Clutch Roller

NOTE: Do not remove solenoid attaching screws.

Remove control valve body and governor pipes.

2 Remove the governor pipes from valve body. Then remove valve body-to-spacer gasket.

REAR SERVO, SOLENOID, VALVE BODY SPACER, AND FRONT SERVO REMOVAL

1 Remove rear servo cover attaching screws, the cover and gasket, then the rear servo assembly from the case.

2 Remove servo accumulator springs.

3 Disconnect the solenoid leads from connector terminal. Withdraw connector and O-ring seal.

4 Remove solenoid attaching screws, solenoid and gasket.

5 Remove valve body spacer plate and gasket.

6 Remove six check balls from cored passages in transmission case. There are seven check balls on Toronado and Eldorado transmissions. An eighth, held in place, should not be removed, unless defective.

7 Remove the front servo assembly.

REAR OIL SEAL AND EXTENSION HOUSING REMOVAL

1 Pry rear oil seal from extension housing.

2 Remove housing attaching bolts, then remove extension housing and housing-to-case oil seal.

Sprocket Cover, Link Assembly, Drive and Driven Sprocket Removal—

ELDORADO AND TORONADO

1 Remove the sprocket cover attaching screws, cover and gasket. Discard the gasket.

2 Remove the snap-rings (as illustrated) from beneath the drive and driven sprockets. Leave the snap-rings in a loose position.

3 Remove the drive and driven sprockets, link assembly, bearings and shafts simultaneously, by alternately pulling up on the drive and driven sprockets until the bearings and shafts are free of the housings.

4 Remove the link assembly from the sprockets.

5 Remove the hook type oil seal ring from the turbine shaft.

6 Be sure that the drive and driven sprocket bearings are not damaged.

7 Inspect the drive sprocket teeth for nicks, burrs, scoring, galling or excessive wear. Inspect the drive sprocket ball bearings for wear or damage.

8 Be sure that the turbine shaft lubrication openings are clear and free.

9 Inspect the splines and ground bushing journals for damage.

10 Inspect the turbine shaft for cracks or distortion.

11 Inspect the link assembly for damage or loose links.

OIL PUMP, FORWARD CLUTCH AND GEAR UNIT REMOVAL

1 Pry front seal from the pump. Then remove pump attaching bolts.

2 With slide hammers attached, remove pump from transmission case. Discard pump-to-case seal ring. Discard pump-to-case gasket.

3 Remove turbine shaft from transmission.

4 Remove forward clutch assembly. Be sure that the bronze thrust washer came out with the clutch housing assembly.

GM TURBO HYDRA-MATIC 400–425–M40

Removing drive sprocket snap ring— Toronado and Eldorado

5 Remove the direct clutch assembly. Remove front band and sun gear shaft.
6 Loosen the jam nut which holds the detent lever to the manual shaft. Then remove detent lever from manual shaft. The manual shaft should not be removed from the case unless replacement is necessary.
7 Remove parking actuator rod and detent lever assembly. Then remove detent lever E-ring and the detent lever.
8 Remove attaching screws and park bracket; then the parking pawl return spring.
9 Remove parking pawl shaft retainer, then the parking pawl shaft, O-ring and pawl.
10 Remove the front band assembly.
11 Remove the sun gear shaft.
12 Remove the case center support-to-case bolt and center support locating screw. 1972 models have no locating screw.
13 Remove intermediate clutch backing plate-to-case snap-ring. Then remove the backing plate, three composition, and three steel clutch plates.
14 Remove the center support-to-case retaining snap-ring.
15 Remove the entire gear unit assembly with slide hammer.
16 Remove the output shaft-to-case thrust washer from the rear of the output shaft, or from inside the case.
17 Remove rear unit selective washer from transmission case.
18 Remove rear band assembly.
19 Remove support to case spacer from inside case.

Component Disassembly, Inspection, and Assembly

GOVERNOR

All components of the governor, except the driven gear, are a select fit and so calibrated. Therefore, service this unit as an assembly.

Clean and inspect all parts for wear or other damage. Check valve opening at feed port with a feeler gauge, holding the

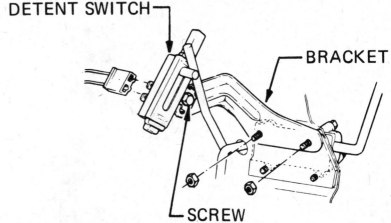

Detent switch-typical

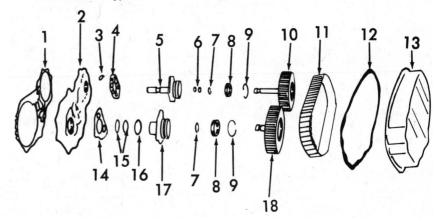

Transmission drive link assy.—425 trans.

1. Gasket
2. Cover plate
3. Check plate
4. Gasket
5. Stator support
6. Oil seals
7. Snap ring
8. Bearing
9. Snap ring
10. Drive gear and link pkg.
11. Drive link
12. Gasket, gear housing
13. Cover and plate
14. Gasket, drive gear support
15. Oil ring support
16. Thrust washer, hsg. to pump cover
17. Support, driven gear
18. Gear, driven w/input shaft

governor with the weights extended completely outward. Check valve opening at exhaust port, holding governor with weights completely inward. If either opening is less than .020 in., replace governor assembly.

If a new governor drive gear is installed, a new pin hole must be drilled 90 degrees from the original hole.

FRONT SERVO INSPECTION

1 Inspect servo pin for damage.
2 Inspect servo piston for damaged oil ring groove, cracks, or porosity. Check freedom of oil seal ring in groove.
3 Check fit of servo pin in piston.

REAR SERVO DISASSEMBLY

1 Remove rear accumulator piston from rear servo piston.
2 Remove E-ring which holds the servo piston to band apply pin.

3 Remove servo piston and seal from band apply pin. Remove second washer from band apply pin.
4 Remove washer, spring, and retainer.

REAR SERVO INSPECTION

1 Check freedom of accumulator rings in piston.
2 Check fit of band apply pin in servo piston. Inspect pin for scores or cracks.
3 Inspect accumulator and servo pistons for cracks or porosity.

REAR SERVO ASSEMBLY

1 Install spring retainer, cup side down, spring, and washer on band apply pin.
2 Install band apply pin, retainer, spring, and washer into bore of servo piston and secure with E-ring.
3 Install oil seal on servo piston. Install outer and inner oil seal rings on accumulator piston and assembly into bore of servo piston.

GM TURBO HYDRA-MATIC 400–425–M40

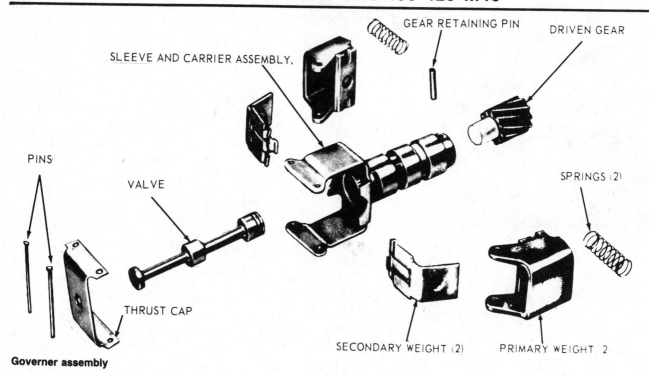

SLEEVE AND CARRIER ASSEMBLY,

GEAR RETAINING PIN

DRIVEN GEAR

PINS

VALVE

SPRINGS (2)

THRUST CAP

SECONDARY WEIGHT (2)

PRIMARY WEIGHT 2

Governer assembly

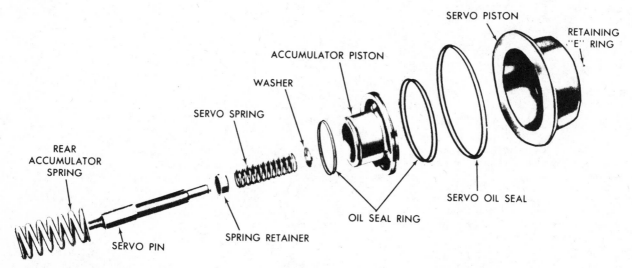

SERVO PISTON

RETAINING "E" RING

ACCUMULATOR PISTON

WASHER

SERVO SPRING

REAR ACCUMULATOR SPRING

SERVO OIL SEAL

OIL SEAL RING

SPRING RETAINER

SERVO PIN

Rear servo and accumulator

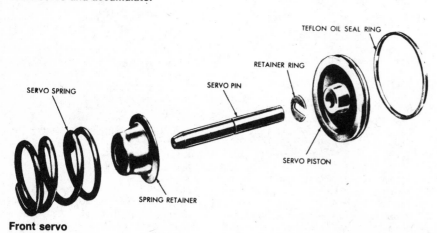

TEFLON OIL SEAL RING

RETAINER RING

SERVO PIN

SERVO SPRING

SERVO PISTON

SPRING RETAINER

Front servo

CONTROL VALVE BODY DISASSEMBLY

When disassembling control valve body, identify all springs so that they can be replaced in their proper locations. Consult illustrations for individual variations of control valve assembly. Be careful to note the exact order of the parts in the various valve bores.

1 Position valve assembly with cored face up and accumulator pocket on bottom.
2 Remove manual valve from upper bore.
3 Compress accumulator piston spring and remove E-ring retainer. Remove accumulator piston and spring.
4 Press out retaining pin from upper right bore. Remove 1–2 modulator

GM TURBO HYDRA-MATIC 400–425–M40

1. MANUAL VALVE
2. DETENT SOLENOID
3. GASKET
4. FRT. ACC. SPRING
5. OIL RING
6. ACC. PISTON
7. E RING
8. 3-2 VALVE PIN
9. 3-2 PIN VALVE SPRING
10. 3-2 VALVE
11. 3-2 BORE PLUG
12. RETAINER PIN
13. 2-3 VALVE
14. 3-2 INTERMEDIATE SPRING

15. 2-3 MODULATOR VALVE
16. 2-3 MODULATOR BUSHING
17. 2-3 VALVE SPRING
18. RETAINER PIN
19. 1-2 SHIFT VALVE
20. 1-2 DETENT VALVE
21. 1-2 REGULATOR SPRING
22. 1-2 REGULATOR VALVE
23. 1-2 MODULATOR VALVE BUSHING
24. RETAINER PIN
25. DETENT REGULATOR PIN
26. DETENT REGULATOR SPRING
27. DETENT REGULATOR VALVE
28. DETENT VALVE
29. VALVE BORE PLUG
30. RETAINER PIN
31. VALVE BORE PLUG
32. RETAINER PIN
33. 1-2 PRIM. ACCUM. SPRING
34. 1-2 ACCUMULATOR VALVE
35. 1-2 ACCUM. VALVE BORE PLUG
36. RETAINING PIN

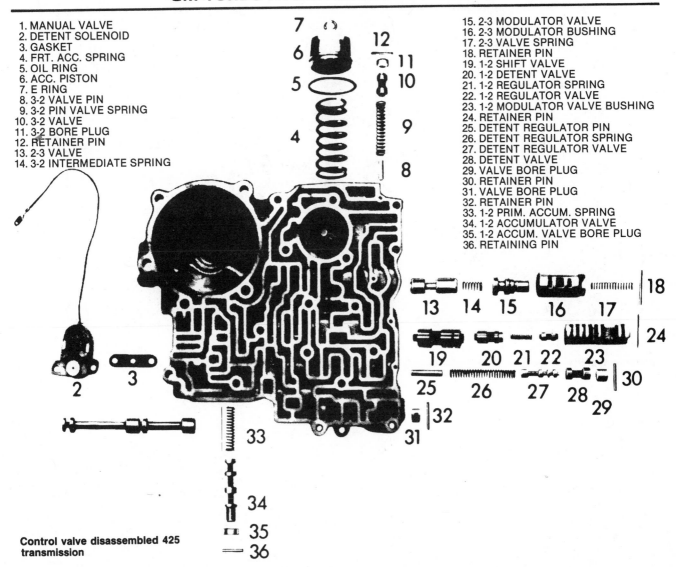

Control valve disassembled 425 transmission

bushing, 1–2 regulator valve and spring, 1–2 detent valve, and 1–2 shift valve. 1–2 regulator valve and spring may be inside of 1–2 modulator bushing.

5 Press out retaining pin from center right bore. Remove 2–3 modulator bushing, 2–3 shift valve spring, 2–3 modulator valve, and 1–2 shift valve. 1–2 and 2–3 shift valve. 2–3 modulator valve will be inside of 2–3 modulator bushing.

6 Press out retaining pin from lower right bore. Hold hand over bore, as plug may pop out. Remove plug, 3–2 valve spring, spacer, and 3–2 valve.

7 Holding hand over bore, press out retainer pin from upper left bore. Remove bore plug, detent valve, detent regulator valve, spacer, and detent regulator valve spring.

8 Pry out grooved retainer ring from lower left bore with long nose pliers. Remove bore plug, 1–2 accumulator bushing, 1–2 accumulator valve, secondary spring, primary 1–2 ac-

cumulator valve, and spring.

9 Remove governor oil feed screen from oil feed hole in valve body.

CONTROL VALVE BODY INSPECTION

1 Wash control valve body, valves, and other parts in solvent. Do not allow valves to bump together. Air dry parts and blow out all passages.

2 Inspect all valves and bushings carefully. Burrs may be removed with a fine stone or fine crocus cloth and light oil. Be careful not to round off shoulders of valves.

3 Test all valves and bushings for free movement in their bores. All valves should fall freely of their own weight.

4 The manual valve is the only valve that can be serviced separately. If any of the other valves are defective or damaged, install a new control valve assembly.

5 Inspect body for cracks or scored bores. Check all springs for distortion or collapsed coils.

CONTROL VALVE BODY ASSEMBLY

1 Replace front accumulator spring and piston into valve body. Compress spring and piston, assuring that piston pin is correctly aligned with hole in piston and that oil seal ring does not catch on lip of bore when installing piston. Secure piston and spring with E-ring retainer.

2 Install 1–2 accumulator primary spring, 1–2 primary valve, and 1–2 accumulator bushing into lower left bore. Place 1–2 accumulator secondary valve, stem end out, into the 1–2 accumulator bushing. Place 1–2 accumulator secondary spring over stem end of valve. Replace bore plug and retaining pin.

3 Install detent regulator valve spring and spacer into upper left bore, making certain that spring seats in bottom of bore. Compress spring and hold with a small screwdriver between end of spring and wall on cored side of valve body. Insert detent regulator

GM TURBO HYDRA-MATIC 400–425–M40

1. MANUAL VALVE
2. RETAINING PIN
3. BORE PLUG
4. DETENT VALVE
5. DETENT REGULATOR VALVE
6. SPACER PIN
7. DETENT REGULATOR SPRING
8. 1-2 SHIFT VALVE
9. 1-2 DETENT VALVE

9A. 1-2 MODULATOR VALVE
10. 1-2 REGULATOR SPRING
10A. 1-2 MODULATOR VALVE SPRING
11. 1-2 REGULATOR VALVE

12. 1-2 MODULATOR BUSHING
13. RETAINING PIN
14. GROOVED RETAINING PIN
15. BORE PLUG
15A. 1-2 ACCUMULATOR SECONDARY SPRING
16. 1-2 ACCUMULATOR VALVE
17. 1-2 ACCUMULATOR PRIMARY SPRING
18. 2-3 SHIFT VALVE

19. 3-2 INTERMEDIATE SPRING
20. 2-3 MODULATOR VALVE
21. 2-3 VALVE SPRING
22. 2-3 MODULATOR BUSHING
23. RETAINING PIN
24. 3-2 VALVE
25. SPACER PIN
26. 3-2 VALVE SPRING
27. BORE PLUG
28. RETAINING PIN

Disassembled control valve body 400 transmission

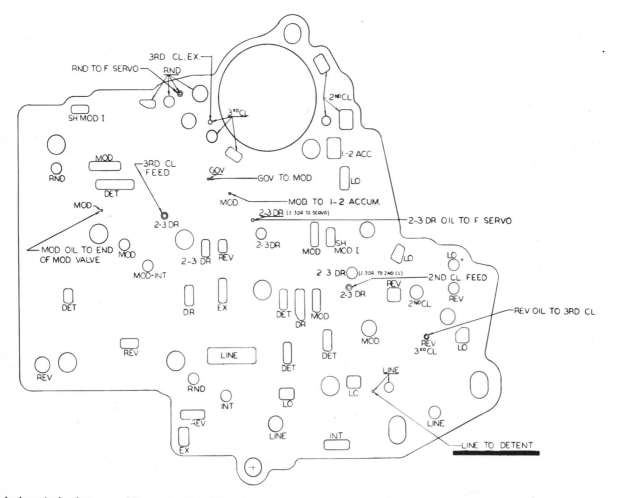

Typical control valve assembly spacer plate 400 transmission

GM TURBO HYDRA-MATIC 400–425–M40

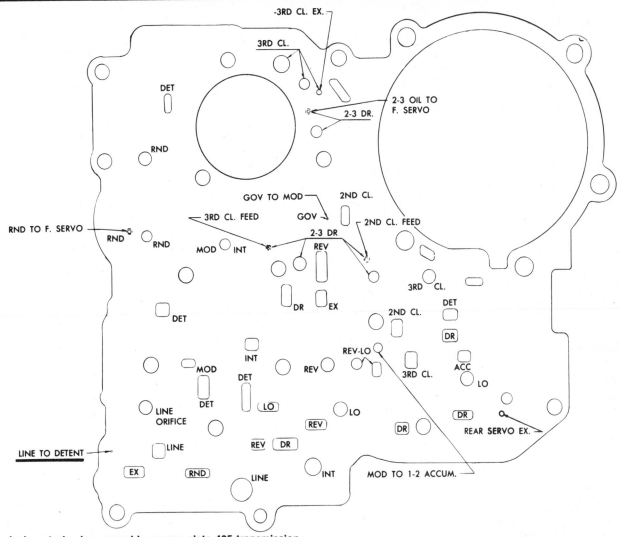

Typical control valve assembly spacer plate 425 transmission

valve, stem end out, and detent valve, small land first. Insert bore plug, press inward, remove screwdriver, and install retaining pin.

4 Insert 3–2 valve in bottom right bore. Place spacer inside 3–2 valve spring and insert spacer and spring in bore. Install bore plug and retaining pin.

5 Install 3–2 intermediate spring on stem end of 2–3 shift valve. Install valve and spring, valve first, into center right bore. Be sure that valve seats in bottom of bore. Place 2–3 modulator valve, hole end first, into 2–3 modulator bushing. Install valve and bushing in bore. Install 2–3 shift valve spring into hole in 2–3 modulator valve. Secure with retaining pin.

6 Seat 1–2 shift valve, stem end out, in bottom of upper right bore. Install 1–2 regulator valve, larger stem first, and spring and 1–2 detent valve, hole end first, into 1–2 modulator bushing. Align spring in bore of 1–2 detent valve. Install assembly into upper right bore of control valve body. Install retaining pin.

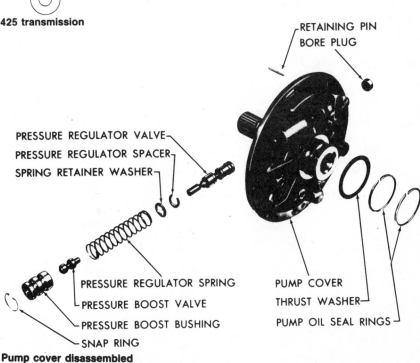

Pump cover disassembled

Component Removal and Overhaul

GM TURBO HYDRA-MATIC 400–425–M40

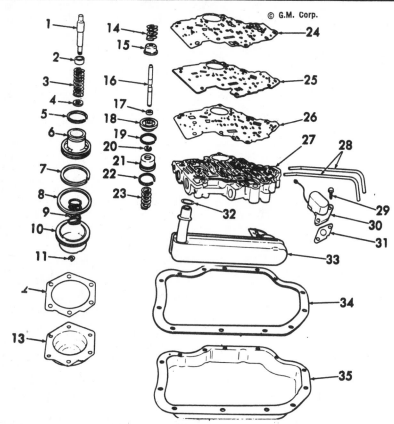

© G.M. Corp.

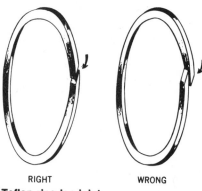

RIGHT   WRONG

Teflon ring lap joint

Servos control valve—400 and 425 trans.

1. Band apply pin
2. Spring retainer
3. Piston spring
4. Washer
5. Seal ring
6. Accumulator Piston
7. Piston ring
8. Seal
9. Accumulator spring
10. Rear piston
11. Snap ring
12. Gasket

13. Cover
14. Piston spring
15. Spring retainer
16. Piston pin
17. Washer
18. Front piston
19. Seal ring
20. Snap ring
21. Accumulator piston
22. Seal ring
23. Piston spring

24. Gasket
25. Spacer
26. Gasket
27. Control valve body
28. Governor pipe
29. Bolt
30. Solenoid
31. Gasket
32. Seal
33. Oil strainer
34. Pan gasket
35. Oil pan

7 Replace governor oil feed screen assembly in governor oil feed hole.
8 Install manual valve with detent pin groove to the right.

OIL PUMP DISASSEMBLY

1 Place the pump over a hole in the bench, shaft down, cover up.
2 Compress regulator boost valve bushing against the pressure regulator spring and remove the snap-ring.

CHILTON CAUTION: *Pressure regulator spring is under extreme pressure.*

3 Remove the boost valve bushing and valve, then the spring.
4 Remove valve spring retainer and spacer/s, if present, and regulator valve.
5 Remove pump cover to body attaching bolts, then remove the cover.
6 Remove the retaining pin and bore plug from the pressure regulator bore.
7 Remove the two hook-type oil rings from the pump cover.

8 Remove pump to forward clutch housing selective washer.
9 Mark drive and driven gears for reassembly, then remove the gears.
10 To disassemble Eldorado and Toronado oil pumps: matchmark gears; remove the pump body; remove the gears and discard the pump body to case O-ring seal.

OIL PUMP INSPECTION

1 Inspect drive gear, driven gear, gear pocket, and crescent for scoring, galling, or other damage.
2 Replace pump gears in pump and check pump body face to face gear clearance. The clearance should be .0008–.0035 in.
3 Check pump body face for scoring or nicks. Check oil passages in pump body for roughness or obstructions. Check condition of cover bolt attaching threads. Check for flatness of pump body face. Check pump body

bushing for scores or nicks. If a new bushing is installed, drive it in flush to .010 in. below the gear pocket face.
4 Replace pump attaching bolt seals if necessary.
5 Check pump cover face for flatness. Check for scoring or chips in pressure regulator bore. Check that all passages are unobstructed. Check for scoring or damage at pump gear face. Check that breather hole in pump cover is open.
6 Check condition of stator shaft splines and bushings.
7 Check oil ring grooves for damage or wear. Check selective thrust washer for wear.
8 Make sure that pressure regulator valve and boost valve operate freely.

OIL PUMP ASSEMBLY

1 Install drive and driven gears into the pump body, alignment marks up and in proper index, (drive gear with drive tangs up).
2 Install pressure regulator spring spacer/s if required, retainer and spring into the pressure regulator bore.
3 Install pressure regulator valve from the opposite end of the bore, stem end first.
4 Install boost valve into bushing, stem end out, and install both parts into the pump cover by compressing the bushing against the spring. Install the snap-ring.
5 Install the regulator valve bore plug and retaining pin into opposite end of bore.
6 Install the front unit selective thrust washer over the pump cover delivery sleeve.
7 Install two hook-type oil seal rings.
8 Assemble pump cover to pump body with attaching bolts. (Leave bolts one turn loose at this time.)
9 Place pump aligning strap over pump body and cover, then tighten tool.
10 Install pump cover bolts. Remove clamp and install pump-to-case O-ring seal and gasket.
11 Assemble Toronado and Eldorado oil pumps as follows: install new pump to case O-ring seal; install gears in the body (drive gear counterbore down) and align matchmarks.

GM TURBO HYDRA-MATIC 400–425–M40

FORWARD CLUTCH DISASSEMBLY

1 Remove forward clutch housing-to-direct clutch hub snap-ring.
2 Remove direct clutch hub.
3 Remove forward clutch hub and thrust washers.
4 Remove five composition and five steel clutch plates. Press out the turbine shaft.
5 Compress the spring retainer and remove the snap-ring.
6 Remove the compressor, snap-ring, spring retainer and 16 release springs, then lift out the piston.
7 Remove inner and outer piston seals.
8 Remove center piston seal from the forward clutch housing.

FORWARD CLUTCH INSPECTION

1 Inspect clutch plates for burning, scoring, or wear.
2 Inspect springs for collapsed coils or distortion.
3 Check clutch hubs for worn splines, thrust faces, and open lubrication holes.
4 Check piston for cracks.
5 Check clutch housing for wear, scoring, open oil passages, and free operation of ball check.
6 Inspect turbine shaft for:
   a. Open lubrication passages at each end.
   b. Spline damage.
   c. Damage to ground bushing journals.
   d. Cracks or distortion of shaft.

FORWARD CLUTCH ASSEMBLY

1 Place new inner and outer oil seals on clutch piston, lips away from spring pockets.
2 Place a new center seal on the clutch housing, lip faces up.
3 Place a seal protector (thimble) tool into clutch drum and install the piston.
4 Install 16 clutch release springs into pockets in the piston.
5 Lay the spring retainer and the snap-ring on the springs.
6 Compress springs using compressor, and install snap-ring. Press in turbine shaft.
7 Install the direct clutch hub washers; retain with petroleum jelly.
8 Place forward clutch hub into forward clutch housing.
9 Oil and install 5 composition, 4 flat steel and 1 waved steel clutch plate (plate with U-notches) starting with waved steel and alternating composition and steel clutch plates.
10 Install clutch hub and retaining snap-ring.
11 Place forward clutch housing on pump delivery sleeve and air-check clutch operation.

DIRECT CLUTCH AND INTERMEDIATE SPRAG DISASSEMBLY

1 Remove intermediate clutch retainer snap-ring, then the retainer.
2 Remove clutch outer race, bushings and sprag assembly.
3 Invert the unit and remove backing plate to clutch housing snap-ring.
4 Remove direct clutch backing plate and the steel and composition clutch plates.
5 Using clutch compressor tool compress the spring retainer and remove the snap-ring.
6 Remove retainer and 14 piston release springs (16 on Chevrolet and Cadillac).
7 Remove direct clutch piston, then remove the outer and inner seals from the piston.
8 Remove center piston seal from the direct clutch housing.

DIRECT CLUTCH AND INTERMEDIATE SPRAG INSPECTION

1 Check for popped or loose sprags.
2 Check sprag bushings for distortion or wear.
3 Inspect inner and outer races for scratches or wear.
4 Check clutch housing for cracks, wear, proper opening of oil passages, and wear on clutch plate drive lugs.
5 Check clutch plates for wear or burning.
6 Check backing plate for scratches or damage.
7 Check clutch piston for cracks and free operation of ball check.

DIRECT CLUTCH AND INTERMEDIATE SPRAG ASSEMBLY

1 Install a new clutch piston seal onto the piston, lips facing away from spring pockets. Apply transmission fluid to oil seals.
2 Install a new outer piston seal, and a new center seal onto the clutch housing, lip of seal facing up.
3 Place seal protectors over hub and clutch housing, then install clutch piston.
4 Install springs into piston, place retainer and snap-ring on retainer.
5 With clutch compressor, compress the clutch and install snap-ring.
6 Install composition and steel clutch plates, starting with steel and alternating composition and steel.

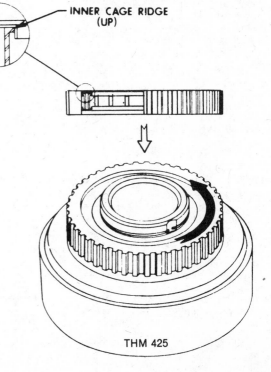

INNER CAGE RIDGE (UP)

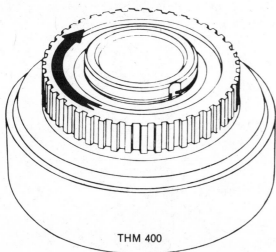

THM 400

THM 425

Correct directions of rotation for direct clutches

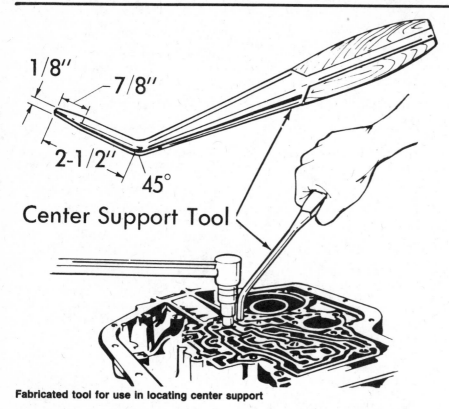

1/8″ 7/8″
2-1/2″ 45°

Center Support Tool

Fabricated tool for use in locating center support

NOTE: Do not use radial grooved plates here.

7 Install clutch backing plate, then install the backing plate snap-ring.
8 Invert the unit and install one sprag bushing, cup side up, over the inner race.
9 Install sprag assembly into outer race. On 1974–75 models install the roller assembly of the intermediate clutch.
10 With ridge on inner cage facing down, start sprag and outer race over inner race with clockwise turning motion.

NOTE: Outer race should not turn counterclockwise.

11 Install sprag bushing over sprag cup side down.
12 Install clutch retainer and snap-ring.
13 Place direct clutch assembly over center support and air check operation of direct clutch.

NOTE: It is normal for air applied to reverse passage to escape from direct clutch passage. Air applied to direct clutch passage should move direct clutch.

Eldorado and Toronado

APPLICATIONS
NOTE: Due to front-wheel drive application, the transmission output shaft turns in the opposite direction from that required by a conventional, rear-wheel drive vehicle.

Failure of the direct clutch to function after reconditioning an Eldorado or Toronado transmission may result from incorrect installation of the intermediate sprag clutch assembly.

In the Eldorado or Toronado transmission, the intermediate sprag clutch assembly must be installed with the identifying inner cage ridge up. When assembled, the outer race should turn counterclockwise only.

On all cars, except Eldorado and Toronado, this assembly is installed with the identifying inner cage ridge down. When assembled, the outer race should turn clockwise only. The same intermediate sprag clutch assembly is used on all cars.

CENTER SUPPORT DISASSEMBLY
1 Remove hook-type oil rings from center support. Do not remove the Teflon oil seal unless replacement is required.
2 Compress the spring retainer and remove the snap-ring.
3 Remove spring retainer and clutch release springs.
4 Remove intermediate clutch piston.
5 Remove inner piston seal.

NOTE: Do not remove the three screws holding the roller clutch inner race to the center support.

6 Remove outer piston seal.

CENTER SUPPORT INSPECTION
1 Inspect roller clutch inner race for scratches or identations. Check that lubrication hole is open.
2 Check bushing for scoring, wear, or galling. If bushing is replaced, drive new bushing into bore until it is flush to .010 in. below the top of the oil delivery sleeve.
3 Check oil seal rings and ring grooves in the center support tower for damage.
4 Make air pressure check of oil passages to be sure they are not interconnected.
5 Inspect piston sealing surfaces for scratches. Inspect piston seal grooves for damage. Check piston for cracks or porosity.
6 Check release springs for distortion or collapsed coils.
7 Check support to case spacer for burrs or raised edges. Repair with a stone or fine sand paper.

CENTER SUPPORT ASSEMBLY
1 Install new inner and outer seals on the piston, lip on inner seal facing away from the spring pocket.
2 Install inner spring protector tool onto the center support hub; lubricate the seal and install the piston.
3 Install release springs into the piston, spaced equally.
4 Place spring retainer and snap-ring over the springs.
5 Using the clutch spring compressor, compress the springs and install the snap-ring.
6 Install hook-type oil rings.
7 Air check the operation of intermediate clutch piston.

TORQUE CONVERTER INSPECTION
The torque converter is a welded assembly and must be serviced as a unit. If converter output shaft has more than .050 in. end-play, renew the unit.
Check for leaks as follows:
1 Install converter leak test fixture and tighten.
2 Fill converter with air, 80 psi.
3 Submerge in water and check for bubbles.

PLANETARY GEAR UNIT DISASSEMBLY
1 Remove center support assembly.
2 Remove center support to reaction carrier thrust washer.
3 Remove center support to sun gear races and thrust bearing. One race may already have been removed with the center support.
4 Remove reaction carrier and roller clutch assembly.
5 Remove front internal gear ring from output carrier assembly.
6 Remove sun gear.
7 Take off reaction carrier to output carrier thrust washer.
8 Turn carrier assembly over. Remove output shaft to output carrier snapring.
9 Output shaft may now be removed.
10 Remove nylon speedometer drive gear by depressing retaining clip and sliding gear off output shaft. Remove steel speedometer drive gear with a suitable puller.
11 Remove output shaft to rear internal

GM TURBO HYDRA-MATIC 400–425–M40

gear thrust bearing and two races.

12 Remove rear internal gear and mainshaft. Remove rear internal gear to sun gear thrust bearing and two races. Remove rear internal gear to mainshaft snap ring. Remove mainshaft.

PINION REPLACEMENT

1 Support the carrier assembly on its front face.
2 Using a ½ in. drill, eliminate the stake marks from the end of the pinion pin.

CHILTON CAUTION: *Do not allow drill to remove any stock from the carrier as this will weaken the part and future failure would be probable.*

3 Drive or press the pinion pins out of the carrier.
4 Remove the pinions, thrust washers and roller needle bearings.
5 Inspect the pinion pocket thrust faces for burrs and remove if present.
6 Install 18 needle bearings into each pinion, using petroleum jelly to hold the bearings in place. Use the pinion pin as a guide.
7 Place a bronze and steel washer on each side of the pinion so that the steel washer is against the pinion and hold them in place with petroleum jelly.
8 Place the pinion assembly in position in the carrier and install a pilot shaft through the rear face of the assembly to hold the parts in place.
9 Drive a new pinion pin into place while rotating its pinion, being sure that headed end is flush or below the face of the carrier.
10 Back up the opposite end of the new pinion shaft while staking the shaft in place.

NOTE: Both ends of the pinion pins must lie below the face of the carrier or interference may occur.

INSPECTION OF REACTION CARRIER, ROLLER CLUTCH, AND OUTPUT CARRIER ASSEMBLY

1 Inspect band surface reaction carrier for burning or scoring.
2 Check roller clutch outer race for scoring or wear. Check thrust washer for wear.
3 Check bushing for damage. If bushing is damaged, replace reaction carrier.
4 Check reaction carrier pinions for damage, rough bearings, or excessive tilt. Check pinion end play. It should be .009–.024 in. Pinions may be replaced if necessary.
5 Check roller clutch for damaged members. Check roller clutch cage for damage.
6 Inspect front internal gear (output carrier) for damaged teeth. Inspect output carrier pinions for damage, rough bearings, and excessive tilt. Check pinion end play. It should be .009–.024 in.
7 Inspect parking pawl lugs for cracks

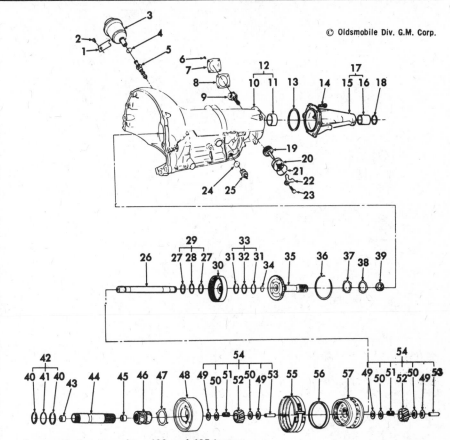

© Oldsmobile Div. G.M. Corp.

Case and planet carrier—400 and 425 trans.

1. Retainer	20. Sleeve and seal	39. Drive gear
2. Bolt	21. Seal	40. Thrust bearing
3. Vacuum modulator	22. Sleeve retainer	41. Thrust bearing
4. Seal	23. bolt	42. Thrust bearing
5. Modulator valve	24. Seal	43. Bushing
6. Bolt	25. Connector	44. Sun gear shaft
7. Cover	26. Mainshaft	45. Bushing
8. Gasket	27. Thrust bearing (kit)	46. Sun gear
9. Governor assy.	28. Thrust bearing (kit)	47. Thrust washer
10. Transmission case	29. Thrust bearing (kit)	48. Reaction carrier
11. Bushing	30. Rear internal gear	49. Pinion carrier kit
12. Case assembly	31. Thrust bearing (kit)	50. Pinion carrier kit
13. Seal	32. Thrust bearing (kit)	51. Pinion carrier kit
14. Bolt	33. Thrust bearing (kit)	52. Pinion carrier kit
15. Case extension	34. Snap ring	53. Pinion carrier kit
16. Bushing	35. Output shaft	54. Pinion carrier kit
17. Case extension	36. Snap ring	55. Rear band
18. Oil seal	37. Thrust washer	56. Internal gear ring
19. Driven gear	38. Thrust washer	57. Output carrier

or damage. Inspect output locating splines for damage.
8 Check front internal gear ring for flaking.

PLANETARY GEAR UNIT ASSEMBLY

1 Install rear internal gear onto end of mainshaft, then install the snap-ring.
2 Install sun gear to internal gear thrust races and bearings against inner face of rear internal gear as follows:
  a. Place large race against the internal gear, with flange facing up.
  b. Place thrust bearing against race.

  c. Place small race against the bearing, with inner flange facing the bearing, or down.
3 Install the output carrier over the mainshaft so that the pinions mesh with the rear internal gear.
4 Place the above portion of the assembly through a hole in the bench so that the mainshaft hangs downward.
5 Install the rear internal gear to output shaft thrust races and bearings as follows:

  a. Small diameter race against internal gear, center flange facing up.
  b. Bearing onto the race.

GM TURBO HYDRA-MATIC 400–425–M40

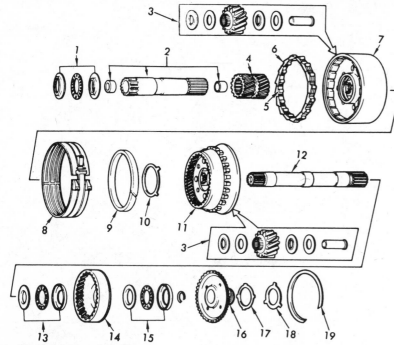

Reaction carriers and shafts

1. BEARING PKG.
2. SUN GEAR SHAFT
3. PINION PKG.
4. SUN GEAR—REAR
5. LOW CLUTCH ROLLER PKG.
6. LOW CLUTCH ROLLER
7. REACTION CARRIER
8. REAR BAND
9. SILENCER RING

10. THRUST WASHER
11. OUTPUT CARRIER
12. MAIN SHAFT
13. BEARING PKG.
14. MAIN SHAFT GEAR
15. BEARING PKG.
16. FLANGE
17. THRUST WASHER
18. THRUST WASHER
19. SNAP RING

c. Second race onto the bearing, outer flange cupped over the bearing.
6. Install output shaft into the output carrier assembly.
7. Install output shaft to output carrier snap-ring. If used, install the O-ring.
8. Turn assembly over and support it so that the output shaft hangs downward.
9. Install the reaction carrier to output carrier thrust washer, tabs facing down and in their pockets.
10. Install sun gear, splines, chamfer down. Install gear ring over output carrier.
11. Install sun gear shaft (long splined end down), then the reaction carrier.
12. Install the center support to sun gear thrust races and bearing as follows:
    a. Install the large race, center flange up over the sun gear shaft.
    b. Install thrust bearing.
    c. Install the second race, center flange up.
13. Install roller clutch into reaction carrier outer race. Install the center support to reaction carrier thrust washer into the recess in the support. Retain with petroleum jelly.
14. Install center support into reaction carrier and roller clutch assembly. With reaction carrier held, center support should only turn counterclockwise.

15. Install a gear unit assembly holding tool to hold units in place. Install output shaft to case thrust washer tabs in pockets and retain with petroleum jelly.

Transmission Assembly

PARKING MECHANISM, REAR BAND AND THE COMPLETE GEAR ASSEMBLY INSTALLATION

1. Install O-ring seal onto parking pawl shaft, then install parking pawl, tooth toward inside of case.
2. Install the pawl shaft retaining clip and the return spring, square-end hooked on the pawl.
3. Install parking brake bracket guides over pawl, using two attaching bolts.
4. Install rear band assembly so that the two lugs index with the two anchor pins. Install support to case spacer with ring gap adjacent to band anchor pin.
5. Install rear selective washer into slots provided inside rear of transmission case. Dip washer in transmission fluid.
6. Install the complete gear unit assembly into the case.
7. Lubricate and install center support to case snap-ring. Install bevel side up.

NOTE: The support to case spacer is .040 in. thick and is flat on both sides. The

center support to case snap-ring has one side beveled. The intermediate clutch backing plate to case snap ring is .093 in. thick and is flat on both sides.

On later models, install the case to center support bolt by placing a center support locating tool or equivalent tool into case direct clutch passage with handle of tool pointing to right as viewed from front of transmission and parallel to bell housing mounting face. Apply pressure downward on tool handle to hold the center support firmly against the case splines. Torque the case to center support bolt to 22 ft./lbs.

8. Install three steel and three composition clutch plates. Alternate the plates, starting with steel.
9. Install the backing plate, ridge up, then the snap-ring. Locate snap-ring gap opposite band anchor pin.
10. Check rear end-play as follows:
    a. Install a threaded rod or a long bolt into an extension housing attaching bolt hole.
    b. Mount dial indicator on the rod and index it with the end of the output shaft.
    c. Move the output shaft in and out. Read the end-play. End-play should be .007–.019 in. The selective washer controlling this end-play is the steel washer, having three lugs, that is located between the thrust washer and the rear face of the transmission case.

If a different washer thickness is required to obtain proper end-play, it can be selected from the following chart.

Thickness	Notches and/or Numeral		
.074–.078 in.	None		1
.082–.086 in.	1 Tab	Side	2
.090–.094 in.	2 Tabs	Side	3
.098–.102 in.	1 Tab	O.D.	4
.106–.110 in.	2 Tabs	O.D.	5
.114–.118 in.	3 Tabs	O.D.	6

MANUAL LINKAGE INSTALLATION

1. Install new manual shift shaft seal into the case.
2. Insert actuator rod into the manual detent lever from the side opposite the pin.
3. Install actuator rod plunger under the parking bracket and over the pawl.
4. Install manual lever and shaft through the case and detent lever, then lock with hex nut on the manual shift shaft. (Be sure detent retaining nut is tight.) Install retaining pin.

FRONT BAND, DIRECT CLUTCH AND FORWARD CLUTCH INSTALLATION

1. Install front band, with band anchor hole placed over the band anchor pin, and apply lug facing the servo hole.

GM TURBO HYDRA-MATIC 400–425–M40

2 Install the direct clutch and intermediate sprag assembly. (Removal of direct clutch plates may help.)

3 Install forward clutch hub to direct housing bronze thrust washer onto the forward clutch hub. Retain with petroleum jelly.

4 Install forward clutch assembly, indexing the direct clutch hub so the end of the mainshaft will be flush with the end of the forward clutch hub. Use turbine shaft as a tool.

5 Install turbine shaft; end with the short spline goes into the forward clutch housing.

6 Install pump-to-case gasket onto the case face and install the front pump assembly and all but one attaching bolt and seal. Torque to 18 ft./lbs.

NOTE: If turbine shaft cannot be rotated as pump is being pulled into place, forward or direct clutch housing has not been properly installed to index with all clutch plates. This condition must be corrected before pump is fully pulled into place.

7 Drive in a new front seal.

8 Check front unit end-play as follows:
  a. Install one rod of slide hammer,

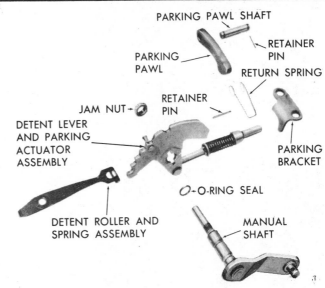

Manual and parking linkage 425 transmission

with 5/16–18 in. thread, into the empty pump assembly attaching bolt hole.

b. Mount dial indicator on the rod and adjust the indicator probe to contact end of turbine shaft.

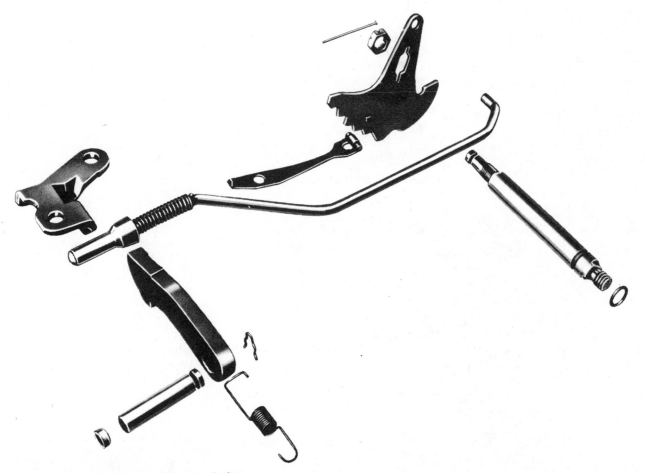

Manual and parking linkage 400 transmission

GM TURBO HYDRA-MATIC 400–425–M40

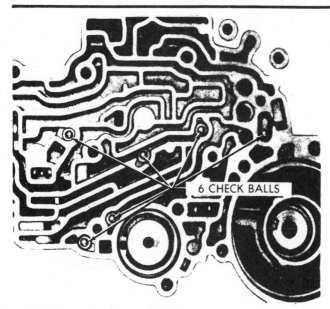

Check ball location in 400 transmission

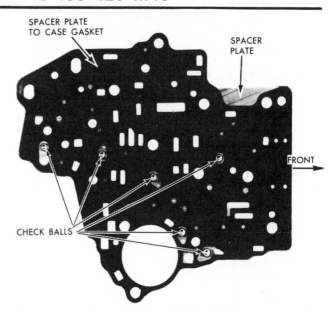

Location of check balls—in-vehicle installation

c. Hold output shaft forward while pushing turbine shaft rearward to its stop.
d. Set dial indicator to zero.
e. Pull turbine shaft forward.
f. Read end-play, as registered on dial. The reading should be .003–.024 in.

The selective washer controlling this end-play is located between the pump cover and the forward clutch housing. If more, or less, washer thickness is required to bring end-play within specifications, make selection from the thickness-color chart.

Thickness	Color
.060–.064 in.	Yellow
.071–.075 in.	Blue
.082–.086 in.	Red
.093–.097 in.	Brown
.104–.108 in.	Green
.115–.119 in.	Black
.126–.130 in.	Purple

9 Remove dial indicator and install the remaining front pump attaching bolt and seal. Torque to 18 ft./lbs.

REAR EXTENSION HOUSING ASSEMBLY INSTALLATION

1 Install extension housing-to-case O-ring seal onto extension housing.
2 Attach extension housing to transmission case. Torque to 22 ft./lbs.
3 Drive in a new extension housing rear seal.

Sprocket Cover, Link Assembly, Drive and Driven Sprocket Installation—

ELDORADO AND TORONADO

1 Position the drive and driven sprockets on a workbench with the shafts up, drive sprocket closest to the transmission.
2 Install the link assembly over the sprockets, with the colored guide link (with etched numerals) down.
3 Lift the sprockets and link assembly, and rotate it until the drive sprocket is up.
4 Allow the driven sprocket to hang freely, and start the turbine shaft into the pump housing until it will support the weight of the assembly.
5 Start the input shaft into the driven support housing. Alternately, push the shafts inward until the sprockets are installed.
6 With a plastic mallet, gently tap the sprockets to seat the sprocket bearing assemblies in the support housing.
7 Install the sprocket bearing retainer snap-rings, located under the drive and driven sprockets.
8 Install a new sprocket cover to case gasket and position the sprocket cover on the transmission case. Torque the attaching bolts to 8 ft./lbs.

NOTE: One sprocket cover attaching bolt is ¼ in. longer. This bolt must be installed in the tapped hole directly over the oiler cooler fittings on the transmission case.

CHECK BALLS, FRONT SERVO GASKETS, SPACER AND SOLENOID INSTALLATION

1 Install front servo spring and retainer into transmission case.

Check ball location 425 transmissions

GM TURBO HYDRA-MATIC 400–425–M40

2 Install flat washer on front servo pin, on end opposite taper. Install pin and washer into case so that tapered end of pin is contacting band.

3 Install oil seal ring on front servo piston, and install on apply pin so that identification numbers on shoulders are exposed. The piston should move freely in the bore.

4 Install check balls into transmission case pockets.

5 Install valve body spacer to case gasket and spacer plate. Install detent solenoid and gasket, with connector facing outer edge of case. Do not tighten bolts at this time.

6 Install O-ring seal on electrical connector. Lubricate and install electrical connector with locator tab in notch on side of case. Connect detent wire and lead wire to electrical connector. Be sure to install electrical wire clip.

REAR SERVO INSTALLATION

1 Before installing the rear servo assembly, check the band apply pin, using rear band apply fixture as follows:

   a. Attach band apply pin selection gauge, to the transmission case with attaching screws.

   b. Apply 25 ft./lbs. torque and select proper servo pin to be used from scale on the tool.

   c. Remove tool and make note of proper pin to be used during assembly.

There are three selective pins:

Pin Identification	*Pin Size*
One ring	Short
Two rings	Medium
Three rings	Long

The identification consists of a ring located on the band lug end of the pin. Selecting the proper pin is equivalent to adjusting the band.

2 Install rear accumulator spring.

3 Install servo assembly, then the gasket and cover. Torque bolts to 18 ft./lbs.

CONTROL VALVE AND GOVERNOR PIPE INSTALLATION

1 Install control valve-to-spacer gasket, then install governor pipes into the control valve body assembly.

2 Install two guide pins, then install the control valve body and governor pipe assembly into the transmission.

NOTE: Be sure the manual valve is properly indexed with the pin on the manual detent lever.

3 Remove guide pins and install valve assembly attaching bolts and manual detent and roller assembly.

4 Tighten detent solenoid and control valve attaching bolts to 8 ft./lbs. torque.

STRAINER AND INTAKE PIPE INSTALLATION

1 Install case-to-intake pipe O-ring onto strainer and intake pipe assembly.

2 Install strainer and pipe assembly. Install new filter on models so equipped.

3 Install new pan gasket, then install the pan. Torque to 12 ft./lbs.

4 Install modulator shield and all pan attaching screws. Torque pan attaching screws.

Modulator Assembly

INSTALLATION

1 Install modulator valve into the case, stem end out.

2 Install O-ring seal onto the vacuum modulator, then install assembly into the case.

3 Install modulator retainer and attaching bolt. Torque to 18 ft./lbs.

GOVERNOR AND SPEEDOMETER DRIVEN GEAR INSTALLATION

1 Install governor assembly.

2 Attach governor cover and gasket with four bolts. Torque to 18 ft./lbs.

3 Install speedometer driven gear assembly. Install retainer and attaching bolt.

CONVERTER INSTALLATION

1 Place transmission in cradle or portable jack.

2 Install converter assembly to pump assembly, making certain that the converter hub drive slots are fully engaged with the pump drive gear tangs and that the converter is installed all the way toward the rear of the transmission.

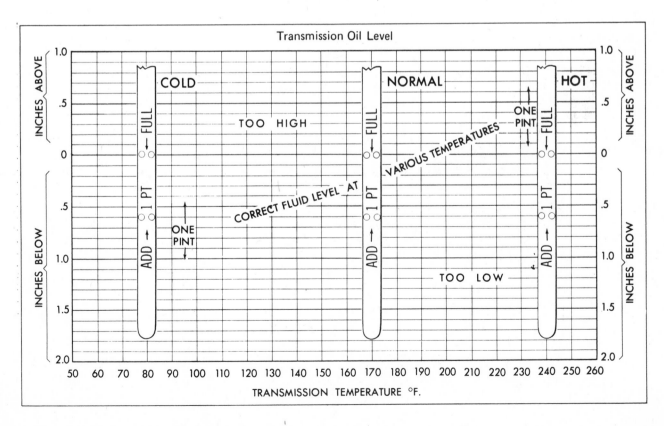

GM TURBO HYDRA-MATIC 400–425–M40

G.M. Type 400–425 Fluid Specifications

Oil capacity, transmission and converter—approx. 22 pints
Capacity between marks on dip stick—1 pint
Type of Oil—Dexron® Automatic transmission fluid Type A
Drain and refill—24,000 miles

G.M. Type 400–425 Torque Specifications

	Ft./Lbs.
Extension-to-case attaching bolts	22–23
Detent solenoid bolts	8
Control valve body bolts	8
Bottom pan attaching screws	12
Modulator retainer bolt	18
Governor cover bolts	18
Manual lever-to-manual shaft nut	20
Linkage swivel clamp screw	20
Transmission-to-engine mounting bolts	40
Rear mount-to-transmission bolts	40
Oil cooler line	16
Filter retainer bolt	10
Pressure switch assembly	8

G.M. Type 400 Control Pressures

Model	Control Pressures (psi) ①					
	Drive④ Brakes Applied @ 1,000 rpm	Super⑤ Brakes Applied @ 1,000 rpm	Reverse —Brakes Applied @ 1,000 rpm	Neutral —Brakes Applied @ 1,000 rpm	Drive— Idle Set engine to spec.	Drive— 30 mph closed throttle
Cadillac, Eldorado	60-90	135-160	95-150	55-70	60-85	55-70
Buick ②	60	150	220-260	150	N.A.	60
Oldsmobile	60-90	135-160	95-150	55-70	60-85	N.A.⑥
Pontiac ②	60	150	107	60	60	N.A.⑦
Chevrolet	60	150	107	60	60	N.A.

N.A. —Not applicable.
① —Can also be checked on a lift with transmission in Drive, brakes applied and engine at 2,000 rpm. Check pressure before rpm falls below 1,200.
② —@ 1,200 rpm.
④ —Drive Left on Cadillac.
⑤ —Drive Right on Cadillac—Super on Oldsmobile—D2 on Chevrolet.
⑥ —55 to 70 on 1974–75 models.
⑦ —60 on 1974–75 models.

TRANSMISSION IDENTIFICATION

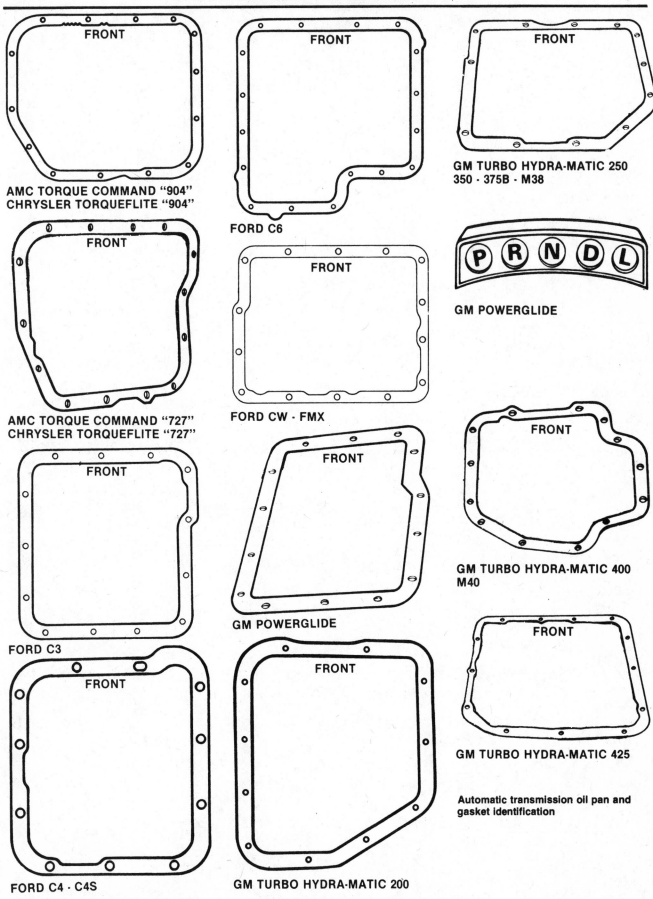

FRONT
AMC TORQUE COMMAND "904"
CHRYSLER TORQUEFLITE "904"

FRONT
AMC TORQUE COMMAND "727"
CHRYSLER TORQUEFLITE "727"

FRONT
FORD C3

FRONT
FORD C4 · C4S

FRONT
FORD C6

FRONT
FORD CW · FMX

FRONT
GM POWERGLIDE

FRONT
GM TURBO HYDRA-MATIC 200

FRONT
GM TURBO HYDRA-MATIC 250
350 · 375B · M38

P R N D L
GM POWERGLIDE

FRONT
GM TURBO HYDRA-MATIC 400
M40

FRONT
GM TURBO HYDRA-MATIC 425

Automatic transmission oil pan and
gasket identification